NUGGETS TO NEUTRINOS

NUGGETS TO NEUTRINOS

THE HOMESTAKE STORY

Steven T. Mitchell

STEVEN T. MITCHELL

Copyright © 2009 by Steven T. Mitchell.

Library of Congress Control Number:	2009911769
ISBN: Hardcover	978-1-4415-7067-3
Softcover	978-1-4415-7066-6

All rights reserved. No part of this book may be reproduced or transmitted in any form or by any means, electronic or mechanical, including photocopying, recording, or by any information storage and retrieval system, without permission in writing from the copyright owner.

The contents of this book are intended as general historical information only and may contain errors or omissions. The author makes no representation or warranty, expressed or implied, as to the accuracy, meaning, completeness, or use of the information, descriptions, and data depicted hereon or contained herein, and the author has no liability or responsibility for the accuracy, meaning, completeness, or use of the information, descriptions, and data.

Many of the historical sites referenced herein are located on private property. Additionally, mine sites and other geographic sites referenced herein may have inherent risks that could result in personal injury or death to persons who visit such sites and properties. Therefore, no attempt should be made to enter real property of others without first obtaining the identities of and required permissions from the legal owners, holders, or lessees of the property and obtaining an understanding of all the risks and requirements in accessing and entering such property. The author recommends that no attempt be made to approach or enter mine shafts, tunnels, adits, quarries, structures, or other sites. The author has no liability or responsibility for claims of injuries, death, damages, losses, including claims of others, which may result from persons accessing or entering such properties.

This book was printed in the United States of America.

To order additional copies of this book, contact:
Xlibris Corporation
1-888-795-4274
www.Xlibris.com
Orders@Xlibris.com

Contents

Preface 29
Acknowledgments 31
Introduction 33

Chapter 1—The Geology 35
    Black Hills Stratigraphy 35
    The Lead Window 41
    Fossil and Recent Gold Placers 48

Chapter 2—The Westward Expansion 51
    Migration of the Plains Indians 51
    Early-Day Visitors and Prospectors 56
    Emigrant Protection 63
    Fort Laramie Treaty of 1851 65
    Grattan and Ash Hollow (Bluewater) Massacres 68

Chapter 3—The Great Reconnaissance 70
    Army Appropriation Act of 1853 70
    Warren-Hayden Expedition of 1857 73
    Raynolds Expedition of 1859-60 75
    Dakota Territory 77
    Hayden Expedition of 1866 79
    Red Cloud's War 80
    Fort Laramie Treaty of 1868 82
    Custer Expedition of 1874 90

Chapter 4—The Gold Rush 97
    Collins-Russell Expedition of 1874-75 97
    Newton-Jenney Expedition of 1875 99
    Miners Are Ordered Out of the Black Hills 103
    Gold Is Discovered in the Northern Black Hills 105

Chapter 5—The Expropriation of the Black Hills 113
    Decision to Extinguish Indian Title to the Black Hills 113
    Great Sioux War of 1876-77 118
    Manypenny Agreement (Black Hills Treaty) 128
    The Sioux Split Up, but Refuse to Give Up 130

Chapter 6—The Homestake Discovery 134
    Locators of the Homestake Mine 134
    Wolf Mountain Stampede 136
    The Need for Water and a Mill 137
    Custom Stamp Mills in Lead City 145

Chapter 7—The Homestake Capitalists and Their Strategies 149
    Hearst, Haggin, Tevis, and Company 149
    Homestake Mining Company 154
    Other Belt Companies of the Homestake Capitalists 165
        Old Abe Mining Company 165
        Giant and Old Abe Mining Company 166
        Highland Mining Company 167
        Deadwood Mining Company 170
        Golden Terra Mining Company 172
        Deadwood-Terra Mining Company 174
        Father DeSmet Mining Company 176
        Father DeSmet Consolidated Gold Mining Company.......177
        Caledonia Gold Mining Company 180
    The Homestake Mergers 182

Chapter 8—Water Rights and Water Fights 184
    Water Rights and Water Ditches 184
    Wyoming and Dakota Ditch, Flume, and Mining Company
        and the Spearfish Ditch..187
    Wyoming and Dakota Water Company 188
        DeSmet, Bowie, and Palmer Ditch 189
        Little Elk Creek Ditch 189
    Black Hills Canal and Water Company 190
        Pioneer Ditch 191
        French Boys Ditch 192
        Montana Ditch 192
        Foster Ditch 193
    The Deadwood Water Question 193
        Boulder Ditch 194
        New French Boys (Old Abe) Ditch 194
    J. B. Haggin Maneuvers to Gain Control 203
    Other Acquisitions of the Black Hills Canal and
        Water Company 205
        Columbia Water Company and Columbia Ditch 206
        Hearst Ditch 206

  Peake Ditch 207
  Little Rapid Ditch 207
 Homestake Water Collection System..208

Chapter 9—Early Mining and Haulage Methods..213
 Mining by Open Cuts and Drawholes...213
 Open Square-Set Method...217
 Homestake System of Stoping...222
 Surface and Underground Haulage 227

Chapter 10—Arrastras, Stamp Mills, and Amalgamation 237
 The Arrastra and Amalgamation 237
 The Stamp Mill and Amalgamation 239
 The Large Stamp Mills along the Homestake Belt.......................................242
  Homestake Mill 245
  Golden Star Mill 247
  Highland Mill 252
  Operation, Standardization, and Optimization.................................254
  The Mills under Homestake's Ownership..263
 Retorting and Refining...264

Chapter 11—The Need for Cordwood, Mine Timber, and a Railroad 267
 Justification for the Railroad...267
 Along the Route..271
  Lead City to Terraville..273
  Lead City to Reno Gulch 273
  Attempted Train Robbery at Reno Gulch 273
  Reno Gulch to Buck's Landing..277
  Buck's Landing to Piedmont via Elk Creek Canyon...........................280
  The Town of Piedmont 283
  Buck's Landing to Este 289
 The First Federal Timber Sale..291
 The Burlington Group Assumes Control 293
 Flood of 1907 Forces Abandonment of the
  Line through Elk Creek Canyon 295
 The Larger Homestake Sawmills..297

Chapter 12—The Lead Townsite Development 303
 Occupants of the Townsite versus the Mineral Claimants...........................303
 The Gushurst and Manuel Connection...311
 First Murder in Lead City..313

Schools..314
A Few Other Early Lead Businesses...317
Churches...321
Celebrations..323

Chapter 13—Sister Cities Near the Homestake Belt..................327
Mining Brings Mining Camps 327
Gayville..329
Blacktail, Lancaster City, and South Bend..................................332
Central City..333
Lincoln and Terraville 341
Golden Gate, Anchor City, and Poorman Gulch........................344

Chapter 14—Cyanide Charlie's Contribution 345
The Advent of Cyanidation...345
Cyanide Plant No. 1...349
Cyanide Plant No. 2...354
Slime Plant...355
Regrind Plant..363
Homestake's Tungsten Mill 364
South Mill..365
Precipitation and Refining...370

Chapter 15—The Conversion to Electricity................................371
Edison Dynamos...371
Englewood Hydroelectric Plant...373
Hydroelectric Plant No. 1 375
Lead Substation..384
Central Steam Plant...387
Auxiliary Steam-Electric Plant...390
Hydroelectric Plant No. 2 391
Communications Systems..395
Kirk Power Plant...396

Chapter 16—The Lockout of 1909-1910 400
Lead Miners' Union 400
The Unions Pressure for an Eight-hour Day and a Closed Shop..401
Grier's Lockout Precludes a Strike 403
The Return to Work with Nonunion Employees.........................406
Subsequent Efforts to Organize the Workforce..........................407

Chapter 17—Wages, Benefits, and Community Support 409
    The Early Workforce 409
    Wages 411
    Homestake Hospitals 413
    Hearst Mercantile Company 416
    Homestake Veterans Association 419
    Homestake Baseball 419
    Homestake Aid Association 421
    Mine Safety 422
    Recreation Building and Opera House 425
    U.S. Commission on Industrial Relations 425
    Lead Country Club 428
    Homestake Band 428
    Employee Housing 429
    The *Sharp Bits* Magazine 431
    Nicknames 431

Chapter 18—Mine Fires and the Subsidence Issue 433
    The Need for a New Shaft 433
    Mine Fire of 1907 437
    Surface Cave-Ins 439
    Mine Fire of 1919 440
    B&M No. 2 as a Replacement Shaft 443
    Milliken (No. 1) Winze 446
    Ellison Fire of 1930 446
    Introduction of Underground Sandfill 448

Chapter 19—Later Acquisitions Along the Homestake Belt 452
    Columbus Gold Mining Company 452
    Columbus Consolidated Gold Mining Company 453
    Hidden Fortune Mining Company 455
    Homestake South Extension Mining Company 460
    Oro Hondo Mining Company 461

Chapter 20—What Depression? 464
    Ross Shaft 464
    Cyanide Plant No. 3 467
    South Dakota Ore Tax 469
    No. 2 and No. 3 Winzes 471
    Yates Shaft 474
    South Mill Expansion 477

Chapter 21—The Deep-Level Projects 479
    No. 4 Winze..479
    No. 5 Shaft 483
    Oro Hondo Shaft..486
    Bored Raise Drilling...487
    Diamond Drilling 489
    No. 6 Winze..491
    No. 7 Winze..494
    Deep-Level Ventilation Project...495
    North Homestake Exploration Project 501

Chapter 22—The End of Amalgamation 503
    Grizzly Gulch Tailing Storage Facility...503
    Carbon-in-Pulp Process..506
    The Wastewater Treatment Plant..510
    Expansions of the Grizzly Gulch Tailing Storage Facility 511
    Mill Optimization Projects...512

Chapter 23—Mechanization of Mining 515
    Square-Set Cut-and-Fill Stoping...515
    Open Cut-and-Fill Stoping 517
    Blasthole Sublevel Stoping..524
    Vertical Crater Retreat Mining...525
    Uphole and Bench Mining...527
    Mechanized Cut-and-Fill Stoping 528

Chapter 24—Modern Mining in the Open Cut......................................530
    Exploration Work 530
    Terraville Test Pit..531
    Pit Expansions..531
    Reclamation of Waste Rock Facilities 535

Chapter 25—How Other Work Was Performed 537
    Mine Surveying...537
    Sampling and Assaying...541
    Smelting and Refining..544
    Sand Casting..548
    Brookhaven Solar Neutrino Experiment 553

Chapter 26—Production Statistics 558
    Production and Revenue History 558
    The Decision to Close the Mine 561

Chapter 27—Beyond Closure 564
    Mine Decommissioning and Postclosure Reclamation Work 564
    Sanford Underground Science and Engineering Laboratory 568
    Deep Underground Science and Engineering Laboratory 573

Chapter 28—Some Early Homestake VIPs 577
    Fred and Moses Manuel 577
    Henry Clay Harney and Alexander Engh 581
    George Hearst 583
    Phoebe Apperson Hearst 588
    James Ben Ali Haggin 591
    Lloyd Tevis 592
    Charles Washington Merrill 596
    Famous VIPs as Visitors 600

Epilogue 603
Appendix
    A: Water rights acquired by Homestake Mining Company 605
    B: Superintendents and general managers of the
       Homestake Mine 609
    C: Mining Law of 1872 611

Notes 613
Glossary of Geologic and Mining Terms 665
Bibliography 677
Index 697

Illustrations

1.1.	The Tertiary rhyolite dike swarm in Homestake's Open Cut, 2008	38
1.2.	Geologic map of the Black Hills	39
1.3.	Geologic cross section A-A'	40
1.4.	The hydrogeologic setting of the Black Hills	40
1.5.	Proterozoic stratigraphy around the city of Lead, South Dakota	41
1.6.	Map of the structure of the Precambrian rocks in the Lead District	44
1.7.	Idealized geologic cross section of the Homestake Mine	45
1.8.	Names and relative positions of the ore-bearing folds in the Open Cut area	46
1.9.	Generalized plan of the 300-foot level of the Homestake Mine	48
1.10.	Generalized section showing Paleo (Fossil or Cement) Placers and Recent Placers in the northern Black Hills	49
2.1.	Migration patterns of northern plains tribes (1650-1820)	53
2.2.	Territorial acquisitions of the United States	55
2.3.	Emigrant trails of the west	64
2.4.	Territory of the Sioux Nation pursuant to the Fort Laramie Treaty of 1851	67
3.1.	Spotted Tail, or Sinte Gleska, chief of the Brulé tribe of the Teton Sioux	71
3.2.	Red Cloud, chief of the Oglala tribe of the Lakota Sioux	72
3.3.	Sitting Bull, chief of the Hunkpapa tribe of the Lakota Sioux	72
3.4.	Map of Dakota Territory, 1861	78
3.5.	Negotiations for the Fort Laramie Treaty of 1868	83
3.6.	Map of the Great Sioux Reservation of 1868 compared to the Sioux Nation of 1851	83
3.7.	A portion of Captain William Ludlow's map of the Black Hills (Custer) Expedition, 1874	92
3.8.	W. H. Illingworth's photo entitled, "Our first grizzly, killed by Gen. Custer and Col. Ludlow."	95
4.1.	Members of the Gordon Party (Collins-Russell Expedition) in Cheyenne, Wyoming Territory, after they were escorted out of the Black Hills in 1875	99
4.2.	"Mining crew drifting for gold below discovery point," by Stanley J. Morrow, 1876	108

4.3.	The Wheeler Ground or Placer Claim No. 2 below the Discovery claim in Deadwood Gulch, 1876	109
5.1.	General Sheridan's Three-Pronged Campaign against the Sioux, 1876	119
5.2.	"Stretcher with wounded man from the Battle of Slim Buttes," Stanley J. Morrow, 1876	125
5.3.	Soldiers of General George Crook at the Gordon Stockade, 1876	127
5.4.	Further reductions of the Great Sioux Reservation	129
6.1.	Map by Louis Janin and Walter P. Jenney of the Old Abe, Homestake, and Golden Star claims and mine workings	138
6.2.	The Hidden Treasure Mill on Deadwood Creek near Hidden Treasure Gulch, 1876	139
6.3.	The Jones and Pinney Mill at Golden Gate, 1878	140
6.4.	A water-powered arrastra	141
6.5.	Map showing the tunnels driven by Samuel Gwinn (left) and Cyrus Enos (middle)	143
6.6.	Cyrus Enos's Tunnel and Lead City, 1877	144
6.7.	The Old Abe Shaft, Highland Mill, and Davenport Mill, ca. 1880	144
6.8.	Lead's Main Street, 1877	147
6.9.	Custom stamp mills of Lead City, 1878	147
7.1.	Samuel McMaster	155
7.2.	Map showing the mining companies along the Homestake Belt (1876-1902)	158
7.3.	J. D. McIntyre's mineral survey plat of the Homestake, Golden Star, Giant, and Gold Run lode claims	160
7.4.	Sanborn map showing the Homestake works, October 1885	163
7.5.	Thomas J. Grier, superintendent of the Homestake Mine (1884-1914)	164
7.6.	Golden Prospect (Highland) Shaft of the Highland Mining Company	170
7.7.	The Deadwood Mine Open Cut and underground workings, 1902	172
7.8.	The Old Brig (Terra) Shaft in Terraville	175
7.9.	The 100-stamp Father DeSmet Mill, ca. 1888	179
8.1.	Early water ditches in the Lead-Deadwood Area	186
8.2.	The water ditches south of Lead	186
8.3.	The Homestake water collection system, 1994	208
8.4.	The Hanna Pump Station	209

8.5.	The Reno Bridge, 2007	211
9.1.	Double-jack drilling in an open cut	214
9.2.	Hand mucking in an open cut	214
9.3.	Breaking rock on a grizzly	215
9.4.	Drawing chute on the haulage level	216
9.5.	Hauling ore on the Highland Tramway	216
9.6.	Square-set mining on the Comstock	218
9.7.	A timber stope in the Pierce Ledge, 200-foot level	219
9.8.	A solid timber crib	220
9.9.	The drawhole method of mining	221
9.10.	Hand mucking in a shrinkage stope, 1922	224
9.11.	Development of a shrinkage stope	225
9.12.	Drilling a breast cut from the top of the rock pile in a shrinkage stope, ca. 1920	225
9.13.	Old and new way of drilling	226
9.14.	The J. B. Haggin, Homestake's first steam locomotive, purchased in 1879	227
9.15.	The I. C. Stump steam locomotive	228
9.16.	Old Smokey pulling a train of 1-ton ore cars on the 400-foot level of the Homestake Mine	229
9.17.	Illustration by Buck O'Donnell showing the procedure for lowering horses and mules down a mine shaft	229
9.18.	Lowering a horse down the shaft to the 900-foot level	230
9.19.	Teddy the mule	230
9.20.	Compressed-air locomotive No. 1 and a train of ore cars on the Ellison Tramway	232
9.21.	The three-track train crossing in Lead, South Dakota	232
9.22.	Compressed-air locomotives from the H. K. Porter Company.	233
9.23.	A compressed-air locomotive and cars at the ore dump	234
9.24.	Charging an air locomotive	235
9.25.	A typical "camelback" ore dump	236
10.1.	Illustration by Buck O'Donnell of a water-powered arrastra	238
10.2.	The Joshua Hendy 10-stamp mill	240
10.3.	Cross section of a Joshua Hendy stamp mill	241
10.4.	The process of recovering gold and silver from amalgam using a mercury retort	242
10.5.	The Homestake, Golden Star, and Highland mills	244
10.6.	The Deadwood, Golden Terra, and Caledonia mills in Terraville	245
10.7.	Cross section of the Highland Mill by Fraser and Chalmers showing the ore bins and drives for the two opposing lines of stamp batteries	253

10.8.	Longitudinal section by Fraser and Chalmers of the 120-stamp Highland Mill	253
10.9.	Cleanup day at the Homestake 80-stamp mill, 1888	259
10.10.	Dressing the plates at the Amicus Mill, 1921	259
10.11.	Balls of gold amalgam	260
10.12.	The Homestake Assay Office, 1896	265
10.13.	Amalgam retorts in the old assay office	265
10.14.	Sacking gold bars, Harry J. Teer, bullion guard	266
10.15.	Wells Fargo Express Company guards transferring $250,000 of gold bullion from the Homestake Mine, 1890	266
11.1.	Installing the cow catcher on the George Hearst locomotive	270
11.2.	The train crew of the George Hearst locomotive	270
11.3.	The George Hearst delivers the first load of wood on the Black Hills and Fort Pierre Railroad, December 1881	271
11.5.	Part of the Black Hills and Fort Pierre Railroad train crew who were on board when the Homestake payroll train was held up on October 11, 1888	275
11.6.	The Uncle Sam Mine at the town of Roubaix	279
11.7.	Homestake's limekilns and quarry at Calcite	282
11.8.	The former Hearst Mercantile Company store in Piedmont, 2008	285
11.9.	The Black Hills and Fort Pierre narrow-gauge train at Giant Bluff in Elk Creek Canyon, ca. 1890	286
11.10.	Timetable of the Black Hills and Fort Pierre Railroad, 1896	287
11.11.	The mining camp of Greenwood	290
11.12.	The Safe Investment Mine near Greenwood	290
11.13.	Fireman Bob Leeper with Black Hills and Fort Pierre engine No. 536 at Homestake's lumber and timber camp in Merritt	293
11.14.	Engine No. 488 on the turntable in front of the Black Hills and Fort Pierre Roundhouse in Lead	294
11.15.	The Hearst Mercantile Store in Nemo, 1921	297
11.16.	A Homestake logging crew, ca. 1888-89	298
11.17.	A load of wood on the Black Hills and Fort Pierre Railroad at Nemo	299
11.18.	The Linn Tractor-Trailer, introduced into the Homestake logging operations in about 1930	299
11.19.	The Moskee Sawmill and forest fire of August 9, 1936	300
11.20.	Homestake's sawmill at Camp 5, which was operated from 1934 through 1937	301
11.21.	Homestake's Spearfish Sawmill	301
12.1.	Lead City and Washington, 1876-77	303
12.2.	North Mill Street, Lead, 1878	304

12.3.	Lead City, 1889	309
12.4.	The original Townsite of Lead, South Dakota per Frank Morris's plat, December 27, 1898	310
12.5.	Peter A. Gushurst's Big Horn Store in Deadwood, Dakota Territory, Stanley J. Morrow, 1876	311
12.6.	Dismantling the Gushurst store in Lead, 1899 *of the Dakotahs*	312
12.7.	The Lead City District School on Wall Street	315
12.8.	The school buildings of Lead's Central Campus	316
12.9.	The South Lead School	317
12.10.	Medicine bottles of Lead's early-day druggists.	317
12.11.	A pencil sketch by Buck O'Donnell entitled, *Where All the Mining Was Done*	318
12.12.	The Homestake Hotel, July 4, 1910	319
12.13.	The Smead Hotel, constructed at the northwest corner of Pine and North Mill streets in 1901	320
12.14.	Ivy Baldwin's tightrope walk over the Open Cut on July 4, 1916	323
12.15.	Ivy Baldwin's parachute jump from a hot-air balloon over the city (Mountain Top) ballpark on July 4, 1916	324
12.16.	Lead's Labor Day Parade, 1961	325
12.17.	The "cage ride" at Lead's Labor Day Celebration	326
12.18.	The "mine train" ride at a typical Labor Day Celebration in Lead	326
13.1.	The mining camps and towns located near Deadwood Creek in upper Deadwood Gulch.	328
13.2.	"Bird's-eye view of Gayville, 1877" by Stanley J. Morrow	329
13.3.	View of Gayville showing the Gibbs, Cook, and Parker Mill at far left, Cyanide Plant No. 2 (upper center), and the Columbus Mill (formerly the Baltimore and Deadwood Mill) at lower right.	332
13.4.	Pollock and Boyden photo of Central City and custom stamp mills, July 1880	334
13.5.	View of Central City from above the Father DeSmet Mill	335
13.6.	Central City near the mouth of Hidden Treasure Gulch	335
13.7.	The Black Hills Brewing Company	338
13.8.	The bottling works of the Black Hills Brewing Company	339
13.9.	Central City, shortly after the fire of April 25, 1888	340
13.10.	The town of Terraville	341
13.11.	The Terraville Tunnel	343
14.1.	The Rossiter Cyanide Plant in Deadwood	346
14.2.	Cyanide Plant No. 1 in Lead	349

14.3. W. B. Perkins photo of the Cyanide Plant No. 1
after its first major upgrade 350
14.4. General plan showing location of Homestake
Cyanide Plant No. 1 and other facilities, 1903..................350
14.5. Butters and Mein distributor filling a redwood sand-leaching vat 351
14.6. Cyanide Plant No. 2 at Gayville, 1908.................................354
14.7. The Slime Plant in Deadwood......................................357
14.8. Plan view of Homestake's stamp mills, cyanide plants,
and slime-treatment plant, ca. 1931..................................358
14.9. The slime pipeline between the Cyanide Plant No. 2 in
Gayville and the Slime Plant in Deadwood, 2001 359
14.10. General arrangement drawing of the
Slime Plant in Deadwood, 1907 360
14.11. Filter presses in the Slime Plant......................................361
14.12. Tube mills in the original Regrind Plant 363
14.13. Flow sheet showing Homestake's amalgamation,
cyanidation, and refining processes, 1915 364
14.14. The South Mill (upper left) and
Cyanide Plant No. 1 (lower left) 365
14.15. General section through the South Mill............................366
14.16. The cam floor of the South Battery of the South Mill.................366
14.17. The original rod mills in the South Mill, 1922 367
14.18. Construction of the plate floor at the South Mill, 1922 368
14.19. Flow sheet of mills and cyanide plants at Lead,
South Dakota, and vicinity, 1931 369
15.1. The Edison 75-lamp dynamo that provided the first electrical
lighting at the Hearst Mercantile Store and part of the 80-stamp
Homestake Mill in 1888..371
15.2. Two 60-kilowatt Edison dynamos at the Old Abe
compressor room, ca. 895 372
15.3. Homestake's first hydroelectric plant, located approximately
one-half mile above the mouth of Spearfish Canyon 373
15.4. The inside of the Englewood Hydroelectric Plant 374
15.5. The Maurice Intake Dam for Hydroelectric Plant No. 1...............376
15.6. The 23,862-foot-long diversion tunnel for
Hydroelectric Plant No. 1 377
15.7. The Forebay, located at the downstream end of the
diversion tunnel for Hydroelectric Plant No. 1 377
15.8. The standpipes above Hydroelectric Plant No. 1..................378
15.9. Homestake's Hydroelectric Plant No. 1..............................378

15.10. Excavation for the penstocks between the standpipes and the
       Hydroelectric Plant No. 1, ca. 1910..................................379
15.11. Westinghouse generators at the Hydroelectric Plant No. 1............380
15.12. Drilling a drift round in Tunnel No. 3 using electric
       Templeton drills, September 19, 1910.................................382
15.13. The completed tunnel with concrete arch between the
       Maurice Intake and the Hydroelectric Plant No. 1, 1911...............383
15.14. The cam floor of the Golden Star Mill................................384
15.15. The Cornish Pump...385
15.16. The B&M pump room on the 1,100-foot level............................386
15.17. The Central Steam (Boiler) Plant and other
       surface facilities of the B&M Group..................................387
15.18. The drum shaft for the new steam-powered ore hoist that was
       installed at the B&M Shaft in 1915...................................388
15.19. The steam-powered ore hoist at the B&M Shaft, 1917...................388
15.20. James Meddaugh photo of the Ingersoll-Sergeant five-drill steam
       compressor installed at the B&M Shaft in 1895........................389
15.21. The Lilly Hoist Controller, similar to the one developed
       by William J. Lilly, in about 1914...................................390
15.22. The Westinghouse-Parsons steam turbines inside the
       Auxiliary Steam-Electric Plant, 1932.................................391
15.23. Homestake's Hydroelectric Plant No. 2, 2008..........................392
15.24. Installation of an electric-powered, high-pressure Ingersoll-Rand
       compressor at the Ellison hoisting works, 1922.......................393
15.25. The 1,400-horsepower, two-drum electric ore hoist at the
       Ellison Hoist Room, 1932...393
15.26. The Wyodak Mine near Gillette, Wyoming, 1928.........................394
15.27. Homestake's first 100-drop magneto switchboard
       installed in 1906..395
15.28. The Kirk Power Plant, 1935...397
17.1. The first Homestake Hospital, constructed in April 1879..............414
17.2. Homestake's second hospital, constructed in 1886
       at the northeast corner of Siever and Main streets...................414
17.3. The Homestake Hospital Staff with the
       company's 1916 ambulance...415
17.4. The Homestake ambulance crew, 1931...................................415
17.5. Homestake's new brick hospital, 1922.................................416
17.6. The Hearst Mercantile Company's Brick Store on
       North Mill Street in Lead, 1898......................................417
17.7. The Hearst Mercantile Store on Lead's Main Street, 1934..............418
17.8. The $15,000 Mountain Top ball park, 1909.............................420

17.9.	Homestake's Mine Rescue team, 1927	422
17.10.	The Homestake Band, 1931	429
17.11.	The Homestake superintendent's mansion on Nob Hill	430
17.12.	The Homestake Mansion on Fairview Avenue in Lead	430
18.1.	The Golden Prospect (Highland) Shaft of the Highland Mining Company	433
18.2.	The Homestake Mine, ca. mid-1880s	434
18.3.	The Homestake Vertical Shaft	434
18.4.	Unloading 1-ton ore cars at the Ellison Shaft	435
18.5.	The Ellison Tramway, July 1, 1920	436
18.6.	Some of the Ellison support facilities, 1923	436
18.7.	One of the fire teams at the Star Shaft, 1907	437
18.8.	Diverting water from Whitewood Creek into the Savage Tunnel, 1907	438
18.9.	Conveying rock into the cave-in on South Gold Street, March 1911	440
18.10.	A typical underground crushing and skip-loading station at the Ellison Shaft	442
18.11.	The 36×48-inch Traylor jaw crusher on the 2,000-foot level at the Ellison Shaft, 1923	443
18.12.	The B&M No. 2 Shaft, ca. 1925	444
18.13.	The caving and subsidence that was plaguing Lead's business district north of Main Street, ca. 1920s	445
18.14.	The original hoist chamber for the Milliken Winze	446
18.15.	The Ellison fire, 1930	447
18.16.	The aftermath of the Ellison fire, 1930	447
18.17.	The sand dam and cement silo	449
19.1.	The Ruth and Lardner 10-stamp mill in Sawpit Gulch near Central City	453
19.2.	The Columbus Consolidated Mill in Deadwood Gulch, ca. 1902	454
19.3.	The famous Hidden Fortune gold discovery site, 1899	455
19.4.	Eastern investors inspect the Hidden Fortune Mine discovery, 1902	456
19.5.	The Bingham Shafthouse of the Hidden Fortune Gold Mining Company.	457
19.6.	A portion of the 1904 "Map of the Ore District of the Northern Black Hills," by Frank S. Peck, showing the claim holdings of the Hidden Fortune Gold Mining Company and the Columbus Consolidated Gold Mining Company	458
19.7.	The Homestake South Extension Company located along the northwest side of Whitewood Creek	460

19.8.	The shafthouse of the Oro Hondo Mining Company	462
20.1.	The Ross Shaft Complex	465
20.2.	A schematic of the mine pumping system	467
20.3.	Cyanide Plant No. 3, 1933	468
20.4.	Homestake's gold cleanup operation at the Pocahontas Mill site in Terraville, 1937	469
20.5.	The supply raise between the 4,100- and 4,700-foot levels	472
20.6.	The Golden Stairway that extended between the 1,550- and 4,100-foot levels	473
20.7.	The new 700-horsepower Jeffrey Aerodyne fan installed over the Oro Hondo Shaft, 1941	474
20.8.	The Yates, Ellison, and South Mill complexes, ca. 1942	475
20.9.	The Yates ore hoist (top) and Yates cage hoist (bottom)	476
20.10.	Deepening the Yates Shaft, 1954	477
21.1.	No. 4 Winze shaft-sinking jumbo	479
21.2.	Mucking the bottom of No. 4 Winze using sinking buckets and a hydro-mucker clamshell (not shown), 1962	480
21.3.	Hanging a set of shaft steel at the 6,350-foot level in No. 4 Winze, 1962	481
21.4.	The newly installed ore hoist at No. 4 Winze, 1962	482
21.5.	Final inspection of the newly installed sheave wheel and rope at No. 4 Winze, 1962	482
21.6.	Slushing ore in the skip-loader trench on the 6,800-foot level at No. 4 Winze, 1966	483
21.7.	No. 5 Shaft and hoist house, 2006	484
21.8.	Alimak full-face method for driving the No. 5 Air Shaft extension	485
21.9.	The first borehole raise drill at the Homestake Mine, May 1967	487
21.10.	Installing the borehole reaming head and track for mucking the drill "cuttings." This photo, taken in 1967, shows the first 6-foot-diameter reaming head that was used to bore raises at the Homestake Mine	489
21.11.	Cross section showing diamond drill holes, geologic structure, mine development, and open cut-and-fill stoping	490
21.12.	A Longyear electric-hydraulic drill with a wireline hoist	491
21.13.	Concept drawing dated July 11, 1969, by Tony Seiler, Homestake mining engineer, showing the proposed location and configuration for No. 6 Winze relative to the Ross Shaft	492
21.14.	Idealized section of the Homestake Mine	494

21.15.	The two 700-horsepower Kirk fans and isolation doors	497
21.16.	A portable refrigeration unit called a spot cooler	498
21.17.	The 3,000-horsepower American Davidson centrifugal fan, *left*, and the 1,250-horsepower Jeffrey Aerodyne backup fan, *right*, over the Oro Hondo Shaft	499
21.18.	The 2,300-ton-capacity refrigeration plant on the 6,950-foot level	500
21.19.	The three main ventilation circuits of the Homestake Mine	501
22.1.	The Grizzly Gulch Tailing Storage Facility	506
22.2.	A simplified flow sheet showing the overall gold-recovery process used at the Homestake Mine after the Slime Plant was decommissioned in 1973	507
22.3.	The 125-foot-diameter Dorr slime thickeners	509
22.4.	The dissolution agitator tanks at the Carbon-in-Pulp Plant	509
22.5.	Aerial view of Homestake's Wastewater Treatment Plant	510
22.6.	Plan illustration of Homestake's Wastewater Treatment Plant	511
22.7.	Carpco-Humphrey spirals	513
23.1.	Illustration showing the square-set cut-and-fill method	516
23.2.	Illustration showing the overhand open cut-and-fill method	518
23.3.	Drilling out a "slab round" in an open cut-and-fill stope using some of the first "jackleg" drills with telescopic legs	519
23.4.	"45-ing" the back of an open cut-and-fill stope using jackleg drills	520
23.5.	"45-ing" the back of an open cut-and-fill stope using stope jumbos designed by Homestake	520
23.6.	Slushing ore to the binline in an open cut-and-fill stope	521
23.7.	An open cut-and-fill stope ready for sandfill	522
23.8.	A typical "doghouse" used by an underground shift boss on the 2,750-foot level, ca. 1941	523
23.9.	The first vertical crater retreat drill at Homestake, 1978.	525
23.10.	Illustration showing the vertical crater retreat (VCR) mining method	526
23.11.	A Tamrock-Secoma electric-hydraulic longhole drill with diesel-tramming capability	527
23.12.	A 2-boom, Jarvis Clark MJM-20B diesel-powered drift jumbo equipped with pneumatic drills	528
23.13.	An Elphinstone R1300, 3.5-cubic-yard load-haul-dump (LHD)	529
24.1.	The Open Cut crushing plant and crushed ore stockpile, 1998	532
24.2.	The 6,300-foot-long Japanese Pipe Conveyor	532
24.3.	An aerial view of the Open Cut area, ca. 2001	533

24.4.	Open Cut Reclamation Summary, 1999	535
24.5.	The top of the reclaimed East Waste Rock Facility	536
24.6.	A portion of the reclaimed Sawpit Waste Rock Facility featuring a wildlife habitat area	536
25.1.	Transferring the survey control at No. 3 Winze	539
25.2.	Illustration showing how the survey control was transferred from the surface to an underground level or from one underground level to another	539
25.3.	Channel sampling at the Homestake Mine, 1923	541
25.4.	One of the chlorine-parting furnaces that was utilized to part silver from the crude bullion	545
25.5.	A gold pour at the Homestake Refinery	546
25.6.	Gold bars at the Homestake Refinery, 1966	547
25.7.	Gerald Powers making a silver pour using an electric tilting furnace	547
25.8.	Flow sheet showing the various refining processes used at the refinery	548
25.9.	Inside view of the foundry	549
25.10.	Jim Brosnahan and Lee Wright making a pour from a cupola, 1954	550
25.11.	Removing a metal casting from the sand mold	551
25.12.	Grinding the bases of hand grenades cast at Homestake Foundry during World War II	552
25.13.	The Brookhaven solar neutrino experiment	554
25.14.	Dr. Ray Davis, senior scientist in charge of the Brookhaven-Homestake solar neutrino experiment, inspects progress on assembly of the chlorine tank in the tank chamber, 1965	555
25.15.	Dr. Ray Davis takes a dip in the water surrounding the tank of perchloroethylene in the tank chamber, 1971	555
25.16.	Dr. Ray Davis assembles the argon gas extraction and purification system in the control room	556
27.1.	Representatives of Homestake and the South Dakota Department of Environment and Natural Resources who made an inspection of Homestake's underground decommissioning and closure work on June 13, 2003	565
27.2.	The City Dump, ca. 1999	566
27.3.	The City Dump site on July 9, 2002, a year after it was reclaimed by Homestake at a cost of $1.5 million	566
27.4.	The Yates Waste Rock Facility, November 12, 2004	567

27.5.	The Yates Waste Rock Facility in July 2009, after it was reclaimed by Homestake at a cost of $3.5 million	567
27.6.	Aerial view showing the 186 acres of land that Barrick Gold Corporation donated to the South Dakota Science and Technology Authority in 2006	571
27.7.	The proposed laboratory module at the 4,850-foot level for the Deep Underground Science and Engineering Laboratory at Homestake	575
27.8.	The proposed laboratory module at the 7,400-foot level for the Deep Underground Science and Engineering Laboratory at Homestake	576
28.1.	Fred Manuel	577
28.2.	Moses Manuel	578
28.3.	Henry Clay "Hank" Harney	582
28.4.	Alexander "Alf" Engh	582
28.5.	George Hearst, a partner in Hearst, Haggin, Tevis, and Company	583
28.6.	Phoebe Apperson Hearst	588
28.7.	James Ben Ali Haggin	591
28.8.	Lloyd Tevis	593
28.9.	Charles Washington Merrill	596
28.10.	President William Howard Taft's visit to the Homestake Mine on October 21, 1911	601
28.11.	Gutzon Borglum's and Mary Garden's visit to the Homestake Mine, October 20, 1931	602

Tables

1.1.	Stratigraphy of the Black Hills and surrounding area	35
2.1.	Divisions of the Sioux Indians	52
6.1.	Stamp mills along Gold Run Gulch in Lead City (1877-78)	145
7.1.	Mining companies along the Homestake Belt (Golden Gate to Lead)	157
7.2.	Operating costs for the Homestake Mine (May 1, 1878-Aug. 31, 1880)	162
9.1.	Cost of mining 200 tons of ore per shift using the Homestake drawhole method of mining	217
9.2.	Compressed-air locomotives used at the Homestake Mine	234
10.1.	The large stamp mills along the Homestake Belt	243
10.2.	Plate arrangement and approximate distribution of recovery at the Amicus Mill	261
10.3.	Distribution of gold recovery in the Homestake stamp mills (May-August 1910)	262
10.4.	Mill production in tons through the Homestake stamp mills (1917-1920)	264
11.1.	Narrow-gauge locomotives of the Black Hills & Fort Pierre Railroad Company	295
14.1.	Typical sand treatment cycle (1930)	352
15.1.	Hydroelectric Plant No. 1 stream diversion tunnel (Maurice Dam to Forebay)	382
17.1.	Wage scale at the Homestake Mine (1877-1917)	411
17.2.	Mine fatalities along the Homestake Belt (1876-2003)	424
24.1.	Open Cut surface mining production	534
26.1.	Production statistics for the Homestake Mine (1878-2002)	558
26.2.	Product revenue and dividends of the other Homestake Belt mines through May 31, 1901	559
26.3.	Product revenue from the Homestake, Highland, and Deadwood-Terra Mining companies (1878-1894)	560
26.4.	Historic ore production in tons by level and ledge (1878-2001)	562

Dedication

To all the men and women whose skill, ability, dedication, and perseverance transformed a world-class ore deposit into a world-class mine and who are now working to help create a world-class science and engineering laboratory.

Epigraph

The Mission of Gold

Gold is the bait whereby nature invites civilization. Whenever the old mother of us all has an especially choice bit of land, teeming with wealth of all kinds, but unknown and unvalued, she throws around the edges of this favored region the gleam of the yellow metal, and men rush in droves to secure it. For gold, they imperil their lives, their health, their family relations, their present happiness, sometimes their souls, and old Mother Nature sits by, watching and content, knowing that in the pursuit of the treasure they will find the other and more valued possibilities of her land. Just as money, in itself is of no value, but is merely a medium of exchange, so are the gold discoveries not so valuable for what they will produce of wealth as for the other avenues of trade which they will open up.

—*The Deadwood Daily Pioneer-Times,* April 1, 1909

Preface

The cover photo, courtesy of the Museum of the American West, Autry National Center, portrays the oil painting *American Progress*, which was artistically created by John Gast in 1872 and marketed as a chromolithograph by George A. Crofutt in 1873. The painting depicts the essence of the Manifest Destiny—the belief that the United States was destined and certain to expand its civilized world to the Pacific Ocean. At the dawning of this civilization is the goddess of progress who personifies the Manifest Destiny. With the star of empire as her vision and the book of wisdom and knowledge in her hand, she breezes across the western lands stringing the wires of communication that will soon bind the virtues of the east with the incompleteness of the west.

The goddess brings forth the light of civilization to the seemingly dark and cloudy lands of the western frontier. Accompanying her is the emigrant wagon train, stagecoach, prospector, farmer, and rancher. Not far behind are the railroads that bring the goods, supplies, and more fancied people of the east. Fleeing from progress, all the while, are the victims of the Manifest Destiny—the American Indians, who have their own vision of the American Dream. The Indians, with youngster and elder in tow, are forced to abandon their favored homelands with much despair and reluctance, as do the buffalo, bear, and other life forms indigenous to the western frontier.

Such was the case with the discovery and development of the famous Homestake Mine—a small, but important facet of the Westward Expansion. The gold mine was discovered in the Black Hills in 1876 when the Hills were still a part of the Great Sioux Reservation that was established under the Fort Laramie Treaty of 1868.

Much has been written about the Black Hills (Custer) Expedition of 1874 or about some of the more famous battles of the Great Sioux War of 1876-1877. A lesser amount has been written about the rich mining history of the Black Hills. Only a handful of authors, such as Edward Lazarus or Martin Luschei, have described how the Black Hills were expropriated from the Great Sioux Reservation for the benefit of the miners, emigrants, and the entire United States.

Unfortunately, the entire Homestake story has not been told to date, primarily because of the longevity of the mine and the seemingly abstruse and arcane nature of much of the historical information. The story would also be diminished if it didn't include the bearing and relationship to the Great Sioux War of 1876-77, the expropriation of the Black Hills from the Great Sioux Reservation, and the related sacrifices of many people.

The Homestake legacy and story continue to evolve. Although the final chapter about Homestake's mining history has been closed, a new chapter about science has been opened, based on current work to convert the mine to an underground science and engineering laboratory. Hence, the title of the book, *Nuggets to Neutrinos: The Homestake Story.*

Acknowledgments

Knowingly or unknowingly, a fairly large number of knowledgeable people have documented much of the history of the Black Hills and the Homestake Mine over the last 135 years or so. These priceless tidbits of information—sometimes quite obscure—can be found in family histories, personal notes, letters, memoirs, reports, technical papers, photographs, illustrations, plats, maps, deeds, affidavits, agreements, litigation, newspapers, and books. To the originators of these works, the author is grateful.

A special thank-you is extended to Todd Duex and Margie Winsel Boorda of Barrick Gold Corporation, the parent company of Homestake Mining Company. Todd, who is the site manager for Homestake and Barrick at Lead, South Dakota, granted the author access to and use of photos and records still in Homestake's possession. Margie, who is director of Corporate Land Management for Barrick and former corporate land manager for Homestake, was especially helpful in taking the time to locate and furnish large amounts of historical information relative to chains of title on mining claims and water rights.

The author is also quite appreciative of the efforts of Carolyn Weber of the Homestake Adams Research and Cultural Center and Cyndi Fisher and Tom Nelson of the Black Hills Mining Museum for allowing use of historic photos and illustrations from the Homestake Collections.

Kathy Hart, Todd Hubbard, Gary Lillehaug, Evelyn (Schnitzel) Murdy, Bob Otto, Tom Regan, and Mark Zwaschka, all former employees of Homestake, provided photos from their personal collections. Charles Mosley, son of the late Fred Mosley, who was foreman of the Homestake Refinery, furnished photos from Fred's collection. Jerry Bryant, Gary Richards, Keith Shostrom, Colette (Flormann) Bonstead, and James W. Harney, great-grandson of Henry Clay "Hank" Harney, provided images of rare photos from their personal collections. John R. Fielder, son of Mildred Fielder, and Mike Green, stepson of Joel Waterland, gave approval to use photos from the Fielder and Waterland Collections, which were made available to the author by Donna Neal at the Devereaux Library, South Dakota School of Mines and Technology. Shirley Anderson and Larry Bradley of the W. H. Over Museum in Vermillion, South Dakota, provided early photos of Lead and Gayville from the Stanley Morrow Collection. Jane Koropsak and Mona Rowe approved use of photos of the Brookhaven Solar Neutrino Experiment, courtesy of the Brookhaven National Laboratory. Syd DeVries provided illustrations for the Deep Underground

Science and Engineering Laboratory at Homestake, courtesy of the Lawrence Berkeley National Laboratory.

Ken Stewart and Matthew Reitzel were particularly helpful in locating and scanning photos and other information at the South Dakota State Historical Society in Pierre. Janet Carter and Daniel Driscoll, of the U.S. Geological Survey, provided excellent illustrations of Black Hills geology and hydrology. LeeAnn Paananen and Juliet Heltibridle helped locate and scan photos from collections at the Phoebe Apperson Hearst Library in Lead. The Deadwood Public Library has an excellent search engine and facilities for researching older newspapers that the library has on microfilm. Mel Schmidt of the MS Book and Mineral Company provided excellent illustrations of the Joshua Hendy Stamp Mill. Wayne Paananen provided superb aerial photos of Lead and the Open Cut, courtesy of Historical Footprints. Marilyn Kim, the collections assistant for the Museum of the American West, Autry National Center, approved use of an image of the 1872 oil painting *American Progress* by John Gast. Thanks to each of these individuals and organizations for their efforts and contributions.

The Web sites for the Library of Congress, National Archives and Records Administration, and National Atlas of the United States were quite useful for sourcing and allowing use of historic photos and maps. The *All Topo Maps* from iGage Mapping Corporation was also useful for the creation of topographic illustrations using personal Global Positioning System (GPS) information.

Charles Tesch Sr., a longtime employee of Homestake, was most helpful in making his personal files available and answering questions. Mark Zwaschka and Bob Otto, former Homestake geologists, provided information and helped edit the chapter on Black Hills and mine geology. Kirby Denton, Alvin Dyer, Ron Enderby, Greg King, John Marks, Jeri Mykleby, LeEtta Shaffner, Larry Trautman, and Ron Waterland, former Homestake employees, also provided detailed information and helped answer questions. Mike Cepak, Natural Resources engineering director for the Minerals and Mining Program of the South Dakota Department of Natural Resources, provided public information and reports relative to Homestake from the state's files. Ariel Larson and Jerry Bryant provided historical information about the Black Hills and Fort Pierre Railroad, as well as camps along the route.

Sheree Green, Roxana Feterl, and Kari Podoll of the Lawrence County Register of Deeds Office were most helpful in directing the author to the proper index books and records. Attorney Max Main helped with legal questions.

Wendy Pitlick of Quill and Scroll Writing and Editing Services and Chile Gadingan of Xlibris Publishing Company deserve a sincere thank-you for their copyediting work.

Last, but not least, the author would like to thank his wife, Cindy, for her support, love, and understanding during the last two years when writing the manuscript or conducting research was the ostensible priority in our lives. Not every wife will tramp through the woods helping trace out long-abandoned railroad lines and water flumes or patiently wait in the car a "few more minutes" while her husband procures "one more GPS location" or searches for "one more record" at the library or courthouse. Thanks, Cindy.

Introduction

Very few mines in the world ever produced gold continuously for more than one hundred years. The Homestake Mine was one that did, producing 40 million ounces of gold from 1876 through 2001, when the quest for the yellow metal was brought to an end for good. Over the next few years after the mine was shut down, tens of thousands of ounces in additional gold were recovered as mine facilities were systematically decommissioned, and the mill site was reclaimed and converted to an open-air museum.

For more than 125 years, the Homestake Mine helped support the livelihoods of countless numbers of people who were directly or indirectly affiliated with the mine. Sadly, some of these people lost their lives or were physically impaired while working at the mine or in support of the mine. Fortunately, a lasting legacy evolved from the dedication, loyalty, and perseverance of each of these people and every other person who was ever associated with the mine. This living legacy continues to evolve with the transformation of the mine into a deep underground science and engineering laboratory.

The Homestake legacy began to unfold in August and September 1875 when the Bryant, Blanchard, Smith, Gay, and Lardner parties discovered rich gold placers in Deadwood Gulch. What they found was mostly Homestake gold, weathered and worn to "nuggets" and "dust." Fred and Moses Manuel, along with their partners, Henry C. "Hank" Harney and Alexander "Alf" Engh, were latecomers to Deadwood Gulch, arriving in February 1876. For the most part, these four men were more interested in finding the source of the placer gold or the "lode gold." Their prowess and diligence paid off. On April 9, 1876, Moses Manuel and Hank Harney discovered a rich quartz outcrop upon which all four men located the Homestake lode claim.

The Black Hills was still a part of the Great Sioux Reservation then, pursuant to the Fort Laramie treaties of 1851 and 1868. It wasn't until the Manypenny Agreement was signed on September 26, 1876, and ratified by Congress on February 28, 1877, that the boundaries of the Great Sioux Reservation were modified, thereby excluding the Black Hills from the reservation and allowing the miners to have a "legal" presence in the Black Hills. The Teton Sioux, also known as the Lakota, probably weren't the first American Indians to have a presence in and around the Black Hills. Notwithstanding, the Fort Laramie treaties specified the boundaries for the Great Sioux Reservation and the Black Hills were included within that description.

Toward the latter part of 1877, the California capitalists George Hearst, J. B. Haggin, and Lloyd Tevis acquired the Homestake and Golden Terry mining claims from the Manuel brothers, Harney, and Engh. From that point forward, the California capitalists and their various other investment partners engaged themselves to try and acquire most all of the mining claims along the Homestake Belt, providing there was good ore and the price was right.

Their acquisition strategies included such methods as outright force, costly court battles litigated by the best lawyers, acquisition and control of precious water rights through separate companies, fair land purchases, creation or consolidation of mining companies, and acquisition and control of competing companies through accumulation of company stock. In other cases, the Homestake capitalists prevailed by simply waiting until the other operators went broke or some other opportunity presented itself to allow acquisition at a bargain price. Aided by their money, skill, and shrewdness, the Homestake capitalists were very successful in fulfilling their passions and paving the roadway for future generations at the Homestake Mine.

Chapter 1
The Geology

Black Hills Stratigraphy

Geologists classify the age of the rocks according to four eras of time: Precambrian, Paleozoic, Mesozoic, and Cenozoic (see table 1.1). The Precambrian era, which spans about four billion years, begins with the creation of the Earth some 4.6 billion years ago and ends with the start of the Paleozoic era some 570 million years ago. The oldest Precambrian rocks in the Black Hills are the granites that are exposed in Little Elk Creek Canyon, west of Tilford, and at Bear Mountain, northwest of Custer. These early granites are approximately 2.5 billion years old and were formed during the Archean period of the Precambrian era.

Table 1.1 Stratigraphy of the Black Hills and surrounding area

Era	Period	(millions of years) Epoch	Formation	Comments
Cenozoic (age of mammals)	Quaternary	Holocene	Alluvium and Colluvium	0-50 ft.; Contain gold placers
		0.01-2.0 Pleistocene	Gravel Deposits	Stratified paleochannels, pediments, and stream terraces.
	Tertiary	2-5 Pliocene	Gravel Deposits	Heterogeneous gravels from igneous & metamorphic rocks
		5-24 Miocene	Arikaree Sandstone	Deposition of clays and sands
		24-38 Oligocene	White River Group	Greenish clays, bentonites, sands, silt, volcanic ash; fossils
		38-55 Eocene	Igneous Intrusive Rocks	Rhyolite, trachyte, & phonolites in dikes, sills, laccoliths
		55-63 Paleocene		

Era	Period	Age (Ma)	Formation	Description
Mesozoic (age of reptiles)	Cretaceous	63-138	Hell Creek	Dark clays/lignites with last of dinosaur bones; 0-450 ft.
			Fox Hills	Greenish-gray shales; 0-200 ft.
			Pierre Shale	Gray shales 1,200-2,700 ft.
			Niobrara	Chalky limestones/clay; 80-300'
			Carlile Shale	Gray shales with shark's teeth
			Greenhorn	Slabby, buff limestone to 750 ft.
			Belle Fourche Shale	Gray shales; to 150-850 ft.
			Mowry Shale	Dark gray shales; 125-230 ft.
			Newcastle Sandstone	Sandstone; 0-150 ft.
			Skull Creek Shale	Gray/black shales; 150-270 ft.
			Fall River	Massive sandstone; 10-200 ft.
			Lakota	Brown sandstone; 35-700 ft.
	Jurassic	138-205	Morrison	Shale, sandstone, gypsum; dinosaur rem; 0-200 ft.
			Unkpapa Sandstone	Varicolored sandstone; 0-225 ft.
			Sundance	Yellow sandstone; 250-450 ft.
	Triassic	205-240	Spearfish	Red, silty shales/sandstone; gypsum beds; 375-800 ft.
Paleozoic (age of fishes)	Permian	240-290	Minnekahta Limestone	Fine-grained, purple to gray laminated limestone; 25-65 ft.
			Opeche	Red siltstone, gypsum, shale; 25-150 ft.
	Pennsylvanian	290-330	Minnelusa	Yellow/red stratified sandstone, limestone, dolomite, & red shale at base; 375-1,175 ft.
	Mississippian	330-365	Madison (Paha Sapa) Limestone	Gray to buff, cavernous limestone and dolomite; cliffs; gold bearing; 200-1,000 ft.
	Silurian and Devonian	410-435	Englewood	Pink, shaly dolomite; gray shale at base; few fossils; 30-60 ft.
	Ordovician	435-500	Whitewood	Limestone & dolomite; 0-235 ft.
			Roughlock	Dolomitic sandstone; 0-30 ft.
			Winnipeg/Icebox	Green shale/siltstone; 0-150 ft.
	Cambrian	500-570	Deadwood	Brown to light-gray sandstone, shale, limestone; local basal conglomerate. 0-500 ft.
Precambrian		570-1,715		Long Period of Erosion
	Proterozoic	1,715-2,500	Harney Peak Granite	(southern Black Hills)
			Grizzly	Schist/phyllite; to 2,000 ft.
			Flag Rock	Basalt/chert; 4,000-5,000 ft.
			Northwestern	Mica schist/phyllite; 4,000 ft.
			Ellison	Quartzites, mica schist; to 4,500'
			Homestake	Iron-rich, siderite/grunerite schist; gold bearing; 50-300 ft.
			Poorman and Yates Member	Slates/schists; 1,000-2,000 ft.
	Archean	2,500+	Elk Cr./Bear Mtn. Granites	Gneiss/granite

Sources: John Paul Gries, *Roadside Geology of South Dakota* (Missoula: Mountain Press Publishing Company, 1996); U. S. Department of the Interior, *Atlas of Water Resources in the Black Hills Area, South Dakota* by J. M. Carter, D. G. Driscoll, J. E. Williamson, and V. A. Lindquist. Hydrologic Investigations Atlas HA-747, U. S. Geological Survey, 2002.

The next oldest Precambrian rocks are the metamorphic quartzites, slates, schists, and igneous granites prevalent in the central and southern Black Hills. These metamorphic rocks were formed during the Proterozoic period some 1.715 to 2.5 million years ago. Today, Proterozoic rocks extend over an area roughly 60 miles long, north and south, by about 25 miles, east and west. The Homestake Mine in the northern Black Hills and the vast majority of mines in the central and southern Black Hills area are located in rock formations that were created within the Precambrian era. The youngest Precambrian rocks in the Black Hills are the Harney Peak granites that comprise Harney Peak and Mount Rushmore.[1]

The Paleozoic era is known as the Age of Fishes. This era extends from about 570 million years ago to 240 million years ago. The Paleozoic era begins with the appearance of marine invertebrate faunas and includes the time in which primitive fish appeared. Later, land, plants, and animals appeared, forests appeared, and finally primitive reptiles began to appear. The Deadwood Formation is the earliest rock formation in the Paleozoic era. The massive limestone formations of Spearfish Canyon, Whitewood Canyon, Little Spearfish, Bear Butte Creek, False Bottom Creek, and others were formed during the Paleozoic era. Except for the Homestake Mine and a few others, the majority of mines in the northern Black Hills are located in rock formations that were formed during the Paleozoic era.[2]

The Mesozoic era, known as the Age of Reptiles, represents the period from about 240 million years ago to 63 million years ago. Birds first appeared during this era, along with primitive mammals. Reptiles spread across the continents, and dinosaurs began to appear. The dinosaurs diversified, spread, and flourished, but suddenly became extinct when some major event occurred near the end of the Cretaceous period about 65 million years ago. Some scientists theorize that a clue to the dinosaurs' sudden extinction relates to a thin band of clay that exists between the Cretaceous and Tertiary sediments on all continents. Scientists believe that the boundary clay, as it is called, is a byproduct of some catastrophic event such as a supervolcanic eruption or asteroid that struck the earth and caused a large cloud of dust and ash to settle upon the earth.

Much erosion and deposition of sediments occurred throughout the Mesozoic era, forming most of the sandstone and shale formations that exist throughout much of present-day South Dakota. By the end of the Cretaceous period, some 7,500 feet of Paleozoic and Mesozoic sediments covered the Precambrian igneous granites and metamorphic schists and slates of the Black Hills.[3]

The Cenozoic Era is known as the Age of Mammals. During this 63-million-year era, mammals began to develop and diversify from a few simple life forms into a wide array of terrestrial and marine vertebrates. Grasses became abundant, as did horses, elephants, and other herbivores.[4] Large carnivores became dominant during the Pliocene epoch two to five million years ago. Finally, man himself appeared during the late Pleistocene epoch.[5]

The Laramide orogeny occurred near the start of the Cenozoic era some 62 million years ago and lasted for approximately 20-25 million years. An orogeny is a geologic process whereby mountains are formed by the tectonic forces from the thrusting and folding of the earth's crust. The Laramide orogeny uplifted and formed the Black Hills, Big Horn, and Laramie mountain ranges. Concurrently, the Williston and Powder River basins were thrust downward into their present configurations. In the Black Hills, the Proterozoic Harney Peak granites were thrust upward along with the 7,500 feet of Paleozoic and Mesozoic sedimentary rock formations.[6]

Between about 50 and 58 million years ago, from the late Paleocene epoch through much of the Eocene epoch, new magmas associated with the Black Hills uplift intruded the Precambrian rock formations and the overlying sedimentary formations. The magmas formed many of the buttes and peaks that dominate the northern Black Hills, including Bear Butte, Custer Peak, Terry Peak, and Pillar Peak. In a geological sense, these buttes and peaks are called laccoliths and are comprised of one or more of the felsitic igneous rocks such as rhyolite, phonolite, and porphyry.

Magmas intruded the Precambrian rocks of the Homestake Mine, as can be seen in the Open Cut in Lead (see fig. 1.1). The magmas were emplaced as dikes along the foliation of the steeply dipping Precambrian schists and phyllites and transitioned to form sills parallel to the strata of the Deadwood Formation, which lies unconformably over the Precambrian rocks. The dike swarms strike north to northwest and dip east to northeast. Dikes and sills of rhyolite and phonolite crosscut all of the Proterozoic rock units in the Lead Window and have been exposed throughout the underground workings of the Homestake Mine.[7] Concurrent with this igneous activity, the Precambrian rock units were subjected to a brittle phase of deformation that resulted in prominent joint sets.[8]

Fig. 1.1. The Tertiary rhyolite dike swarm in Homestake's Open Cut, 2008. The tan-to-buff-colored "ribbons" of rock shown in the center of this photo are Tertiary rhyolite dikes that have intruded the Precambrian rocks of the Homestake Mine. Modern surface mining was conducted in the Open Cut from October 1983 to September 1998.—Photo by author.

By about 37 million years ago, most of the uplifting of the Black Hills was completed. The period of uplifting was accompanied and followed by a long period of erosion that lasted through the Oligocene and Miocene epochs. As a result, the overlying softer sedimentary rock formations were slowly eroded away, exposing the uplifted Precambrian rocks and creating an appearance of concentric rings of Paleozoic and Mesozoic rocks around the core (see figs. 1.2 and 1.3). Further erosion formed the gold placers throughout the streams and valleys of the Black Hills.[9]

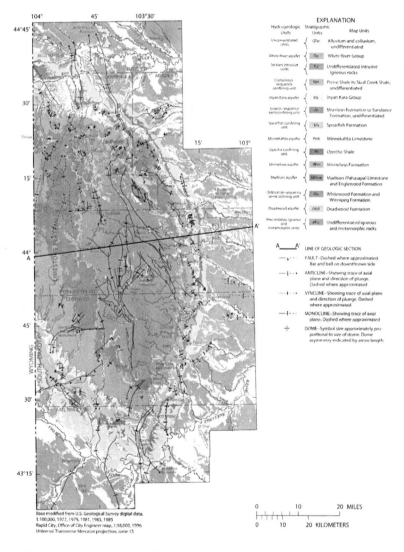

Fig. 1.2. Geologic map of the Black Hills. Refer also to cross section A-A'.—Reproduced by permission from Janet M. Carter and Daniel G. Driscoll, *Hydrology of the Black Hills Area, South Dakota*, Water-Resources Investigations Report 02-4094, U.S. Geological Survey, 2002.

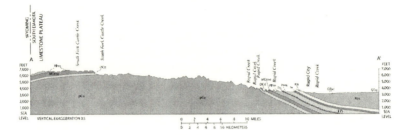

Fig. 1.3. Geologic cross section A-A'. This cross section illustrates the present-day geologic setting of the Black Hills. —Reproduced by permission from Janet M. Carter and Daniel G. Driscoll, *Hydrology of the Black Hills Area, South Dakota*, Water-Resources Investigations Report 02-4094, U.S. Geological Survey, 2002.

The sedimentary rock units, particularly the Deadwood, Madison (Paha Sapa), Minnelusa, Lakota, and Fall River formations, contain the main aquifers of the Black Hills (see fig. 1.4). Most of the groundwater within the Madison Formation enters the formation in areas where it is exposed, particularly around the headwaters of Spearfish Canyon near O'Neill Pass. Water wells drilled in the eastern portions of the Black Hills and prairie areas are often artesian (i.e., a flowing wellhead) if the weight of the water in the saturation zone "above" the bottom of the well is great enough to overcome the depth of the well and any friction losses within the aquifer.[10]

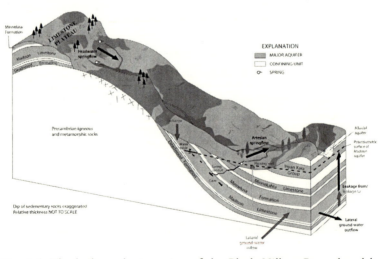

Fig. 1.4. The hydrogeologic setting of the Black Hills.—Reproduced by permission from Janet M. Carter and Daniel G. Driscoll, *Hydrology of the Black Hills Area, South Dakota*, Water-Resources Investigations Report 02-4094, U.S Geological Survey, 2002.

The Lead Window

Because much of the content of this book is directed toward the Homestake Mine and other mines along the Homestake Belt, discussion is limited to rock units of the northern Black Hills that formed during the Proterozoic and Cambrian periods. The Homestake Belt generally includes the trend of the Homestake Formation from Sawpit Gulch, near Central City, to Whitewood Creek and Kirk Road on the southeast side of Lead. The Homestake Mine and the other large lode mines along the Homestake Belt were located because of the ore-grade mineralization that was discovered within the "leads" or "ledge" substructures of the Precambrian Homestake Formation. Other mines along the belt were located because of ore-grade mineralization discovered within the overlying Cambrian Deadwood Formation.

Several Proterozoic rock units exist in the northern Black Hills around the city of Lead (see fig. 1.5). Collectively, the area encompassed by these rock units is referred to as the Lead Window. Because of localized conditions during the Precambrian era, some geographic areas experienced heavy deposition of sands, silts, or shale, whereas other areas were devoid of such depositions. Therefore, not all of the formations in the northern Black Hills exist in other areas of the Black Hills where Precambrian rocks are exposed.

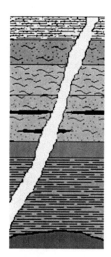

Grizzly Formation - Metagraywacke, sericite-biotite schist (3,300 ft.)

Flag Rock Formation - Biotite-sericite schist, graphitic phyllite (5,300 ft.)

Northwestern Formation - Biotite-quartz-sericite-garnet schist (0-4,300 ft.)

Ellison Formation - Sericite-biotite schist and phyllite with interbedded impure quartzite and minor amphibolite; quartzities (4,900 ft.)

Homestake Formation - Grunerite-siderite schist and chert with chlorite-rich and biotite-rich phyllite or schist; locally sulfide rich (0-400 ft.)

Poorman Formation - Thin, well-banded sericite-biotite carbonate phyllite; graphitic phyllite (700-3,300 ft.)

Yates Member - Hornblende-plagioclase schist (2,000-3,900 ft.)

Fig. 1.5. Proterozoic stratigraphy around the city of Lead, South Dakota.—Modified from U.S. Department of the Interior, "*The Homestake Gold Mine, An Early Proterozoic Iron-Formation-Hosted Gold Deposit, Lawrence County, South Dakota*" by S. W. Caddey et al., Bulletin 1857-J, *Geology and Resources of Gold in the United States*, U.S. Geological Survey, 1991.

The primary Proterozoic rocks exposed within the Homestake Mine include, from oldest to youngest, the Poorman, Homestake, and Ellison formations. These formations were formed by the deposition of sediments in a shallow sea basin from about 1.9 to 2.2 billion years ago. Thereafter, from approximately 1.7 to 1.8 billion years ago, the sedimentary formations were subjected to intense heat and pressure, metamorphosing the rocks into slates, schists, and quartzites. Additionally, the formations were subjected to several events that resulted in tight folding and very complex structural geology within the mine.

The oldest part of the Poorman Formation, called the Yates Member, is comprised of a large body of amphibolite. The Yates Member is believed to have been formed from a deep-water sequence of sediment and rocks dominated by tholeiitic basalt flows and volcaniclastic debris. The basalt was transformed to amphibolite through metamorphism. The Yates Member of the Poorman is a relatively competent rock unit that contains hornblende, plagioclase, calcite, dolomite, and ankerite. Minor amounts of ilmenite, magnetite, titanite, pyrrhotite, and pyrite also occur. Coarse-grained grunerite-quartz-biotite-chlorite schist is locally embedded within the Yates Member amphibolite.[11]

The youngest portion of the Poorman Formation is well foliated and consists of dark gray carbonaceous phyllites with minor amounts of micaceous tuffs. The phyllites contain sericite, biotite, graphite, carbonate, and quartz with localized areas of chlorite and garnets. Chlorite content is greatest in areas adjacent to the Homestake Formation.[12]

The Homestake Formation is an iron formation that ranges in width from about 10 feet to 300 feet. The strike of the formation is about North 35° 40' West. The dip varies but is generally to the northeast. The formation grades from a carbonate facies (siderite) in the western regions of the mine to a silicate facies (grunerite) in the eastern areas. The gradation is based on the degree of metamorphism, which generally reduced the iron carbonate content and increased the iron silicate content of the formation. The Homestake Formation is a light gray to greenish schist, depending on the amount of biotite and chlorite present. The formation consists of cummingtonite-grunerite, siderite, quartz, chlorite, graphite, and muscovite minerals. The Homestake Formation also contains localized fine-grained pyrrhotite, arsenopyrite, biotite and/or chlorite, chert, quartz, and gold. Pyrrhotite ranges between 8 percent and 30 percent. Arsenopyrite ranges between 1 percent and 12 percent. All of the economic gold values within the Homestake Mine exist within the Homestake Formation.

The Ellison Formation consists of quartzites, phyllites, and micaceous tuffs. Foliation is well developed in the phyllites, which closely resemble those of the Poorman Formation. The phyllites consist of sericite, biotite, quartz, and graphite. The Ellison Formation is light gray to dark gray, depending on the amount of graphite. The micaceous tuffs are composed of muscovite, chlorite, biotite, and carbonate.

The Tertiary rhyolite dikes in the Homestake Mine represent aphanitic to porphyritic magmas that preferentially flowed along the foliation of the schists in an upward direction until they reached the Paleozoic unconformity. The magmas then propagated laterally, forming sills. The rhyolites typically exhibit phenocrysts of feldspar within a matrix of feldspar, pyroxene, and quartz. The younger phonolite dikes crosscut the rhyolite dikes. The phonolite dikes tend to strike northeast and dip to the northwest.

Most all of the gold extracted from the Homestake Mine was sourced from the Homestake Formation. Associated ore minerals within the Homestake Formation include siderite, chlorite, quartz, pyrrhotite, and arsenopyrite, although not all of these were always present or indicative of ore. The majority of the ore-grade mineralization within the mine occurred during the Precambrian era.[13]

The Lead Syncline, Lead Anticline, and Poorman Anticline constitute the three main fold structures within the Lead Window (see figs. 1.6 and 1.7). Within these fold structures are substructures that are locally referred to as "ledges." Main Ledge, an anticlinal structure, hosted the original discovery of the Homestake Mine where the ledge cropped out in an area presently occupied by the Open Cut. The synclinal ledges are designated by odd numbers from east to west (i.e., 3, 5, 7, 9, 11, and so on). The anticlinal ledges are designated by even numbers and are also numbered from east to west. Diamond drilling has shown that the Homestake Formation projects more than 10,000 feet below the surface and extends more than five miles along its strike. From 1876 through 2001, ore was mined from the Homestake Formation to a depth of 8,000 feet.[14]

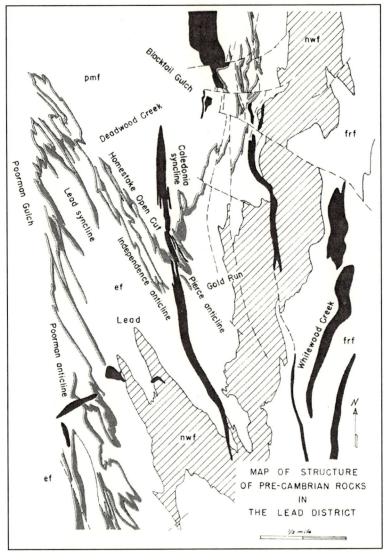

Fig. 1.6. Map of the structure of the Precambrian rocks in the Lead District. This geologic plan map illustrates the structural complexities of the Precambrian rock formations around Lead, South Dakota. The formations represented include the Flag Rock (frf), Northwestern (nwf), Ellison (ef), Homestake (hmf), and Poorman (pmf). The Tertiary rhyolite and phonolite intrusive rocks are also shown cutting the Precambrian formations.—Source: Homestake Mining Company, *Homestake Centennial: 1876-1976*, (Lead: Homestake Mining Company, 1976).

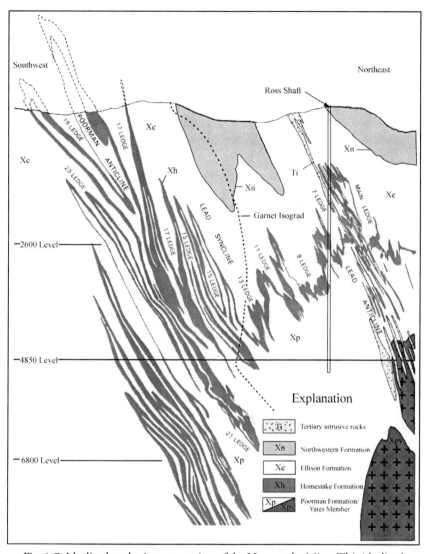

Fig. 1.7. Idealized geologic cross section of the Homestake Mine. This idealized cross section of the Homestake Mine illustrates the predominant Precambrian and Tertiary rock formations exposed in the mine and the complexity of fold structures. The substructures of the Homestake Formation represent the "ledges" from which mining operations were conducted.—Modified from U.S. Department of the Interior, "*The Homestake Gold Mine, An Early Proterozoic Iron-Formation-Hosted Gold Deposit, Lawrence County, South Dakota*" by S. W. Caddey et al., Bulletin 1857-J, *Geology and Resources of Gold in the United States*, U.S. Geological Survey, 1991.

Sidney Paige mapped the structural geology within the Lead Window in about 1923,[15] and Homestake geologists did the same thing in 1929 and 1930[16] (see fig. 1.8). The easternmost structure was named the Caledonia Syncline, which included the Caledonia Ledge and the Amicus Ledge. The axial trend of the Caledonia Syncline is slightly northeast, whereas all of the other structures trend northwest. The Amicus Ledge represents the western flank of the Caledonia Syncline. Two small open cuts, the Clara and Hercules, were situated along the Amicus Ledge between the Caledonia Mine and the Homestake Mine.

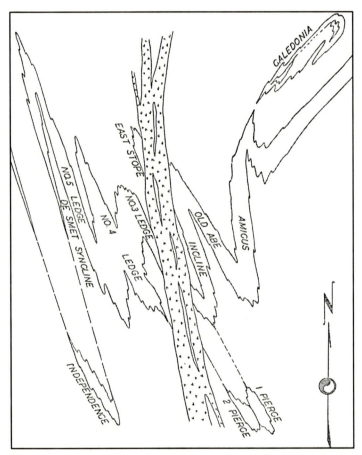

Fig. 1.8. Names and relative positions of the ore-bearing folds in the Open Cut area. This plan view illustrates the fold substructures and "ledge" designations that were mapped by Homestake geologists in 1930. The area is mostly within the present-day Open Cut. Stippled areas represent Tertiary rhyolite dikes that can be seen in the east wall of the Open Cut.—Source: A. B. Yates, "Structure of the Homestake Ore Body" (PhD diss., Harvard Engineering School, March 1931).

The Amicus Ledge also represents the eastern flank of the Old Abe Anticline, which is situated immediately east of the Pierce Anticline. The western limb of the Old Abe Anticline is referred to as the Old Abe Ledge. Ore was mined from this ledge between the 900- and 1,100-foot levels. The dip of the Amicus Ledge is approximately 60 degrees east, whereas the Old Abe Ledge dips about 60 degrees west. The Old Abe Anticline plunges southeast at about 35 degrees.[17]

The Old Abe Ledge and the Incline Ledge represent the limbs of the Incline Syncline. The nose of this structure is intersected by the rhyolite dike swarm that is exposed in the east wall of the Open Cut. That portion of the fold nose situated on the west side of the rhyolite dike swarm was called the East Stope. Abundant rich ore was mined in the Incline Ledge and the East Stope. The axial plane of the Incline Syncline generally defined the eastern limit of ore within Main Ledge.

The Pierce Anticline, sometimes referred to as the Incline Anticline, has the Incline Ledge as its east limb and No. 3 Ledge as its west limb. The No. 3 Ledge is also intersected by the Tertiary rhyolite dikes. The Pierce Anticline plunges about 35 degrees southeast. The limb represented by the Incline Ledge dips at about 65 degrees to the east. The Pierce Ledge consists of a double-fold substructure that is designated as No. 1 Pierce on the east and No. 2 Pierce on the west. The latter anticlinal substructure has been sheared off and detached from the east flank of the No. 4 Anticline. The early-day miners found that the No. 2 Pierce substructure contained continuous ore, whereas No. 1 Pierce was barren to at least the 1,850-foot level. The No. 1 Pierce appears to be sheared and detached from the main Pierce Anticline. Below about the 2,150-foot level, the No. 1 Pierce converges with the main Pierce Anticline.[18]

The north end of No. 3 Ledge forms the No. 3 Syncline to the west of the rhyolite dike swarm. The next fold structures to the west are named the No. 4 Anticline and No. 5 Syncline. The latter is also referred to as the DeSmet Syncline. Collectively, these three fold structures comprise the large Main Ledge ore body. Above the 1,850-foot level, the three ledges were mined from hanging wall to footwall without regard to the narrow ribbons of barren slate or phyllite that separated the tightly compressed fold structures. Below the 1,850-foot level, No. 5 Ledge separates from the Main Ledge Anticline. The single limbs represented by No. 3 Ledge and the No. 5 or DeSmet Ledge dip almost vertical to the east. The limb represented by No. 4 Ledge dips from 60-70 degrees to the east. The Main Ledge Anticline plunges at about 39 degrees southeast between the 800- and 1,850-foot levels.

The west limb of the DeSmet Syncline has been intensely sheared and was observed to exist as pinched and detached lenticular bodies throughout much of the Open Cut surface mine area. A very distinct shear zone, related to the intense folding that occurred, was found to exist over much of the trace of the west limb of this syncline (see fig. 1.9). To the south, the west trace of No. 5 Syncline merges into the fold nose of the Independence Anticline. The Independence, or No. 6 Ledge, represents the western flank of the Independence Anticline.[19] Historically, only a minor amount of underground mining was conducted in the Independence Ledge.

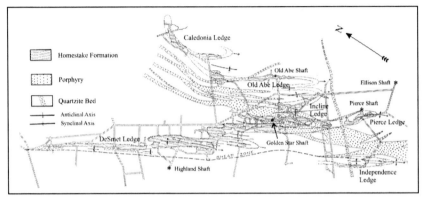

Fig. 1.9. Generalized plan of the 300-foot level of the Homestake Mine.—Source: U.S. Department of the Interior, *Description of the Central Black Hills*, N. H. Darton and Sidney Paige, Folio No. 219, Geologic Atlas of the United States: Central Black Hills Folio, South Dakota, U.S. Geological Survey, 1925.

To the west of the Independence Anticline is the large structure known as the Lead Syncline. Within this syncline are several substructures, including 7, 9, 11, and 13 ledges. These ledges, particularly 9 Ledge, yielded significant amounts of ore over the life of the Homestake Mine. West of the Lead Syncline is the Poorman Anticline, which contains the West Ledge substructures. The Poorman Anticline includes 15, 17, 19, and 21 ledges, all of which yielded ore from the lower levels of the mine.

Fossil and Recent Gold Placers

Overlying the Precambrian formations is the Cambrian Deadwood Formation. Starting at about 1.7 billion years ago, the upturned edges of the Precambrian slates and schists were subjected to a long period of weathering and erosion. At about 570 million years ago, and for the next 70-80 million years, the sediments were deposited over the eroded schists, forming the Deadwood Formation. The relatively flat, undulating horizon between the schists and the sandstones of the Deadwood Formation is referred to as the Cambrian-Precambrian unconformity. This unconformity marks the beginning of the Cambrian period of the Paleozoic era. The unconformity is evident along Houston Street in the city of Lead.

Throughout much of the late Proterozoic and early Cambrian periods, the Precambrian rocks were mostly submerged in shallow seawater. In some cases, reefs or topographic high areas protruded through the ever-deepening seawater. Wave action washed sands and gravels into depressions in the Precambrian surface and concentrated the gold that had been liberated from vein croppings of the Homestake Formation. Weathering continued to liberate additional gold from the vein croppings. Gradually, the gold was mobilized and concentrated, forming placers along Paleozoic stream channels above the unconformity. The sands and gravels and localized gold placers were eventually "cemented" with fine silica and calcium carbonate through the oxidation of pyrite, forming a conglomerate near the bottom of the Deadwood Formation. This conglomerate typically contains pebbles of quartz, quartzite, amphibolite, schist, hematite, and other detrital material that originated from the Precambrian rocks.[20]

As erosion continued into the Pleistocene and Holocene periods, the hills and valleys of the Black Hills became much more steep-walled, causing the streams to be cut down into the Precambrian rocks with force. This rapid cutting action resulted in continued erosion and liberation of gold from the Precambrian and Cambrian rocks. In some cases, the streams were completely rerouted, causing remnant placers to be left high up on the hillsides or on divides between the valleys. Gravels associated with the Old River around Tinton[21] or the famous bench placers on the south side of Spring Creek Valley at Rockerville are examples of bench and hillside placers. The historic gold placers that were discovered along Deadwood Gulch during the gold rush of 1875-76 primarily represent placers formed primarily during the Quaternary period (see fig. 1.10).

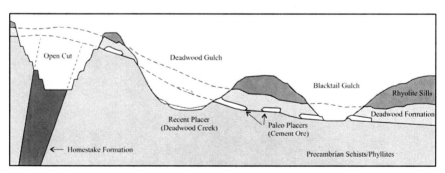

Fig. 1.10. Generalized section showing Paleo (Fossil or Cement) Placers and Recent Placers in the northern Black Hills.—Modified from Joseph P. Connolly, *The Tertiary Mineralization of the Black Hills*, (Rapid City: South Dakota School of Mines, Bulletin 15, September 1927), 56.

As the placer miners began exploiting the various gold placers in 1875-76, the lode miners quickly began searching the flanks of Deadwood and Gold Run gulches and the associated tributary gulches in an effort to locate the sources of gold. Upon sampling the outcrops of the Deadwood Formation near the ridgetops, the miners soon discovered the paleo placers along the former Cambrian stream channels, which they called the "fossil placers" or "cement ore." Because of the location and hardness of the fossil placers, the miners had to resort to underground mining methods to exploit the cement ore.

With a few exceptions, most of the early-day mines discovered around the north end of the Homestake Belt were located for the "cement ore" or paleo placers within the basal conglomerate of the Deadwood Formation. Most of these mines were discovered in areas where the lower reaches of the formation cropped out along the flanks of the gulches. Such mines included the Hidden Treasure, Alpha, Gustin, Minerva, Keets, Wooley-Peacho, Gentle Annie, Hawkeye, Pluma, Monitor, Hercules, Deadwood, Golden Terra, Big Missouri, Hidden Fortune, and others.[22]

Chapter 2
The Westward Expansion

Migration of the Plains Indians

Members of the Crow tribe are believed to have been the primary inhabitants around the Black Hills in the early 1700s. The Crow people roamed the foothills of the Black Hills and inhabited areas as far west as the Little Missouri, Powder River, and Tongue River. During this same time frame, the Kiowa and Apache people inhabited areas southwest of the Black Hills.[1] The Padouca and Comanche bands occupied portions of the present-day Wind Cave National Park area from about AD 1500 to 1700.

Around 1760, the Kiowa, Arapaho, Plains Apache, and Crow tribes began to occupy areas between the Platte River and the southern Black Hills. Members of the Suhati, Omisis Cheyenne, Kiowa, and Arapaho tribes occupied the northern fringes of the Black Hills. Areas between the Cheyenne River and the Black Hills were occupied by the Oglala, Sicangu, Minniconjou, and Itazipco Lakota, Arikara, and Ponca tribes. These tribes conducted seasonal hunting excursions to the eastern and southern portions of the Black Hills.[2]

The Arikara or Ree people lived along the Missouri River and its tributaries near present-day Pierre in the 1600s and 1700s. These people resided in circular earthen lodges dug into the ground as a cellar.[3] By 1760, approximately twenty thousand Arikara people inhabited areas along the Missouri River below Big Bend.[4] Throughout the 1700s, bands of Arikara from the Missouri River area often ventured out on buffalo hunts as far west as the Black Hills. There, the Arikara met with the Comanche, Arapaho, Kiowa, and Cheyenne, and they traded goods for horses. The Arikara also traded for horses with the Teton (Lakota) tribe of the Sioux, who had been slowly migrating southwest from the headwaters of the Mississippi River since about 1670.[5]

Although the Sioux people belonged to one linguistic group, there were three major divisions or tribes based on certain dialectic, geographic, historical, and cultural distinctions. The Santee Sioux, who spoke the Dakota dialect, represented the eastern division. The Yankton Sioux represented the middle division, speaking the Nakota dialect. The Teton Sioux, who spoke the Lakota dialect, represented the western division. Within each of the three main tribes were subgroups called bands or clans (see table 2.1). Each band or clan was subdivided into family groups called Tiyospayes, who selected their own chief and were largely independent of the band and tribe.[6]

Table 2.1 Divisions of the Sioux Indians

(Lakota Dialect) Teton	(Nakota Dialect) Yankton	(Dakota Dialect) Santee
Oglala	Yankton	Mdewakantonwon
Brulé (Sicangu)	Upper Yanktonai	Wahpeton
Hunkpapa	Lower Yanktonai	Wahpekute
Blackfeet (Sihasapa)		Sisseton
Sans Arc (Itazipacola)		
Minniconjou		
Two Kettle (Oohenupa)		

Source: Sven G. Froiland, *Natural History of the Black Hills* (Iowa: Graphic Publishing Company, 1978), 62.

During the early 1600s, the majority of the Lakota and Dakota Sioux inhabited the area from the mouth of the Wisconsin River to central Minnesota. A few bands inhabited areas as far north as Lake Winnipeg and as far west as the Rocky Mountains. Gradually, the people of the Sioux Nation were pushed westward by the Ojibwa, also known as the Chippewa. The Ojibwa were stronger in battle, having previously acquired guns from the French.[7]

The Sioux inhabited north central Minnesota around the headwaters of the Mississippi River in the early 1600s. By about 1670, the Yanktons and the Tetons began migrating to south central Minnesota. The Santee Sioux also migrated to this area after 1735. Gradually, the various bands of Sioux continued their southward movement to the mouth of the Blue Earth River on the Minnesota River at present-day Mankato (see fig. 2.1). Here, the various bands faced a decision whether to cross the Minnesota River into the open prairie country or remain on the north side of the river in the wooded country. Those who crossed the river at the Oiyuwega, or "crossing place," accepted a new way of life from a woodlands tradition to a nomadic, prairie-plains culture.

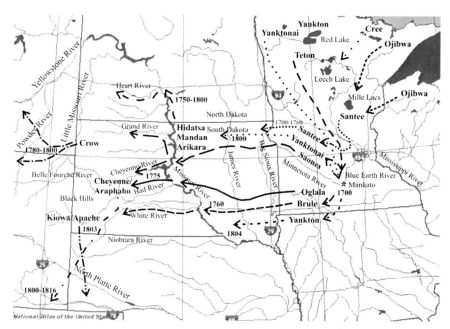

Fig. 2.1. Migration patterns of northern plains tribes (1650-1820). The Sioux, who were forced to move west by the Ojibwa, reached the Missouri River in present-day South Dakota by about 1750. Standing Bear and his band reached the Black Hills in about 1776.—Modified from Sven G. Froiland, *Natural History of the Black Hills* (Sioux Falls: The Center for Western Studies, 1978); Fort Laramie Natural Historic Site and Museum, Fort Laramie, Wyoming.

The Yankton crossed the river and eventually settled along the James River Basin. The Yanktonai elected to follow the Minnesota River upstream, settling in northeastern South Dakota. By about 1720, the Teton tribe also split into two groups at the river crossing. The Brulé and Oglala Tetons crossed the Minnesota River and started a slow migration to the west. The remaining five bands of Teton, referred to as Saone, elected to stay on the north side of the Minnesota River and migrated upstream to the Lake Traverse area.[8]

As the Ojibwa forced the Sioux westward across the Mississippi and south to the Minnesota River, the Teton Sioux forced the Mandan, Hidatsa, and Cheyenne tribes to the west. The Mandan and Hidatsa reached the upper Missouri River from the northeast ahead of the Cheyenne and Teton. The westward migration of the Cheyenne was checked by the Mandan at the river. After a period of hostility, the two tribes became friendly, which allowed the Cheyenne to cross the river and continue their migration to the west. Near Bear Butte, the Cheyenne encountered the Crow. Eventually, the Cheyenne defeated the Crow, Kiowa, and Arapaho and pushed them out of present-day western South Dakota.[9]

By 1750, the five bands comprising the Saone-Teton had migrated to the east bank of the Missouri River. The Oglala and Brulé Tetons followed about ten years later. The Teton soon learned that the Missouri River area had much to offer in terms of water, rich grasses, and buffalo. Snowfall along the river was relatively light compared to what the Teton were accustomed to. These were reasons enough for the Teton to want to live along the east side of the river, at least initially.[10]

The Cheyenne, Arikara, Mandan, and Hidatsa tribes occupied the west side of the Missouri River. These tribes prevented the Teton Sioux from crossing the river. However, when a smallpox epidemic and other diseases nearly decimated the west river tribes between 1772 and 1780, the Oglala and Brulé took advantage and moved across the river below Big Bend in 1775. The remaining Arikara were forced to move northward and relocated along the Grand River. The Mandan and Hidatsa also fled north and settled in present-day North Dakota.

The Oglala took up residence between the Bad and Cheyenne rivers. The Brulé settled along the White River. Some bands of Oglala continued west. One such band included a war party led by Chief Standing Bull, who reached the Black Hills area in 1776. Over the next several years, the Oglala defeated the Cheyenne and took control of the areas that the Cheyenne had taken from the Crow and the Kiowa.[11]

Between about 1782 and 1806, the Oglala Lakota tribe dominated most of the major tributaries on the eastern side of the Black Hills. The Oglala, along with the Brulé, continued to occupy tributaries of the White and Bad rivers. The Kiowa, Crow, Arapaho, Plains Apache, and Comanche tribes occupied areas between the southern Black Hills and the tributaries of the White and Bad rivers. The Cheyenne occupied the Cheyenne River area and southeastern flanks of the Black Hills. Between 1814 and 1816, the Lakota tribe began forcing the Kiowa, Cheyenne, and Arapaho people out of present-day South Dakota. After these tribes relocated to the headwaters of the Platte, Arkansas, and Red rivers, the Lakota dominated most of the area between the Missouri River and the Black Hills.[12]

Following the decimation of an Arikara village on the Missouri River in 1823,[13] the U.S. government decided it should negotiate treaties with the various tribes of the Great Plains to help ensure the safety of the fur traders traveling along the Missouri River. In 1825, a group under the direction of General Henry Atkinson and Indian agent Benjamin O'Fallon traveled up the Missouri River in nine keelboats to begin negotiating treaties with the various tribes. Several different treaties were signed, including the June 22, 1825, Treaty with the Teton Sioux, the July 12, 1825, Treaty with the Sioune and Oglala Tribes, the July 16, 1825, Treaty with the Hunkpapa Band of the Sioux Tribe, and the July 18, 1825, Treaty with the Arikara Tribe. Collectively, these treaties became known as the Atkinson and O'Fallon Treaty of 1825.

The Atkinson and O'Fallon Treaty called for "perpetuating the friendship which has heretofore existed" between the United States and its citizens and the bands of the Sioux tribe of Indians. The treaty emphasized the friendship and supremacy of the United States over Sioux country and specified that the tribes would admit the right of the United States to regulate all trade and intercourse with them and that the United States would extend to the tribes "from time to time, such benefits and acts of kindness as may be convenient and seem just and proper to the president of the United States." The treaty required the tribes to protect the persons and property of the traders and the persons legally employed by them. It also required the tribes to allow safe passage to all people authorized by the United States to pass through the territories of the tribes and to protect the agents and other people sent by the United States to temporarily reside among them. The treaty also limited the trading privileges of the tribes to persons authorized by the United States and required the tribes to apprehend and turn over to the nearest military post, any foreign or other unauthorized traders.[14]

Between 1803 and 1849, the United States significantly expanded its boundaries west of the Mississippi River. The government consummated the Louisiana Purchase in 1803, British Cession in 1819, Spanish Cession in 1819, Oregon Country in 1846, Texas Annexation in 1845, and Mexican Cession in 1848 (see fig. 2.2). Although little was known about any of these areas, it was becoming clear to government officials and the general public that the western frontier had much to offer. Many of these people, including trappers, traders, and prospectors began forming their own small expeditions to find out for themselves what the West had to offer.

Fig. 2.2. Territorial acquisitions of the United States.—Courtesy of the National Atlas of the United States, U.S. Geological Survey.

Early-Day Visitors and Prospectors

There are many stories and accounts about the existence of gold and presence of prospectors in the Black Hills long before the gold rush of 1874-1876. One such story involves Regis Loisel, a fur trader who had a post on Cedar Island, some 35 miles south of present-day Pierre, South Dakota. While returning to St. Louis in the spring of 1804, Loisel encountered Lewis and Clark, who were sailing up the Missouri River. Loisel continued on to New Orleans. In a letter dated May 28, 1804, to Carlos de Hault de Lassus, the last Spanish lieutenant governor of Upper Louisiana, Loisel cautioned the still-lingering Spanish authority about the intent of the aggressive Americans, who he feared would soon take over the fur trade and invade the Spanish territory near Santa Fe. Loisel described how adept the Americans were at bribing the Native American tribes with presents in order to incite them into war or peace or to obtain their fur trade.

Loisel described the value of the various tributaries of the Missouri River, including those that connected to Spanish and Mexican territories to the southwest. In one section of his letter, Loisel described gold nuggets he claimed could be found in the Costa Negra (Black Hills):

> *Ascending the Misury [Missouri] one hundred and thirty leagues above the mouth of the Rio Chato [Platte River], one comes to the Rio Que Corre [Niobrara River or White River]. Its direction is the same, and it rises in the first mountains known under the name of the Costa Negra [Black Hills]. That name was doubtless given these mountains because of the color of the earth. Under that earth are hidden precious minerals, as is declared by the tribes who frequent them. They are so abundant that they are found in nuggets, scattered here and there both in various places upon the Rio Chato and upon this river.*
>
> *Reascending the Misury to a distance of 450 leagues from San Luis [St. Louis], one comes to the River Chayennes, or as it is called the Courche or Braso. It offers the same means of communication with the Nuevo Mexico to the west by crossing the Costa Negra of which we have spoken above.*[15]

Another story dating back to 1811 involves natives who traveled to a trading post of the American Fur Company located at the confluence of the north and south forks of the Cheyenne River, about 25 miles from the Black Hills. Here, the natives traded fine nuggets of gold for sugar, coffee, trinkets, and whiskey. The natives claimed the gold came from the Black Hills country. Supposedly, an old French trapper for the American Fur Company knew of the exact location where the nuggets were found but died in 1855 before he could pass on the information to anyone else.[16]

Louis Thoen lived near the base of Lookout Mountain, now within present-day Spearfish, South Dakota. Louis was a farmer and stonemason who gathered and cut stone for homes and businesses around Spearfish. On March 14, 1887, Louis and his brother, Ivan, were gathering slabs of sandstone from a draw on the southwest side of Lookout Mountain. Louis spotted a nice, flat-looking stone. While digging the rock out, he noticed that it was inscribed. After rubbing off some of the dirt, he deciphered the words, "Indians hunting me." He then laid the slab aside and finished the day gathering sandstone rocks.

That night, Louis finished cleaning the dirt off the 8×10×3-inch rock and discovered it had writing on both sides. One side was inscribed as follows: "Came to these hills in 1833 seven of us De Lacompt, Ezra Kind, G W Wood, T. Brown, R Kent, Wm King, Indian Crow, al ded but me Ezra Kind. Killed by Ind beyond the high hill got our gold June 1834." The other side of the stone was inscribed as follows: "Got all of the gold we could carry our ponys all got by the Indians I hav lost my gun and nothing to eat and Indians hunting me."[17]

One researcher of the Thoen Stone, as it is called, was Frank Thomson of Spearfish, South Dakota, who spent a considerable part of his life doing research on the Thoen Stone and the seven prospectors. In 1951, Thomson established contact with Louise Thoen Courtney, one of Louis Thoen's daughters. Courtney communicated the following to Thomson in a letter dated February 28, 1951:

> *I was a little girl and lacked two days of being seven years old, when my father found the stone tablet on March 14, 1887, in a draw at the foot of Lookout Mountain. My father and Uncle Ivan Thoen were getting out some stone for sale. They saw the corner of a large suitable flat stone sticking out of the dirt where the soil had been washed away . . . so they dug it out. Under the stone was the tablet. It looked like it was worm eaten . . . they rubbed some of the damp sand off and found the words, "Indians hunting me." The stone was laid aside and taken to the house, and by lamp light, they rubbed off more sand and found the words, "Got all of the gold we could carry," which made us all think it would tell where to find the gold so we would be rich. That was my first thought, too, so I could get a pony. The next morning the entire inscription was read and later that day the stone was sent to John Cashner, as he had a little curio store and was interested in prehistoric things and liked to investigate them . . . . Father had some pictures of the stone taken and sold a few to tourists (tourists were scarce as hens' eggs in those days).*[18]

Thomson also received information about the stone from Florence Keats Bettleheim, a young neighbor of the Thoens at the time. Thomson relates how even though Bettleheim was seventy-five years old at the time he communicated with her, she was able to recall exactly where each person was sitting at the breakfast table the morning Thoen rushed into their house to describe what he had found. Excerpts from Bettleheim's letter to Thomson dated March 22, 1951, are as follows:

> Mr. Louis Thoen had a farm at the foot of Lookout, and we all knew him very well. I recall very clearly the day he rushed into our kitchen to tell my father what he had found, and to ask him to come at once to see it. My father, a pioneer of 1875-76 and greatly interested in all that pertained to the entire history of the west, and particularly the Black Hills, examined the stone very carefully, and also the exact place where it was found. He decided it had lain there many years, and was the true record of the fate of some unfortunate group of gold seekers.
>
> Mr. Thoen was a quiet, conservative man, but was naturally deeply excited over his find and exhibited it at once. My old friends, John Cashner and his wife, and I examined the stone many times and thought it should be carefully preserved in some museum.[19]

As part of his ongoing research, Thomson was able to obtain genealogical information and family histories about men who may very well have been members of the Ezra Kind party. In communicating with descendants of Andrew Jackson Brown, Thomson learned that a Tom Brown and a man named Kent had joined a party headed for the Black Hills in the late 1820s or early 1830s. The men were never heard from again. Descendants of Kent said their ancestors told stories about receiving a letter from him in the early 1830s saying he and members of his party had found all the gold they wanted and that he would be home soon.[20] Thomson's genealogical research about the seven prospectors lends credibility to the authenticity of the stone. The original Thoen Stone is in the possession of the Adams Museum in Deadwood, South Dakota.

Pierre-Jean De Smet was an immigrant from Belgium who began ministering to the Indian tribes in America after taking his vows for the priesthood at Florissant, Missouri, in 1827. Over the next twenty to twenty-five years, De Smet gained the trust and respect of the tribes west of the Mississippi River. For this reason, he was called upon by the tribes and the military to act as an intermediary during the negotiations of the Fort Laramie treaties of 1851 and 1868.[21]

In 1848, De Smet spent four months in and around the Black Hills. Although he was well versed in the science of geology, De Smet rarely spoke about the presence of gold in the Black Hills or anywhere else. He visited the various tribes in and around the Black Hills again in 1851 and 1864 and made his last trip there during the months of June and July in 1870 with Father Panken. In a letter dated October 12, 1887, to Father Peter Rosen, a Catholic priest residing in the Black Hills, Panken mentioned that on his return trip through Vermillion in 1870, there were "some pioneers ready to go over to the Black Hills country in search of gold."[22]

Over De Smet's lifetime, there were many people who believed he knew of the precise location of a rich deposit of gold in the Black Hills. De Smet never would divulge any of this information to others. He feared he would compromise the respect and trust he had developed with the tribes who entrusted him with such secrets. In 1870, he made the comment that "in the Black Hills of Dakota, beyond the ken of the white man, and where his feet have never trod, there is gold enough with which to pay off the debt of the nation, and for that matter, the entire debt of the civilized world, and yet these deposits would scarcely have been drawn upon."[23]

In his book, *Pa-ha-Sa-Pah: Or the Black Hills of South Dakota*, written in 1895, the Rev. Peter Rosen describes a story that he says appeared in the *St. Louis Globe-Democrat*. The story relates to a dinner given at the St. Louis University in the 1860s to honor Father De Smet. During the dinner, the subject of mineral deposits came up. Rosen says Father De Smet shared a story about how, during his missionary work in the Black Hills, he discovered a most extraordinary gold deposit while digging into a mountain. After viewing the precious ore, he carefully covered it up. Chicago congressman Barney Caulfield also attended the dinner. Caulfield later stated that those people present used every effort to induce De Smet to describe the location of his find. De Smet firmly declined to do so, claiming he did not want his children to be disturbed.

In April 1875, it was reported that similar conversations had taken place between Father De Smet and other people, such as Colonel Stewart Van Vliet in the 1850s and John B. Stolley, in 1864.[24]

The Indians themselves knew of the existence of gold in the Black Hills. Red Cloud, an Oglala chief born in 1822, said his people knew as early as 1800 that there was gold in the Black Hills. Red Cloud recalled that when he was a boy, his people killed three white men close to the Hills.[25] In the book *Black Elk Speaks*, Black Elk discusses Custer's expedition into the Black Hills where Pahuska [Custer] "found there much of the yellow metal that makes the Wasichus [white people] crazy." Black Elk goes on to say, "Our people knew there was yellow metal in little chunks up there; but they did not bother with it, because it was not good for anything."[26]

In 1952, Frank B. Bryant, a mining engineer, wrote a letter dated June 10 to Frank Thomson, researcher of the Thoen Stone. Bryant shared a story about his father, Frank Bryant Sr., who had a contract in 1873-74 at Rollo, Nebraska, to cut cottonwood ties for the Union Pacific Railroad. While there, the elder Bryant met Jim Pierman, who had an Indian wife and had lived with the Indians for many years. The parents of Pierman's wife had been going to the Black Hills for many years to cut tepee poles and hunt for game. On one of those trips, the wife's parents met another Indian party from Wyoming who had just "cleaned out" an emigrant wagon train.

The elder Bryant says that Father De Smet, who was camped in the Black Hills with a party of seven other men, was met by two groups of Indians. One young Indian showed De Smet a six-shooter that the Indian had just acquired along with several canisters of powder from the emigrant train. The Indian said he wanted to trade the six-shooter for a cap-and-ball gun, since he didn't have any cartridges for the six-shooter, but did have ammunition for a cap-and-ball gun. De Smet asked the Indian where he could get ammunition for a cap-and-ball gun. The Indian then left De Smet's camp. Upon his return, the Indian showed De Smet a pouch full of gold nuggets and asked if the nuggets would work all right as ball-bullets. De Smet cautioned the Indians to never go near the place where the nuggets were obtained, warning that the white men would kill every Indian on the plains if they found out about the gold.[27]

There are other stories and reports of prospectors in the Black Hills prior to the gold rush of 1874-1876. Shortly after arriving in the Black Hills in 1874, the Gordon Party discovered remnants of old sluice boxes, an old cabin, and a grave marked J. M., 1846.[28] In 1878, the *Black Hills Telegraphic Herald* reported how "every few months the miner or the adventurous prospector brings to light some fresh evidences of early mining operations in the Hills." The newspaper described instances where miners had found mining implements buried several feet below the surface, an old chain partially embedded in a large tree, and an old miner's shack, complete with tools.[29]

The *Telegraphic Herald* told of two Frenchmen by the name of Le Fevre, who discovered two human skeletons while hunting in the Bear Lodge Mountains. The skeletons were found behind a crude breastwork constructed of rocks and logs. One of the men had apparently been shot through the skull. The other man had an arrowhead firmly embedded in his femur. No traces of clothing were found among the skeletons. The Le Fevre brothers also found part of a kettle, a broken rifle stock made of fancy knotted wood, and the remnants of a leather-bound diary. The only writing that could be deciphered in the weathered diary was the figures "1-2-52."[30]

In the summer of 1852, an expedition of about 300 prospectors, led by Captain Douglas of St. Joseph Valley, Michigan, departed Council Bluffs to partake in the California gold strike. After reaching Fort Laramie, the group met a French trapper who told them it was unnecessary to travel as far as California to search for gold. The trapper said he could show the prospectors a place in the Black Hills where they could get all the gold they would want. Thirty of the prospectors decided to take the trapper up on his offer. It was decided that if the smaller party ended up finding gold, they would catch up with the main party at the Humboldt River and inform them of their find.

After actually finding gold in the Black Hills, eight of the thirty prospectors departed the Black Hills, per the agreement, and caught up with the main party at the Humboldt River. There, the eight prospectors reported that they had indeed found gold on two different streams in the Black Hills. Because of the amount of water and depth of the gravels, the prospectors stated they were unable to reach bedrock where they thought the best gold would be found. The prospectors related how they then prospected in an area north of the two streams and found gold there also. Because it was November, and in consideration of reports of unrest with the Indians in the Black Hills, the prospectors decided not to return to the Hills and proceeded instead to California. The remaining twenty-two prospectors in the Black Hills were never heard from again.[31]

In 1852, a man named Hale also relayed another story about prospectors in the Black Hills. Hale told of a "pitiable wreck of a man" named Thomas Renshaw, from Cincinnati, who came hobbling into a Mormon hunter's camp on Green River in October 1852. After receiving some nourishment, Renshaw stated his party had been on their way to California. On their way, they met some friendly Indians who told the party gold could be found in the Black Hills. Renshaw and his party decided to do some prospecting in the Black Hills before continuing on to California. Upon entering the southern Black Hills, they cached their wagons and placed their gear on the backs of their oxen. Renshaw stated that they didn't find much of any gold until they reached the northern Black Hills. One day, while the rest of his party was placer mining, Renshaw went out in search of game. When he returned late that day, he discovered that all of his companions had been massacred and that their camp had been burned.[32]

Another story involves a party of French Canadians who made a successful strike on Castle Creek in the Black Hills in 1854. Unfortunately, they were supposedly caught and massacred. Indians buried their gold, which they placed in a jug.[33]

In 1855, Jim Bridger and Jeremiah Proteau guided an English sportsman named Sir Saint George Gore on a tour of the West. One of their stops was in the Black Hills, where Proteau found handfuls of glittering dust on Rapid Creek. After Proteau finished stuffing his shirt full of the gold dust, Gore laughed, claiming the glitter was merely mica. Later, Bridger told Proteau that Gore admitted that the glitter was indeed gold.[34]

In April 1856, the hunting expedition arrived at some of the headwater tributaries of the Belle Fourche River. Here, Proteau discovered gold again. Gore told Proteau it was only fool's gold, but Proteau put some of the material in two of his black bottles anyhow. After the expedition departed the Black Hills, Bridger informed Proteau that Sir George said the material in the bottles was gold. Lamourie, another scout, was told that if he prospected at the head of Swift Creek, he would find rich gold there.[35]

Late in 1863, thirteen prospectors, including G. T. Lee from Missouri, traveled from Fort Laramie to the Black Hills. Over three days, members of the party recovered $180 worth of gold but were forced to leave the Hills because of a large snowfall. Lee didn't return to the Black Hills until 1876. Unfortunately, he was never able to find the location he successfully prospected in 1863.[36]

Toussaint Kensler, part Sioux, was a prospector who worked the gold placers of Adler Gulch in Montana in 1863-64. While there, he was arrested on a charge of murder but managed to escape the authorities. Kensler was not heard from again until he appeared at an Indian Agency on the Missouri River in Dakota Territory in 1866. There, he showed the people five goose quills filled with placer gold that he claimed were found along a tributary to the south fork of the Cheyenne River. Toussaint also showed the group a fossilized skull, which he said he found on his way to the agency. Toussaint was arrested again on the murder charge and put in jail. He later produced a map showing the location of his gold find. Interestingly, Walter P. Jenney used Kensler's map as a reference when he explored the Black Hills in 1875 as part of the government-sponsored Newton-Jenney Expedition. Jenney believed Kensler might have found the gold along Amphibious Creek or possibly French Creek. Jenney referenced Kensler's map in his 1876 report to the Department of Interior.[37]

In 1866, a Sioux hunting party led by Chief Yellow Bear entered the Black Hills by way of Makes-the-Ghost Hill. The Sioux party included a son of John Richards Sr., a French fur trader who was married to an Indian woman. One version of the story says that somewhere in Deadwood Gulch, the Sioux hunters came upon seven prospectors. The prospectors opened fire, killing one Indian and wounding another. The Sioux chased the prospectors up Blacktail Gulch and down False Bottom Creek to Lookout Mountain. There, five of the prospectors were killed in a surprise attack and two escaped. Later, the two prospectors were surrounded on top of a mountain near Sundance, Wyoming, and killed. The younger Richards supposedly gave the gold to his brother, John Jr.[38]

Charles Richards, a nephew of John Richards Jr., told Thoen researcher, Frank Thomson, a slightly different version of the story. Charles said a party of Sioux, which included his father and uncle, came upon a group of prospectors led by Sioux guides. The prospectors were bringing a pack train of mining equipment into the Black Hills. The Sioux warriors proceeded to kill all of the prospectors and the Sioux guides. Charles Richards also shared what he called the Three Whiskers story, whereupon a party of Sioux came upon three elderly long-whiskered men who were mining in the Black Hills. The Indians killed these prospectors also.[39]

Alice Brown, a cousin of Charles Richards and a granddaughter of John Richards Jr., described a story to Thomson that is similar to the first version about the seven prospectors. Brown claimed that one of the Sioux Indians who was killed was a seventeen-year-old nephew of Sitting Bull. The other Indian was American Horse, who was wounded, but not killed. In 1925, Thomson also learned of a story from one of his employees, George Acker. Acker was part

Sioux, and he grew up on the Cheyenne Indian Reservation. One day, while the two were looking at Lookout Mountain, Acker exclaimed to Thomson, "There is much gold hid over there, on that mountain." Acker recounted how an old woman once told him that when she was a young girl, the band of Indians she was with killed four white men who had horses. Three of the horses were loaded with gold, and the fourth held their camp gear. The Indians buried the gold on the mountain. In 1950, Thomson learned that the old woman was Acker's grandmother, Mollie Hodgkiss Landreau.[40]

Frank Thomson continued his research on the story of the seven prospectors. Thomson's research indicates that John Richards Jr. did receive the gold from his brother. After selling the gold, Richards started a fur-trading post on Deer Creek, a tributary of the Platte River. Richards was licensed to trade with the Sioux in an area between the Black Hills and the Rocky Mountains. In 1870, Richards killed a soldier by the name of Lieutenant Morrison at Fort Fetterman. Richards subsequently fled to the Red Cloud Agency, where he became the leader of a band of Sioux.

Richards departed Fort Laramie in 1872 and arrived at the Red Cloud Agency to see his brother-in-law, Chief Yellow Bear. While there, on May 17, 1872, Richards proceeded to Chief Yellow Bear's tent, followed by twelve or fifteen of Yellow Bear's people. Richards and Chief Yellow Bear got into a heated argument over family matters and the gold that was taken from the prospectors in 1866. Richards then shot and killed Chief Yellow Bear. Immediately afterward, other Sioux who were present attacked Richards, and he was "cut to pieces."[41]

Emigrant Protection

The early-day fur traders who traveled to the Yellowstone River country via the Black Hills were not much of a problem for the Indians. What irritated the Indians were the emigrants and prospectors who, by the thousands, began establishing trails through the plains areas of the American Indians from the late 1840s through the early 1870s. The situation worsened as the invaders began decimating and driving off the buffalo herds.

The term "Manifest Destiny" was coined by politicians and others in the 1840s to rationalize the nation's desire to expand its boundaries to the Pacific Ocean. Some people believed the Manifest Destiny was God's will that Americans should spread over the entire continent and Christianize the land. For whatever reasons, the nation's people quickly began to embrace the Westward Expansion in pursuit of the American Dream. Primary transportation routes such as the Oregon Trail, California Trail, Mormon Trail, Bozeman Trail, Santa Fe Trail, and others were developed along existing Indian trails to facilitate emigrant travel (see fig. 2.3).[42]

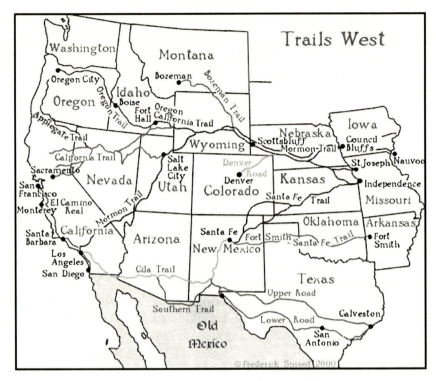

Fig. 2.3. Emigrant trails of the west.—Courtesy of Frederick Smoot, TnGenWeb Project.

A massive amount of emigrant travel took place across the Great Plains in the 1840s in the quest for fertile lands in the Willamette Valley of Oregon, a new homeland in the west for the Mormons, or for gold in California. Eventually the Indians became angered and enraged as they observed their buffalo herds being pushed around and slaughtered for food and sport. Grasslands along the emigrant trails were reduced to dust. The Indians also realized they were being subjected to the diseases that the emigrants were spreading across the plains. As the number of altercations increased between the tribes and the emigrants, the U.S. government began to recognize the need to provide support and protection for the emigrants on the trails. Such protection was soon provided by the military whose troops were based out of forts strategically positioned along the main trails.[43]

Most of the forts that existed along rivers of the Great Plains prior to about 1845 were merely fur-trading posts established by the trading companies for the purpose of exchanging goods and furs. In 1834, traders Robert Campbell and William Sublette established Fort William at the confluence of the North Platte and Laramie rivers to facilitate trade to the north. The fort was sold to the Rocky Mountain Fur Company in 1835. A few years later, the fort was sold to the American Fur Company, who changed the name to Fort John. The fort was eventually sold to the U.S. government in 1849 at which time it was converted to a military post called Fort Laramie. Over the next forty years, Fort Laramie became the crossroads and staging grounds for emigrant protection, Indian wars, and military reconnaissance investigations, all of which were deemed necessary to facilitate the Manifest Destiny and associated Westward Expansion.[44]

Fort Laramie Treaty of 1851

In 1793, the U.S. government passed legislation giving the president the authority to appoint Indian agents and assign them to broad geographic areas to oversee the government's trading houses. The trading houses served to increase the Indians' dependence on manufactured goods, since the trading houses were only permitted to accept trade goods, such as furs, hides, and raw materials from the Indians. The Indian agent was responsible for ensuring that the trading policies were followed within his area of jurisdiction. Any violations of the trade laws were to be reported to the governor of the territory, a local military commander, or the War Department. Agents were also responsible for distributing annuities from the government to the chiefs who, in turn, distributed the annuities to the members of his tribe.[45]

One such Indian agent was Thomas Fitzpatrick, who was appointed agent for the Upper Platte and Arkansas River area tribes in 1846. Fitzpatrick was an Irish fur trader who previously had worked for William Ashley and participated in some of the Upper Missouri area trading expeditions with Jedediah Smith, Jim Bridger, and others. Many of the tribal members already knew Fitzpatrick from his trapping and trading days and respected him. They called him Broken Hand because of an accident with a rifle that shattered his wrist.

During the same year, the Oglala and Brulé (Sicangu) Teton decided to seek help from President James Polk. The chiefs petitioned the president for help, saying the emigrants were chasing the buffalo from the Indians' hunting grounds. As a consequence, the Indians claimed their tribes were being forced to go into the country of their enemies to hunt for game. The government issued a response to the Teton chiefs, saying, "It is the nature of the Buffalo and all other kinds of game to recede before the approach of civilization, and the injury complained of is but one of those inconveniences to which every people are subjected by the changing and constantly progressive spirit of the age."[46]

Empathetic with the plea for help from the Oglala and Brulé (Sicangu) chiefs and fearing a large Indian retaliation, Fitzpatrick proposed a treaty to David Dawson Mitchell, superintendent of the western region. Under the treaty, the Plains Indians would be provided with annual payments for the losses they were experiencing. Mitchell agreed and said a general treaty council meeting should be held with the tribes of the Plains Indians. After overcoming some political resistance, Fitzpatrick and Mitchell eventually persuaded Congress in February

1851 to approve an appropriation of $100,000 for a general treaty council. Couriers were sent out to all the tribes of the Plains Indians notifying them of a council that would convene on September 1, 1851.

About ten thousand Indians, including the Lakota, Assiniboine, Cheyenne, Arapaho, Arikara, Gros Ventre, Mandan, Hidatsa, Shoshone, and Crow attended the council. The Mandan, Hidatsa, Arikara, and Assiniboine delegation departed Fort Union for Fort Laramie on July 31 accompanied by their Indian agent, Alexander Culbertson, and Father De Smet. Upon reaching the Oregon Trail, the delegation was amazed at the condition of the trail. "This immense avenue," wrote Father De Smet, "resembles a wind-swept surface, worn bare by the perpetual march of Europeans and Americans marching to California." De Smet continued, "The Indians, familiar with only the paths of the chase, thought, upon beholding this beaten track, that the entire nation of the white man had traversed it and that the country of the rising sun must be deserted."[47]

The talks were held on Horse Creek, which is a tributary of the North Platte River near the Nebraska-Wyoming border. Other plains tribes such as the Comanche, Kiowa, and Apache refused to attend for fear of having their horses stolen by other tribes. Entire villages of people traveled to the council, including elders, women, children, dogs, and horses. The large turnout of Indians was partially attributable to the promise of the U.S. government to provide several weeks' worth of provisions and presents to those gathered. However, much to the dismay of the chiefs, the wagon train of provisions was delayed in getting to Fort Laramie. The chiefs then expressed concern about whether they should even agree to meet until the provisions arrived. The concern was alleviated after Broken Hand assured the Indians that the provisions and presents would soon arrive.[48]

The first day of the Big Talk was held on September 12. The Indians were dressed in their best attire. Superintendent Mitchell and Thomas Fitzpatrick were seated in the center of the large gathering. Mitchell opened the meeting with words of goodwill, which were followed with the passing of the peace pipe. Mitchell informed the Indians that the Great Father in Washington "has heard and is aware that your buffalo and game are driven off, and your grass and timber consumed by the opening of roads and the passing of emigrants through your countries. For these losses, he desires to compensate you."[49]

Mitchell conceded to the chiefs that the Indian nations generally acknowledged certain tracts of land as their respective territories. Mitchell proposed that definite boundaries be established for each territory. He also told the chiefs that they would not have to abandon any rights or claims to lands outside their respective territories, including the privilege of hunting, fishing, or passing over any of the territories described. In consideration of the treaty stipulations and for the damages which had already occurred, or might occur to the Indian nations, Mitchell informed the chiefs that the United States would deliver to the nations $50,000 per annum for fifty years.[50]

Mitchell and Fitzpatrick then commenced to draft the treaty while the tribes talked, danced, and debated the boundaries of their respective territories. Each tribe, of course, wanted to claim more territory than the other tribes would allow. The detail of describing metes and bounds descriptions for the respective territories was assigned to Jim Bridger, Robert Campbell, and Thomas Fitzpatrick, who were quite familiar with the lands being discussed. The Oglala objected to the boundaries assigned to the Sioux Nation (see fig. 2.4). Black Hawk, an Oglala chief, exclaimed, "You have split the country, and I do not like it. These lands once belonged to the Kiowas and the Crows, but we whipped these nations out of them, and in this we did what the white men do when they want the lands of the Indians."[51]

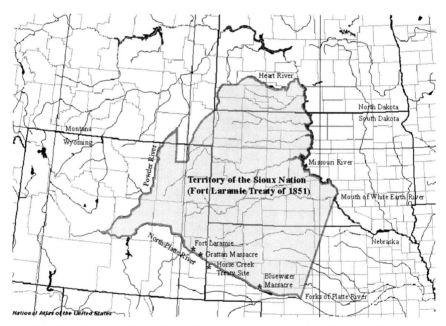

Fig. 2.4. Territory of the Sioux Nation pursuant to the Fort Laramie Treaty of 1851. The Sioux or Dahcotah Nation was created under the Fort Laramie Treaty of 1851. Territories were also established for the other Indian Nations attending the council. The other territories are not shown.—Illustration by author.

On September 17, 1851, all of the chiefs of the tribes, with the exception of the Oglala, signed the treaty by "touching the pen" and making their mark. Four days later the cannon sounded, signifying the arrival of the provisions and gifts. Each of the chiefs was presented with a colorful uniform, a gold-plated sword, a certificate from the Great Father, and a medal to hang around their neck. Over the next two days, the chiefs distributed the tobacco, cloths, beads, knives, and blankets to their respective people.[52]

At the conclusion of the Great Council, the people were satisfied and confident of the promises and kindness offered by Superintendent Mitchell. Father De Smet shared in the confidence, saying, "This Council will be the beginning of a new era for the redskins; an era of peace when travelers will be able to cross the desert unmolested, and the Indians in turn will have nothing to fear from the white man." On September 24, the tribes began their departure from the Great Council.[53]

Grattan and Ash Hollow (Bluewater) Massacres

The Great Council, pursuant to the Treaty of 1851, created an era of peace. Unfortunately, the peace only lasted for about three years. On August 18, 1854, a Mormon wagon train moving west along the Platte River passed by a circle of Brulé, who were camping and waiting for receipt of their annual annuity from Fort Laramie. One particular cow lagged behind the emigrants' train and wandered into the Brulé camp. The stray cow ended up getting shot by High Forehead, a Minniconjou who was visiting the Brulé camp. The emigrant who owned the cow promptly registered a complaint at Fort Laramie with Lieutenant Hugh B. Fleming, commander of the post.[54]

Conquering Bear, a Brulé chief, hurried to Fort Laramie and offered $10 or a horse from his own herd of ponies as restitution for the cow. The emigrant rejected the offer and demanded $25. The counteroffer was considered unreasonable to Conquering Bear, and he returned to camp.[55] The following morning, Man-Afraid-of-His-Horses, a Brulé chief himself, went to Fort Laramie to try and resolve the matter. His efforts were futile.

While Man-Afraid was at the fort, Lieutenant John L. Grattan requested permission from Fleming to apprehend and arrest the guilty Indian. Fleming approved the request. Grattan, along with Sgt. William Faver, Cpl. Charles McNulty, twenty-five privates, two musicians, and Lucien Auguste, the interpreter, then departed Fort Laramie with a mountain gun and a howitzer. Man-Afraid, who left the fort at the same time, cautioned the officer several times about the large number of Sioux. The officer replied, "Yes, that is good." Lucien Auguste, the interpreter, turned to Man-Afraid and said, "I am ready but must have something to drink before I die."

By the time the detachment arrived at the lodges, Auguste was intoxicated. He began riding his horse at full gallop around the Brulé lodges shouting obscenities. Chief Conquering Bear then arrived, accompanied by Little Thunder, Big Partisan, and Man-Afraid-of-His-Horses. Grattan insisted that High Forehead be surrendered, but Conquering Bear denied the request on the basis that he had no authority over the visiting Minniconjou. Over the next forty-five minutes or so, Conquering Bear tried to convince High Forehead to surrender but was unsuccessful in his efforts.

At about this time, Red Cloud and his band of Oglala warriors arrived at the camp. Other Indians concealed themselves in the willow thickets around Grattan's troops. Grattan then signaled his troops with a command that the Indians did not understand. A shot or shots were fired, and the howitzer was fired in return. Chief Conquering Bear fell to the ground, mortally wounded. Members of the camp were immediately enraged over the death of their chief. Over the next few minutes, Grattan's entire detachment was annihilated, except for Pvt. John Cuddy, who died from his injuries a few days later at Fort Laramie.

That evening, and again over the next two days, the Indians attacked the trading post formerly known as Fort John and the Gratiot Houses. The government annuities were delivered to the Gratiot Houses for temporary storage until Major J. W. Whitfield, agent for the Platte Agency, could arrive to distribute the annuity goods. The Indians proceeded to seize the annuity goods, along with goods that the traders owned.[56]

Following the Grattan Massacre, Colonel William Harney, who was in Paris where his wife and daughter lived, received orders to take command of a retaliatory expedition against the Sioux. Harney boarded ship on Christmas Eve, 1854, and sailed for St. Louis. After preparations were completed, a detachment under the command of Colonel William Harney was sent out in August 1855 with a force of thirteen hundred troops to secure the Platte Road and punish and discipline those who were involved with the Grattan Massacre. Harney learned that the guilty people were members of Chief Little Thunder's Brulé tribe, who were camped along Bluewater Creek, northwest of Ash Hollow.

On September 3, despite attempts of the tribe to surrender, Harney's men proceeded to massacre eighty-six members of the tribe, including women and children. The soldiers showed great remorse for what they had just done, and they took good care of the prisoners. The following spring, Harney secured a peace treaty with the Sioux and Chief Little Thunder at Fort Pierre. Although the U.S. Senate never ratified Harney's treaty, it did serve to improve relations between the military and the Indian tribes for the next decade. The Grattan and Bluewater massacres precipitated the beginning of a long series of Indian Wars across the Great Plains.[57]

Chapter 3
The Great Reconnaissance

Army Appropriation Act of 1853

Following the Mexican Cession and Oregon Territory acquisitions in the late 1840s, huge domains of additional ground became the possession of the U.S. government. People flocked to Oregon, California, and Utah in search of the American Dream as part of the Westward Expansion. Unfortunately, the eastern politicians had only a mental image of what the western land looked like or had to offer. As a result, the government decided that it needed to formally investigate the western territories to determine the geography and resources and identify possible routes for extending roads and railroads west of the Mississippi River.

In 1845, Congress began debating the need for a railroad that might extend from the Mississippi River to the Pacific Ocean, thus providing a transcontinental railway. Congress subsequently passed the Army Appropriations Act on March 3, 1853, allotting $455,000 for the U.S. Army to conduct surveys of four possible routes for such a railroad. Collectively, these resource and topographical surveys became known as the Great Reconnaissance. Two of the proposed routes would pass through Nebraska Territory, which at the time included the Black Hills. The resource and topographical surveys through Nebraska Territory included the Harney-Warren Expedition of 1855, Warren Expedition of 1856, Warren-Hayden Expedition of 1857, and Raynolds-Hayden Expedition of 1859.

The Great Reconnaissance expeditions were largely scientific undertakings with the objective of collecting data, information, and physical and cultural specimens. The U.S. Army was charged with assessing settlement possibilities and performing boundary, wagon-road, natural resource, railroad route, and waterway surveys. To provide protection and general support, the Army Corps of Topographical Engineers was assigned responsibility for the scientists and their investigations. Some of the most talented civilian scientists, including topographers, geologists, botanists, zoologists, and meteorologists contracted with the government to investigate the western lands. The importance of these Great Reconnaissance expeditions is supported by the fact that the federal funding for such expeditions accounted for almost one-third of the annual federal budget between 1843 and 1856.[1]

Concurrent with the Great Reconnaissance, members of the Sioux gathered for a great council of the Seven Campfires at Lake Traverse in June 1857. Another council was held at Bear Butte in present-day South Dakota in August 1857. As many as seven thousand members of the Oglala, Minniconjou, Sans Arc, Blackfeet, Two Kettles, and Hunkpapa tribes attended the council. Bands of Sioux were camped all along the banks of the Belle Fourche River and Bear Butte Creek. Spotted Tail (see fig. 3.1) and his band of Brulé did not attend since he was imprisoned at Fort Laramie for stealing $10,000 during the Kincaid Coach Raid. Several other warriors and emerging leaders were in attendance, including Red Cloud (see fig. 3.2), Crazy Horse, and Sitting Bull (see fig. 3.3).

Fig. 3.1. Spotted Tail, or Sinte Gleska, chief of the Brulé tribe of the Teton Sioux.—Library of Congress (LC-USZ62-131515).

Fig. 3.2. Red Cloud, chief of the Oglala tribe of the Lakota Sioux.—Library of Congress (LC-USZ62-91032).

Fig. 3.3. Sitting Bull, chief of the Hunkpapa tribe of the Lakota Sioux.—Library of Congress (LC-USZ62-111147).

Crazy Horse, at the age of sixteen, witnessed the Grattan and Bluewater massacres. He vowed to his father that he would battle the whites for the rest of his life. At the Bear Butte Council, the Sioux chiefs resolved to put a stop to the encroachment of emigrants into the Sioux Nation provided for under the Fort Laramie Treaty of 1851.[2] Furthermore, the Sioux resolved to execute any tribesman who revealed the existence of gold in the Black Hills to the whites. Based on these resolutions, warriors from Bear's Rib tribe would prevent the Warren-Hayden Expedition from conducting a full investigation of the Black Hills that fall.[3]

Warren-Hayden Expedition of 1857

G. K. Warren, a lieutenant in the U.S. Army's Corps of Topographical Engineers, was given an order to complete the necessary examinations to determine the best route for extending the military road westward from Sioux City to Fort Laramie and South Pass in Wyoming Territory. Afterward, he was to examine the Black Hills and return by way of the Niobrara River valley to Omaha. Warren's expeditions were completed in 1855, 1856, and 1857. He was assisted on the 1857 expedition by J. H. Snowden and P. M. Engel, topographers; Dr. Ferdinand V. Hayden, geologist; W. P. C. Carrington, meteorologist; Dr. S. Moffit, surgeon; and Lieutenant James McMillan as commander of the escort. The escort consisted of twenty-seven enlisted men and three noncommissioned officers of the Second Infantry.

Warren's expedition to the Black Hills departed Fort Laramie on September 4, 1857. Members of the expedition mapped the area through Raw Hide Butte, Old Woman Creek, and the south fork of the Cheyenne River. From there, they proceeded into the Black Hills via Beaver Creek. They continued northwest until they came to Inyan Kara. Here, a large force of Hunkpapa met them and made serious threats against the party entering the Black Hills. Warren recorded later that an attack would have taken place with success had General Harney not taught the Indians a lesson at the Blue Water Massacre in 1855.

Warren determined that the main objection from the Hunkpapa involved a large herd of buffalo they were following and herding, waiting for the winter hair of the buffalo to get sufficiently long before slaughtering part of the herd. None of the expedition members was allowed to kill any of the buffalo for fear of stampeding and scattering the entire herd. The Hunkpapa informed Warren that the treaty made with General Harney allowed the whites to travel along the Platte River to Fort Laramie, and up and down the Missouri River in boats, but did not provide for any travel elsewhere within the Sioux Nation. The Hunkpapa suspected that Warren was trying to assess the value of the Black Hills for the whites, determine routes, and select places for military posts. They made it clear to Warren that the Black Hills were to be left wholly for the Indians.

Warren reached an agreement with the Indians that he would camp near Inyan Kara for three days and await the arrival of their chief, Bear's Rib. Bear's Rib was the person appointed in Harney's treaty to be the first chief for the tribe. While Warren's group was camped near the forty lodges of the Hunkpapa, a larger group of Hunkpapa and Sihasapa joined the camp. The camp became quite unpleasant for the members of the expedition, considering their company and the fact that it snowed for two days and two nights. After three days, Bear's Rib still had not arrived, so Warren decided to break camp. The expedition traveled backward 40 miles along the trail on which they had come and then continued eastward through the southern Black Hills.

After two days, Bear's Rib and one other Indian caught up with Warren. Bear's Rib repeated what the other chiefs had said and informed Warren that no guarantee could be made to prevent his expedition from being destroyed if he proceeded any farther. Warren replied that he intended to proceed to Bear Butte at which point he would leave the Black Hills. After a whole day of deliberations, Bear's Rib agreed, but only if he accompanied the expedition for part of the way. Warren also agreed to tell the president that no one else would be allowed to travel through their country. Bear's Rib further informed Warren that he did not want any presents delivered in the future to procure a right of entry into the Black Hills or to preclude a war with the Crow. Bear's Rib told Warren that the annuities were hardly worth going after, and if they weren't available when the Indians brought their robes to the trading posts along the Missouri River, they did not want them.

After Bear's Rib departed, the expedition completed their route through the eastern half of the Black Hills, taking time to map each of the streams and drainages along the way. After concluding their visit to the Black Hills, the expedition members spent some time investigating the north fork of the Cheyenne River (the Belle Fourche River) before heading down the south fork of the Cheyenne River to the mouth of Sage Creek. The main party departed Fort Randall on November 7 and mapped their way down the Missouri River to Sioux City. Here, the expedition ended on November 16, 1857, and Warren returned to Washington, via Fort Leavenworth and St. Louis.[4]

In his final report of the expedition, Warren wrote, "In these mountain formations, which border the great plains on the west, are to be found beautiful flowing streams and small rich valleys covered over with fine grass for hay, and susceptible of cultivation by means of irrigation. Fine timber for fuel and lumber, limestone and good stone for building purposes are here abundant. Gold has been found in many places in valuable quantities, and without doubt the more common and useful minerals will be discovered when more minute examinations are made."[5]

Warren framed the future possibilities for development of the West:

> As I before stated, an irreclaimable desert of two hundred to four hundred miles in width separates the points capable of settlement in the east from those on the mountains in the west. Without doubt these mountain regions will yet be inhabited by civilized men, and the communication with the east will require railroads, independent of the want of an interior overland route to the Pacific. For this purpose the valley of the Platte offers a route not surpassed for natural gradients by any in the world, and very little more is to be done west of the Missouri than to make the superstructure. A cheap road for light trains and engines could easily be built, and when settlements are formed in the mountains will become profitable; and the gold that has been discovered there in valuable quantities may produce this result much sooner than we anticipate. The Niobrara apparently presents a more short and direct route to the interior than the Platte, but its natural features are not so favorable. The direct route from Sioux City to Fort Laramie by the Niobrara would be, for a railroad, about forty miles shorter than by way of the Platte and Fort Kearney.[6]

Some eighteen years later on August 20, 1875, A. A. Humphreys, brigadier general and chief of engineers, wrote a letter to the Honorable William W. Belknap, secretary of War. In his letter, Humphreys informed Belknap of the need to distribute Warren's report of his reconnaissance expedition to the Black Hills:

> Sir: In 1857 an exploration of the Black Hills was made by Lieut. G. K. Warren, Topographical Engineers. As this report was printed with the documents accompanying the President's annual message to Congress in December 1858, it can always be referred to for official purposes, but is not accessible to those specially interested in the region of which it treats. Recent developments in the Black Hills country have awakened a great interest in that region, and there are constant inquiries for the report I referred to. I would therefore recommend that this report be reprinted at the Government Printing Office, and that 2,000 copies be furnished to this Office, upon the usual requisition.[7]

Raynolds Expedition of 1859-60

On April 13, 1859, Captain W. F. Raynolds received orders from the War Department, Office of Explorations and Surveys, to explore the regions of the upper Yellowstone and Missouri River areas over two seasons. The objective of the expedition was to ascertain the numbers, habits, and disposition of the Indians inhabiting the area. The expedition was to also take note of the agricultural and mineralogical resources, the climate, navigability of streams, topographical features, and any obstacles that might present a problem relative to the construction of roads and railroads, "either to meet the wants of military operations or those of emigration through, or settlement in, the country."

Raynolds's orders included some general instructions: "To accomplish these objects most effectually the expedition should proceed by the Missouri River to Fort Pierre. Here a large number of the Dakotas will be assembled to receive their annuities, and overtures should be made to obtain their assent to your proceeding to the source of the Powder River by the Shayenne and its north fork, by which a new route leading west from the Missouri River would be examined. To aid you in accomplishing this object, the clothing, et cetera, to be given to the Dakotas by the government, under the treaty made with them by General Harney, will be turned over to you by the Indian Bureau for distribution."

Captain Raynolds's orders were fairly explicit. He was also authorized to employ eight assistants, including topographers, a geologist and naturalist, an astronomer, a meteorologist, and a physician. The average salary for each person was not to exceed $125 per month. Raynolds assigned Lieutenant H. E. Maynadier as his assistant and hired J. D. Hutton as topographer and assistant artist. J. H. Snowden was hired as topographer, H. C. Fillebrown, as meteorologist and assistant artist, Dr. F. V. Hayden, as naturalist and surgeon, Dr. M. C. Hines, as surgeon and assistant naturalist, and George Wallace, as timekeeper and computer.[8]

The expedition was within sight of the Black Hills on July 7. On July 11, they traveled along the divide between Bear Butte Creek and Cottonwood Creek, whereupon they made camp on the north fork of Bear Butte Creek. After dinner, followed by getting drenched in a rainstorm, a large party climbed to the top of Bear Butte, which was about a mile from camp. The expedition subsequently continued mapping along the northeastern fringe of the Black Hills until about July 16 when they moved on toward the Yellowstone River.

In his report, Raynolds made some interesting comments about the Black Hills: "Civilized life could find no home in this region," Raynolds's wrote, "and if the savage desires its continued possession, I can see no present reason for its disputing." He went on to write, "The mountains will furnish a sufficient supply of pine lumber for ordinary uses, and, although timber is very scarce in the region as a whole, yet the Black Hills will fully supply this great deficiency in the district immediately adjoining."

More interesting are his comments about gold in the Black Hills:

> *Very decided evidences of the existence of gold were discovered both in the valley of the Madison and in the Big Horn Mountains, and we found some indications of its presence also in the Black Hills, between the forks of the Shayenne. The very nature of the case, however, forbade that an extensive or thorough search for the precious metals should be made by an expedition such as I conducted through this country. The party was*

> composed in the main of irresponsible adventurers, who recognized no moral obligation resting upon them. They were all furnished with arms and ammunition, while we were abundantly supplied with picks and shovels, and carried with us a partial stock of provisions. Thus the whole outfit differed in no essential respect from that which would be required if the object of the expedition had only been prospecting for gold. The powder would serve for blasting and the picks and shovels were amply sufficient for the primitive mining of the gold pioneer, while the arms would be equally useful for defence and in purveying for the commissariat. It is thus evident that if gold had been discovered in any considerable quantity the party would have at once disregarded all the authority and entreaties of the officers in charge and have been converted into a band of gold miners, leaving the former the disagreeable option of joining them in their abandonment of duty, or of returning across the plains alone, through innumerable perils. It was for these reasons that the search for gold was at all times discouraged, yet still it was often difficult to restrain the disposition to "prospect," and there were moments when it was feared that some of the party would defy all restraint.[9]

In 1877, while Jim Bridger was living in Jackson County, Missouri, he mentioned to historian Robert Strahorn that while on the Raynolds Expedition in the Black Hills in 1859, he found gold in one of the streams. After showing the gold to Captain Raynolds, Bridger was ordered to throw the nuggets away to prevent the rest of the men from learning of his find. Bridger added, "Since my discovery of gold, I have found the same metal in that country while trading with the Indians though not in such abundance as the first."[10]

Based on hearsay from the Great Reconnaissance surveys and other sources, the Black Hills Exploring and Mining Association was formed in January 1861 by Byron M. Smith, William P. Lyman, Moses K. Armstrong, Wilmont W. Brookings, Newton Edmunds, and J. Shaw Gregory. This organization received strong support from the citizens of Yankton, who were in the process of forming Dakota Territory. More than half the town's adult male population were enrolled as members of the organization. The association wasted little time in submitting a request to Congress for funding to conduct additional surveys of the Black Hills.[11]

Dakota Territory

After the Yankton Treaty was signed with the Yankton band of the Sioux in 1858, early settlers in the area began to form their own unofficial provisional government. The settlers lobbied congressional leaders for territorial status, but were unsuccessful. Three years later, J. B. S. Todd, an in-law of Abraham Lincoln, successfully lobbied for territorial status. Under the Organic Act, Dakota Territory was created during the second session of the Thirty-sixth Congress in

February 1861. President Buchanan signed the legislation into law on March 2, 1861. Dakota Territory was formed from a portion of Nebraska Territory and a portion of Minnesota Territory that was abandoned when Minnesota became a state in 1858. Dakota Territory embraced some 326,902 square miles over what is now South Dakota, North Dakota, a very small portion of Nebraska, most of Montana, and the northeastern portion of Wyoming to the continental divide (see fig. 3.4). An office of surveyor general was opened on July 1, 1861, in Yankton. Yankton served as the territorial capital of Dakota from 1861 to 1883, when the capital was moved to Bismarck.

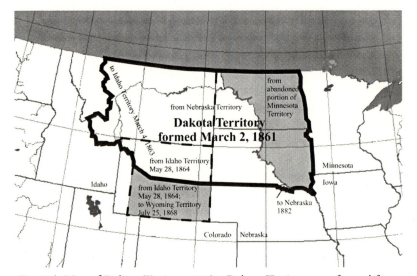

Fig. 3.4. Map of Dakota Territory, 1861. Dakota Territory was formed from Nebraska Territory and an abandoned portion of Minnesota Territory on March 2, 1861. Dakota Territory consisted of present-day South Dakota, North Dakota, most of Montana, and the northeastern portion of Wyoming. South Dakota and North Dakota were admitted to the Union as states on November 2, 1889.—Illustration by author.

The northeastern portion of future Wyoming and all of the area comprising future Montana were transferred from Dakota Territory to Idaho Territory when it was organized on March 4, 1863. Nebraska Territory transferred the southeastern portion of future Wyoming to Idaho Territory at about the same time. After Montana achieved territorial status on May 28, 1864, the southeastern portion of Idaho Territory, consisting of most of future Wyoming, was transferred to Dakota Territory.

Dakota subsequently created four new counties in future Wyoming in 1867-68, including Laramie, Carter (Sweetwater), Carbon, and Albany. These became the first counties of Wyoming when it achieved its territorial status on July 25, 1868.[12] The territory of Dakota maintained jurisdiction over Wyoming

Territory until May 19, 1869, because of delays in the appointment of officers in Wyoming Territory.[13]

As legislation to create Dakota Territory was being finalized in 1861, two acts of Congress were considered to facilitate the Westward Expansion. The first was the Homestead Act, which was passed on May 20, 1862, and became effective on January 1, 1863. The act entitled an adult citizen of the United States to file for up to 160 acres of unappropriated public land for his or her exclusive use for the purposes of settlement and cultivation. After five years, upon filing an affidavit with signatures from two credible witnesses, the person could then apply for a patent, or title, to the land.[14]

The second piece of legislation addressed the need for a transcontinental railroad. As a result of the railroad reconnaissance surveys that were completed in the 1850s, President Abraham Lincoln signed the Pacific Railway Act into law on July 1, 1862. This act authorized the Union Pacific Railroad and the Central Pacific Railroad to complete the first transcontinental railway. Actual work began in 1863 with the Union Pacific commencing at Omaha, Nebraska, and the Central Pacific commencing at Sacramento, California. The golden spike was driven on May 10, 1869, at Promontory, Utah.[15]

Hayden Expedition of 1866

In August 1866, Dr. Ferdinand V. Hayden of the Smithsonian Institution, who previously accompanied Lieutenant G. K. Warren to the Black Hills in 1857 and Captain W. F. Raynolds in 1859, agreed to head up a private expedition to the Hills. The government consented to provide a small detachment of soldiers from Fort Randall as an escort. After the expedition was completed, Hayden gave a talk before the Dakota Historical Society on October 6, 1866. A portion of his talk included the following comments:

> *Intermingled with these rocks and in the layers above, are found the gold bearing formations which are developed in the Black Hills. Little particles or grains of gold can be found in almost any little stream in the vicinity of these hills. But gold is not always found in paying quantities where "color" is raised. While there is every indication of rich gold deposits in these hills, my explorations have been more for the purpose of collecting old fossil remains than glittering dust.*[16]

Such information stirred additional interest among members of the Black Hills Exploring and Mining Association, who proceeded to arrange for their own expedition to the Black Hills to prospect for gold. By June 1867, a large number of people, including one hundred former soldiers, were camped in tents near Yankton, eager to depart for the Black Hills. The association convinced Brevet Major General A. B. Dyer, chief of the Ordnance Office in Washington, D. C., to provide two mountain howitzers and two hundred rounds of ammunition, which Dyer said could be picked up at Fort Randall.

When Lieutenant-General William Sherman learned of the proposed expedition, however, he promptly agreed with General Alfred H. Terry that the association should be kept out of the Black Hills. "You may therefore, forbid all white people going there at present," Sherman said, "and warn all who go in spite of your prohibition, that the United States will not protect them now, or until public notice is given that the Indian title is extinguished." The expedition was subsequently cancelled.

In November 1867, Captain P. B. Davy of Blue Earth, Minnesota, formed another expedition. Davy had passed near the Black Hills in 1866 while traveling from Fort Abercrombie to Montana. He proposed an expedition that would leave Yankton and travel to the Black Hills in the spring of 1868. By spring, approximately three hundred people had gathered in Yankton to participate in the expedition. Unfortunately, the nature of the expedition posed a potential problem for the U.S. government relative to its planned negotiations for a new treaty. As a result, Brevet Major General D. S. Stanley, commanding officer for the District of Southeast Dakota, received orders from the Department of Dakota "to prevent the proposed expedition, using force if necessary." Davy agreed to abandon the expedition.[17]

Red Cloud's War

Meanwhile, Colonel Henry B. Carrington was dispatched from Fort Laramie with his soldiers in June 1866 to establish a new fort by which emigrant protection would be provided along the Bozeman Trail. At a point where the Bozeman Trail crossed Big Piney Creek, Carrington's troops and civilian contractors constructed Fort Phil Kearny. The fort consisted of an 800×600-foot stockade that was 11 feet high. The walls were constructed of pine logs buried 3 feet into the ground. The inside of the stockade included barracks for the troops and their families, mess and hospital facilities, and a munitions magazine. A separate area contained shops and stables. Upon learning of the fort and its purpose, Chief Red Cloud was infuriated.

On December 21, 1866, a small band of Oglala Sioux, led by Crazy Horse, attacked a supply train leaving the newly established fort. The attack was typical of many that had recently been occurring along the trail, wherein the Indians would swoop down on the wagons and pick off a few of the soldiers or emigrants. Colonel Carrington quickly dispatched a contingent of eighty-one soldiers, which included Lieutenant Colonel William Fetterman with forty-four soldiers, Lieutenant George Grummond with twenty-seven soldiers, and several volunteers. Fetterman was instructed to pursue the warriors as far as Lodge Trail Ridge.

After Crazy Horse saw the detachment approaching, he charged toward the soldiers, then turned and abruptly retreated, whereupon the troops gave chase over the ridge. Much to Fetterman's surprise, Red Cloud's warriors were hidden beyond the ridge. The warriors proceeded to quickly massacre every soldier within the Fetterman-Grummond detachment. Upon investigating what had happened, Colonel Carrington found a gruesome scene. He later reported, "Eyes torn out and laid on rocks; noses cut off; chins hewn off; teeth chopped out . . . private parts severed and indecently placed on the person."[18]

Upon receiving word of the Fetterman massacre, General Sherman was outraged. In a wire to General Grant in Washington, Sherman was emphatic: "We must act with vindictive earnestness against the Sioux, even to their extermination, men, women, and children. Nothing less will reach the root of the case." As commander of the west, it was Sherman's job to ensure the safety of the emigrants and military posts along the trails. In Washington, the Interior Department took more of a humanitarian view and disagreed with the notions of the War Department. The Interior Department advocated separating the Indians from the emigrants and placing the Indians on newly created reservations, thereby saving them from extermination. It was proposed that the Indians be trained "in the arts of western civilization" until they could learn to support themselves and receive full rights of U.S. citizenship.

By spring 1867, Congress began debating whether to adopt the War Department's approach or that of the Interior Department. A special commission charged with investigating the Fetterman Massacre concluded that the government had erred in establishing a road in the Powder River Basin and that peace would not be achieved unless the road was abandoned. They also stated that defending the road was becoming cost prohibitive. The commission became convinced that it was the moral and Christian duty of the United States "to save the Indians from destruction and elevate them from their degraded, barbarous state."[19]

In an effort to resolve the situation, Congress passed a bill on July 20, 1867, authorizing the creation of a new commission to make peace with the plains tribes. An amendment was passed, however, that authorized the secretary of War "to conquer a peace" if peace was not achieved through negotiations. Senator Henderson, commissioner of Indian Affairs, Taylor, John Sanborn, Samuel Tappan, General William Sherman, General William Harney, and Alfred Terry were named to the commission.[20] Before the commission had a chance to meet with the Sioux, another skirmish called the Wagon Box incident took place on August 2, 1867.[21]

In September 1867, the peace commission, under Commissioner Taylor, met with nine bands of the Sioux at North Platte, Nebraska. Very few of the hostile tribes from the Powder River Basin attended this council. Commissioner Taylor opened the meeting, saying, "We are sent here to inquire of you and find out what the trouble is between the white men and you . . . . Whatever is wrong we will make right.... We ought to bury the tomahawk, smoke the pipe of peace and henceforth live in friendship forever like brothers of one family." Spotted Tail spoke first for the tribes, saying, "We object to the Powder River Road. The country which we live in is cut up by the white men, who drive away all the game. That is the cause of our troubles." After other chiefs had spoken in a similar manner, the meeting was adjourned for the day.

The following morning, the council gathered again whereupon General Sherman spoke. He said, "The United States cannot abandon its road across your country, but it will pay for any damage done by the Bozeman Trail." Sherman went on to say, "You must learn to live like white men and we will help you for as long as you need—just as we helped the Cherokee, Creeks, and Choctaws. We will set aside land for you, land for farming, housing and cattle-raising and we will teach your children to read and write."

Sherman concluded his remarks by warning them, "The railroads are coming and you cannot stop [them] any more than you can stop the sun or the moon. You must decide; you must submit. This is not a peace commission only; it is also a war commission. Without peace, the Great Father who, out of love for you, withheld his soldiers, will let loose his young men and you will be swept away."[22]

The council then ended with the understanding that the parties would reconvene in November. However, when the commissioners returned in November, none of the Indians showed up. Red Cloud sent a message on behalf of the Sioux, saying, "If the Great Father kept white men out of my country, then peace would last forever. The Great Spirit has raised me in this land and has raised you in another land. What I have said I mean. I mean to keep this land."

After debating its next move, the commission decided to take a more friendly approach with the Indians. In its report, the commission recommended closing the Powder River Road. The commissioners believed that closing the road would not impose much of a hardship, since the Indians had not allowed safe travel over the trail for the last two years anyhow. Moreover, the Northern Pacific Railroad had progressed to the point where the Bozeman Trail would soon become obsolete. The government accepted the recommendations.[23]

Fort Laramie Treaty of 1868

In April 1868, the Peace Commission returned to Fort Laramie armed with a new proposal for a treaty (see fig. 3.5). The proposal contained everything Red Cloud wanted. Unfortunately, the commission learned that Red Cloud wasn't going to show up. Negotiations proceeded with Chief Spotted Tail and Iron Shell representing the Brulé. The new treaty defined the Great Sioux Reservation, which would extend from the Missouri River to the 104 degree of longitude on the west side (see fig. 3.6). The reservation would be bounded by the forty-sixth parallel of latitude on the north side and the Platte River on the south side. The area would embrace the Black Hills.

Fig. 3.5. Negotiations for the Fort Laramie Treaty of 1868. The Peace Commission who participated in the Fort Laramie Treaty of 1868 included General William T. Sherman, General William S. Harney, General Alfred H. Terry, General Christopher C. Augur, Senator John B. Henderson, commissioner of Indian Affairs, Nathaniel G. Taylor, John G. Sanborn, and Samuel F. Tappan. Spotted Tail, Iron Shell, Man-Afraid-of-His-Horses, Sitting Bull, American Horse, and numerous other chiefs represented the Sioux.—National Archives (111-SC-95986).

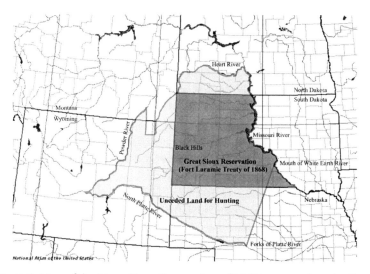

Fig. 3.6. Map of the Great Sioux Reservation of 1868 compared to the Sioux Nation of 1851. Pursuant to the Fort Laramie Treaty of 1868, the area of the Great Sioux Reservation was greatly reduced compared to the Sioux Nation established under the Fort Laramie Treaty of 1851.—Illustration by author.

The proposal specified that "no persons, except those herein designated and authorized so to do, and except officers, agents, and employees of the government as may be authorized to enter upon Indian reservations in discharge of duties enjoined by law, shall ever be permitted to pass over, settle upon, or reside in the territory described in this article, or in such territory as may be added to this reservation for the use of said Indians, and henceforth they will and do hereby relinquish all claims or right in and to any portion of the United States or its Territories."[24]

Article XI of the Fort Laramie Treaty of 1868 required the tribes to relinquish all right to permanently occupy the territory outside the reservation, but provided them with the right to hunt on any lands north of the Platte River and on the Republican Fork of the Smoky Hill River "so long as the buffalo may range thereon in such numbers as to justify the chase." Under the proposed treaty, the tribes would withdraw all their opposition to the railroads being built on the plains and permit peaceful construction of any railroad not passing over their reservation. Additionally, the Indians would refrain from attacking any persons at home or travel, refrain from molesting wagon trains, coaches, mules, or cattle belonging to people of the United States, and never capture or carry off white women or children or scalp white men or do harm to them. The proposal specifically required the Indians to withdraw their opposition to the construction of the railroad being built from the Platte River to the Pacific Ocean and withdraw all opposition to the military posts or roads now established south of the North Platte River.

In return, the United States proposed to furnish physicians, teachers, carpenters, farmers, blacksmiths, and others for the education of the Indians. Any head of a family who desired to commence farming was eligible to select a tract of land not exceeding 320 acres, which would be recorded in a land book. After three years, the person could apply for a patent for up to 160 acres of the land he was occupying and farming. The land outside the proposed reservation, but essentially within the boundaries provided for under the 1851 Treaty of Fort Laramie, would be considered "unceded land."

Article XII specified that no new treaty for the cession of any portion or part of the proposed reservation would "be of any validity or force . . . unless executed and signed by at least three-fourths of all the adult male Indians occupying or interested in the same." Article XVI of the treaty proposal specified that no white person would be permitted to settle upon or occupy any portion of the unceded lands and could not pass through the same without prior consent from the Indians. Furthermore, within ninety days after signing, the military posts that had already been established within the unceded lands would be abandoned, along with the road leading to them and to the settlements in the territory of Montana.

Spotted Tail, Iron Shell, and other headmen of the Brulé band of the Sioux signed the treaty by "touching the pen" on April 29, 1868.[25] Upon signing, Chief Iron Shell of the Brulé remarked, "I will always sign any treaty you ask me to do, but you have always made away with them—broken them." The Peace Commission needed Red Cloud's signature. After being notified of the terms of the treaty, the Oglala chief exclaimed, "When we see the soldiers move away and the forts abandoned, then I will come down and talk." After the last of the troops vacated Fort Phil Kearny toward the end of summer, Red Cloud's warriors torched the abandoned posts and burned them to the ground.

On May 25, Man-Afraid-of-His-Horses, Sitting Bull, American Horse, and other chiefs and headmen signed for the Oglala, as did the chiefs and headmen representing other bands of Sioux. Unfortunately, the commissioners still didn't have Chief Red Cloud's signature. Eventually, the peace commissioners got tired of waiting for Red Cloud and left the treaty with Major William Dye, commander of Fort Laramie. Red Cloud finally appeared at Fort Laramie on November 4. Charged with getting a signature, Major Dye proceeded to explain each article of the treaty to Red Cloud. When Major Dye got to the sections pertaining to settling on the reservation and farming, Red Cloud interrupted. He said he had heard all he cared to hear from others and that he saw no reason to give up hunting for farming on a reservation.[26] After unsuccessfully attempting to negotiate for weapons to "fight the Crow," Red Cloud, Thunder Man, High Eagle, and Thunder Flying Running signed the treaty on November 6, 1868.[27]

Immediately after signing the treaty, Red Cloud announced that the Oglala would spend the winter along the Powder River among the remaining buffalo and that his people intended to make war with the Crow. Red Cloud said his people did not care to learn to farm and would stay with the game as long as possible. He also said he expected the people at Fort Laramie to resume trading since the wars and winters had impoverished his people. Red Cloud wasn't too concerned, though, saying his people had plenty of hides to make the Indians rich again. He was happy that the Bozeman Trail would be closed, but he was far from being ready to become a farmer.

Spotted Tail, more of a friend of the government, echoed similar feelings: "Now we want to live as our fathers have lived. There is plenty of game in our country at present and we cannot go farming until all that is gone. When that time comes, I will let my grandfather know it." Crazy Horse and Sitting Bull did not sign, rejecting the treaty. Instead, they showed great resolve in wanting to maintain their way of life in the Powder River Basin. "Now you tell us to work for a living," Crazy Horse informed the commission, "but the Great Spirit did not make us to work but to live by hunting. You white men can work if you want to. We do not interfere with you, and again you say, why do you not become civilized? We do not want your civilization. We would live as our fathers lived, and their fathers before them."

Very quickly, it became apparent that the Treaty of 1868 was not going to motivate the Sioux to move to the reservation, or transform them from hunters to farmers. General Sherman could see that the provisions of the treaty that relate to unceded lands and hunting grounds served only to create obscure boundaries that would delay getting the Indians to relocate to the newly defined reservation. In his report to General Sheridan, Sherman wrote that it would be better "to invite all the sportsmen of England and America this fall for a Great Buffalo Hunt and make a grand sweep of them all."[28]

By spring 1869, less than half of the Lakota and Nakota Sioux had moved to the Great Sioux Reservation. Those who did settled along the Missouri River close to the established forts and newly established agencies. Spotted Tail kept his Brulé band about 30 miles above the mouth of the White River, where they would be more isolated from the new farm plots and the alcohol of the white man. Red Cloud and his Oglala people refused to leave the Powder River Basin despite the hard winter their people were experiencing. This hardship was exacerbated by the fact the Indians were not allowed to trade at Fort Laramie. Notwithstanding, the majority of the bands of Sioux were generally happy with the freedom they believed they had achieved under the treaty. However, they had absolutely no reason or desire to become farmers.[29]

In spring 1870, Red Cloud and Spotted Tail consented to a request by Commissioner Ely Parker of the Federal Indian Bureau to travel to Washington, D. C., to discuss the Treaty of 1868 and the Great Sioux Reservation. When Red Cloud and his Oglala delegation arrived in Washington in June, Spotted Tail and his delegation of "friendly" Brulé from the reservation were already there. One evening, Red Cloud and Spotted Tail were taken to the White House, where they met President and Mrs. Ulysses Grant in a room filled with chandeliers. While eating dinner, which included a dessert of strawberries and ice cream in the State Dining Room, Spotted Tail commented that the whites had many more good things to eat than what they send to the Indians.

The next day, Secretary Cox opened the conference by saying the president had authorized General Sherman to negotiate the recent treaty because he wished to find a place where the Indians could live undisturbed. Secretary Cox concluded his opening comments by saying, "The first thing we want to say to is that they must keep the peace and then we will try to do for them as is right."[30] Red Cloud walked to the head table, shook hands, and sat on the floor in preparation for giving his remarks. Through an interpreter, he pleaded his case:

> *What I have to say to you, and to these men and to my Great Father is this: Look at me. I was raised where the sun rises and I come from where he sets. Whose voice was first heard in this land? It was the red people who used the bow. The Great Father may be good and kind, but I can't see it. The Great Father has sent his people out there and left me nothing but an island. Our nation is melting away like the snow on the side of the hills where the sun is warm; while your people are like the blades of grass in spring when summer is coming. I don't want to see the white people making roads in our country . . . I have got two mountains in that country—Black Hill[s] and Big Horn. I want no roads there. There have been stakes driven in that country, and I want them removed. I have told these things three times, and I now have come here to tell them for the fourth time.*[31]

In discussing the Treaty of 1868, Red Cloud exclaimed, "This is the first time I have ever heard of it and do not mean to follow it." Secretary Cox already knew that Major Dye had previously attempted to go over the treaty, section by section, with Red Cloud, but was interrupted before getting to the end. Red Cloud countered by saying, "I do not say that the Commissioners lied, but the interpreters were wrong." When Secretary Cox offered to give Red Cloud a copy of the treaty, Red Cloud angrily replied, "I will not take the paper with me. It is all lies."[32]

Red Cloud then departed Washington, D. C., for a visit to New York. There he made a scheduled appearance at the Cooper Institute and was welcomed with an ovation. Speaking through an interpreter, Red Cloud gave a short speech. Nearly every sentence was followed with applause from the audience. The following is an extract from his speech:

> *The Great Spirit made us poor and ignorant. He made you rich and wise. In 1868 men came out and brought papers. We could not read them. When I reached Washington the Great Father explained to me what the treaty was, and showed me that the interpreters had deceived me. I was brought up among the traders, and those who came out there in the early times treated me well and I had a good time with them. But, by and by, the Great Father sent out a different kind of men; men who cheated and drank whiskey; men who were so bad that the Great Father could not keep them at home and so sent them out there. I came to speak to you myself; and now I am going away to my home.*[33]

As Red Cloud was returning home, editors of newspapers such as the *Yankton Union and Dakotaian* wrote that "a dose of terrible war" was the only solution to the Indian problem. An officer of the Seventh Cavalry reported that many bands of Indians were turning to Sitting Bull, chief of the Hunkpapa. In reflecting the sentiment of many of the Sioux, the officer reported that "Red Cloud saw too much. The Indians say that these things cannot be true; that the white people must have put bad medicine over Red Cloud's eye to make him see everything and anything that they please."[34]

Red Cloud was experiencing change. After being invited to meet with Sitting Bull, Red Cloud declined but replied with a message: "I shall not go to war anymore with the whites. I shall do as my Great Father says and make my people listen. Make no trouble for our Great Father. His heart is good. Be friends to him and he will provide for you."[35] Sitting Bull, however, was gaining more and more popularity with the Sioux based on his stance against the ways of the white man. "Look at me and see if I am poor, or my people either . . . ," he said, "you are fools to make yourselves slaves to a piece of fat bacon, some hardtack, and a little sugar and coffee."[36]

By early 1872, formalized efforts began to emerge for the purpose of prospecting the Black Hills. On February 27, 1872, Charles Collins, editor of the *Sioux City*

Weekly Times, organized the Black Hills Mining and Exploring Association in Sioux City. Several prominent businessmen supported the effort, including Thomas H. Russell, Charles D. Soule of the Northwestern Transportation Company, Dan Scott, editor of the *Sioux City Journal*, and General A. C. Dawes, passenger agent for the Kansas City and St. Joe Railway. Russell traveled to the towns of Dakota Territory, distributing pamphlets that Collins had printed in his campaign to promote development of the Black Hills.[37]

"Ho! for the Black Hills! Off to the New Eldorado!" exclaimed the headline of the pamphlet. The pamphlet included directions to the Black Hills and specified the cost for outfitting a party of five. Participants were advised to bring one rifle and one revolver per man. Except for personal items, most everything else would be provided. The pamphlet guaranteed freedom from care and responsibility and participants could expect "a good time during the whole trip." The effort culminated in an expedition that would tentatively depart for the Black Hills on September 1, 1872.[38]

On March 16, 1872, Moses Armstrong, another member of the Black Hills Mining and Exploring Association and territorial delegate to Congress, appeared before Columbus Delano, secretary of the Department of Interior, Francis Walker, commissioner of Indian Affairs, and Ferdinand V. Hayden, director of the U.S. Geological Survey. Hayden had participated in several expeditions to the Black Hills in the 1850s and 1860s. Armstrong pleaded with the group to amend the Treaty of 1868 and allow development of the Black Hills for mining, lumbering, and settlement. After considering the plea, Delano refused to support the proposed amendment, although he said he was not opposed to negotiating with the Sioux for mineral and timber rights.[39]

Hearing all of the pleas and plans for privately sponsored expeditions to the Black Hills, Major General W. S. Hancock, U.S. Army, issued the following press release on March 26, 1872, to newspapers throughout Dakota:

> *Letters are being received at these headquarters from various parts of the United States, making inquiries in regard to the reputed gold discoveries in the section of the country west of the Missouri River known as the Black Hills of Dakota, and asking if expeditions, presumed to be now in the process of organization, will be permitted to penetrate that region.*
>
> *The section of country referred to is set apart as an Indian reservation, by treaty with the Sioux, and the faith of the Government is understood to be pledged to protect it from the encroachments of, or occupation by, the whites. Accordingly, any parties or expeditions which may organize for the purpose of visiting or "prospecting" the region in question, will be engaging in an unlawful enterprise, the consummation of which it will be my duty, under the law and my instructions, to prevent, by the use, if necessary, of the troops at my disposal. In this connection, I may mention that I am just in receipt of an official letter from General Stanley, in command, subordinate*

to me, on the Missouri River, in which he refers incidentally to the Black Hills gold reports, in which he says no gold has been found there.

If you will give publicity in your columns to the statements herein contained, I do not doubt it will be the means of saving many worthy people from incurring useless expense.[40]

Collins, however, was not ready to give up on his business plan for the Black Hills. Collins proposed a "stumpage corporation," which would pay the Lakota a royalty for cutting timber in the Black Hills. Collins figured the Hills could be investigated for minerals during timber-cutting operations. Once a gold strike had been declared, the Black Hills could be seized. "Favored friends of the administration" could quickly "secure the choicest tracks in the Hills." Unfortunately, Collins's plan was ill received, and it was soon abandoned.[41]

Meanwhile, Secretary Delano and possibly other influential people inside and outside of the federal government, such as Dr. Ferdinand Hayden, had second thoughts relative to the Black Hills. In a letter dated March 28, 1872, from Secretary Delano to owner and publisher Charles Collins, Delano wrote,

I am unable to express an intelligent opinion, now as to the propriety of immediate efforts to extinguish the Indian title in the pine forests of the Black Hills. I am inclined to think that the occupation of this region of country is not necessary to the happiness and prosperity of the Indians, and it is supposed to be rich in minerals and lumber it is deemed important to have it freed as early as possible from Indian occupancy.

I shall, therefore, not oppose any policy, which looks, first, to a careful examination of the subject upon the basis indicated in this letter. If such examination leads to the conclusion that the country is not necessary or useful to the Indians, I should then deem it advisable that steps be taken to extinguish the claim of the Indians and open the territory to the occupation of the whites.[42]

Following the signing of the Fort Laramie Treaty of 1868, Red Cloud and his followers gave up the fight and moved to the mouth of Horse Creek on the Platte River, just west of present-day Scottsbluff, Nebraska. This area, located within the unceded area but outside of the reservation boundary specified in the treaty, became known as the Red Cloud Agency. By September 1872, there were approximately 6,320 Oglala and Upper Brulé Sioux, 1,515 Cheyenne, and 1,342 Arapaho registered at the Red Cloud Agency. The agency was moved to a site east of Fort Robinson, the Soldier's Town, in August 1873.[43]

The followers of Crazy Horse and Sitting Bull refused to move to the Great Sioux Reservation and continued to follow the remaining herds of buffalo throughout the Powder River Basin. Red Cloud's people became known as Hangs-Around-the-Fort by the followers of Crazy Horse and Sitting Bull. When winters were harsh, some of

Crazy Horse's followers, such as Black Elk and his parents, would spend the winter at the Soldier's Town. When spring arrived, they migrated back to the Powder River Basin or spent time hunting in the Black Hills.[44]

Custer Expedition of 1874

In the midst of the growing dissension over the Treaty of 1868, President Ulysses S. Grant signed the Mining Law of 1872 on May 16, 1872. The Mining Law of 1872 was one of a number of public land laws passed by Congress in the late 1800s to promote the development or settlement of public domain lands in the western United States. The Mining Law of 1872 provided (and still provides) citizens of the United States with the right to explore for, discover, and exploit certain valuable mineral deposits on those federal lands opened for these purposes. The Mining Law specified the means for (1) discovery of a valuable mineral deposit, (2) location of mining claims and sites, (3) recordation of mining claims and sites, (4) maintenance of mining claims and sites, and (5) mineral patents (see Appendix C).[45]

With all of the opportunities to stake mining claims, mine gold, and establish homesteads, the business and territorial leaders of Dakota Territory were by no means resting on their laurels regarding the Black Hills. On January 14, 1873, the Legislative Assembly of Dakota Territory appeared before Congress. Here, members of the assembly asked for a scientific exploration of the Black Hills. Because the Indians only used the Black Hills as a retreat from their hostile actions, the members argued, the government should confine the Indians to a smaller area of the Great Sioux Reservation and open the Black Hills to settlement.[46]

By spring 1874, General Philip Sheridan had concluded that many of the "hostile" Sioux were not living up to the terms of the treaty. To remedy this situation, Sheridan believed a new and more centrally located military post was needed. Sheridan proposed the Black Hills for the preferred location of such a post, based on its geographical location and findings of previous reconnaissance expeditions that revealed an abundance of wood, water, and grass with which to support a new fort. From this location, Sheridan believed, the military could monitor the activities of the more hostile Lakota Sioux who resided to the south and west of the Black Hills, as well as those who migrated between the Powder River Basin and the agencies to receive their annuities. Moreover, Sheridan knew that the officials of the Northern Pacific Railroad wanted the Indians subdued so the railroad could safely be continued to the west. In a similar fashion, businessmen, farmers, and ranchers also pressured the government for access to the resources rumored to exist in the Black Hills.

Within a short time, General Sheridan received permission from President Grant and General William Sherman to conduct a reconnaissance mission to the Black Hills to search for a site for a fort and explore the area for possible communication routes. Perhaps, President Grant was the driving force that caused

Secretary Delano and other governmental officials to change the government's position about conducting another investigation of the Black Hills. The size and breadth of the scientific and civilian corps that was invited to participate suggests that there may have been other reasons for the expedition.[47]

On June 8, 1874, Lieutenant Colonel George Armstrong Custer received Special Order No. 117 from General Alfred Terry to lead an expedition from Fort Abraham Lincoln, Dakota Territory, to Bear Butte in the Black Hills. From Bear Butte, Custer was to explore the country to the south, southeast, and southwest for possible communications routes. The order stated that the expedition would consist of six companies of the Seventh Cavalry stationed at Fort Abraham Lincoln, four companies of the Seventh Cavalry stationed at Fort Rice, Company I, Twentieth Infantry, and Company G, Seventeenth Infantry. By the time the expedition was finally assembled, it included approximately 1,000 men, 1 woman, 110 wagons and ambulances, 3 Gatling guns, and one 3-inch gun.[48]

A large complement of military and civilian individuals was assigned to the expedition. These people included Lieutenant Colonel George A. Custer, commander; Lieutenant Colonel George A. Forsythe, aide to General Sheridan; Captain William Ludlow, chief engineer; James Calhoun, soldier (and brother-in-law to Custer); Lieutenant Colonel Frederic Dent Gant (President Grant's son); and Bloody Knife, scout. A scientific corps included Newton H. Winchell, civilian geologist from the University of Minnesota; William H. Illingworth, civilian photographer; George Bird Grinnell, civilian paleontologist and botanist from Yale; Charley Reynolds, civilian guide; Horatio N. Ross and William T. McKay, civilian miners; Aris B. Donaldson, civilian reporter and botanist; William Curtis, civilian reporter; Felix Vinaterri, civilian band director; Michael Smith, civilian wagon master; and Sarah Campbell, also known as Aunt Sally, a black female cook. The civilian corps included five newspaper reporters who represented newspapers in New York, Chicago, St. Paul, and Bismarck. The entourage also included one hundred Arikara scouts, a sixteen-member marching band, a doctor, and three hundred head of cattle that would provide food for all members of the expedition.[49]

As the expedition was being assembled, news stories were already being published informing the public about the mysterious Black Hills. "The press has praised the Black Hills country to the skies," wrote Lieutenant James Calhoun in his diary. "We are informed in glowing terms 'that it is believed to be a land of ambrosial luxury—flowing with milk and honey.' In fact so much has been circulated regarding this section of the country, that thousands are wild with curiosity—longing to see it."[50] An editorial in the *Bismarck Tribune* promised nuggets of "pure gold as large as walnuts." Private Theodore Ewert was one who showed remorse. Reflecting on the stated and unstated purposes of the expedition, Private Ewert wrote in his journal that the United States "forgot its honor, forgot the sacred treaty in force between itself and the Dakota Sioux, forgot its integrity, and ordered the organization of an expedition for the invasion of the Black Hills."[51]

At 8:10 a.m. on July 2, 1874, the signal to "advance" sounded and the mile-long procession under the command of General George A. Custer departed Fort Abraham Lincoln from Bismarck. After traveling to Inyan Kara, the expedition moved through the western Black Hills, reaching French Creek on July 30 (see fig. 3.7).[52]

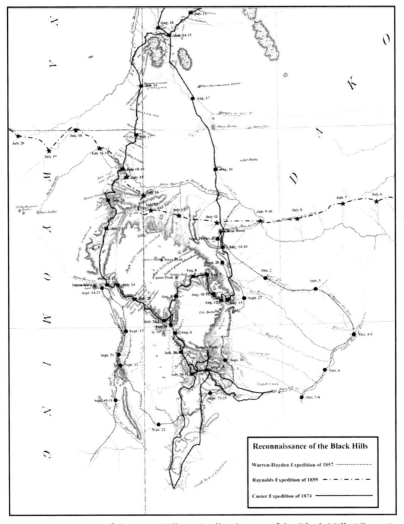

Fig. 3.7. A portion of Captain William Ludlow's map of the Black Hills (Custer) Expedition, 1874. The map also shows the routes of the Warren-Hayden Expedition of 1857 and the Raynolds Expedition of 1859.—Modified from Captain William Ludlow, *Report of a Reconnaissance of the Black Hills of Dakota, Made in the Summer of 1874*, Washington, D. C.: Government Printing Office, 1875.

Horatio Ross wasted no time panning for gold along French Creek. The correspondent from the *New York Tribune* accompanying the expedition reported the following: "There was gold there but it was merely a color requiring careful manipulation and an experienced eye to find it. The few glittering grains, with a slight residue of earth were carefully wrapped up in a small piece of paper and put in the miner's pocketbook. It was simply an earnest of what was to come. The discovery announced created a good deal of interest but little commotion. The next day, the expedition remained in camp and the miners had a chance to renew their search. The result was the discovery of a good bar, yielding from five to seven cents per pan."[53]

On August 1, Custer moved the camp approximately 2½ miles down French Creek where permanent camp was set up for the next five days. Almost immediately, Horatio Ross and William McKay began finding gold in their pans. Soon, many other people in the camp, including Aunt Sally, became afflicted with gold fever and engaged themselves in panning for gold, regardless of whether they knew what they were doing or not. The next day, August 2, Custer dispatched Charley Reynolds to Fort Laramie to deliver the news. "Lonesome Charley" Reynolds completed the 115-mile trip in four days.

Custer's notice was telegraphed to General Terry, who immediately released a message to the newspapers: "Gold has been found at several places and it is the belief of those who are giving their attention to this subject that it will be found in paying quantities. I have upon my table forty or fifty small particles of pure gold, in size averaging that of a small pinhead, and most of it obtained today (August 2) from one panful of earth. As we have never remained longer at our camp than one day, it will be readily understood that there is no opportunity to make a satisfactory examination in regard to deposits. Veins of what the geologists term gold-bearing quartz crop out on almost every hillside."[54]

Captain William Ludlow's final report described similar findings: "August 2—There is much talk of gold, and industrious search for it is making. I saw in General Custer's tent what the miner said he had obtained during the day. Under a strong reading-glass it resembled small pinheads, and fine scales of irregular shape, perhaps thirty in number. The miners expressed themselves quite confident that if they could reach bedrock in the valleys at a favorable place, plenty could be obtained by use of the pan."[55]

Immediately the newspapers were abuzz with news of the discovery of gold in the Black Hills. "Count me in for the Black Hills," read the *Bismarck Tribune* on August 12. Headlines of the August 27 issue of the *Chicago Inter-Ocean* read, "Gold! The Land of Promise—Stirring News from the Black Hills." The *Yankton Daily Press and Dakotaian* announced, "Struck it at Last . . . Gold Expected to fall 10 percent . . . The National Debt to be Paid when Custer Returns."[56]

The *New York Tribune* was much more realistic, cautioning its readers that the region belonged to the Sioux and that the value of the gold fields was yet to be determined. The newspaper reported that "speculators in St. Paul and Bismarck, in the interest of the Northern Pacific Railroad, are actively engaged in fomenting the gold fever . . . and that all the gold actually discovered thus far could be put in a thimble."[57]

While sitting around the campfire in the permanent camp on French Creek on August 5, members of the expedition decided to form the first mining company in the Black Hills. They named it the Custer Park Mining Company. A discovery notice was written on the inside of a hardtack box cover and posted at the discovery site. The notice read as follows:

> *District No. 1, Custer Park Mining Company, Custer's Gulch, Black Hills, D. T., August 5, 1874.*
>
> *Notice is hereby given that we the undersigned claimants, do claim four thousand (4,000) feet commencing at number eight (8) above, and running down to number twelve (12) below discovery, for mining purposes, and intend to work the same as soon as peaceable possession can be had of this portion of Dakota Territory by the General Government, and we do hereby locate the above claim in accordance with the laws of Dakota Territory governing mining districts.*[58]

Private Creighton, assigned to assist Ross and McKay, said this about placer mining:

> *I was detailed to accompany prospectors as one of a guard. This was a fine trip if you think looking for gold is fun. Try it sometime for your own enjoyment. Dig a hole four or five feet deep through gravel and sand until you come to what is called "bed-rock," then you commence to wash the gravel, and after having worked all day, you find no gold. If this doesn't satisfy you, repeat the process the next day and perhaps you will succeed in finding, by extra hard work, gold paying from four to five dollars; the next day nothing, and so on. It is fun if you like it, and it's fun if you don't.*[59]

On August 6, the Custer Expedition departed permanent camp and began making their way to the north, following parts of the trail they had previously established. Their route took them across Gillette Prairie and down Gold Run Gulch to Castle Creek at present-day Deerfield Lake. From there, they traveled north across Reynolds' Prairie, down the South Fork of Rapid Creek and up Dump Draw to an area along the North Fork of Rapid Creek approximately 1½ miles south of Nahant. It was here that General Custer, Captain Ludlow, Private Noonan, and Bloody Knife shot a grizzly bear on August 7 (see fig. 3.8).

Fig. 3.8. W. H. Illingworth's photo entitled, "Our first grizzly, killed by Gen. Custer and Col. Ludlow." Pictured are Bloody Knife, Custer's chief scout, General Custer, Private John Noonan, and Captain William Ludlow, all of whom shot at the bear.—National Archives (77-HQ-264-847).

From there, Custer experienced some difficulty finding a good route out of the Black Hills. The expedition made its way down Middle Boxelder Creek to a point near present-day Roubaix Lake and then across the North Fork of Boxelder Creek and Corral Creek to Hay Creek, where they made camp on August 8 at present-day Reausaw Lake. From here, the group made their way along Little Elk Creek, cut south to present-day Nemo, and followed Boxelder Creek to the mouth of Estes Creek, where they made camp on August 10 and 11. After that, Custer followed Boxelder Creek and the present-day Norris Peak Road to Piedmont Valley. From here, he proceeded north, passing Bear Butte on its east side. The expedition returned to Fort Abraham Lincoln on August 30, 1874.[60]

Elizabeth Custer provided a vivid description of the arrival in Fort Lincoln:

> *When the day of their return came, I was simply wild with joy. I hid behind the door as the command rode into garrison, ashamed to be seen crying and laughing and dancing up and down with excitement. I tried to remain there and receive the general, screened from the eyes of*

outsiders. It was impossible. When we could take time to look everyone over, they were all amusing enough. Many, like the general, had grown heavy beards. All were sun-burnt, their hair faded, and their clothes so patched that the original blue of the uniform was scarcely visible.

The boots were out at the toes, and the clothing of some were so beyond repairing that the officers wanted to escape observation by slipping, with their tattered rags, into the kitchen-door. The instruments of the band were jammed and tarnished, but they still produced enough music for us to recognize the old tune of "Garryowen," to which the regiment always returned.

By and by the long wagon-train appeared. Many of the covers had elk horns strapped to them, until they looked like strange bristling animals as they drew near. Some of the antlers were brought to us as presents. Besides them we had skins, specimens of gold and mica, and petrified shells of iridescent colors, snake rattles, pressed flowers, and petrified wood. My husband brought me a keg of the most delicious water from a mountain-stream. It was almost my only look at clear water for years.[61]

In September 1874, Lieutenant Colonel George Forsythe returned to Chicago to update General Sheridan. While there, a reporter from the *Chicago Tribune* asked Forsythe if the rumors about gold in the Black Hills were true. Forsythe replied, "There is no doubt that there is a great deal of gold there."[62]

In St. Louis, General William Sherman remarked, "It's the same old story, the story of Adam and Eve and the forbidden fruit." Edwin A. Curley, an English journalist who was present when Sherman made the remark, then asked, "Is it not true that in practical effect the officers of this government were sent to pluck the forbidden fruit, and to show it to the people, saying, 'This is most excellent food?'"[63]

Fueled with new energy, the *Yankton Daily Press and Dakotaian* began to issue a daily flier called the *Yankton Black Hiller*. The flier promoted Yankton as the most logical jumping-off point for outfitters and prospectors headed for the gold country on the basis that the capital of Dakota was now connected with the Dakota Southern Railroad and had riverboat service to Fort Pierre.[64]

The gold rush was on.

Chapter 4

The Gold Rush

Collins-Russell Expedition of 1874-75

The announcements by General George Armstrong Custer in 1874 that gold had been discovered in the Black Hills were the catalyst that precipitated the gold rush of 1875-1876. For decades, prospectors and venture capitalists such as Charles Collins, editor of the *Sioux City Times*, had heard tales about early-day prospectors, fur traders, Father De Smet, and the Indians who had found or had some knowledge of gold in the Black Hills. Moreover, hearsay and reports from the Warren-Hayden Expedition to the Black Hills in 1857, the Raynolds Expedition of 1859-1860 to the Yellowstone, and the Custer Expedition of 1874 pointed to the richness of the Black Hills and what the Hills might offer in terms of fulfilling the American Dream.

Collins and his partner, Thomas H. Russell, had been promoting expeditions to the Black Hills since February 27, 1872, when they formed the Black Hills Mining and Exploring Association of Sioux City. After Lieutenant General Philip Sheridan became aware of the promotion, he promptly issued an order prohibiting any travel within the Great Sioux Reservation. Collins and Russell shelved their plan.[1] However, following Custer's announcement of a gold discovery, Collins and Russell renewed their interest in the Black Hills and began organizing another expedition. Some two hundred confidential letters were sent to those prospective adventurers who previously indicated a desire to participate in an expedition to the Black Hills. "If you can raise $300, can handle a rifle, and mean business, be at Sioux City on or about the middle of September," the letter stated. By September 3, 1874, about one hundred adventurers had arrived in Sioux City and were getting ready for the expedition.[2]

Hearing rumors once again of private expeditions to the Black Hills, Lieutenant General Sheridan sent a written order to Brigadier General Alfred H. Terry on September 3, 1874:

> *Sir: Should the companies now organizing at Sioux City and Yankton trespass on the Sioux Indian reservation, you are hereby directed*

> to use the force at your command to burn the wagon trains, destroy the outfits and arrest the leaders, confining them at the nearest military post in the Indian country. Should they succeed in reaching the interior, you are directed to send such force of cavalry in pursuit as will accomplish the purposes above named. Should Congress open the country to settlement by extinguishing the treaty rights of the Indians, the undersigned will give cordial support to the settlement of the Black Hills. A duplicate of these instructions has been sent out to General Ord, commanding the Department of the Platte.[3]

On the afternoon of October 6, 1874, the Collins-Russell Expedition, also known as the Gordon Party after its leader, John Gordon, departed the west side of the Missouri River near Sioux City. The party consisted of twenty-eight people, including Captain Thomas H. Russell, John Gordon, David G. Tallent and his wife, Annie D. Tallent, and their nine-year-old son, Robert E. Tallent. The Gordon Party reached the Black Hills on December 9, 1874, near Tilford. Here, they spotted Custer's trail and followed it to French Creek, arriving there on December 23. Immediately, members of the party started panning and found gold in every pan.[4]

The Gordon Party embarked on the task of building a stockade for protection against the winter weather and any other danger. On February 23, 1875, a meeting was held at the stockade for purposes of forming a mining district and drafting the first mining laws. A. F. Long acted as chairman and Thomas H. Russell served as secretary. A committee consisting of B. B. Logan, Thomas Russell, Lyman Lamb, D. G. Tallent, and R. B. Whitney was appointed to draft the mining laws. At a meeting on March 6, the first mining district in the Black Hills was organized as the Custer Mining District, and Angus W. McDonald was elected district recorder.

Things were going good at the stockade until April 6, 1875, when a military detachment under the command of Captain John Mix, Company M, Second U.S. Cavalry, arrived at the Gordon Stockade with twenty-five mules to escort the Gordon Party out of the Black Hills. The detachment escorted the Gordon Party out of the Black Hills on April 8 bound for Fort Laramie via the Red Cloud Agency. Upon reaching Fort Laramie, the Gordon Party was detained for two days and released. Most of the party subsequently went to Cheyenne (see fig. 4.1), where Charlie Collins, who had traveled from Sioux City to welcome the group home, met them. Some of the party remained in Cheyenne and others returned to Sioux City.[5]

Fig. 4.1. Members of the Gordon Party (Collins-Russell Expedition) in Cheyenne, Wyoming Territory, after they were escorted out of the Black Hills in 1875. From left to right, back row: Tommy Quinner, David Aken, Angus McDonald, Lyman Lamb, "Red Dan" McDonald, and John Boyle. *Front row*: Jim Dempser, Dempster McDonald, R. R. Whitney, and B. B. Logan.—Photo courtesy of the State Archives of the South Dakota State Historical Society.

Newton-Jenney Expedition of 1875

By spring 1875, Secretary of the Interior Columbus Delano and President Ulysses S. Grant were convinced that the Black Hills needed to be opened up to settlement. It now became a question of what the Black Hills were worth. In a letter to the secretary of War, Secretary Delano, wrote, "To settle this question satisfactorily, the Department has decided, under the advice of the President, to send a competent Geologist to explore that region." Professor Walter P. Jenney of the Columbia School of Mines was selected as chief geologist to head the expedition.[6]

On April 18, Secretary Delano approved a request by Edward P. Smith, commissioner of Indian Affairs of the Indian Bureau, to add a topographer and an astronomer to the Black Hills geological expedition. Smith stated that it would be necessary to determine with certainty whether the locations of any findings of the upcoming expedition might actually be located in Dakota Territory, and subject to the Fort Laramie Treaty of 1868, or in Wyoming Territory, which was public domain. Delano also informed the *New York Times* that the Sioux chiefs would arrive in Washington, D. C., around May 1 to negotiate the sale of the reservation.[7]

Spotted Tail and Red Cloud, who previously had established friendly relations with the government, were invited to be the primary representatives of the Sioux at the meeting in Washington. On May 26, 1875, the two chiefs and seventeen other members of their delegation met with President Ulysses S. Grant and Secretary of the Interior Columbus Delano. President Grant informed the chiefs that it was going to be difficult to keep prospectors from entering the Black Hills and that strong efforts might not be made to keep them out.

The chiefs informed President Grant and Delano that they were infuriated that prospectors had already trespassed into the Black Hills. They also expressed frustration that the administration of their agencies was being handled so poorly. Delano reminded the chiefs that under the Treaty of 1868, the government was no longer required to continue providing food rations at the agencies. He suggested that they relinquish the Black Hills. At the conclusion of the meeting, the chiefs agreed to sell their hunting rights along the Platte and Republican rivers for $25,000 but adamantly refused to give up their right to the Black Hills.[8]

The Newton-Jenney Expedition departed Fort Laramie on May 25, 1875, with Professor Walter P. Jenney heading a scientific party of seventeen people. Henry Newton was hired as assistant geologist and Dr. Valentine T. McGillycuddy as topographer. Lieutenant Colonel Richard I. Dodge commanded the military escort, which consisted of six companies of cavalry, two companies of infantry, seventy-one wagons, and more than one hundred cattle. Five news correspondents participated in the expedition.[9]

The expedition reached French Creek on June 14. Here, Colonel Dodge spotted a miner running into a gorge and sent one of his men to investigate. After a short while, when the soldier failed to return, Dodge went to the gorge to investigate. There, he found the soldier engaged in a conversation with six miners. After learning that there were actually twenty miners prospecting along the creek, Dodge informed the six miners he didn't intend on arresting them. In return, however, Dodge persuaded the miners to give him a sizeable quantity of gold to send to General Crook.[10]

Colonel Dodge wasted no time in sending a dispatch to the nearest telegraph office advising General Crook of the gold find. Almost immediately, the message was released to newspapers across the country.

> *A dispatch received at the War Department today from Assistant Adjt. Gen Whipple of Gen. Sherman's staff, dated St. Louis, Mo., June 23, announces that a dispatch just received there from Col. Dodge, dated Harney's Peak, June 17, reports that gold was found in paying quantities on French Creek. Custer's report is confirmed in every particular. Col. Dodge reports the command is well and in fine condition. He has written fully on the discovery, and will forward his reports by mail.*[11]

Not to be outdone by the military and to possibly couch the reports of Dodge and Custer, Walter Jenney sent his own message to E. P. Smith, commissioner of Indian Affairs, Washington, D. C.:

Camp On French Creek, June 17, 1875.

> *I have discovered gold in small quantities on the north bend of Castle creek, in terraces of bars and quartz gravel. Arrived here yesterday. About fifteen men have located claims on the creek above here and have commenced working. Gold is found southward to French creek at this point. The region has not been fully explored, but the yield of gold is small and the richness of the gravel has been greatly exaggerated. The prospect, at present, is not such as to warrant extensive operations in mining.*[12]

The expedition continued on to Spring Creek. There, according to Colonel Dodge, gold was found in "very considerable quantities" with "ample water for sluices." Dodge proclaimed that "there is no doubt but that the gold on this creek will richly pay the hydraulic miner." He said the area "is a lovely dairy country, full of beautiful sites for residences, timber and building-stone being convenient and abundant, and everywhere is the most ample supply of cold, pure water."

After investigating Castle Creek, Dodge found that the miners regarded this stream as the best in the Black Hills. The expedition then moved on to Rapid Creek about two miles upstream from where it exits the Black Hills. In describing that particular location, Dodge wrote, "The soil of the bottom is rich loam, covered with the finest grass. The water is pure and excellent, and the stream amply large for sawmills . . . It is one of the choicest spots in the Black Hills for the settler. Unfortunately, there is gold enough in the sands and soil of the creek bottom to induce some Vandal to put the whole of it through sluice boxes, leaving the now lovely valley a desert of rocks and sand . . . There is abundant timber for building purposes, fencing, and fuel and unlimited limestone for building or other purposes." Dodge alluded to the value of timber several times and regarded the timber of the Black Hills as "one of the principal sources of its future wealth. Scattered through the young forests are old and large trees which by some means escaped the fires that destroyed others of their age."[13]

While visiting the area around Rapid Creek, the Newton-Jenney Expedition encountered the Hinman Expedition led by Reverend Samuel D. Hinman. The stated objective of Hinman's expedition was to find a more suitable location for the Spotted Tail Agency. A secondary objective of the expedition, as Hinman later communicated to chiefs Spotted Tail and Two Strike, was to disprove the exaggerated rumors of gold and fertility in the Black Hills and to protect the Indians' rights to the Black Hills.[14] Red Dog, a Lakota from the Red Cloud Agency and part of the Hinman Expedition, asked Colonel Dodge why the soldiers were so friendly with the miners and weren't escorting the miners from the Black Hills. Dodge replied, "This is a matter for the chiefs above me."[15]

In writing about the gold found on the expedition, Colonel Dodge commented, "In the valleys of the streams gold is found almost everywhere, in the bars, in the gravel and sand of the beds, even in the grass roots." He went on to write, "Every stream within the area yielded gold, but I saw no single spot where the quantity would pay the ordinary pan miner." According to Dodge, "Pan mining will pay nowhere in the Hills. Sluice and hydraulic mining will pay on French Creek; it will pay well on Spring Creek, and best on Rapid Creek." Dodge explained that "Castle Creek, the principal tributary of Rapid, was the favorite among the miners, but Mr. Jenney preferred the Spring." Dodge theorized that "money is to be made here by men who have sufficient capital to buy up many claims along a creek, sufficient to warrant the expense of dams, ditches, and all the works necessary for hydraulic mining."[16]

Colonel Dodge speculated that "the Black Hills will, sooner or later, be opened to the miner." Regarding the order to escort miners out of the Black Hills, Dodge explained how a miner was often captured by his troops, taken to the nearest military post, turned over to the civilian authorities, and set free, whereupon they "immediately start again for the Black Hills." Dodge spoke of one miner who stated he had been captured and removed from the Hills four different times. The miner bragged to Colonel Dodge that each time, he would give the soldiers more trouble in being caught. "I guess I can stand it as long as they can," the miner concluded. By July 20, 1875, as many as six hundred miners were engaged in prospecting and mining in the Black Hills, according to Dodge. He noted that General Crook himself came to the Hills and "mingled and freely conversed with these men."[17]

Throughout his book, Dodge wrote as though he understood the basics of geology and mining. He quipped that "placer mining is the poor man's diggings, while quartz mining is only for the rich." Dodge believed the mineral wealth of the Black Hills would be derived from development of the "leads." If gold or silver did exist in the veins, he said, "the value of the Hills is incalculable." He went on to say, "Appearances are . . . so favorable that it is hard to believe that the Black Hills will not yet furnish its 'Big Bonanza.'" Dodge couched his thoughts by saying, "Of each twenty men who will rush to the Black Hills as miners, nineteen would have been better off if they had remained home. I am aware that this statement will deter no man from going, as the American people are so constituted that each man expects himself to be the twentieth."[18]

Colonel Dodge reported that he saw little evidence of Indian lodges anywhere in the Black Hills, except for one location near the head of Castle Creek where a few temporary lodges were observed. Based on this and the people he talked to, Dodge did not believe the Black Hills ever served as a permanent home for Indians. While in the Black Hills, the expedition encountered several small parties of Indians who were either on hunting excursions or had ventured into the area because they were curious as to why the soldiers were in the Black Hills. One Indian Dodge encountered was Robe Raiser, who was quite communicative through Dodge's interpreter. Robe Raiser said he was fifty years old and had lived just outside the Black Hills most of his life.

Robe Raiser told Dodge that small bands of Indians generally did not venture very far into the Black Hills except to cut lodgepoles or to hunt a little. Robe Raiser gave several reasons for this: First, he said, the Black Hills were "bad medicine" and many spirits resided there. Second, although lodgepoles are plentiful, game was scarce and more difficult to kill than on the plains. Third, he said, the pine areas were so thick that the Indians tended to lose their ponies if they were turned loose. If the ponies were tied up, the flies tormented them. Fourth, he said, it rained quite frequently in the Black Hills, and the Indians did not like the rainstorms that are usually accompanied by terrible thunder and lightning that "tears trees to pieces and sets the woods on fire." For these reasons, he said, the Indians never lived in the Black Hills and would not live there now. Robe Raiser told Dodge that the Indians would have sold the Black Hills long ago, but were now being urged to make a "big fuss" so they could get a "big price" for the country.[19]

The Newton-Jenney Expedition ended on October 14, 1875, upon their arrival at Fort Laramie. In November, representatives of the party convened in Washington, D. C., to prepare their reports. A preliminary report by Professor Jenney, along with a map prepared by Dr. McGillycuddy, was included in the annual report of the commissioner of Indian Affairs for 1875.

Miners Are Ordered Out of the Black Hills

Following President Grant's meeting with Red Cloud and Spotted Tail in Washington, D. C., it became apparent to the government that as long as there were still miners occupying the Black Hills, the government was in no position to bargain with the Sioux for extinguishment of title to the Black Hills. To remedy the situation, Brigadier General George Crook was sent to the Black Hills in July 1875 to expel all unauthorized people. Upon arriving at the miner's camp on French Creek on July 27, 1875, Crook canvassed much of the Black Hills and reported that there were as many as twelve hundred prospectors there.[20]

Toward the end of July, D. W. Flick departed French Creek and traveled up Laughing Water Creek to do some prospecting work on a quartz claim. One afternoon, after completing his work for the day, Flick returned to his tent to get something to eat. While he was in the process of frying up some flapjacks, a stranger appeared at the tent and struck up a conversation with him. "What do you think of the order old Crook has issued, directing us to get out of the Black Hills?" Flick asked the stranger. Before the stranger could answer, Flick continued,

"I don't know what you think about it, but I don't intend to stand it. I believe I can raise a hundred men about here, and I propose to form a company to stand old Crook off, if he tries to drive us out. I am here to fight, if necessary. It cost a lot of trouble and hardship to get in here and I don't propose to be run out."

The stranger replied that he didn't completely agree and that the Indians had a right to complain about white men occupying their reservation and hunting grounds. The stranger indicated it would be better to comply with the order until a treaty could be made with the Indians. He seemed to think a new treaty could be made fairly soon. As Flick and the stranger started to engage in a more heated discussion, Samuel Shankland and some other men arrived at the tent. Immediately, Shankland recognized the stranger, saluted him, and introduced him as General Crook to the other men. Flick nearly choked to death on the flapjack he was eating.[21]

On August 10, 1875, General Crook held a meeting with the miners at the Gordon Stockade. Crook informed them that they were being ordered to leave the Black Hills immediately and would not be able to return until a settlement agreement was reached with the Sioux. The miners, who knew they were trespassers, asked for five days to allow them time to lay out a townsite and otherwise prepare to leave. Additionally, the miners asked that six or more of their men be allowed to remain in the Black Hills to guard their claims and equipment. Crook, who probably felt he had achieved the impossible, agreed to the requests of the miners.

The following morning, August 11, the miners organized and laid out a townsite, which they initially called Stonewall. The name was later changed to Custer City. Some twelve hundred lots were platted and marked out on the ground. A drawing was held with about two hundred miners present to determine who would own which lot. The results were entrusted to the eight men selected to remain in the Black Hills. These men included Samuel Shankland, Thomas Hooper, Alvah D. Trask, Robert Kenyon, W. H. Wood, Alexander Thompson, Alfred Gay, and H. F. Hull.

The miners who were still in the Black Hills with the Newton-Jenney Expedition were not affected by General Crook's order. These miners included John W. Allen, Brown, Carlin, Flarida, Warren, and possibly a few others. A mass exodus of miners began leaving the Black Hills on August 15, 1875. Within a few days, almost all the miners had departed, except for the eight people who were selected to remain, the miners on the Newton-Jenney Expedition, and the numerous others who may or may not have received orders to leave.[22]

The *New York Times* provided its readers with a glowing report of the rich gold finds and General Crook's order for miners to leave the Black Hills:

> *Cheyenne, Wyoming Territory, Aug. 10. Gen. Crook and Col. Stanton returned here today from the Black Hills. The miners were preparing to leave, covering up the richest lodes to prevent their becoming*

known until such time as they can return. The country is considered rich in gold. The mountains are full of quartz. Capital and skilled labor will develop mines equal to those of Colorado or Nevada. There were about 1,500 miners in the Hills, and a great deal of preliminary work has been done by them in the way of ditches and sluices. There were no Indians in that region, and but few have been seen. Those at the agencies are still demanding that the miners be driven out.

Some gold was panned out in the presence of the party, which yielded seventy-five cents to the pan. There is abundance of water and grass, and also timber for building, but the pine is not of the highest merchantable quality. A town called Stonewall has been laid out on Custer's Gulch, in the vicinity of which some rich diggings are located. The whole country will be well adapted to grazing and farming. The troops are now en route to establish a temporary post near Stonewall for the purpose of keeping out the miners. Prof. Jenney's party were still exploring the Hills, and will probably remain until the middle of October. Gen. Crook and party had fine hunting on Spring, Rapid, Elk, and Box Elder Creeks, taking a large number of deer, elk, and mountain sheep. Gen. Crook leaves tomorrow for Omaha.[23]

Gold Is Discovered in the Northern Black Hills

After the prospectors and miners were ordered out of the Black Hills on August 15, 1875, some never left and others were just arriving. Much of the prospecting activity after August 15 took place in the northern Black Hills away from General Crook's troops, who mostly monitored areas in the southern Black Hills.

A considerable difference of opinion exists among historians as to which party of prospectors was the "first" to discover gold in 1875 along Deadwood Creek and Whitewood Creek. Notwithstanding, it appears there were at least three different parties of prospectors in the area at about the same time but the parties may not have realized it until they encountered each other. The first groups of prospectors to arrive in the area in August were the Bryant and Pearson Party, who arrived from Fort Pierre; the Blanchard Party, from Custer City; and the Smith Party, from Helena. Toward the end of September, the Gay Party arrived, followed by the Lardner Party in October.

The Bryant Party arrived in the Black Hills at about the time everyone else was ordered to leave. This party included Frank S. Bryant, J. Pearson, Thomas Moore, Richard Lowe, James Pierman, Samuel Blodgett, and George Hauser. Disguising themselves as Indians, the men left the Missouri River on August 1, 1875. Bryant had in his possession a map that was given to him by an Indian friend in Rulo, Nebraska.[24] Tom Labarge, Charley De Gray, and Lephiere Narcouter, former employees of the American Fur Company, reportedly prepared the map. These three individuals and Bryant's Indian friend apparently had some knowledge of where gold could be found in the northern Black Hills, based on their trapping and trading experiences.[25]

Upon arriving in the Black Hills near Bear Butte, Bryant and his party entered the Hills through Elk Creek Canyon, where they did some prospecting. From there, they moved to the head of Spruce Gulch, which Bryant himself named, and traveled down the gulch to Whitewood Creek in present-day Deadwood. They arrived at this location on the afternoon of August 10, 1875. According to their map, this was the location where gold could be found. On August 11, Blodgett, Pearson, and Bryant continued prospecting up Whitewood Creek and subsequently found small amounts of gold on Deadwood Creek just above the mouth of City Creek. Encouraged by what they found, the men returned to their campsite just below the mouth of Spruce Gulch, constructed several sluice boxes, built a crude cabin, and began washing the gravels along Whitewood Creek. About two weeks later, the Lardner Party met the Bryant Party on Whitewood Creek but elected to move farther up Deadwood Gulch toward what later became Gayville.[26]

Around the middle of September 1875, Frank Bryant's party moved away from their cabin at the mouth of Spruce Gulch and prospected along present-day Nevada and Whitetail Gulches. From here they tried their luck on Spearfish Creek. They moved down the creek and ended up camping near the mouth of Spearfish Canyon on September 13. The following day, members of the party spotted some tents of Colonel Dodge in the distance. The prospectors hastily broke camp, headed up False Bottom Creek, which Bryant also named, and returned to their cabin below Spruce Gulch.[27] A few days later most of the Bryant Party, except for Bryant and Pearson, departed the Black Hills. Moore and Lowe followed the Newton-Jenney Expedition past Bear Butte and returned to their homes in Missouri. Blodgett and Hauser joined the soldiers at Custer City and returned to Fort Laramie.[28]

Bryant and Pearson decided to move up Deadwood Gulch to where the Lardner and Blanchard parties were camped. Bryant located a placer claim at the mouth of Bobtail Gulch. After the two men ran low on supplies, they too left the Hills and headed to Fort Laramie to get additional supplies. By late fall, Bryant and Pearson returned to Deadwood Creek and continued prospecting. Needing a winter shelter, they constructed a cabin in what is now Central City.[29]

The Blanchard Party made their discovery on Deadwood Creek near the mouth of Blacktail Gulch on September 6, 1875. Members of the party included A. S. Blanchard, Tom Patterson, H. A. Albien, James Verpont, and one other person who was a member of Custer's expedition to the Black Hills in 1874. On April 28, 1876, Blanchard sold his Claim No. 1 above Discovery to B. E. Murphy for $1,500. James Verpont, who found the first gold along Deadwood Creek on September 6, owned the Discovery claim. Hildebrand owned Claim No. 1, below Discovery.[30]

The Smith Party left Cave Gulch, Montana, in August 1875 with their saddle horses, pack animals, tents, rifles, and about four months' worth of provisions. The party consisted of William Smith, John Kane, and three other men. On September 5, they arrived in the Black Hills and made camp near the site that later became Crook City. From there they proceeded up Whitewood Creek as far as they were able, diverted away from Whitewood Canyon, and made their way back to the creek via Smith Gulch. Upon reaching an area that had abundant white trees just below the mouth of City Creek in present-day Deadwood, the men named the area Whitewood. They then made camp near the mouth of Spruce Gulch. The Smith Party discovered gold along Whitewood Creek on September 7. The men staked their first claims about one-half mile below the confluence of Deadwood and Whitewood creeks.

Two days later, the men decided to prospect their way up Deadwood Creek. Travel was exceedingly difficult because of the large amount of deadfall timber and thick underbrush. Upon reaching the mouth of Blacktail Gulch, the Smith Party encountered the Blanchard Party, whose members were busy placer mining on Deadwood Creek at that location. Both parties joined together and staked additional 300-foot-long claims in accordance with the mining laws of Montana Territory. Members of the Blanchard Party were given preference in the selection of claims since they had made the first discovery.[31]

Toward the latter part of September 1875, the Gay Party arrived and located some claims on Deadwood Creek just below the mouth of Blacktail Gulch. Their camp became known as Gayville. In October, the Lardner Party arrived and joined the Gay Party.[32] The Lardner Party included William Lardner, Ed McKay, M. Joe Ingaldsby, Hilan "Pat" Hulin, James Hicks, William Gay, Alfred Gay, E. D. Haggard, and two of Frank Bryant's old partners, John B. Pearson and Dan Muckler. The party had previously prospected along Little Rapid, Whitewood, Whitetail, and Little Spearfish creeks. They then crossed over the divide near Bald Mountain to Deadwood Creek. Just below the mouth of Blacktail Gulch, the party discovered rich gold on Deadwood Creek and proceeded to locate their first claims on November 9, 1875.

The party immediately staked nine additional claims upstream of the Discovery claim. The Claim No. 4 above Discovery belonged to William Gay. The Claim No. 9 above Discovery belonged to Lardner. None of the party staked any claims below Discovery. The area below Discovery was covered with a dense growth of underbrush, beaver dams, and deadwood that was strewn all over the place. This was probably the reason no one from the Lardner Party staked any claims below Discovery.

By December, the Blanchard, Smith, Bryant, and Lardner parties in upper Deadwood Gulch held their first miner's meeting and organized the Lost Mining District. William Lardner was elected recorder of the district. Mining laws were established, and a fee of $1.50 per claim was charged for the right to record a mining claim.

In late December, Lardner made a trip back to Little Rapid Creek to retrieve a cache of mining equipment. There, he met J. J. Williams, W. H. Babcock, Eugene Smith, and another person named Jackson who had been prospecting on Castle Creek. Lardner mentioned his rich gold find to them. The next day, the four prospectors packed up their belongings and followed Lardner back to Deadwood Gulch. Around the first of January, the Williams Party began staking claims below the Lardner Party's Discovery claim. Jackson staked Claim No. 1 below Discovery on Deadwood Creek (see fig. 4.2). Later, that claim was sold to Montana miners, Hildebrand and Harding.[33]

Fig. 4.2. "Mining crew drifting for gold below discovery point," by Stanley J. Morrow, 1876. Shallow shafts and pits were often sunk 20-30 feet deep to access the better gold-bearing gravels that were lying on the bedrock. The gravels were excavated by driving small drifts, hoisted to the surface in buckets, and washed in rockers or sluice boxes.—Photo courtesy of the National Archives (165-FF-2F-10).

W. H. Babcock staked Claim No. 2 below Discovery and sold it to the Wheeler brothers. This claim was reportedly the richest placer claim in Deadwood Gulch (see fig. 4.3). Over a fifty-day period, working two sluice boxes day and night, the Wheeler brothers recovered $43,000 worth of gold. Over four months, the total amount of gold they recovered amounted to $150,000 (roughly equivalent to more than 7,250 troy ounces of gold at $20.67 per troy ounce). Claims No. 4 and No. 5 below Discovery were staked by the Chisholm brothers and were later sold to Robert Neill. Claim No. 6 below Discovery yielded $2,300 in one day. Jack McAleer located Claim No. 9 below and recovered $40,000 in gold.[34] At the end of the placer mining season in 1876, a party of thirty of the miners banded together and personally escorted their placer gold to the First National Bank in Cheyenne. The value of their cargo was estimated at $800,000 to $900,000.[35]

Fig. 4.3. The Wheeler Ground or Placer Claim No. 2 below the Discovery claim in Deadwood Gulch, 1876. The claim was located downstream from the mouth of Blacktail Gulch. The Wheeler brothers recovered more than $150,000 worth of placer gold over a four-month period in 1876.—Photo courtesy of the Black Hills Mining Museum.

Robert Kenyon, one of the eight miners allowed to stay in the Black Hills after August 15, 1875, held Claim Nos. 14 and 15. J. J. Williams located Claim No. 22 below the Discovery claim on Deadwood Creek. Over a 3-month period, Claim No. 22 yielded $27,000 in gold. Williams then sold that claim and acquired Claim No. 14 above Discovery on Whitewood Creek, which yielded $35,000 in gold.[36] By mid-January 1876, virtually all of the land along Deadwood Creek and Whitewood Creek was staked with placer claims.[37]

On December 11, 1875, John B. Pearson and his party located the first lode claim in the Black Hills, which they named the Blacktail claim.[38] John B. Pearson,

William Gay, M. J. Ingaldsby, Alfred H. Gay, Ed Haggard, and D. Muckler then located the Giant Lode No. 115 and the Gold Run Lode No. 116 on January 7, 1876.[39] It turns out that the Giant and Gold Run claims were located above the famous Homestake ore body at the site of the present-day Open Cut. The men obtained good specimens of rock from the Giant lode but apparently failed to develop the lode sufficiently to fully understand the value of their claim.[40]

On March 31, 1876, William Gay wrote a letter from his residence on Deadwood Creek, Lost District, to Captain Jack Crawford of Omaha advising him of the ever-increasing number of prospectors who had returned to the Black Hills and the claims his party had staked:

> *I arrived here yesterday and found everything all right. We had a very hard trip. The snow is about three feet deep and still snowing. There is quite a change here since I left. Instead of eighteen men there are several hundred. There has been a number of discoveries made since I left, both in placer and quartz. The weather has been so cold that prospecting has been very difficult, but where it has been tried they got good prospects. Prospects on Deadwood and Whitewood [creeks] are from ten to twenty-five cents, and as high as $1.30 to the pan, while on some of the side gulches emptying into Deadwood there has been as much as $5.00 to the pan. There have been a number of claims sold at prices varying from $500.00 to $4,000.*
>
> *When I got here the town sites I spoke to you about had all been taken up and all the lots taken. One is at the mouth of Whitewood, and is called Creek City; the other is a short distance below the mouth of Deadwood, on Whitewood Creek, and is called Whitewood City. We are going to lay off a town here on Deadwood to be called Gay City. I will reserve a lot for you. Tell Curran that if he had come over with me he could have had a chance to get a good claim on shares, but they are all taken now. Old Dan had sold his discovery claim before I got here. We have particularly developed some of the ledges we had staked, and two of them proved to be very good. I send you a specimen from the "Giant" lode, owned by Wm. Gay, A. H. Gay, M. J. Ingelsby, E. D. Haggard, D. Meckles and J. B. Pearson. We have another ledge that we consider better, called the "Blacktail," owned by the same parties. We propose to start on a prospecting expedition as soon as the snow leaves sufficiently, and if I find anything I will try to give you a show.*[41]

In October 1875, a party including John R. Brennan, George W. Stokes, N. H. Hawley, and George Ashton departed Denver, Colorado, for the Black Hills. Upon reaching Cheyenne, the party determined that the four of them only had a combined total of $20 with which to purchase supplies. Not too worried, the party purchased the supplies they could afford and continued on. At the Platte River, they joined up with a party of forty-five prospectors who were also headed for the Black Hills. In this group were California Joe, Dick King, George Palmer, John Argue, Robert Ralston, and James Hepburn. Accompanying Hepburn was his wife, who became the fourth nonnative woman to enter the Black Hills behind Sarah "Aunt Sally" Campbell with the 1874 Custer Expedition, Annie Tallent with the Collins-Russell Expedition in December 1874, and Martha "Calamity Jane" Cannary-Burke, who accompanied the military escort of the Newton-Jenney Expedition in 1875.

The prospectors traveled up Red Canyon and reached Custer on November 12, 1875. Here, the party split up. Brennan, Stokes, Palmer, Hawley, Byron, and Argue prospected their way to Palmer Gulch, where they discovered gold in early December. The prospectors decided to construct three cabins and stay for the winter. On December 20, the Palmer Gulch Mining District was organized at a miner's meeting held at the cabin of John R. Brennan, George W. Stokes, and George Palmer.[42]

In October 1875, Thomas H. Mallory and his party of thirty men returned to the Iron Creek area. Mallory's party had previously prospected on Iron Creek but left the Black Hills in August in accordance with General Crook's order. On their return, these prospectors made major discoveries on Deer Creek, Beaver Creek, Bear Gulch, Sand Creek, Potato Creek, and Mallory Gulch.[43] Over the next few months, the very richest placers on these streams were staked and controlled by nine men, including Mallory, Noah Kipp, D. S. Lunt, Frank "Portigee" George, H. Bisslinghoff, John Duffy, J. Sauders, Robert Neill, and C. H. Fry.[44]

George W. Corey wrote a letter dated December 20, 1875, to Professor Walter P. Jenney, who had recently completed his scientific expedition to the Black Hills. In his letter, Corey informed Jenney, "I have some fine specimens of Black Hills gold that I will send to you. They were taken out of Bear Gulch by Frank George de Oliver, who would not sell them, but let me take them to send to you. He took out with a rocker in eight and one-half days, $165. One lump, which I have, weighs one and one-half ounces."[45]

Thomas H. Mallory sent his own letter to Professor Jenney on January 5, 1876. Mallory informed Jenney about his recent prospecting results and knowledge of the Black Hills as follows:

> Iron Creek runs into Spearfish above Crow Peak. Bear Creek runs into Spearfish lower down and near the buttes. Sand Creek runs into Red Water. These streams head nearly together, and they with their smaller tributaries make a large mining district. They are richer too than anything on this side of the Hills. One pan of dirt on the discovery of Bear Creek contained $27. I have a report that another small stream running into Spearfish from the southwest, is still better than anything

> *yet found. How much truth there is in it, Bottsford and I will know in a few days. I send you a little gold from Sand Creek. It has been retorted and does not look bright.*[46]

By mid-January 1876, there were more than four thousand prospectors and miners throughout the Black Hills. Custer City had about one thousand residents. The area around Hill City had about five hundred residents.[47] However, after rumors got out about the rich strikes in the northern Black Hills, most of the prospecting activity in the southern Black Hills was temporarily abandoned and the gold rush moved to areas around Deadwood Gulch and Bear Gulch. By April 1876, four new mining districts were organized. The Buckeye District was established in August 1875 on Iron Creek. After the Mallory Party returned to the Black Hills in October 1875, the Hurricane District, comprised of Sand Creek and Mallory Gulch, was organized. The Rawlings District followed in February 1876 and comprised Bear Creek in Bear Gulch. The Beaver Creek District included Beaver Creek and possibly Potato Creek.[48]

By spring 1876, nearly every foot of every creek in the Black Hills was staked with placer claims. Lode claims were being staked in large numbers outward from the placer claims. Nothing would be able to stop the Black Hills gold rush now.

Chapter 5
The Expropriation of the Black Hills

Decision to Extinguish Indian Title to the Black Hills

On January 11, 1875, the U.S. House of Representatives received a petition from the Legislative Assembly for the territory of Dakota. The petition was referred to the Committee on Indian Affairs. The petition asked Congress to "abrogate the treaty now in force, or, if such action be deemed unjust to the Indians, in lieu thereof extinguish the Indian title to that portion of the reservation known as the Black Hills of Dakota, so that the nation may receive the benefit of its great wealth." The petition listed reasons why Congress should take such action and recommended timely legislation "to prevent a repetition of the bloody scenes in California between the miners and the Indians, which resulted in the almost total destruction of the latter."[1]

General William T. Sherman was interviewed by the *New York Times* on March 11, 1875, relative to the private gold-hunting expeditions to the Black Hills. Sherman reaffirmed that the integrity of the 1868 Treaty of Fort Laramie would be "maintained at all hazards" and that as soon as the weather permitted, troops would be sent in to bring out the [Gordon] party now in the Hills. He emphasized that any other expeditions would be prevented from entering the region. Any others found to already be there would also be driven out.[2]

On March 18, 1875, the *New York Times* reported that President Ulysses S. Grant had transmitted a number of documents from the War and Interior departments to the Senate pursuant to a Senate resolution that had recently been passed. The news article stated that secretary of the Interior, Columbus Delano, communicated the articles of the Treaty of 1868 to President Grant. Delano confirmed that measures had been adopted by the secretary of War to prevent further intrusion of the portion of the reservation in question and remove persons already there. Delano assured the president that efforts were being made "to arrange for the extinguishment of the Indian title" and that all proper means would be used to accomplish that end. Delano told the president that if the steps taken toward the opening of the country to settlement failed, "those persons at present within that territory without authority" would be expelled.[3]

Delano went on to explain to the president that "this department has taken steps to bring to this city a delegation of the Sioux, parties to the treaty, for the purpose of negotiating for the extinguishment of their right to the reservation embracing the Black Hills country, with a view to opening up the same to settlement, and, until such an arrangement has been effected, it is the intention of this department, with the cooperation of the War Department, to protect the rights of the Indians as guaranteed to them by the Treaty of 1868, and to prevent any further infraction of those rights. It is also the intention of this department to use every effort possible to extinguish the Indian title of the Black Hills country, and open the same to settlement and explorations of mineral wealth at the earliest day practicable."[4]

Needing more information, General Sherman sent a telegram to Lieutenant General Philip H. Sheridan asking, "What do you know of the Black Hills?" Sheridan responded to Sherman's request with a letter dated March 25, 1875. Oddly, Sheridan's letter was published in the *New York Times* two days later. General Sheridan explained that his first knowledge of the Black Hills was based on discussions he had with Father Pierre Jean De Smet, when the two of them met many years ago on the Columbia River in Oregon. While there, De Smet told him about an Indian romance of a mountain of gold in the Black Hills. Sheridan explained that at that time, reference to the Black Hills included all of the various mountain ranges within the country "bounded on the east by longitude 102°, on the south by the Sweet Water and Laramie Rivers, on the west by the Big Horn and Wind Rivers, and on the north by the Yellowstone River."[5]

Sheridan explained that many people believed that the Black Hills of the Cheyenne River, which General George A. Custer visited the previous year, was the place that contained Father De Smet's mountain of gold. Sheridan said Father De Smet was shown gold nuggets that had been obtained from the beds of many rivers within the various mountain ranges that comprised the entire country of the Black Hills. One mountain supposedly contained enough of the yellow metal to pay off the national debt, according to De Smet. The Black Hills of the Cheyenne River, as described by Custer, were situated between the north and south forks of the Cheyenne. The north fork, according to Captain William Ludlow's map of the 1874 Custer Expedition, was the Belle Fourche River.

In his letter, Sheridan assured Sherman that he had the utmost confidence in the statements of General Custer that gold had been found near Harney's Peak. Sheridan couched his remarks, though, by stating that Custer's gold find did not mean that gold existed in great quantities. Sheridan proposed sending military expeditions in the spring of 1875 to "open up the Yellowstone River" and investigate the areas around the mouths of the Big Horn, Powder, Tongue, Rosebud, and Wind Rivers. Sheridan informed Sherman that "this country is as yet entirely unexplored, and the expedition may develop a very valuable auriferous section, and make the Father De Smet story to some extent true, but

I am of the belief that a mountain of mica has not changed to gold." Sheridan concluded his letter promising to prevent trespassers from entering the Black Hills, in accordance with his orders dated September 3, 1874.[6]

On June 18, 1875, a day after Colonel Richard I. Dodge and Professor Walter P. Jenney reported their gold finds, Secretary of the Interior Delano appointed a commission to negotiate with the Sioux for relinquishment of the Black Hills. Senator William B. Allison, President Grant's friend from Iowa, was selected to head the commission. Immediately, the commission commenced work to form a "grand council" through which mining rights would be secured to the Black Hills. Additionally, the Big Horn Mountains in Wyoming Territory would be purchased.

On September 20, 1875, the first meeting was held on the White River, some 8 miles from the Red Cloud Agency. As many as twenty thousand members of the Oglala, Minniconjou, Brulé, Hunkpapa, Blackfeet, Sans Arc, Yankton, Santee, Cheyenne, and Arapaho tribes attended the meeting. Red Cloud, Crazy Horse, and Sitting Bull boycotted the meeting. Some 120 cavalry men stood by in case of trouble.[7]

Senator Allison opened the council by saying to the chiefs, "We have come to ask you if you are willing to give our people the right to mine in the Black Hills as long as gold or other minerals are found, for a fair and just sum." He continued, "If you are so willing, we will make a bargain with you for these rights. You should bow to the wishes of the Government which supports you. Gold is useless to you and there will be fighting unless you give it up." The chiefs were confused over the various options discussed, and the council was adjourned to allow the chiefs time to contemplate their options.[8]

Over the next two days, negotiations were moved to the Red Cloud Agency. Twenty of the leading chiefs of the Sioux, Arapaho, and Cheyenne participated. Red Cloud expressed his thoughts to the commissioners, saying, "These Hills out here to the northwest we look upon as the head chief of the land. My intention was that my children depend on these for the future . . . I think that the Black Hills are worth more than all the wild beasts and all the tame beasts in the possession of the white people. I know it well, and you can see it plain enough, that God Almighty placed these Hills here for my wealth, but now you want to take them from me and make me poor, so I ask so much that I won't be poor."[9]

In exchange for everything within the "racetrack" around the Black Hills, Spotted Tail informed the commissioners he wanted steers, flour, coffee, sugar, tea, a wagon, horses, cattle, sow, houses, and furniture for every family for seven generations. Little Wolf, a Cheyenne, exclaimed, "My people own an interest in these hills that you speak of buying . . . . We want to be made rich too." The chiefs then submitted a demand of $70 million from the government for the Black Hills.

After considering their options, the commissioners presented six different options for the Indians to consider. One option gave the government a license to mine, grow stock, and cultivate the soil for $100,000 per year. Another option offered $6,000,000, in fifteen equal installments, for an outright purchase of the Black Hills. In addition, the proposal required the Indians to allow three routes to be designated across the reservation to the Black Hills. Any agreement would require approval of Congress and approval from three-fourths of the adult males of the Great Sioux Nation, pursuant to the Fort Laramie Treaty of 1868. The Indians stood firm with their demand. Recognizing that the negotiations weren't going anywhere, the commissioners ended their talks with the Indians on September 29, 1875, without an agreement.[10]

On November 1, the Annual Report of the Commissioner of Indian Affairs was released. In the section entitled "Black Hills," Commissioner Edward Smith reported, "The occupation of the Black Hills by white men seems now inevitable." Referring to the Sioux, the commissioner stated, "They are not now capable of self-support; they are absolute pensioners of the Government in the sum of a million and a quarter of dollars annually above all amounts specified in treaty-stipulations. A failure to receive Government rations for a single reason would reduce them to starvation."[11]

Commissioner Smith, in referencing the results of the Newton-Jenney Expedition, remarked,

> *The report confirms, in large degree, the statements of travelers and explorers and the reports of General Custer's military expedition of last year, and shows a gold-field with an area of eight-hundred square miles, and around this gold region, principally to the north, an additional area within the Black Hills country of three thousand square miles of arable lands, and this latter embracing along its streams an area equal to two hundred square miles finely adapted to agriculture, while the hillsides and elevations contiguous thereto are equally adapted to purposes of grazing, making the whole area of three thousand square miles of timber, grazing, and arable land of great value for agricultural purposes . . . . So long as gold exists in the same region, the agricultural country surrounding the gold-fields will be largely required to support the miners, and to attempt to bring the wild Sioux into proximity to the settlers and miners would be to invite provocations and bloody hostility.*

The commissioner summarized by saying, "The fact that these Indians are making but little use of the Black Hills has no bearing upon the question of what is a fair equivalent for the surrender of these rare facilities for farming and grazing."[12]

On November 3, 1875, President Ulysses Grant met with Secretary of War William Belknap, Secretary of the Interior Zacharia Chandler, Assistant Interior Secretary R. B. Cowan, Commissioner of Indian Affairs Edward Smith, General

Sheridan, and General Crook. At this meeting, notwithstanding provisions of the 1868 Treaty, the president announced that "no further resistance shall be made to miners going into [the Black Hills]"[13]

General Sherman later shared his thoughts on the matter. He said, "In as much as the Sioux have not lived at peace, I think Congress has the perfect right to abrogate the whole treaty or any part of it."[14] On November 9, General Sherman sent written correspondence to Brigadier General Alfred Terry informing him that the president had decided the military would offer no further resistance to the occupation of the Black Hills by miners, "it being his belief that such resistance only increased their desire and complicated the troubles." Sherman instructed Terry to enforce these orders quietly and to keep the president's decision confidential.[15]

The government had some knowledge that Sitting Bull and the Hunkpapa were wintering on the Yellowstone River and the Minniconjou and Sans Arc on the Tongue River, and that Crazy Horse, Black Elk, and other Oglala were elsewhere in the Powder River Basin. On December 3, the secretary of the Interior instructed the commissioner of Indian Affairs to deliver an ultimatum to the Indians in the Yellowstone and Powder River valleys. The ultimatum stated that unless the Indians abandoned their hunting grounds in the unceded areas and moved to the Great Sioux Reservation by January 31, 1876, they would be considered "hostile" and the government would deal with them accordingly. Upon learning of the decision, the military officers probably recognized that it would be extremely difficult for the Indians to comply with the ultimatum in the dead of winter.[16]

Black Elk, one of the Oglala Sioux wintering in the Powder River Basin, recounted later, "During the winter, runners came from the Wasichus and told us we must come into the Soldier's Town right away or there would be bad trouble. But it was foolish to say that, because it was very cold and many of our people and ponies would have died in the snow. Also, we were in our own country and were doing no harm. Late in the 'Moon of the Dark Red Calves' [February], there was a big thaw, and our little band started for the Soldier's Town, but it was very cold again before we got there. Crazy Horse stayed with about a hundred tepees on Powder."[17]

At about the same time, President Ulysses Grant presented his Seventh Annual Message to the Senate and House of Representatives on December 7, 1875. In his message, the president informed Congress as follows:

> *The discovery of gold in the Black Hills, a portion of the Sioux Reservation, has had the effect to induce a large emigration of miners to that point. Thus far the effort to protect the treaty rights of the Indians to that section has been successful, but the next year will certainly witness a large increase of emigration. The negotiations for the relinquishment of the gold*

> *fields having failed, it will be necessary for Congress to adopt some measures to relieve the embarrassment growing out of the causes named. The Secretary of the Interior suggests that the supplies now appropriated for the sustenance of that people, being no longer obligatory under the treaty of 1868, but simply a gratuity, may be issued or withheld at his discretion.*[18]

Great Sioux War of 1876-77

The Great Sioux Reservation and the Powder River Basin, where the "hostile" Sioux were located, both fell within the jurisdiction of Lieutenant General Philip H. Sheridan, head of the U.S. Army's Division of the Missouri. The Department of Dakota and Department of the Platte were two of four departments under the Division of the Missouri. General Sheridan involved both departments in his military efforts, primarily because the Black Hills straddled the boundary between the two departments.

The Department of Dakota was headquartered in St. Paul under the command of Brigadier General Alfred H. Terry. This department had jurisdiction over present-day Minnesota, Montana, North Dakota, parts of Idaho and South Dakota, as well as the Yellowstone portion of Wyoming. It was subdivided into districts, two of which included the southern district at Fort Randall and the middle district at Fort Abraham Lincoln in Bismarck.

The Department of the Platte was headquartered in Omaha under the command of Brigadier General Edward O. C. Ord. General Ord was replaced by Brigadier General George Crook in May 1875. The Department of the Platte had jurisdiction over Colorado, Iowa, Nebraska, Wyoming (except Yellowstone), and the territory of Utah. One of the districts within the Department of the Platte was the District of the Black Hills. This district was headquartered at Fort Laramie and included Fort Fetterman, Camp Robinson, and Camp Sheridan.[19]

On February 1, 1876, the secretary of the Interior notified the secretary of War that many bands of Sioux had not complied with the order to relocate to the Great Sioux Reservation by January 31. These Indians, who were residing mostly in the areas of the Yellowstone River and Powder River Basin, were now considered "hostile" and subject to "appropriate action by the Army." Some of the bands of Sioux, such as those under Crazy Horse and Sitting Bull, refused to relocate from their unceded hunting grounds to the reservation. These two chiefs claimed to never have been a party to any treaty with the United States. It didn't make any difference, though. The matter was now in the hands of the War Department. As a result, some twenty-two battles and skirmishes, known as the Great Sioux War of 1876-77, were fought across much of the Great Plains—land that the American Indians still claimed as theirs.[20]

General Sheridan previously tried to persuade the government to allow him to conduct a winter campaign to catch the hostiles off guard. However, Sheridan's original suggestion was called off, based on a suggestion from the commissioner of Indian Affairs to give the Sioux until January 31, 1876, to relocate to the reservation. By February, Sheridan ordered Brigadier General George Crook and General Alfred H. Terry to begin preparing to subdue the hostiles.

Sheridan's overall strategy was to use three columns of soldiers who would march from three separate directions and surround the area where he believed the Sioux were camped (see fig. 5.1). Two columns were derived from Terry's command and one from Crook's command. Sheridan and his two commanders believed only one column would be necessary to subdue the hostiles, based on the Indian Bureau's estimate of five hundred to eight hundred hostile warriors.[21]

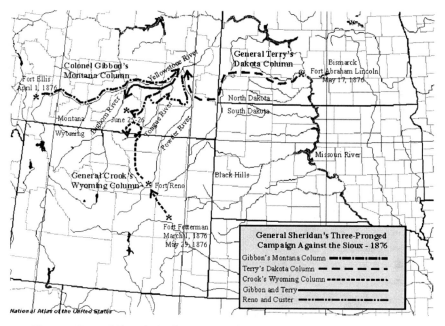

Fig. 5.1. General Sheridan's Three-Pronged Campaign against the Sioux, 1876. General Sheridan's Campaign against the "hostile" Indians resulted in numerous battles in 1876-1877, including the battles of the Powder River, Rosebud Creek, Little Big Horn River, Warbonnet Creek, Slim Buttes, Spring Creek, Cedar Creek, Dull Knife, and Wolf Mountain.—Illustration by author.

Because of bad weather, General Terry's two columns were delayed in departing Fort Abraham Lincoln. However, General Crook had already departed Fort Fetterman on March 1. After marching in temperatures reaching 40-50 degrees below zero, Crook may have gained a better appreciation for the reason very few of the Sioux attempted to move to the Great Sioux Reservation by January 31.

On March 17, under orders from General Crook, Colonel Joseph J. Reynolds and six cavalry companies staged an attack at dawn on what was thought to be

Crazy Horse's band of Oglala on the Powder River. Much to Reynold's chagrin, the camp turned out to be a village of Cheyenne under Chief Two Moon. Crazy Horse's Oglala friend, He Dog, was also at the camp. In total, there were approximately 735 people residing in 120 lodges in the camp, including 210 warriors. Reynold's soldiers proceeded to burn the village, forcing the survivors to flee in the cold to Sitting Bull's Hunkpapa camp in the Blue Mountains on Beaver Creek. Sitting Bull took the Cheyenne in and provided them with horses, robes, and food. He even directed his people to double up in their lodges so that some of the lodges could be turned over to the Cheyenne.[22]

The attack on March 17 did little more than incent Sitting Bull and the other Indian leaders toward a stronger alliance and allegiance among one another. At a council meeting on the Tongue River, Sitting Bull announced, "We are an island of Indians in a lake of whites. We must stand together, or they will rub us out separately. These soldiers have come shooting; they want war. All right, we'll give it to them!" Indian runners were then sent out to all the Sioux, Cheyenne, and Arapaho agencies west of the Missouri River. The messengers rode hard and fast. Upon arriving at each agency, they immediately went to each chief and reported what Sitting Bull had told them to say: "It is war. Come to my camp at the Big Bend of the Rosebud. Let's all get together and have one big fight with the soldiers!"[23]

The word spread like wildfire. The Indians at the agencies were already starving because the rations had essentially run out in February. This situation, coupled with reports of the mass influx of prospectors into the Black Hills, the report of the March 17 attack on the Powder River Village, and now the plea from Sitting Bull, caused a large number of Sioux and Cheyenne Indians to begin moving to the Powder River Basin to hunt for food and prepare to fight the whites.

In May, the Indian agent at Red Cloud Agency commented to Red Cloud that General Crook had conducted an inept winter campaign against the Indians. Upon learning of the comment, Crook himself made a visit to the agency and accused the agent of helping the hostiles. While at the agency, Red Cloud warned Crook about the Lakotas, "They are not afraid of the soldiers or of their chief . . . and they will say of the Great Father's dogs, 'Let them come!'"[24]

On May 28, 1876, as many as seven hundred to eight hundred warriors were spotted by a lieutenant who was returning from the Powder River. Similarly, Colonel Wesley Merritt reported that as many as two thousand Indians had left the Red Cloud Agency in May. By early June, as many as fifteen thousand Sioux and Cheyenne were camped along the banks of the Rosebud River, including thousands of warriors who were ready to die defending their homelands.[25]

After all had gathered, Sitting Bull called the chiefs together in the council. The Cheyenne named Two Moon as their chief. The Sioux chose Sitting Bull as their chief. Two Moon then suggested to all the chiefs that Sitting Bull should lead the war campaign for all the tribes. "You already have the right man—Sitting Bull. He has called us all together. He is your chief, and you always listen to him.

I can see no reason for another choice." With that said, the vote was unanimous among the chiefs. Once the warriors were informed, they too were satisfied.[26]

Within a few days, the Sun Dance was conducted by Black Moon, as Intercessor, on the banks of the Rosebud. Sitting Bull was chief of the dancers. The dance became known as Sitting Bull's Sun Dance. Sitting Bull proceeded to the sacred pole, sat down with his back against it, and offered one hundred pieces of his flesh, which were cut out by Jumping Bull. For the next eighteen hours, Sitting Bull performed the Sun Dance. Around noon the next day, he appeared faint and fell asleep.

Upon awakening, Sitting Bull exclaimed to the others that his offering had been accepted and his prayers were answered. He said he heard a voice from above saying, "I give you these because they have no ears." Furthermore, he said, "[I] looked up and saw soldiers and some Indians on horseback coming down like grasshoppers, with their heads down and their hats falling off. They were falling right into our camp." Upon hearing this revelation, all the Indians rejoiced, and the Sun Dance ended on June 14.

A few days later, Sitting Bull warned the warriors, "These dead soldiers who are coming are the gifts of God. Kill them, but do not take their guns or horses. Do not touch the spoils. If you set your hearts upon the goods of the white man, it will prove a curse to this nation." Chiefs Ice and Two Moon, Crazy Horse and Gall believed in Sitting Bull's vision because they had heard him prophesy before and knew that his prophecies always came true.[27]

On May 29, 1876, General Sheridan's three-pronged campaign got underway once again after General Crook had time to regroup at Fort Fetterman. Crook departed Fort Fetterman with 1,051 soldiers who comprised the Wyoming column. Crook was supposed to meet up with his guide, Grouard, and a contingent of Crow and Shoshone scouts at the ruins of old Fort Reno. However, when Crook arrived at the designated meeting place, Grouard wasn't there. Crook then proceeded toward Goose Creek (near present-day Sheridan, Wyoming) but ended up getting lost in the absence of his guide.

Crook finally reached Goose Creek and established camp on June 11. Scout Grouard arrived at Crook's camp three days later with 261 Crow and Shoshone allies. Grouard reported that he was late because of trouble getting the Crow to agree to join in as allies against the Sioux and Cheyenne. Within a week, approximately eighty civilians joined ranks with Crook. Most of these people were miners bound for the Black Hills.

Brigadier General Alfred H. Terry and General George A. Custer departed Fort Abraham Lincoln on May 17. Terry's Dakota column consisted of about 925 men from twelve companies of the Seventh Cavalry, two companies of the Seventeenth Infantry, and one company of the Sixth Infantry. Colonel John Gibbon had already departed Fort Ellis on April 1 with his Montana column and was awaiting instructions from his camp at the mouth of Tullock Creek on the Yellowstone River. Gibbon's column consisted of about 450 soldiers from the Second Cavalry and Seventh Infantry regiments. While Gibbon's column was camped at Tullock Creek, small bands of Indians began stealing horses from camp. On May 16, while tracking the horse thieves, Gibbon's chief of scouts, Lieutenant James H. Bradley, discovered a large village of Sioux on the Rosebud River. Immediately, Gibbon sent a dispatch to Terry informing him that he had discovered some Sioux. Gibbon provided no other details. Terry responded with instructions for Gibbon's column to march to the mouth of the Powder River. Upon meeting Gibbon on June 6, Terry learned in more detail about the large contingent of Sioux camped on the Rosebud.[28]

All the while, Sitting Bull had as many as sixty scouts of his own on the lookout for soldiers each day. On June 16, 1876, the scouts returned to the camp saying the valley of the Rosebud was black with "Three Stars" Crook and his soldiers. That evening, Sitting Bull assembled about one thousand warriors, including some one hundred Cheyenne. Crazy Horse led the Oglala and Sitting Bull led the Hunkpapa.

After an informal war dance was held that evening, the warriors proceeded up the south fork of Reno Creek and down Corral Creek, stopping short of Crook's encampment at daybreak on June 17. At around 8:30 a.m., as Crook's column was halted for a coffee break, the Indians attacked. The Battle of Rosebud Creek lasted six hours. In the end, ten of Crook's men died and twenty-one were wounded. Thoroughly shaken by the surprise attack, Crook retreated from the area and failed to fulfill his part of the three-pronged attack as planned.[29]

After the attack, Sitting Bull decided to move the tribes to the north in order to put some distance between his people and Crook's people. Sitting Bull's camp now consisted of more than two thousand lodges, each with a family of one to three warriors. Camp was reestablished along the valley of the Greasy Grass or the Little Big Horn River. Each of the various bands of Indians occupied a specific area, which resulted in a multitude of circles within the overall camp. Box Elder, a Cheyenne prophet, announced that the people should keep their horses tied up at their lodges because "in my dream, I saw soldiers coming." On June 24, 1876, a herald announced that soldiers were, in fact, coming and that they would reach the camp the next day.[30]

Meanwhile, Lonesome Charley Reynolds, Bloody Knife, and other scouts were conducting their own reconnaissance. They warned the officers that there were too many Sioux and Cheyenne for the army to handle. The officers didn't listen, though, just as Sitting Bull saw and heard in his vision.

General Terry made new plans, since no one had heard from General Crook. The new plan called for Custer to lead the Seventh Cavalry up the Rosebud River to its head and cross over into the valley of the Little Big Horn River. General Gibbon and General Terry would travel back up the Yellowstone River on the Far West riverboat, move along the Big Horn River to the Little Big

Horn Valley, and block the Indians' northward escape route. Mark Kellogg, a newspaper reporter who accompanied Custer, stated he thought Custer had the best chance of locating the hostiles and therefore chose to travel with him. Each soldier was issued one hundred cartridges for their single-shot Springfield rifle and twenty-four rounds for their Colt revolver.[31]

On the morning of June 25, the hostile camp was located. Bloody Knife reported to Custer that there were more Indians camped along the Little Big Horn River than the soldiers had bullets. Mitch Bouyer, another of Custer's scouts, reported to Custer that the camp was the largest he had ever seen in his thirty years on the northwest plains. Custer wasn't overly concerned. Major Marcus Reno, second in command to Custer, led the charge with Bloody Knife and 112 soldiers into the south end of the Indians' village. Within minutes, a bullet struck Bloody Knife in the head, splattering blood on Reno's face. Reno's troops began fighting from the ground while waiting for Custer to arrive. After a short time, Reno ordered his troops to retreat. The soldiers mounted their horses and galloped up the west side of the river.[32]

Sitting Bull was puzzled by Reno's behavior. "Look out," Sitting Bull yelled to his people, "there must be some trick about this." Sitting Bull, sensing that Reno had been waiting for other troops to arrive, instructed his people not to follow Reno's troops. He was too late. The Indians were already charging after Reno's soldiers. One Indian reported later, "We killed the soldiers easy; it was just like running buffalo. One blow killed them. They were shot in the back; they offered no resistance."[33]

Just as everything began to quiet down, Custer's soldiers were spotted on the bluffs above the village. Custer's men started their charge but suddenly stopped. The Indians, led by Crazy Horse, turned their attention to Custer. With Crazy Horse shouting, "Hoka hey," the Indians charged across the river. Sitting Bull was now suspicious about why Custer stopped his charge, thinking more troops might be in the area. Sitting Bull found out later that day that Gibbon, Terry, and Benteen were indeed in the area.

After Custer's force was annihilated, the Sioux and Northern Cheyenne regrouped to attack Reno and Benteen. The fight continued until dark and for much of the next day. On June 26, upon arrival of General Terry's column from the north, the Indians retreated to the southwest for good.[34] Over the course of the two-day battle, 269 soldiers were killed.[35]

Not long after the Battle of the Little Big Horn, Chief Sitting Bull moved his village to the Big Horn Mountains, where a victory dance was held on the fourth day after the battle. After a few weeks, the Indians traveled back to the Rosebud, whereupon they proceeded north to the Tongue River, Powder River, Beaver Creek, and Little Missouri in search of game. By August, the various bands started to split up again. Sitting Bull moved his one thousand warriors to the Twin Buttes area along the upper Grand River. Other camps were situated close by. Iron Shield, also known as American Horse, Thunder Hoop, and

Looking-for-Enemy had forty lodges of Oglala, Brulé, and Minniconjou at Slim Buttes some 30 miles from Sitting Bull's camp. Many bands of Sioux had finally decided to move to the Great Sioux Reservation.[36]

On August 5, 1876, General Crook began moving his troops northward to join General Terry's column at the confluence of the Powder River and Yellowstone River. Crook ordered that much of their equipment and rations were to be left behind so the troops could travel more quickly. After meeting on August 25, Terry and Crook decided to split up. Terry moved along the Yellowstone River to search for hostiles. Crook moved into Dakota Territory to chase down the bands of hostile Indians that were reportedly moving east. Upon hearing reports of harassments going on in the Black Hills, Crook decided to proceed to the Black Hills instead of returning to Fort Abraham Lincoln.

On September 8, General Crook decided to camp and allow his troops to recuperate. Needing food and medicine, he sent a detachment of 150 troops from the Third Cavalry, under Captain Anson Mills, to the Black Hills to procure the provisions and return to camp. Lieutenants Emmet Crawford, A. H. Von Leuttwitz, Frederick Schwatka, and Scout Frank Grouard accompanied Mills. That evening, Scout Grouard reported there were thirty-nine lodges and as many as two hundred Indians camped at Slim Buttes. Mills decided to make camp for the evening and plan an attack for the following morning.

At three o'clock on the morning of September 9, the soldiers were awakened and received their assignments for the attack. Lieutenants Crawford and Von Leuttwitz were directed to each take fifty men and advance quietly on foot to the left and right sides of Chief American Horse's (Iron Shield) camp. The camp was situated on both sides of Rabbit Creek near present-day Reva Gap. Lieutenant Schwatka was instructed to charge through the camp with twenty-five mounted cavalrymen and take positions on the other side of the camp. The remaining twenty-five men under Lieutenant Bubb were directed to guard the horses and then close in from the near side.

The surprise attack was thwarted when a small herd of Indian ponies was spooked and proceeded to run through the camp. Mills ordered Schwatka to charge and commence firing. The Indians quickly emerged from their teepees and returned fire. Chief American Horse and numerous women and children fled to a deep ravine behind the camp. Lieutenant Von Leuttwitz was shot through a kneecap. Mills then sent George Herman as a dispatch to General Crook, who was about 17 miles away. Upon receiving the dispatch, Crook responded by sending General Carr with one hundred soldiers to the battlefield. General Crook and the rest of his column arrived at Slim Buttes by about 4:00 p.m.

Crook then led a charge into the ravine and another battle ensued. Scout James "Buffalo Chips" White was killed during the charge. Furious over the loss of Buffalo Chips and other soldiers, Crook ordered his soldiers to circle the ravine and commence continuous fire. After some time, Crook ordered a cease-fire.

He then instructed his guides Frank Grouard and Baptiste Pourier to ask the warriors if they wanted to surrender any women and children. Twelve women and six children were surrendered. On Crook's order, the soldiers commenced firing again. After about two hours, Crook called for another cease-fire. Chief American Horse, who was badly wounded, cautiously emerged and requested to surrender before any more of his people were killed. Crook accepted the surrender. The army surgeons tended to American Horse's abdominal wound, but he died later that evening.[37]

Earlier that morning when the battle first commenced, one of Chief American Horse's people was able to escape and rode to Sitting Bull's camp some 8-10 miles away. Sitting Bull rapidly assembled some six hundred to eight hundred warriors, who rode to Slim Buttes. Sitting Bull's people commenced firing on the soldiers following American Horse's surrender but soon decided to retreat. At the end of the day, Crook found he had lost thirty soldiers. Many others sustained disabling injuries and had to be transported in unique ways (see fig. 5.2). The Indians' losses were considered heavy. The village was then examined and salvaged. In the process, the troops found many souvenirs from the Battle of the Little Big Horn, including clothing, letters, money, McClellan saddles, gauntlets, firearms, and a guidon of the Seventh Cavalry.

Fig. 5.2. "Stretcher with wounded man from the Battle of Slim Buttes," Stanley J. Morrow, 1876.—National Archives (111-SC-85704).

On the morning of September 10, most all of General Crook's troops departed for the Black Hills. General Carr's two troops of the Fifth Cavalry remained behind to burn the village and act as the rear guard. Just as Carr's troops were ready to depart, Indians who were hiding in the ravines attacked them. After the captive Indian women and children were released to the warriors, Sitting

Bull called off the attack and the battle ended.[38] Captain Mills and fifty of the healthiest soldiers were sent forward to get rations. Mills arrived in present-day Crook City on September 13 and began gathering supplies and rations.

The day of September 12 became known as the "mud march" or the "horsemeat march" between Owl Creek and the Belle Fourche River. By this time, the soldiers and horses were exhausted and too weak to continue. A large number of horses were too weak to stand and had to be abandoned. Some of the men also gave up in despair, because of weakness and exhaustion, and said they couldn't go on. Somehow, under constant persuasion and with great difficulty, the men found the strength and courage to continue.

On September 13, the soldiers were met near the Belle Fourche River by a wagon train loaded with bacon, flour, coffee, hardtack, supplies, and fifty cattle compliments of the citizens of Crook City and Deadwood. After fording the river, camp was made, and everyone obtained some nourishment and rest. Upon entering the Black Hills on September 14, the troops made camp for several days in Whitewood Valley near present-day Crook City.[39]

General Crook and several of his officers were invited to Deadwood on the evening of September 15. There, they were welcomed and entertained at the Grand Central Hotel by Mayor E. B. Farnum and the city council. During the dinner reception, a large group of citizens gathered outside the hotel and chanted loudly for General Crook. Crook responded to the call from the balcony of the hotel. After being introduced by General A. R. Z. Dawson, Crook spoke to those gathered for the next half hour about the fierce battles and hardships he and his troops had encountered over the spring and summer. Crook expressed hope that the government would continue its efforts to provide protection from the "marauding bands of Indians."

Following the reception, General Crook and his officers were invited to Jack Langrishe's theater, which was filled with citizens of the community. There, Judge Joseph Miller, in an appropriate address, presented General Crook with a petition signed by 635 citizens, lamenting the murder of more than one hundred citizens over the last four to five months. The petition urged the establishment of a military post at some convenient location in the northern Black Hills. General Crook sympathized with the petitioners and suggested the citizens form their own militias. Over the next several days, the people of the Black Hills celebrated with the soldiers by holding mass meetings, dances, and consuming large amounts of food and liquor.[40]

While in Deadwood, General Crook received orders to meet General Sheridan at Fort Laramie. Crook departed about two days later. The troops spent a few more days in Deadwood under the command of Colonel Merritt. They then marched to Custer City and camped there for about two weeks (see fig. 5.3). From there, the troops marched to Camp Robinson, where the mission was concluded on October 24, 1876.[41]

Fig. 5.3. Soldiers of General George Crook at the Gordon Stockade, 1876. After the Battle of Slim Buttes, General Crook's troops returned to the Gordon Stockade and camped along French Creek for about two weeks in late September 1876 while making their way to Camp Robinson.—Photo courtesy of the State Archives of the South Dakota State Historical Society.

Such advice from General Crook to form private militias was not needed; community leaders had already taken action to ensure their own protection. On July 26, the Lawrence County Commission established a reward of $25 for each Indian brought in dead or alive. Soon thereafter, Seth Bullock reported that the reward had been increased to $50, payable "in clean merchantable gold dust."

In August, a man named Brick Pomeroy shot an Indian while the native was attempting to steal some of Pomeroy's horses. A Mexican man decapitated the Indian and brought the head to Deadwood, whereupon he received $66 as a token of appreciation from the citizens. The Mexican was later shot and killed in a Crook City bar by none other than Brick Pomeroy, who had become irate over the Mexican's actions. Apparently, public officials condoned this type of self-protection. In July 1877, the Lawrence County Commission passed a resolution to increase the reward on Indians to $300, based on the Board of Health's belief that a bounty was "conducive to the health of the community."[42]

A militia named the Black Hills Brigade was formed in the northern Hills in late summer, 1876. Some thirty-four citizens enlisted in the militia from the communities of South Bend, Central City, and Golden Gate. In the southern Black Hills, the Minute Men organized in Custer City under Captain Jack Crawford. The Minute Men were so named because they were never more than a minute from town. Notwithstanding the efforts for self-protection, some thirty-five to forty settlers were killed by Indians in 1876, with some estimates as high as one hundred. Most of the skirmishes with Indians occurred in the foothills or on the emigrant trails just outside of the Black Hills.[43]

Manypenny Agreement (Black Hills Treaty)

Following the Custer massacre at the Battle of the Little Big Horn, grief turned to rage across the nation. A notion prevailed throughout much of the United States that the issues associated with the Sioux needed to be resolved once and for all. Nebraska Senator Algernon Paddock introduced a bill to exterminate the Indians. President Grant extended the military's jurisdiction to include the Indian agents and the agencies. In August, Congress approved the Indian Appropriations Act for the entire nation. However, a rider was tacked onto the act that would cut off food and rations to the Sioux until they agreed to cede the Black Hills, cede their hunting rights outside of the Great Sioux Nation, and allow three travel routes across the reservation.

Concurrent with the legislative action, President Grant appointed another commission to meet with the Sioux. The objective of the commission was to fulfill the terms of the rider attached to the Indian Appropriations Act. In effect, the Indians could choose to die from a continued military campaign, die from starvation, or comply with the demands of the government and learn to farm on the Great Sioux Reservation—outside of the Black Hills.

George Manypenny, former commissioner of Indian Affairs, was selected to chair the commission. Other members included Bishop Whipple and Reverend Hinman, both of whom disagreed with the military's actions against the Sioux. The men believed that the Sioux Nation would only survive if they, as a commission, were successful in achieving the government's ultimatum.

The Manypenny Commission convened with representatives of the Sioux at Red Cloud Agency on September 7, 1876. Sitting Bull and Crazy Horse decided not to attend. After Bishop Whipple gave the invocation, Red Cloud remarked, "We are glad to see you. You have come to save us from death." Red Cloud questioned Bishop Whipple whether the God to which he was praying was the same God that the whites had deceived two times previously while making treaties in the name of God. Bishop Whipple responded that it was the hope of the commission that the Indians would visualize and agree that "instead of growing smaller and smaller until the last Indian looks upon his own grave, they might become as the white people have become, a great and powerful people." Manypenny then communicated the three conditions recently established by Congress. He also suggested that a Sioux delegation look over the Indian Territory in Oklahoma. This suggestion greatly infuriated the chiefs, and they left the meeting.[44]

The next day, some of the chiefs stated they didn't understand all of the terms of the agreement. Others said they had no intention of abiding by the treaty. Red Cloud consented to sending a delegation to Oklahoma. He also expressed concern that if his people were forced to go to the Missouri River, they would die. "There are a great many bad men there and bad whiskey," he said, "therefore, I don't want to go there."[45] Standing Elk then spoke, "Your words are like a man knocking me in the head with a stick. What you have spoken has put great fear upon us. Whatever we do, wherever we go, we are expected to say, 'Yes! Yes! Yes! Yes!'—and when we don't agree at once to what you ask of us in council you always say, 'You won't get anything to eat! You won't get anything to eat!'"

After nearly three weeks, most all of the chiefs signed the agreement on September 26, 1876. Under the Manypenny Agreement, also known as the Black Hills Treaty, the Black Hills were ceded from the Great Sioux Reservation, as were the hunting grounds outside of the reservation to the Powder and North Platte rivers in Wyoming Territory and the Niobrara River in the state of Nebraska (see fig. 5.4).

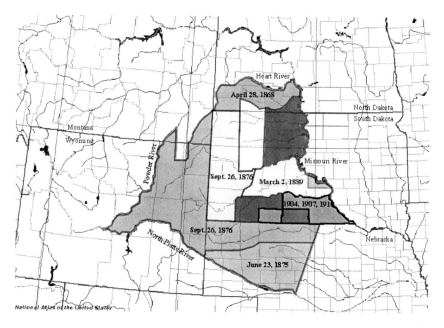

Fig. 5.4. Further reductions of the Great Sioux Reservation. Various tracts of land within the Great Sioux Reservation, as well as the remainder of the Sioux Nation established in 1851, were ceded to the United States on the dates shown. The Manypenny Agreement, which removed the Black Hills of Dakota Territory from the reservation, was signed on September 26, 1876, and ratified by Congress on February 28, 1877. Source information: Charles C. Royce with introduction by Cyrus Thomas, *Indian Land Cessions in the United States*, Smithsonian Institution, Bureau of American Ethnology, Library of Congress Geography and Map Division (Washington, D. C.): 1896-97, 521-997.—Illustration by author.

Over the next few weeks, the commissioners obtained the signatures of approximately 10 percent of the adult males of the Sioux nation, far short of the three-fourths required under the 1868 Treaty of Fort Laramie. Upon their return to Washington, D. C., the commissioners prepared their report, concluding that "our country must forever bear the disgrace and suffer the retribution of its wrongdoing. Our children will tell the story in hushed tones, and wonder how their fathers dare so to trample on justice and trifle with God." Congress ratified the Manypenny Agreement on February 28, 1877.[46]

The Sioux Split Up, but Refuse to Give Up

After the Battle of Slim Buttes, Crazy Horse moved his band of Oglala Lakota to the winter hunting grounds of the Powder River Basin south of the Yellowstone River. Sitting Bull moved a large group of Hunkpapa, Minniconjou, and Sans Arc to hunting grounds around Spring Creek and Cedar Creek north of Glendive on the Yellowstone River. Skirmishes were held with the Sioux at Spring Creek on October 11 and again on October 15-16. The Battle of Cedar Creek was held on October 20-21, 1876, after a council between Colonel Miles and Chief Sitting Bull failed to convince Sitting Bull to give up the fight. Sitting Bull and many of his people subsequently fled to the Fort Peck Indian Reservation. On October 27, about four hundred lodges of Sitting Bull's people surrendered and agreed to return to the Great Sioux Reservation.[47]

Meanwhile, on October 22, 1876, the military entered the Red Cloud Agency, searched the camp, and confiscated all the guns and horses that belonged to the Sioux. The same was done at the Standing Rock and Cheyenne River camps. More than four thousand ponies were seized, which effectively disabled the Sioux from functioning even within their designated reservation. Hundreds of the ponies were given to General Crook's Crow scouts. This infuriated the Sioux, since the Crow were their long-standing adversaries. The rest of the ponies were taken to St. Paul. The ponies that survived the trip, or that were not stolen along the way, were sold for a small amount of money.[48]

On December 7, a detachment under Lieutenant Frank D. Baldwin encountered Sitting Bull's winter camp and drove them across the upper Missouri River, forcing them to abandon their winter supply of meat. Over the next few weeks, Miles pushed Sitting Bull back and forth across Montana Territory.[49] A month later, Crazy Horse, along with five hundred Sioux and Cheyenne warriors, attacked Colonel Miles's detachment near the Tongue River on January 8, 1877. A battle was fought in subzero temperatures in the middle of a raging blizzard. The Indians finally retreated, but Miles followed them for several miles in the snowstorm. The battle became known as the Battle of Wolf Mountain. A few days later, Sitting Bull arrived at Miles's camp and announced that he had given up the fight and intended to take his Hunkpapa to the Great Mother, Canada, in Saskatchewan. Miles decided to allow the move and focus on Crazy Horse.[50]

In the spring of 1877, Spotted Tail, who was Crazy Horse's uncle, went out in search of Crazy Horse to persuade him to give up the fight and come to the reservation. Spotted Tail's attempt was unsuccessful. Upon hearing this, General Crook sent Red Cloud to talk to Crazy Horse. When Red Cloud found Crazy Horse on April 27, 1877, Crazy Horse spread his blanket on the ground for Red Cloud to sit on. Crazy Horse then took off his shirt and gave it to Red Cloud as a symbol of surrender. Crazy Horse, flanked by He Dog and Little Big Man, subsequently led the surrender march to Fort Robinson. The entire procession included about nine hundred of Crazy Horse's followers and stretched for two miles. Several miles from Fort Robinson, Crazy Horse surrendered 900 people, 2,000 horses and mules, and 117 guns.[51]

On May 6, 1877, when the procession was within about two miles of Fort Robinson, post-Commander Lieutenant William Clark rode out to meet the procession. When they met, Crazy Horse dismounted, extended his left hand to Clark, and exclaimed, "Friend, I shake with this hand because my heart is on this side. I want this peace to last forever." With this, Crazy Horse took off his war bonnet. Little Big Man, He Dog, Bad Road, and Little Hawk then offered this same act of peace.[52]

On about July 1, 1877, Crazy Horse was enlisted as the first sergeant of Company E at Fort Robinson. Lieutenant Clark surmised that Crazy Horse would be invaluable as a scout for the army. At first, Crazy Horse agreed to the assignment but probably questioned whether he really wanted to fight again. Convinced that Crook and Miles only wanted him to pursue Sitting Bull, Crazy Horse threatened to take his people north. Crazy Horse was tired of distributing rations at the agency.

Hearing this, Crook put out an order to have Crazy Horse arrested. Crazy Horse decided to flee to Spotted Tail Agency to seek refuge with his uncle, Spotted Tail. Spotted Tail showed no mercy for his nephew and ended up helping the military escort Crazy Horse back to Fort Robinson. As Crazy Horse was being led to his jail cell on September 5, he suddenly made a bold attempt to escape. An Indian policeman, who happened to be Crazy Horse's old friend, Little Big Man, grabbed Crazy Horse and held his arms. In the ensuing confusion, Crazy Horse was stabbed by a soldier and died a short time later.

Throughout 1876 and 1877, small bands of Indians continued to mount raids on settlers in the Black Hills. In 1876 alone, some forty settlers were killed, including Henry Weston "Preacher" Smith. After a series of raids were made on settlers and livestock along False Bottom Creek and Redwater River in early 1877, the residents of Deadwood became fearful that they might be subjected to the same fate that was inflicted upon General Custer and his troops. To mitigate the situation, Mayor E. B. Farnum sent a telegram to Brigadier General George

Crook and Lt. General Philip Sheridan on February 15, 1877: "We are attacked by Indians. All our stock captured. Can you send us immediate relief?" Crook responded immediately, informing Farnum that Second Lieutenant Joseph Cummings and sixty-one members of Company C of the Third Cavalry would be dispatched from Fort Robinson the next day. Despite the winter conditions, the troops arrived in Deadwood five days later.

In the interim, Mayor Farnum became frantic. He sent another telegram to Crook: "Indians seen in every direction." Crook suspected that the Indians would be gone before the troops would arrive in Deadwood. Notwithstanding, he decided to dispatch companies B and L from Fort Robinson under the command of Captain Peter Vroom. Crook's Deadwood Expedition was now supported by three officers and 220 enlisted men from the Third Cavalry.

Throughout the next several weeks, the troops engaged in several skirmishes along False Bottom Creek and Crow Creek. Eventually, the various bands of Indians gave up and fled to the Bear Lodge Mountains. In July, various other raids were made on the ranches along Spearfish Creek and Redwater River. For the most part, however, the Deadwood Expedition was successful in bringing an end to the harassment, and the miners and settlers continued to establish homes in the Black Hills.[53]

In response to the residents' requests for military protection, the government established a small army post called Camp Jack Sturgis in 1877. The camp was situated on the prairie north of Bear Butte. In 1878, Camp Sturgis was abandoned, and the soldiers were relocated to a new post called Camp Ruhlen. This camp was located on the edge of the Black Hills and south of Bear Butte. Almost immediately, Camp Ruhlen was renamed Fort George A. Meade. Many cavalry and infantry units were soon stationed there, including the Fourth Cavalry, Seventh Cavalry, and the Buffalo soldiers of the Tenth Cavalry.[54]

With the hunting grounds outside of the Great Sioux Reservation ceded pursuant to the Manypenny Agreement (treaties were discontinued in 1871), the Red Cloud Agency and the Spotted Tail Agency were now situated outside of the reservation. The government decided the agencies needed to be moved to Dakota Territory. On October 26, 1877, Red Cloud was forced to move his seven thousand members of the Oglala band of Teton Sioux from the Red Cloud Agency near Fort Robinson to the Missouri River in Dakota Territory. Similarly, Spotted Tail had to move his four thousand Brulé band of Teton Sioux from the Spotted Tail Agency near Camp Sheridan (north of present-day Hay Springs, Nebraska) to Dakota Territory.

Spotted Tail and Red Cloud were not at all pleased with having to live along the Missouri River. By early 1878, they successfully persuaded President Rutherford B. Hayes to allow them to move elsewhere in Dakota Territory. New agencies were soon constructed. Spotted Tail moved his Brulé people to Rosebud Creek in Dakota Territory in August 1878. The new agency was called the Rosebud Agency. That winter, Red Cloud moved his Oglala people one last time to the new Pine Ridge Agency along Big White Clay Creek in southwestern Dakota Territory.[55]

Between 1877 and 1880, many other Hunkpapa people, along with some Minniconjou and Sans Arc people, fled the agencies of Dakota and joined Sitting Bull in Saskatchewan. As many as five thousand people established residence in eight hundred lodges there. By 1881, Sitting Bull's people were facing starvation. Running Antelope, Gall, and others pleaded with Sitting Bull to relocate to the Standing Rock Agency in Dakota Territory. Chief Sitting Bull, along with 187 of his followers, finally surrendered on July 19, 1881, at Fort Buford, located at the confluence of the Missouri and Yellowstone rivers.

The next morning, Chief Sitting Bull laid his Winchester on the floor between his feet. He then signaled his son, Crow Foot, to pick up the rifle and hand it to the post commander. Sitting Bull told the commander, "I surrender this rifle to you through my young son, whom I now desire to teach in this manner that he has become a friend of the Americans." Sitting Bull also stated he wanted to be remembered as "the last man of my tribe to surrender my rifle." Over the next nineteen months, Sitting Bull was a prisoner of war at the Standing Rock Agency in Dakota Territory. There, he learned to raise small crops and was successful at doing so over the next few years. In 1885, Sitting Bull joined Buffalo Bill's Wild West Show. Together, the two traveled all around the country and made the show a huge success.[56]

On August 5, 1881, Spotted Tail was shot and killed by Crow Dog, one of Crazy Horse's relatives. The dispute was said to have occurred over an insult Spotted Tail made to Crow Dog's wife. Crow Dog was arrested. In March 1882, he was brought to trial in the district court in Deadwood before Judge G. C. Moody. Crow Dog was found guilty of murder and was sentenced to be hanged. In an appeal that ended up in Supreme Court, Crow Dog was acquitted and set free on the basis that the federal government did not have jurisdiction over crimes involving Indians on Indian lands. Congress acted quickly and passed legislation that extended jurisdiction of the federal court to tribal land over major crimes such as murder or armed robbery.[57]

So ends the saga of how the Black Hills was expropriated from the Great Sioux Reservation. The Manifest Destiny, the Great Reconnaissance, the Panic of 1873, the Custer Expedition, the Newton-Jenney Expedition, the Great Sioux War, and the petition from the Legislative Assembly for the territory of Dakota collectively caused President Grant and Congress to decide in 1875 that the issues associated with the Sioux needed to be resolved in a timely fashion. Based on their decisions and actions, the prospectors, miners, and homesteaders now had a legal right to claim a portion of the Black Hills.

Chapter 6
The Homestake Discovery

Locators of the Homestake Mine

Ferdinand Frederick "Fred" Manuel and Moyse "Mose" Manuel were born in Quebec in 1845 and 1848, respectively. Intrigued by the Montana gold rush, Fred decided to move to Helena, Montana Territory, in 1866 and try his hand at placer mining. Mose, also known as Moses, joined Fred the following year. The brothers decided to prospect for quartz lodes. In no time, they discovered a rich quartz lode west of Helena. After successfully operating this mine for a few years, Fred and Moses decided to go their separate ways in search of the next bonanza.

In the fall of 1874, after reading about Custer's gold find in the Black Hills, Moses returned to Helena to see if Fred might be interested in going to the Hills. Fred was reluctant. However, by the fall of 1875, Fred and Moses had heard enough about the Black Hills and decided to have a look for themselves. Toward the end of December, Fred and Moses arrived in Custer and began prospecting on French Creek. Not finding enough gold to suit them, they worked their way to present-day Hill City and Palmer Gulch. Still finding nothing of great interest, the two headed north.

Upon reaching Boxelder Creek, the Manuels began finding better "color." While there, they met two other prospectors, Henry Clay "Hank" Harney and Alexander "Alf" Engh, who reported that rich placers were being discovered to the north. Together, the four men decided to head in that direction. Upon reaching the northern Black Hills around February 10, they traveled down Spruce Gulch to Whitewood Creek. Here, as Moses later recorded, "The excitement was running high; everybody was reporting rich diggings and new strikes." Right away, the Manuel brothers moved up Whitewood Creek to its confluence with Deadwood Creek. They then traveled up Deadwood Creek to Bobtail Gulch, where several claims were staked on February 21, 1876.[1]

One of the claims staked that day was the Golden Terry, a 300×1,500-foot claim, located by Fred Manuel, C. Harris, and Alexander Engh. Harris sold his interest to Fred Manuel on March 21. Moses Manuel and Henry Harney acquired interests in the claim a short time later.[2] Toward the latter part of March, the four prospectors moved up Bobtail Gulch and crossed over the divide to the north fork of Gold Run Creek. Here, they began finding abundant quartz float, but they couldn't find the outcrop because of deep snow.[3] In his memoir written years later, Moses described the discovery of the Homestake claim as follows:

> *When the snow began to melt, I wanted to go and hunt it again, but my three partners wouldn't look for it, as they did not think it was worth anything. I kept looking every day for nearly a week, and finally the snow melted on the hill and the water ran down a draw, which crossed the lead, and I saw some quartz in the bottom and the water running over it. I took a pick and tried to get some out and found it very solid, but I got some out and took it to camp and pounded it up and panned it and found it very rich. Next day Hank Harney consented to come and locate what we called the Homestake Mine, the 9th of April, 1876. We started to dig a discovery shaft on the side of this little draw, and the first chunk of quartz weighed about 200 pounds and was the richest ever taken out. We came over next day and ran an open cut and found we had a large deposit of a rich grade of ore.*[4]

The Manuels, Harney, and Engh initially located two claims on the north fork of Gold Run Creek, one of which was the Mammoth claim. After staking the claims, Moses and Fred Manuel visited the camp of W. P. Raddick. Raddick mentioned that the country looked good enough to make a "homestake," a term used by prospectors to denote a strike rich enough to enable a man to return home to his family. The Manuel brothers and their two partners decided this was a fitting name for their new discovery, so they filed the other claim as the Homestake.[5] Ironically, the Homestake claim was staked over parts of the Giant and Gold Run claims previously located by William Gay, Alfred H. Gay, M. J. Ingaldsby, E. D. Haggard, D. Muckler and John B. Pearson, known as the Gay and Pearson Party, on January 7, 1876.[6]

Technically, Moses Manuel and Henry C. Harney were the locators of the Homestake claim, as shown on the location certificate dated April 27, 1876. Sometime later, Fred Manuel and Alexander Engh acquired interests in the claim. Deeds reflective of such a transaction were either never issued, never recorded, or were destroyed during the Deadwood Fire of September 26, 1879. Nonetheless, Fred Manuel, Moses Manuel, Henry Harney, and Alexander Engh became equal partners in the Homestake claim, each holding an undivided one-fourth interest.[7]

The Manuels, Harney, and Engh staked the Old Abe claim, Lot No. 117, on the northeast side of the Homestake claim on April 29, 1876.[8] Around May, the Manuels, Harney, and Engh decided to abandon the Mammoth claim, based on their reasoning that it did not cover any vein properly and that the quartz float found on it actually came from the Homestake and Old Abe claims.[9]

The partners now needed water and a mill. Gold Run Creek was the logical choice for water, but its flow was insufficient for milling purposes. After searching the area, the miners decided Whitewood Creek was the next logical choice for a mill site. The men selected a site on Whitewood Creek, a short distance above the mouth of Whitetail Creek and the mining camp of Pennington, known today as Kirk. Upon reaching agreement on this site, Alexander Engh directed Hank Harney to prepare a notice for a water right of 400 miner's inches, include their two names and the Manuel brothers' names as locators, and return to the site to post a location notice for the water right. Engh did so on May 10, 1876, posting the notice on a pine tree on the north side of Whitewood Creek some 700 feet above the confluence of Whitetail and Whitewood creeks.[10]

Wolf Mountain Stampede

In the spring of 1876, Moses Manuel, along with Alec McBatt and a man named Spiegle staked four claims on Whitewood Creek. The men worked the claims during the day and returned to their camp in Bobtail Gulch each night. At the claims, Moses stood guard while the other men prospected. Several times, Moses reported seeing Indians on the hillsides above the men. After realizing that the lode claims in Bobtail Gulch and Gold Run Gulch had better potential, Moses gave his share in the four claims to McBatt and one of his friends. The very next day, while the men were prospecting on the Whitewood Creek claims, a band of Indians swooped down and shot McBatt in the leg.

A few weeks later, while he was recovering, McBatt told Moses and others about some gold he had found near the head of the Little Missouri River while traveling with General George Custer in 1874. At about the same time, a rumor began circulating in Deadwood about a fabulously rich gold discovery near a "bald peak" somewhere around the Big Horn Mountains. Being the adventurers and prospectors that they were, the Manuels rounded up a party of seventy-three prospectors from upper Deadwood Gulch while Spiegle rounded up seventy-two from lower Deadwood Gulch. The Wolf Mountain Stampede—the "maddest of mad stampedes"—was soon underway.[11] Between May 14[12] and May 20[13] the prospectors departed their respective camps in Deadwood Gulch and joined together, as planned, at a temporary camp on Redwater River near Spearfish. From Redwater River, they proceeded to the Little Missouri River as one group.

From there, one-half of the party struck out for Clear Creek in the Big Horns. The other half struck out with Moses toward the head of the Little Missouri River. During their travels, both groups noted several bands of Indians who, it was later learned, were traveling from the various reservations to gather for what resulted in the Battle of the Little Big Horn. Not finding much gold in either location, the members of both parties found their way back to the Black Hills by mid-June.[14]

The Wolf Mountain Stampede is believed by some to have been a cruel hoax started by Red Clark, the owner of a livery stable in Deadwood, who had acquired an abundant supply of horses. Supposedly, Clark paid another man $50 to spread rumors of the Wolf Mountain gold strike. In no time, Clark was able to sell most of his horses to the unsuspecting miners and prospectors. The merchants of Deadwood also realized a near windfall when some $60,000 worth of equipment, supplies, and livestock was purchased by the prospectors and miners as they stampeded out of Deadwood in search of the Wolf Mountain gold discovery.[15]

The *Pioneer News* summarized the stampede as follows:

> *Twenty-three men belonging to the expedition, which left here about the 20th of May on a prospecting tour of the Big Horn Mountains, returned on the 20th inst., making the trip from Clear Creek to Deadwood in nine days. Thirty-five men remained on Clear Creek. The returned prospectors report favorable of the country, and expect to return in a few days. They saw General Crook and command June 1st on Clear Creek, marching north, but up to that time had had no fight with the Indians. His command consists of 1,000 cavalry and 1,000 infantry. They also saw 180 Snake Indians on Powder River, who were on their way to join Crook's command to act as scouts. The party saw no hostile Indians until within 75 miles of Deadwood, on their return, when a small party showed themselves at a distance, but made no attack.*[16]

The Need for Water and a Mill

Unable to find any appreciable amount of gold on the latest stampede, Engh and the Manuels returned to the Hills around June 7, 1876.[17] The Manuel brothers constructed a cabin on the south end of the Homestake claim approximately 75 feet east of where the Homestake 80-stamp mill was later constructed.[18]

On June 16, Hank Harney and Alexander Engh sold H. B. Young "one-hundred feet (undivided) of our interest in a Quartz Ledge situated in Bobtail Creek and recorded as the 'Golden Terry' Ledge; also one-hundred feet (undivided) of our interest in a Quartz Ledge situated on the northeast side of the north fork of Gold Run and recorded as the 'Home Stake' Ledge, for the consideration of six hundred ($600) dollars." Young was a partner in the freighting company of Cuthbertson and Young. He proceeded to mine a shallow open cut to the south of the Homestake discovery shaft, as well as the open cut that was being mined by the four other owners of the Homestake claim (see fig. 6.1).[19]

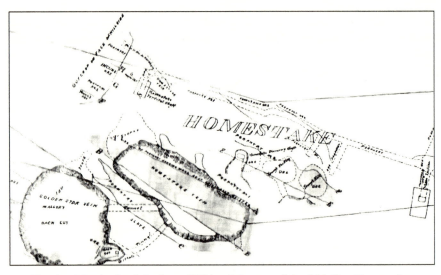

Fig. 6.1. Map by Louis Janin and Walter P. Jenney of the Old Abe, Homestake, and Golden Star claims and mine workings. After H. B. Young purchased his interest of 100 feet in the Homestake claim, he drove two short crosscut drifts from a small open cut and mined two stopes immediately southeast of the original Homestake discovery shaft. Note, also, the Homestake Vertical Shaft (upper left center), Discovery Shaft (center), Incline Shaft (lower right), Homestake Open Cut, and the Mallory Open Cut, which is on the Golden Star claim.—Source: Louis Janin and Walter P. Jenney drawing entitled, *Giant & Old Abe and Homestake Mines*, 1879. Black Hills Mining Museum, Lead, South Dakota.

By August, work on the Homestake claim had come to a standstill "owing to the absence of reduction works." In fact, except for a few arrastras, there weren't any mills in the Black Hills yet. The *Black Hills Pioneer* reported, "There is not less than 50 tons of ore on the dump at this mine, which, when properly reduced, will yield from $50 to $500 per ton." The *Pioneer* went on to say, "There is a large crevice of this ore, and when stamp mills arrive and are in working order, this mine will supply a fifty-stamp mill."[20]

While waiting for the custom stamp mills to arrive, and with reports of rich gold and silver discoveries near the mining camp of Galena, Moses decided to go to Bear Butte Creek with A. G. McShane during the first part of August 1876 to do some prospecting on the Rutherford B. Hayes claim.[21] In the interim, Fred Manuel, Hank Harney, and Alexander Engh decided to mine and stockpile ore on the Homestake and Golden Terry claims.

Toward the end of September 1876, Capt. C. V. Gardner delivered the first quartz mill to the northern Black Hills (see fig. 6.2).[22] The *Black Hills Pioneer* reported on September 30 that "the first quartz mill ever shipped to the Black Hills . . . passed through Deadwood on its way to Gayville last Monday, the

25th."[23] The mill consisted of a small steam boiler and engine, Blake jaw crusher, and a 5-foot-diameter Bolthoff pulverizer that was charged with iron balls. The falling action of the balls pulverized the ore as the mill was rotated on a horizontal shaft. The pulverized ore flowed over a table fitted with copperplates where the gold was amalgamated with mercury.

The Bolthoff Pulverizer, later called the Hidden Treasure Mill, commenced milling cement ore from the Hidden Treasure Mine in October 1876. On October 21, the *Black Hills Pioneer* reported that "results so far are of the most satisfactory."[24] By year's end, some $20,000 in gold had been recovered from the cement ore processed in the little ball mill.[25]

Fig. 6.2. The Hidden Treasure Mill on Deadwood Creek near Hidden Treasure Gulch, 1876. Craven Van Horn Gardner, nicknamed Cap, was a Black Hills Pioneer freighter who delivered the mill to the Hidden Treasure Mine in which he was a partner. Gardner organized and delivered the first 80,000 pounds of commercial freight from Cheyenne to Custer City in the spring of 1876. In May, he brought A. W. Merrick and the first printing press to the Black Hills.—Waterland Collection, Devereaux Library Archives, South Dakota School of Mines and Technology.

In November, Milton E. Pinney of Nebraska City[26] brought the first stamp mill into the Black Hills.[27] The 10-stamp mill commenced operation on December 31, 1876, crushing ore from the Alpha and Omega mines located near Central City (see fig. 6.3).[28] The mill was located on the south side of Deadwood Creek, a short distance east of the Father DeSmet Mill. Pinney was superintendent, William E. Jones, general manager, and A. G. Thorpe, engineer.[29]

Fig. 6.3. The Jones and Pinney Mill at Golden Gate, 1878. The Jones and Pinney Mill was the first stamp mill in the Black Hills. The mill, which was located on the south side of Deadwood Creek, was commissioned on December 31, 1876, with ore from the Alpha Mine.—Photo courtesy of the Black Hills Mining Museum.

Like Pinney, the Homestake men needed their own mill. Around mid-August, when Moses returned from Bear Butte Creek, the four men decided to construct an arrastra on Whitewood Creek and start milling ore from their Homestake claim. An arrastra consisted of a circular, rock-lined pit into which broken ore was placed. The ore was crushed and pulverized by hard dragstones. The dragstones were attached to horizontal arms fastened to a central spindle that was powered by horses, mules, or water. Mercury was added to the arrastra to amalgamate the gold and silver as they were liberated from the crushing and pulverizing action of the dragstones. The average production through an arrastra was less than one ton of ore every twenty-four hours.

Water from the creek was used to power the mill, which was dubbed the French Boys Arrastra (see fig. 6.4). To harness this water power, the men dug a ditch from a point on the creek near their water rights location notice to a point approximately 300 feet above the confluence of Whitewood and Whitetail creeks. At the same time, the men built a road over the hill to the Homestake claim so oxen teams and wagons could haul the ore to the arrastra.[30]

Fig. 6.4. A water-powered arrastra. This water-powered arrastra is probably very similar to the one that Fred and Moses Manuel, Henry Harney, and Alexander Engh constructed on Whitewood Creek approximately 300 feet above the mouth of Whitetail Creek and the mining camp of Pennington.—Waterland Collection, Devereaux Library Archives, South Dakota School of Mines and Technology.

The French Boys Ditch, as it was called, was about 2 feet wide at the bottom, 3 feet wide at the top, and 2 feet deep. Several other men were hired to help complete the work. One of these men was Thomas Jones, who was working the General Custer and Little Maud claims to the west of the Homestake claim. As compensation for his work on the road and ditch, Jones was granted use of half the water right for a six-month period. In addition, the Manuels, Harney, and Engh allowed Jones to construct his own arrastra next to theirs.

The Manuels' Arrastra, or French Boys Arrastra, was started up during the first part of September 1876. It was powered by a waterwheel 12-14 feet in diameter. Most all of the water in Whitewood Creek had to be diverted to operate both arrastras. The French Boys Arrastra was operated almost continuously until just after Christmas when cold weather and snow prevented the haulage of ore and operation of the arrastra. Jones operated his arrastra on ore from the Golden Star claim until about Christmas at which time he sold his arrastra to Fred May. May operated the Jones arrastra on ore from the Lincoln claim until the spring of 1877. He subsequently constructed a 2-stamp mill inside his arrastra.[31] On October 21, 1876, the *Pioneer News* reported that "the 'cleanup' of the arrastra situated upon Whitewood from ore taken out of the Golden Star Mine, yielded $489 after four days' run."

The Manuels, Engh, and Harney were able to resume milling in their arrastra in February 1877.[32] In March, the *Black Hills Pioneer* reported the Manuel brothers were "taking out the richest kind of ore" from the Homestake claim and were in the process of tunneling toward the footwall.[33]

The Manypenny Agreement, which removed the Black Hills of Dakota Territory from the Great Sioux Reservation, was signed on September 26, 1876, and ratified by Congress on February 28, 1877.[34] Upon learning that Congress had ratified the agreement, the Manuels, Harney, and Engh filed supplemental location notices for the Homestake and Old Abe claims on March 15, 1877. This was done to protect their right and title to the mining claims, which were located pursuant to the Mining Law of 1872. The claims might otherwise have been declared invalid since the Black Hills were still a part of the reservation when the claims were originally filed. Many of the other prospectors holding mining claims in the Black Hills also filed supplemental location notices to protect their rights to their claims.[35]

On April 28, the *Black Hills Pioneer* reported that work was nearly complete on construction of the Whitetail and Gold Run ditches. Within a few days, much-needed water from these two ditches was turned into the Pioneer Ditch and made available to placer miners working along the north fork of Gold Run Creek.[36]

Concurrently, mining work was resumed on the Homestake lode claim. Around late May, work had to be suspended when spring runoff and water from the Pioneer Ditch flooded the Homestake adit, which was located just above the bed of the north fork of Gold Run Creek. To compound the problem, spring conditions rendered the road to the arrastra nearly impassable. In view of the operating problems and work interruption, the Manuels, Harney, and Engh became frustrated and decided to lease the French Boys Arrastra and water right to a man named Woodward on or around June 3, 1877.[37] Woodward operated the arrastra through the remainder of 1877 with ore from his Enterprise, Telegraph, and Ready Relief claims.[38]

By late spring 1877, Hank Harney needed more cash. He subsequently made an agreement with Cyrus H. Enos that gave Enos, an agent and superintendent for the Davenport and Black Hills Milling and Mining Company, the right to mine on the Homestake claim pursuant to Harney's undivided interest in the Homestake claim.[39] Enos and the Davenport Company also purchased the north 150 feet of the Golden Star claim.[40] Enos's interest in the north end of these two claims later became known as the Segregated Homestake, Mineral Survey No. 229, and the Segregated Golden Star, Mineral Survey No. 230.[41]

Enos proceeded to drive a tunnel across the north end of the two claims (see figs. 6.5 and 6.6). He opened up a stope and commenced crushing ore in the 10-stamp Davenport Mill in July 1877.[42] The Davenport Mill was located on the south fork of Gold Run Creek on the Lincoln Lode near the west end of present-day Park Avenue (see fig. 6.7).[43]

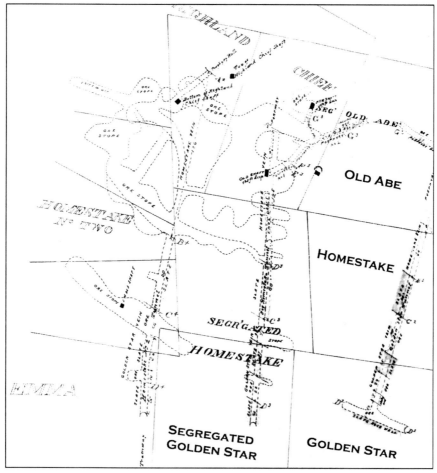

Fig. 6.5. Map showing the tunnels driven by Samuel Gwinn (left) and Cyrus Enos (middle). The Segregated Homestake represented Hank Harney's interest that was purchased by Cyrus Enos on behalf of the Davenport and Black Hills Milling and Mining Company in the spring of 1877. This illustration represents a portion of a drawing entitled *Giant & Old Abe and Homestake Mines* drawn by Louis Janin and Walter P. Jenney in 1879.—The Black Hills Mining Museum, Lead, South Dakota.

Fig. 6.6. Cyrus Enos's Tunnel and Lead City, 1877. This photo was taken from the north fork of Gold Run Gulch looking south to Lead City. The tramway shown in the foreground is likely the tunnel that Cyrus Enos drove across the Segregated Golden Star and Segregated Homestake claims. The photographer may have been standing near the Homestake No. 2 Tunnel that Samuel R. Gwinn drove to the east across the Emma and Homestake No. 2 claims in 1877.—Photo courtesy of the Black Hills Mining Museum.

Fig. 6.7. The Old Abe Shaft, Highland Mill, and Davenport Mill, ca. 1880. This photo shows (top to bottom) the Old Abe Shaft, Highland Mill, and Davenport Mill. The Davenport (Enos) Mill was located at the confluence of the north and south forks of Gold Run Creek near the west end of Park Avenue.—Photo courtesy of the Black Hills Mining Museum.

To the north of the Homestake and Segregated Homestake claims was the Homestake No. 2 claim (see fig. 6.5). The claim, 800 feet long and 300 feet wide, was located on June 20, 1876, by L. H. Mallory and Dan Latham. The location notice was filed on October 25, 1876.[44] In the summer of 1877, Samuel R. Gwinn purchased the Homestake No. 2 claim. He then drove the Homestake No. 2 Tunnel from the north fork of Gold Run Gulch (see fig. 6.5). Finding good ore, Gwinn sank two shafts and completed additional drifting.[45]

Because the Homestake workings were flooded and the arrastra was leased out, Fred Manuel and Alexander Engh decided to take a trip to Chicago on or about July 2, 1877. They didn't return until mid-October.[46] Moses decided to remain in Lead to try and dewater the Homestake workings and resume mining operations. He hired J. W. Kissack to assist with the work. After a short while, the men were successful in dewatering the mine and immediately resumed mining ore from the Homestake. Moses concurrently worked the Old Abe claim and discovered a large ledge of ore on that claim. Based on the good ore he found, Moses arranged for A. D. Bevan and William Filan to sink a shaft 50 feet deep on the Old Abe. As compensation, the men were deeded a 50 percent interest in the shaft and the north 350 feet of the claim, which became known as the Segregated Old Abe, Lot No. 481.[47]

Custom Stamp Mills in Lead City

Approximately, eleven custom stamp mills were operated in Lead in 1877 and 1878 (see table 6.1). The majority of these mills, such as the Racine Mill, Thompson, Carle, and Snell Mill, and Morris and Costello Mill (see figs. 6.8 and 6.9) were situated close to the north and south forks of Gold Run Creek where a limited amount of water was available for milling. The Pioneer Ditch also supplied water to some of the mills. In 1877 and 1878, before the California capitalists established the large stamp mills, the owners of the custom mills enjoyed a thriving business simply because most of the claim owners did not have sufficient capital to construct their own mills. Through February 15, 1878, the stamp mills in Lead City crushed more than 40,000 tons of ore.[48]

Table 6.1 Stamp mills along Gold Run Gulch in Lead City (1877-78)

Name of Mill	Owners/Operators	Location	Date	Stamps
Davenport	Cyrus H. Enos, superintendent; John Noonan and William Mosely, crushers; C. Hopkins, amalgamator	South Fork of Gold Run Creek near west end of Park Avenue	Jan. 15, 1877	10

Racine	W. L. Sackett, George Beemer, E. F. Lane, F. Lane, H. Utley, owners; W. L. Sackett, superintendent	South side Main St.; one block east of original routing of Mill St.	Apr. 11, 1877	20
Lake Superior (Old Abe)	Frank White, manager; Daniel Shane and Malcom Campbell, engineers; William Davis, amalgamator	Old Abe claim	May 2, 1877	20
Gwinn	Samuel R. Gwinn	Present-day Gwinn Ave.	Jun 1877	10
Thompson, Carle, and Snell	W. R. Carle, D. Thompson, J. T. Carle, C. J. Carle, owners; C. J. Carle, superintendent	South of Main St.; near Mill St.	Jun 23, 1877	20
Smith and Pringle	John Cumisky, manager; Peter Rourke, engineer	On South Gold Street between Julius and Addie Streets	Jul 1, 1877	10
Wambold	—	Between Main St. and Julius St. near Siever St.	Aug. 1, 1877	10
Marshman, Lewis, and Brown	Brown, Lewis, A. C. Marshman, Loving, James Black, Joseph Bradley, engineers	South side of Julius St. between Grand and Stone Streets	Aug. 1877	15
M. L. White & Co.	Mr. Kennedy, superintendent; John Delehanty, engineer; M. Longsbreth, amalgamator	Gold Run Gulch	Jan. 1, 1878	20
Welch and Burns	—	Unknown	Jan. 1, 1878	20
Morris and Costello	John Costello and William Morris, owners; D. H. Moore and C. T. Herar, engineers; John Costello, amalgamator	North of Main St.; one block east of N. Mill St.	Feb. 23, 1878	20

Source: *Black Hills Daily Times*, August 11, 1877, and May 18, 1878; Willard Larson, *Early Mills in the Black Hills* (Isanti, Minnesota: Self-Printed), 2001.

Fig. 6.8. Lead's Main Street, 1877. The view in this photo is to the west on Lead's Main Street. The large structure on the lower left is the original Racine Stamp Mill. Above that is the Thompson, Carle, and Snell Stamp Mill marked with an "x." The building on the far right is the Holvey Pharmacy, which was located on the northwest corner of Main and Mill streets.—Photo courtesy of Fielder Collection, Devereaux Library Archives, South Dakota School of Mines and Technology.

Fig. 6.9. Custom stamp mills of Lead City, 1878. This photo, by Stanley Morrow, shows the 20-stamp Racine Mill in the foreground. Above it is the 20-stamp Morris and Costello Mill. The large building on the right is the newly constructed Homestake 80-stamp Mill, which was commissioned on July 12, 1878. The street between the two rows of buildings on the left is North Mill Street. —Photo courtesy of the W. H. Over Museum, Vermillion, South Dakota.

By mid-1877, with several custom stamp mills now operating in Lead City and the French Boys Arrastra under lease to Woodward, Moses Manuel decided to contract his milling requirements to several of the local mills. In July, the 10-stamp Smith and Pringle Mill and the 10-stamp Davenport Mill were contracted to crush ore from the Old Abe claim.[49] The 10-stamp Racine Mill, named after the Wisconsin hometown of its owners,[50] was also contracted in July to crush ore from the Homestake and Golden Star claims.[51] Ten stamps were added to the Racine Mill later in the summer to boost production.[52] The 15-stamp Marshman, Lewis, and Brown Mill began crushing Old Abe ore in August.[53]

Robert "Smoky" Jones staked the Golden Star claim on the west side of the Homestake claim in 1876. The following year, Jones sold interests in his claim to sixteen or seventeen other people, including William P. Lyman, L. H. Mallory, and H. B. Young.[54] In August 1877, Samuel R. Gwinn's original 10-stamp mill and Thompson, Carl, and Snell's 10-stamp mill began crushing ore from the Golden Star claim and Gwinn's Homestake No. 2 claim. In an eleven-day run, 196 ounces of gold were recovered from ore mined on his Homestake No. 2 claim.[55]

On August 11, 1877, the *Black Hills Pioneer* reported that Henry Bolthoff sold a new 20-stamp mill, which he had just received, to Moses Manuel and Samuel R. Gwinn.[56] Moses Manuel then constructed his own 10-stamp mill on the Old Abe claim and purchased a one-half interest in Samuel Gwinn's stamp mill.[57] Gwinn's mill was located in the area of present-day Gwinn Avenue in Lead.

By mid-1877, virtually all of the ground from Golden Gate to Lead City was completely covered with mining claims, claim stakes, and location notices. Undivided interests in mining claims were bought and sold for hundreds of dollars. New claims were staked over existing claims. Disputes arose almost daily over claim boundaries, invalid discoveries, or for most any other reason that might give way to a better land position around the best discoveries. When the California capitalists and their business associates learned of the gold finds in the Black Hills, they too decided they wanted a piece of the action.

Chapter 7
The Homestake Capitalists and Their Strategies

Hearst, Haggin, Tevis, and Company

George Hearst first became acquainted with James Ben Ali Haggin and Lloyd Tevis in San Francisco, California. After discovering that they shared the same types of investment interests, the trio formed a partnership called Hearst, Haggin, Tevis, and Company.[1] James Ben Ali Haggin and Lloyd Tevis were college acquaintances from Kentucky who opened a law office in Sacramento in October 1850. The two men moved to San Francisco in 1854, where they relocated their law office.

Hearst first heard about gold mines in the Black Hills in August 1875. In his memoir written in 1890, he wrote as follows:

> *As to the mine, the way we got on to that was as follows: There was a wild fellow that when broke always used to come to me to get money to go somewhere or other. I believe he wanted to go to Chicago this time but before he went he sent me word of this mine. The first I saw of it was a sample at Salt Lake in which I could see gold. I took it and had it assayed. I think that was in August 1875. Then the next year I thought I would send Chambers out there to look at it. I made arrangements for Chambers to go in September but the Indians got so bad that it was as much as a man's life was worth to go there.*
>
> *Next summer, however, this man Sevenoaks dropped in there and got hold of a little lead mine. I was very sick at the time, but he kept telegraphing me and when I got well I suppose there was about $400 worth of telegrams lying on the table. When I got better I looked over these telegrams. At that time there was a man who I knew who used to hang round the mill and I telegraphed to Chambers to send him to the Black Hills. He fixed him up and started him off. When he got there he did not like the place that Sevenoaks had. He looked round a while and went pretty slow but finally he said he thought it was a good thing. I telegraphed him at once to get a bond, which he did.*[2]

Ludwig D. Kellogg, an experienced mining engineer, was the man Hearst and Chambers dispatched to evaluate the various mines of the Black Hills in June 1877. One of the mines Kellogg inspected was the Homestake. The Manuels and their partners told Kellogg they would take $50,000 for the claim. Kellogg declined, thinking the price was too high.[3]

By the time Kellogg inspected the Homestake Mine, Hank Harney and Alexander Engh had already sold 100 feet of their half interest in the 1,500-foot-long Homestake claim to H. B. Young on June 16, 1876. This left Harney and Engh each with an undivided 325-foot interest in the claim.[4]

Harney was also bound by the agreement he had recently made with Cyrus H. Enos, agent and superintendent of the Davenport and Black Hills Milling and Mining Company. This agreement allowed the Davenport Company to do some tunneling and stoping on the north end of the Homestake claim.[5] By late summer, Enos had driven a tunnel and was mining ore, which was being milled in the 10-stamp Davenport Mill on the south fork of Gold Run Creek.[6] Based on that agreement, Harney decided to sell his remaining 325-foot undivided interest in the Homestake claim to the Davenport and Black Hills Milling and Mining Company around the end of September 1877 for $10,000.[7] The deed was recorded on October 20.[8]

Around October 1, 1877, Kellogg inspected the Homestake claim a second time. Based on the ore that Moses Manuel and Cyrus Enos had proven up over the summer, the Manuels and Engh informed Kellogg that the selling price for the Homestake had increased and that Harney had just sold his remaining interest in the claim to the Davenport Company. Not wanting to risk a higher price later, Kellogg agreed to work out a purchase agreement with the three men.[9]

Fred and Moses Manuel each held their original 375-foot interests in the Homestake claim, which they were willing to sell to Kellogg. Engh was also ready to sell his 325-foot interest. After the price was negotiated, a deal was reached whereby the two Manuels and Engh agreed to sell their 1,075-foot combined interest in the Homestake claim to Kellogg for $60,166.66. A purchase agreement was drawn up on October 3, 1877.[10]

Kellogg immediately sent a telegram to Hearst advising him of the terms of the purchase agreement. Hearst replied with instructions for Kellogg to post a thirty-day bond on the claim and meet him in San Francisco.[11] After the purchase agreement was recorded on October 4, Kellogg took a stage to Sidney, Nebraska, and boarded a Union Pacific train to San Francisco.[12]

Upon his arrival, Kellogg reported on his evaluation of the Homestake to Hearst and Haggin. Haggin was dubious. "It is too far off; I don't want it," he told Hearst. Hearst then went to some other investors, including the Ashe brothers, and told them they could have all but one-fourth of the Homestake claim, which he wanted for himself. The Ashe brothers and the other investors agreed to the deal. Hearst then informed Haggin that he intended to make a trip to the Black Hills with Kellogg to have a firsthand look at the Homestake claim.[13]

The following day, Kellogg and Hearst boarded a train bound for Cheyenne. Accompanying them was Louis Janin, a mining engineer who periodically performed consulting work for Hearst, Haggin, Tevis, and Company.[14]

Upon arriving in the Black Hills in mid-October 1877, Hearst, Kellogg, and Janin completed an assessment of the Golden Terry claim that the Manuels, Harney, and Engh had sold to J. W. Bailey and Thomas Durbin in January. Bailey acquired Durbin's interest on July 20. Bailey then entered into an agreement with a man named Moses Thompson to sell a 9/10 interest in the south 500 feet of the Golden Terry claim to him. Through an assignment of that purchase agreement on October 16, Hearst was able to post a bond on the southerly 500 feet of the Golden Terry claim, less an undivided 50-foot interest. After the purchase was consummated, the Golden Terra Mining Company was formed, and the property was deeded to the new company.[15]

Hearst and Janin also inspected that portion of the Homestake claim still owned by the Manuels and Engh.[16] Encouraged by what he saw, Hearst was able to amend Kellogg's purchase agreement of October 3. The amended agreement, dated October 22, 1877, added a provision requiring a refund for any part of the Homestake claim that might later be found to be invalid because of the Gold Run claim, which was an earlier location.[17] After the Ashe brothers and the other investors failed to come up with their share of the money, Hearst wasted no time in paying for the Homestake claim himself.[18] He subsequently arranged for a Notice of Transfer and Assignment of the two agreements from Ludwig Kellogg to Lloyd Tevis, which was completed on October 25. Tevis, in turn, deeded the Homestake claim to Homestake Mining Company on December 26, 1877, almost two months after the company was incorporated.[19]

In 1949, C. S. T. Farish, son of Homestake's first superintendent, William A. Farish, wrote a letter to Donald H. McLaughlin, president of Homestake Mining Company. The correspondence informed McLaughlin of an interesting letter that William had written to W. W. Farlow of New York City many years ago. Excerpts from William's letter are as follows:

> *In 1877, the late Senator George Hearst went to Mr. John W. Gashwiler, of San Francisco, saying that a mining property in the Black Hills had been brought to the attention of Mr. Haggin and himself, by one named Seven Oaks; that they had their experts out there for several months and Mr. Haggin had, on that day, ordered the one still in the Black Hills, by the name of Kellogg, to throw the whole thing up and go back to Park City, Utah.*
>
> *Hearst told Gashwiler that if he sent me out there to make the examination and I reported favorably upon it, he and Mr. Haggin would join him in the purchase. This Gashwiler did. I reported upon the property favorably, and, on receipt of my report, Mr. Hearst arrived in Deadwood, and, on the next day he paid for the property. I took temporary charge of the property for five months, started development work, and I think that*

the shaft that I started at the time is the principle one in use for operating the mines. In this connection, there had been probably twenty men who turned this property down cold, including Kellogg. As soon, however, as I had reported favorably on the property, the latter gentleman went to San Francisco, and became a very enthusiastic advocate of its purchase, although he did not have the stamina to report it on his own stock.

At the time I made this examination, there were five holes or shafts, in the mine, the deepest being about 20 feet. However, the vein was opened at different points from the south end of the Homestake to the north end of the De Smidt, a distance of about one mile, and in every opening there was paying ore. I recognized, in these developments and the character of the veins, the making of a great mine, if worked on a large scale.

The experts visiting the property before I did, were what can properly be called "conservative" men, and had their opinion been accepted by Senator Hearst, as they were by Mr. Haggin, those gentlemen would never have profited to the extent that they have, by taking any "visionary" advice.[20]

A week after consummating the deal on the Homestake claim and bonding the south end of the Golden Terry claim, Hearst wrote a long letter to J. B. Haggin and Lloyd Tevis updating them of his activities in the Black Hills and the investment potential along the entire Homestake Belt. The "easterners" Hearst refers to were Cyrus Enos's partners in the Davenport and Black Hills Milling and Mining Company, who resided in Davenport, Iowa. H. B. Young was the one who now owned the 100-foot interest. The letter was written in Deadwood City on November 1, 1877, as follows:

We arrived here in good shape and in due time.

Found the mines as good as we expected. Janin and I are getting along first rate since our settlement and it was lucky for them that they settled. We have secured only the interest we had under the bond, owing to the fact that other parties had bought the balance, except about 100 feet, before we got here. The purchasers live east and the owners of the one hundred feet referred to above, are here, but have agreed to act with them. We have been in communication with these eastern people and they are now on their way out to see us. We have offered them the same price for their interest that we paid for what we have got and have proposed that if that is not satisfactory they come into the corporation with us. The owner of the 100 feet is in favor of accepting the money and will recommend the others to do so when they arrive.

As a mine always appears to best advantage when being actively worked, we have shut down all operations and will keep the mine closed until these people are settled with in some way. We hope they will all take the money instead of stock.

In regard to the "Star" mine, we have as yet accomplished nothing. We offered them $50,000 for a quit claim deed to all their interest. They want to reserve a portion of the north end to which we object and in this way the matter now stands.

The Star people contend that they are entitled to all the ground next to us and outside of our surface side lines, while we maintain that our title being the oldest, we are entitled to the full width of our vein without any regard to the surface lines. I want to settle this dispute before I leave and do whatever else is necessary to give us a good, clear title.

I want also to secure an eligible millsite.

I know you will think it strange that 75 feet in surface width does not cover our vein, but it is nevertheless true. The "Homestake" pay vein is more than 100 feet wide at the point where we came in conflict with the "Star" and I am certain it will average from one end of the Homestake to the other, thirty (30) feet. By milling several hundred tons and other thorough investigations the ore has been found to average $14.00 to $15.00 per ton. The ore is easily mined and being very free is easily saved.

Labor, supplies, and everything necessary are as cheap as in Colorado and I think the entire expense per ton for mining and milling will not exceed $4.00 certainly not $5.00.

Sixty to 70 percent of the assay value of the ore can be saved. The amount of dividends would depend, of course, upon the size of the mill. We should not have less than one-hundred stamps and that number of stamps cannot work out the mine in 25 years. Figure it yourselves, how many tons of ore are there in a pile 1,500 feet long, 30 feet wide, and 100 feet deep? These figures are inside of the facts and safe to rely on.

The vein lies between the quartzite and laminated slate, dipping about 70 degrees east with every indication favorable to its continuing to great depth. I think the vein is safe from any disturbances for at least 1,000 feet and probably for 2,000 feet. This big vein on which our claim is located travels the country north of us for about a mile and a half; next to us on that side is the "Homestake No. 2," an 800 feet claim. Near our line they have run a tunnel which goes through one vein being hard rock. The rock in these two veins worked in the mills yields from $9.00 to $17.00 per ton, all of which is corroborated by the sampling we have done.

Next north of us and adjoining the "Homestake No. 2" is our part of the "Golden Terry," or rather the 500 feet of that claim on which we hold a bond. Work on this mine has proven that the vein is about 100 feet wide and that the pay streaks are three in number and ranging from 12 to 20 feet in width.

The ore has not yielded as much per ton as that from the "Homestake," but I am satisfied the reason is that the men who did the sorting did not know the difference between the pay and barren rock. The workings under our administration have gone to prove this and I think the mine will come up to the mark and that we will take it.

North of the "Golden Terry" is the "Golden Terry #2," a small ravine separating the two claims. The vein in the 'Golden Terry #2' is enormous being over one hundred feet in width and they are running a mill and working about thirty tons per day, which amount of ore is extracted by four or five men. Their ore is paying $15.00 per ton, some more and some less. This location runs north in and through the mountain 1,000 feet.

Next to, adjoining the "Golden Terry #2" on the north, is the "Father DeSmet" mine, which is opened on the north slope of the bluff of Deadwood Creek. This mine, over 400 feet wide, has but little waste or worthless material in it. This mine had been, for sometime, and is now having ore worked at the "Elliots Mill" and an old acquaintance of mine informs me that it pays $20.00 per ton. This mine can be opened at a depth of 400 feet by a tunnel from Deadwood Creek ravine.

In this lode, all other veins to the south seem to have come together and the result is such a deposit of ore as I never saw or dreamt of before.

Deadwood Creek is where so much placer gold is found and I think this lode has furnished the supply. This mine, the "Father DeSmet" is, I am sorry to say, a high priced mine and I fear out of our reach as I am informed the owners have refused $700,000 for it. The claim continues down the bluff and across the creek up and on to the foot of the north bluff, facing to the south.

At the foot of the North bluff, the veins again separate and but one pay streak has been opened, say 15 feet. It yields $14.00 per ton. Beyond this point the vein can be seen, but it has not been opened and it is said not to prospect so well on the surface.

I have an arrangement with the superintendent for the "Homestake #2" to get for us three-fourths or at least one-half of the mine at the rate of $50,000 for the whole mine. Neither Janin or anyone else except Kellogg knows. I am also using other means to secure a good trade. I want to close up Homestake matters before I leave as I am afraid some blunder might be made by those that I should have to leave in charge, which would cost us a large sum of money.[21]

Homestake Mining Company

Homestake Mining Company was incorporated as a California corporation on November 5, 1877. The five original directors and subscribers of capital stock included Lloyd Tevis, Henry Janin, M. L. McDonald, George Hearst, and George S. Dodge, all of whom resided in San Francisco, California. The $10 million of capital stock was divided into one hundred thousand shares valued at $100 each.

The amount of capital stock initially subscribed amounted to $2,500, based on five directors holding five shares valued at $500.[22] Homestake was immediately listed on the San Francisco Stock and Mining Exchange. In January 1879, Homestake became the first mining company to be listed on the New York Stock Exchange.[23]

In November 1877, Hearst hired Dan Rathburn as mine foreman and J. W. Kissack as a miner. J. D. McIntyre, a U.S. deputy surveyor, was hired as a surveyor and Jake Siegrist as master mechanic. Despite having missed out on acquisition of the Old Abe claim, Hearst recognized the vastness of the ore systems in the north Gold Run Creek area. His strategy was to acquire a mill and begin negotiating for additional claims and interests, starting with the Homestake claim. Hearst placed an order for an 80-stamp mill from the Union Iron Works of San Francisco in November.[24]

As indicated in his letter dated November 1, 1877, to J. B. Haggin, Hearst was able to arrange for the purchase of H. B. Young's undivided 100-foot interest in the Homestake claim for $10,000. The deed was made out on November 28 to Lloyd Tevis, who, in turn, deeded the interest to Homestake Mining Company on December 26.[25]

Toward the latter part of December 1877, the Manuel brothers departed the Black Hills[26] having sold all their claims at a handsome profit in a mere 1½ years. Hearst himself returned to San Francisco. A short time later, he returned to the Black Hills to finalize the company's management and support team. William A. Farish was hired as superintendent for the Homestake Mine and L. H. Edelen as bookkeeper. Samuel McMaster (see fig. 7.1) was hired to assist with development of the properties. The Deadwood law firm of Corson and Thomas was hired to represent the Homestake Mining Company locally.[27]

Fig. 7.1. Samuel McMaster.—Photo courtesy of the Black Hills Mining Museum.

Having already purchased H. B. Young's 100-foot interest in the Homestake claim, Hearst began negotiating with representatives of the Davenport and Black Hills Milling and Mining Company for its 325-foot interest in the Homestake claim that had been acquired from Hank Harney. The Davenport Company agreed to sell its entire undivided interest to clear up the title but wanted the north 150 feet of the claim deeded back to them, since that was the area where Superintendent Enos was engaged in mining operations. Hearst agreed to the terms of the deal. The Davenport Company deeded their undivided 325-foot interest to Homestake Mining Company on January 17, 1878, for $5,000. In return, Homestake deeded the north 150 feet back to the Davenport Company on January 23. The 150-foot interest became known as the Segregated Homestake claim.[28]

Hearst soon found that litigation was occupying a large percentage of his time. Quite often, it was Hearst who was initiating some legal maneuver to improve or expand the holdings of the Homestake capitalists. The competition was fierce. With all of the discoveries along the Homestake Belt, everyone in the Black Hills and beyond wanted a piece of the action. Claim jumpers staked and filed new claims over abandoned and active claims. Undivided interests in mining claims were bought and sold almost daily to raise money or make an investment, depending on the individual. The Golden Star claim, for example, had fifteen or more owners at one time, each with an undivided interest in the claim.

In view of the flurry of litigation involving mining claims and water rights, Hearst, Haggin, and Tevis began devising and implementing other ways and means to protect their Homestake interest and facilitate expansion of the mine through acquisitions, as Hearst recommended. One such way involved having one or more members of the investment team accumulate stock in a competing company until such time the investors had controlling interest in the company. When convenient, the stock price was driven down by selling large blocks of shares. A larger amount of stock was then repurchased at a reduced price.

Another method involved forming a new company for the purpose of procuring those claims and other business assets that would benefit the new company and Homestake as well. With George Hearst and Samuel McMaster managing the mines and mills and Haggin and Tevis managing the legalities, the California investors found that an entire array of complementary companies could be efficiently managed and held under one investment umbrella. Hearst, Haggin, and Tevis began involving other business associates in some of the mining companies along the Homestake Belt. Two such investors included John T. "Jack" Gilmer and O. J. Salisbury, owners of the Gilmer and Salisbury Stage Company and the Cheyenne and Black Hills Stage and Express Line.[29]

The mines and mining companies along the Homestake Belt that the Homestake capitalists either formed or eventually acquired are listed in table 7.1 and shown in figure 7.2.

Table 7.1 Mining Companies along the Homestake Belt (Golden Gate to Lead)

Company	Formed	Formed By	Acquired by HMC
Homestake Mining Company	Nov. 5, 1877	L. Tevis, H. Janin, M. L. McDonald, G. Hearst, G. S. Dodge	—
Old Abe Mining Company	Jan. 15, 1878	Capt. Richard Uren et al.	Sep. 9, 1878
Father De Smet Consolidated Gold Mining Company	Jan. 23, 1878	L. Graves, A. Borland, A. Bowie,	Dec. 1880[1] Jun 17, 1901[1]
Golden Terra Mining Company	Jan. 23, 1878	J. B. Haggin et al.	Aug. 23, 1880[2]
Caledonia Gold Mining Company	Sep. 2, 1878	J. W. Gashwiler, S. Heydenfeldt, H. D. Bacon, M. L. McDonald, F. Locan	1886[3] Jan. 22, 1901[3]
Giant and Old Abe Mining Company	Sep. 23, 1878	J. B. Haggin, A. E. Head, T. Bell, H. Janin, J. Clark	Jan. 18, 1882
Highland Mining Company	Sep. 30, 1878	J. B. Haggin, G. Hearst, W. Willis, J. Clark, A. E. Head	Feb. 23, 1900
Deadwood Mining Company	Oct. 4, 1878	L. Tevis et al.	Jan. 14, 1881[2]
Deadwood-Terra Mining Company	Aug. 23, 1880	J. B. Haggin, A. L. Hopkins, R. Lounsbery, B. A. Haggin, A. B. Baulis, H. B. Parsons, P. W. Hohnes	Apr. 9, 1902

(1) The Father DeSmet Consolidated Gold Mining Company was formed as a consolidation of the Father DeSmet, Golden Gate, and Justice Mining companies on January 23, 1878. J. B. Haggin and Lloyd Tevis acquired controlling interest in the company in December 1880. Assets of the company were deeded to Homestake Mining Company on June 17, 1901.

(2) Assets of the Golden Terra and Deadwood Mining companies were transferred to the Deadwood-Terra Mining Company between August 23, 1880, and February 8, 1881. Homestake Mining Company acquired assets of the Deadwood-Terra Company on April 9, 1902.

(3) Homestake acquired operating control of the Caledonia Gold Mining Company in about 1886. The assets of the company were deeded to Homestake on January 22, 1901 for consideration of $500,000.

Source: Mildred Fielder, *The Treasure of Homestake Gold* (Aberdeen: North Plains Press, 1970); Homestake Mining Company, "Pre-Patent Title Summaries;" Joel Waterland, *The Spawn & the Mother Lode* (Rapid City: Grelind PhotoGraphics & Typesetters, 1987).

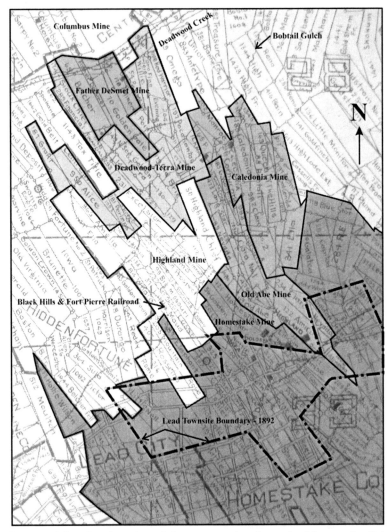

Fig. 7.2. Map showing the mining companies along the Homestake Belt (1876-1902). Note the Lead City townsite boundary that was established by George S. Hopkins on May 11, 1885. On February 5, 1892, a settlement agreement was reached between the occupants of the Lead City townsite and the owners of the mining claims, primarily Homestake, whereby Homestake agreed to grant to the owners of city lots the right to occupy the surface of their lots. The agreement gave the owners of the mining claims the right to continue mining below the surface.—Illustration by author using 1904 "Map of Ore District of the Northern Black Hills" by Frank S. Peck.

Hearst was able to purchase those portions of the Gold Run claim outside the boundaries of the Homestake claim from E. Welch, William Gay, and James

Levy for $1,000 each. Hearst then began negotiating in earnest for the Golden Star claim, which he believed contained a portion of the Homestake ore vein. Moreover, part of the Golden Star claim was needed for the new 80-stamp Homestake Mill.[30]

In February 1878, Hearst purchased Tim O'Leary's 25 feet of the Golden Star claim for $1,000 and 50 feet owned by James M. Woods for $1,666.66. Hearst was then able to procure the remaining 1,375 feet of the claim from numerous other people who had already established businesses, houses, and residential lots on the claim or in some way held a mineral interest.[31] Some of these people included C. F. Thompson, J. W. Dawson, R. H. Girard, J. H. Keyo, Angus McMasters, John Green, Steele and McClung, Richard Nichols, Alexander Engh, Alex McDougal, John Sullivan, L. Cardwell, P. Konsella, and George Graham.[32]

On March 1, 1878, Hearst fired William Farish as superintendent, based on his dissatisfaction with a location survey Farish had made of the Homestake and Giant claims. The Homestake Board of Directors approved Farish's termination on March 5, 1878, and immediately appointed Samuel McMaster as agent and general superintendent of the company's properties.[33]

In May 1878, the "survey error" that Farish had made with respect to the property boundaries of the Homestake and Giant claims was fast becoming a greater concern for Hearst. He wrote to Haggin on May 22 and May 23 claiming that the survey of the Giant did not resemble at all where the stakes were originally set. Hearst claimed that "the survey as made only takes in 150 feet on each side of center of vein and at southeast end does not come within four hundred feet of where the center south end stake was put by the owners when first located on January 7, 1876, the second location on Book of Record." Hearst expressed concern to Haggin that the Giant, as now staked, "leaves out a large amount of the best ground here now claimed by 'Old Abe' and the 'American Flag' and must have been run of the great vein for the benefit of one or both of them at a suggestion of some very accumulating and interested cuss."[34]

Hearst rationalized to Haggin that the Gold Run and Giant claims should originally have been staked 600 feet wide as provided by the Mining Law of 1872. Hearst also expressed concern that because the Black Hills was still part of the Great Sioux Reservation when the Giant and Gold Run claims were originally located, the courts might rule against any legal right to the mining claims. Hearst informed Haggin that he intended to have the Giant and Gold Run claims restaked to a width of 600 feet (see fig. 7.3), but he would refrain from "publishing" the same until he heard back from Haggin. "Telegraph 600×1500 or 1500×300 as you conclude the law is," Hearst wrote. "Both Gold Run and Giant 1500×600 feet or 300×1500 will do, but six is better and as we have to fight, it is a good plan to fight for all."[35]

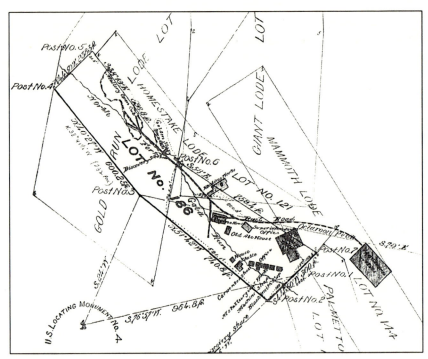

Fig. 7.3. J. D. McIntyre's mineral survey plat of the Homestake, Golden Star, Giant, and Gold Run lode claims. This mineral survey plat, completed on November 15, 1879, illustrates how the Homestake Lode (No. 121) and Golden Star Lode (No. 186) were staked over the top of the Giant Lode (No. 115) and Gold Run Lode (No. 116), based on Hearst's assertion that the latter two claims should have originally been staked 600 feet wide. The Giant and Gold Run claims were first staked by John B. Pearson and his party on January 7, 1876. The Homestake and Golden Star mills are the two structures shown in the lower right.—Source: U.S. Department of the Interior, Bureau of Land Management.

On May 23, 1878, Hearst provided Haggin with another update: "Surveyors and six witnesses on the ground preparing to [stake] and looking up original location of 'Giant.' [Surveyors] will run out center line today, which will fix status of claim," he wrote. Hearst emphasized to Haggin that the dispute was over the south half of the Giant claim, which all the parties recognized was the most valuable area. "I do not say Mr. Farish made the survey wrong willfully," Hearst wrote. "But he was induced to do so from the fights before." Hearst went on to say, "You must wake up Tevis to the importance of protecting our interests here, for while the quartz is not very rich, the amount of quartz that will pay a profit, is truly enormous. You nor your children will never live to see the end of the time when this property will not be worked for a profit. But, as you are aware, it will take time, patience, and money to perfect a title to so large and wide-spread property as this, as we have all kinds of people to contend with, and all on the make. This claim affects a dozen different interests, and several taking pay rock out of the mine."[36]

As a postscript to his letter, Hearst added a note that he had made up his mind to "fight the thing out and the old line at all hazards as it shows the most damnable frauds ever perpetrated in any country. I will hurt a good many people, as I wrote you; if we succeeded in finding out the fraud and maintain our rights there would be more squealing than ever was heard of before. And it is quite possible that I may get killed, but if I should, I can't but lose a few years and all I ask of you is to see that my wife and child gets all that is due them from all sources and that I am not buried in this place. I have made up my mind to survey the claim 1500×600 feet and try to make it stick; if not we can fall back on 1500×300 feet as is the law of D. T. [Dakota Territory] passed in January 1875."[37]

Hearst ordered J. D. McIntyre to conduct a new survey and plat of the Homestake claim, which was completed on June 29, 1878. Excluding the 150-foot Segregated Homestake interest, the Homestake claim, as surveyed, amounted to 4.57 acres. Meanwhile, Hearst was able to purchase interests in the Giant claim from M. V. Boughton, George Fardele, and one other man in June, August, and November 1878. Various other houses and lots were also purchased.[38]

Hearst got a few other breaks. He was able to purchase the Segregated Homestake claim from the Davenport and Black Hills Milling & Mining Company on June 2, 1879. The deed was made out to J. B. Haggin.[39] In consideration, the Davenport Company was paid $80,000 and Cyrus Enos, $26,000 for their interests in the 150-foot-long claim. Hearst was able to purchase the remaining interests in the Gold Run claim on December 24, 1879, from E. Welch, William Gay, and James Levy, who were paid $1,000 each. M. Boughton, George Fardele, and Fred T. Evans were paid amounts ranging from $25 to $150 for their minor interests.[40]

Between May 1, 1878, and August 31, 1880, some 153,372 tons of ore were milled at the Homestake 80-stamp mill; 121,910 tons at the 120-stamp Golden Star Mill; and 3,001 tons at outside mills. Overall mill yield was 113,754 troy ounces of gold-and-silver bullion, called doré, with a net value, after freight charges, of $1,889,283.98, or $6.79 per ton milled (see table 7.2). Total operating costs amounted to $3.45 per ton, which resulted in income of $3.34 per ton. Some 64.6 percent of net income, or $600,000, was paid to shareholders as dividends. Capital additions for the mills, tramway, hoists, and shops amounted to $455,210.10.[41]

Table 7.2 Operating costs for the Homestake Mine (May 1, 1878, through Aug. 31, 1880)

	Total	Cost per Ton Milled[1]
Net Yield	$1,889,283.98	$6.7891
Operating Costs		
Mine	$366,213.32	$1.3160
Shafts	$101,706.42	0.3655
80-Stamp Mill	$199,661.01	0.7175
120-Stamp Mill	$94,726.48	0.3404
Crushing, Shops and Other	<u>$197,518.55</u>	<u>0.7098</u>
Total Operating Costs:	$959,825.78	$3.4492
Net Income:	$929,458.20	$3.3399
Dividends	$600,000	$2.1561
Plant Construction	$455,210.10	$1.6358

(1) Mill unit costs based on 278,283 tons milled from all sources.

Source: Data from Lounsbery and Haggin, Financial Agents, *Report of the Homestake Mining Company from January 1, 1878 to September 1, 1880.*

In March 1879, two donkey steam engines, each of which was 10×12 inches and rated at 50 horsepower, were installed at the Incline Shaft, which was located within the Incline Ledge. Two boilers were also installed, along with a 6-inch Cornish steam pump and a 15×18-inch Burleigh air compressor. The Incline Shaft, which had a skipping capacity of 200 tons per day, was initially sunk to the 100-foot level. Ore was hoisted at the Incline Shaft using ten 1,400-pound ore cars. By September 1, 1880, the Incline Shaft was 150 feet deep.

The two-compartment Homestake Shaft, also referred to as the Vertical, was located approximately 400 feet north of the Incline Shaft. The Vertical Shaft had a capacity of 1,000 tons per day and was equipped with a 12×30-inch, 60-horsepower steam engine. One boiler, 50 inches in diameter by 16 feet long, supplied steam. The Vertical Shaft was initially sunk to a depth of 100 feet in the fall of 1878.[42] In 1879, the Hoskins brothers were awarded a contract to sink the Vertical Shaft to 200 feet.[43] By September 1880, the shaft was 250 feet deep.[44]

Work started on the Golden Star Shaft in 1879. The shaft was located on the Golden Star claim, south of the Homestake Vertical and Incline Shafts on the Homestake claim. By September 1880, the Homestake Mining Company's surface works consisted of the Homestake 80-stamp mill and the 120-stamp Golden Star Mill, the Vertical, Incline, and Golden Star shafts, a blacksmith shop 30×60 feet, a machine shop 20×30 feet, a foundry, a sawmill 20×80 feet, a timber-framing house, and a tramway 1,500 feet long on which the J. B. Haggin steam locomotive was operated (see fig. 7.4). In addition, two carpenter shops, a coal and iron house, a general office, an assay office, the superintendent's residence, the construction engineer's office, and a stable and a storehouse all supported the company.[45]

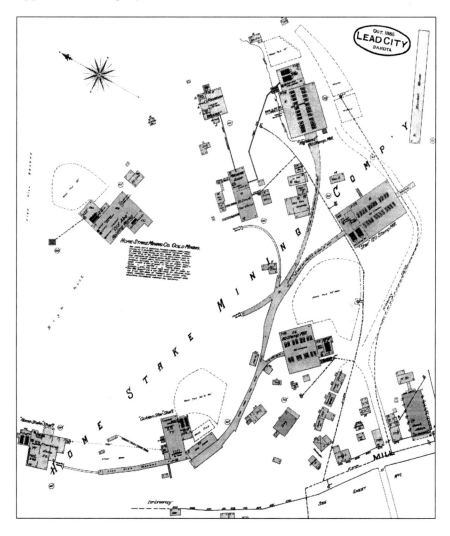

Fig. 7.4. Sanborn map showing the Homestake works, October 1885.—Courtesy of the State Archives of the South Dakota State Historical Society.

On December 15, 1882, after learning that his brother was ill, Samuel McMaster put Harry M. Grigg, superintendent of the Father DeSmet Mine, in charge of the mines and mills in the Lead area. McMaster departed for Tombstone, Arizona, to see his ailing brother and didn't return to Lead until February 15. McMaster himself was having health problems. In May, he suddenly left town again, this time for the West Coast. Grigg was left in charge of more than he could handle but wrote to his brother saying he was happy to have all the responsibility. In order to provide Grigg with some relief, Haggin put George E. Webber in charge of the Father DeSmet operation in 1883.

McMaster didn't return to Lead until May 1884. Unfortunately, his health had deteriorated. He finished his annual report for the period ending June 1, 1884, and left town one final time. On December 23, 1884, Samuel McMaster died in San Francisco. Within a short time, the Homestake Board met and decided to put Thomas J. Grier (see fig. 7.5) in charge of operations at Lead. Grier was no stranger to the Homestake operations, having been working as a telegrapher and accountant under McMaster. Harry Grigg decided to terminate his employment with Homestake and went to work for the Uncle Sam Mine at Roubaix.[46]

Fig. 7.5. Thomas J. Grier, superintendent of the Homestake Mine (1884-1914).—Photo courtesy of the Black Hills Mining Museum.

On January 1, 1888, T. J. Grier reported that there were 936,700 tons of ore blocked out between the surface and 500-foot level of the Homestake Mine. Most of the reserves were located above existing stopes in the Old Abe, Amicus, Farish, Incline, West Stope, and American Flag ledges.[47] Development of the mine was also progressing nicely. In June 1891, it was reported that the Golden Star Shaft was 600 feet deep, the Vertical Shaft 500 feet, and the Old Abe Shaft 700 feet.[48]

Other Belt Companies of the Homestake Capitalists

Old Abe Mining Company

As Homestake Mining Company was getting its feet on the ground, Fred and Moses Manuel, Henry C. Harney, and Alexander Engh sold their remaining interests in the south 1,150 feet of the Old Abe claim to Richard Uren, Henry Richards, Charles Briggs, John Senter, M. S. Delano, W. E. Parnell, John Duncan, Francis Palms, T. G. White, and G. W. Shears on December 12, 1877, for $45,000. A short time later, these men sold their interest in the claim, including their one-half interest in the mill site, to the Old Abe Mining Company on February 15, 1878.[49]

The Old Abe Mining Company was incorporated under the laws of the state of Michigan on January 15, 1878. The company did not have any business affiliation with the California capitalists involved in the Homestake Mine. The amount of capital stock was fixed at $2.5 million and was divided into one hundred thousand shares at $25 each. John M. Curnow of Lead City, Dakota Territory, was appointed agent and superintendent for the Old Abe Mining Company, and Frank G. White was elected president. Capital stock was issued to some twenty-five to thirty shareholders, most of whom resided in Houghton and Calumet, Michigan.[50] At a meeting of the stockholders in early January 1878 at the Calumet House in Michigan, five directors were elected to the Old Abe Mining Company, including John Duncan, Frank G. White, Moses Delano, John Senter, and Richard Uren.[51]

On June 19, 1878, Old Abe Superintendent John Curnow convinced Lawrence County Deputy Sheriff John H. Kehoe to issue an injunction and restraining order against James M. Young, Abram B. Chaplin, and others on the allegation that their mining work on the American Flag and Palmetto claims was encroaching on the Old Abe claim.[52] The dispute was somewhat complicated. Michael Lynch and Michael Heffron had previously purchased the Sunnyside claim on June 25, 1876. Not satisfied with the Sunnyside's discovery and location, Lynch and Heffron located the American Flag claim on ground previously covered by the Sunnyside claim and the Mammoth claim. The Mammoth was originally staked by the Manuels, Harney, and Engh but was abandoned in favor of the Homestake and Old Abe claims.[53] The American Flag and Palmetto claims were then merged as the Palmetto and American Flag Consolidated Mines under the supervision and direction of James M. Young.

Young claimed that the owners of the Old Abe Mining Company had moved the stakes for the Old Abe claim farther up Clara Hill compared to how they were originally set, thereby encroaching on the Palmetto claim. Young maintained that there never was a boundary conflict with the Manuels, Harney, and Engh and that the conflict only came about after the men sold their interests in the Homestake and Old Abe claims. Young also claimed that Henry Richards, foreman of the Old Abe Mining Company, told him that "the Homestake owners have shoved their lines up on us," causing the Old Abe Mining Company to move its south-end stakes farther up the hill.[54]

As the claim disputes continued, the Old Abe Mining Company made an application for patent for the Old Abe claim on July 27, 1878, at the U.S. Land Office in Deadwood, Dakota Territory.[55] While the application for patent was being reviewed, the Old Abe Mining Company and its assets were sold to J. W. Gashwiler on September 20, 1878. That same day, per a preconceived plan, Gashwiler sold the Old Abe Mining Company, its office building, and two 20-stamp mills, known as the Welch and Burns and Gwinn Mills, to J. B. Haggin.[56]

J. W. Gashwiler apparently was not one of George Hearst's best friends. In a letter to J. B. Haggin dated March 6, 1879, Hearst unloaded on Haggin, saying, "I think the advent of Gashwiler into this camp has been the cause of nearly all our troubles, and our compromise with the 'Old Abe' and consequent payment of a large amount of money for a claim that was in a fair way to have been otherwise settled, has given outsiders confidence in thinking they could at any time force us to pay out money and I certainly think the effect of it all has been stupendious and outrageous."[57]

Giant and Old Abe Mining Company

J. B. Haggin deeded the assets of the Old Abe Mining Company to the Giant and Old Abe Mining Company on October 25, 1878.[58] The Giant and Old Abe Mining Company was incorporated under the laws of the state of California on September 23, 1878, after Haggin acquired the assets of the Old Abe Mining Company from J. W. Gashwiler. J. B. Haggin, A. E. Head, Thomas Bell, Henry Janin, and Joseph Clark—all residents of San Francisco, California—were named directors of the company. The amount of capital stock was set at $2 million, which was divided into one hundred thousand shares valued at $20 each. The amount of capital stock that was actually subscribed amounted to $500,000. Of this amount, J. B. Haggin was issued 24,980 shares, valued at $499,600.

A. E. Head, Thomas Bell, Henry Janin, and Joseph Clark were each issued five shares, valued at $100.[59]

Shortly after the Giant and Old Abe Mining Company was formed, the board of directors instructed Samuel McMaster to solicit bids for new hoisting and pumping works and to solicit bids for sinking a new three-compartment shaft to a depth of 200 feet. On December 11, 1879, Fraser & Chalmers—a company specializing in boilers, hoists, and stamp mills—was awarded a $30,200 bid to supply a 26×60-inch Corliss stationary steam engine, 225 feet of 12-inch-diameter pump line, two tubular steam boilers 54 inches in diameter and 16 feet long, and one pair of 18×60-inch hoisting engines with .5×4-inch flat wire rope.[60]

On January 8, 1879, McMaster received eleven bids for sinking and timbering the 200 feet of the new Giant Shaft. Bids ranged from $30 per foot to $65 per foot. The work was awarded to Alex Ballantyne and John McMaster, whose bid was $32.50 per foot for the first 100 feet and $50 per foot for the next 100 feet. The Giant Shaft was subsequently named the Ballantyne and McMaster (B&M) Shaft after the two shaft-sinking contractors.[61] By June 1880, Alex Ballantyne and John McMaster had sunk the B&M Shaft to a depth of 212 feet.[62]

In November 1881, the Homestake Board of Directors passed a resolution to merge the Giant and Old Abe Mining Company with Homestake Mining Company. Homestake shareholders were issued sixteen shares for each ten shares of Homestake stock. Shareholders of the Giant and Old Abe Mining Company received four shares of Homestake stock for every ten shares of the Giant and Old Abe stock.[63] Upon approval from both companies, the assets of the Giant and Old Abe Mining Company were deeded to Homestake Mining Company on January 18, 1882.[64]

Highland Mining Company

Mike Cavanaugh located the Highland Chief claim in September 1877. It was located just to the north of the Old Abe claim.[65] Cavanaugh sold the claim to two men named Brodie and Terry. On January 23, 1880, Brodie and Terry sold the claim for $12,500 to John T. Gilmer and O. J. Salisbury, investment partners with the Homestake syndicate.[66]

At about the same time Cavanaugh located the Highland Chief claim, Samuel R. Gwinn sold the Homestake No. 2 claim to Charles C. Ottinger. Ottinger also owned an interest in the Emma claim, which was situated west of the Homestake No. 2 claim. After Ottinger defaulted on the sale, Gwinn sold the Homestake No. 2 claim to Gilmer and Salisbury. Ottinger later sold his interests in the Emma and Homestake No. 2 claims to Gilmer and Salisbury.[67]

Hearst wrote to J. B. Haggin on March 4, 1878, informing him that "ourselves and others have control of Homestake No. 2 and are in possession of mine and mill and in very favorable times.... We are now running the 30-stamp mill and things look well so far as we can determine."[68] In an update on May 19, Hearst wrote that the Homestake No. 2 Mine "looks well as to the quantity and all things pertaining to its permanency. We run first eight days and cleaned up $9.55 per ton with 55 percent profit. Second run cleaned up yesterday and today and find the ore to pay $12 per ton and think we will be able to improve upon that. Our mill will crush 30 tons per day."[69]

Immediately after the Homestake No. 2 and Emma claims were purchased, the Highland Mining Company was incorporated under the laws of the state of California on September 30, 1878—just one week after the Giant and Old Abe Mining Company was incorporated. J. B. Haggin, George Hearst, William Willis, Joseph Clark, and A. E. Head—all residents of San Francisco, California—were named directors of the Highland Company. The amount of capital stock of the company, set at $2 million, was divided into one hundred thousand shares valued at $20 each. The amount of capital stock actually subscribed amounted to $500. Of this amount, each director was issued five shares valued at $100.[70]

Because the principal shareholders were the same for both companies, Homestake effectively had control over the Highland Mining Company and Homestake's site management personnel operated the mine.[71] Almost immediately, plans were made for sinking a new shaft and constructing a 120-stamp mill that would be patterned after the smaller mills at the Deadwood Mining Company and the Golden Terra Mining Company. On November 19, 1878, Samuel McMaster received seven bids for sinking the Homestake No. 2 Shaft on the Highland property. The bid specification called for sinking 100 feet of shaft and framing the shaft into compartments using 10×10-inch timbers. A contract award was issued, and the shaft was sunk to its initial depth in 1879.[72]

One of Hearst's other acquisitions, on behalf of the Highland Mining Company, was the Prince Oscar mining claim. Hearst informed Haggin on March 9 that "the Highland is doing very well . . . and as you have been informed we have got 'Prince Oscar' which will take some of our surplus. We have not as yet determined what size mill we will build, nor where we will place it, nor will we do so until Jack Gilmer or Salisbury get here, which will be in a few days."[73]

It was in March 1879, on behalf of the Highland Mining Company, that Hearst decided he wanted the Segregated Homestake claim back from Cyrus Enos and the Davenport and Black Hills Milling & Mining Company. He offered $50,000 for the Segregated Homestake, but the offer was declined. Hearst also made an offer to them for the Segregated Golden Star and the Ophir. Again, the offer was declined. Hearst was determined. He took another approach by filing a lawsuit over the Ophir, claiming extralateral rights to the ore he said originated on the Golden Terry claim. The judge ruled that the Ophir claim was invalid, since its discovery shaft was actually on the Golden Terry.[74]

J. B. Haggin questioned the value of the Segregated Homestake. Hearst responded to Haggin's concern on March 23, 1879, as follows:

> McMaster and myself have returned only today from examining the coal fields and on returning found your dispatch of the 21st in relation to the "Segregated Homestake." I telephoned Lounsbery . . . that we did not want any of it at that price as the "Segregated" had allowed the Highland Chief to get patent to the most valuable portion of their mine.
>
> Now it is of some importance that we should own that piece of ground from its position: lying between two companies that we control it becomes necessary that we should spring some plan to so depreciate it in value that we can buy it at a fair price. First, the Highland Chief will bring suit to recover that portion of the ground covered by their patent. Secondly, the Homestake has the "Gold Run" title, which is an older and perhaps a better title to the ground held by the "Segregated Homestake" and its lines, and I think we can show it to be a better title. You understand that the Homestake has guaranteed to the "Segregated Homestake" a title, to the extent of $50,000, now can we sue them on the "Gold Run" title and if we get a patent to the "Gold Run" what figure does that guarantee cut? Or can we then pay them the $50,000 and take the ground? Or had we better convey the title to the "Gold Run" to an outside party? Or can we do that after putting it into the Homestake without being bound by it? Please look into this carefully and see what is the best course to pursue as some scheme must be gotten up to bulldoze these fellows.[75]

On behalf of the Highland Mining Company, Hearst ended up settling the matter in June 1879 by paying Cyrus Enos $26,000 for his interest in the claims and $80,000 to the Davenport Company for its interests.[76] The Highland Mining Company now had ownership of the Homestake No. 2, Highland Chief, Segregated Homestake, Segregated Golden Star, and the Golden Prospect claims. Legal title to the Segregated Homestake and Segregated Golden Star claims was transferred to J. B. Haggin on August 17, 1880.[77] The White 30-stamp mill, which had been on lease for two years, was also purchased for the benefit of the Highland Mine and the Homestake No. 2 claim.

Ore reserves for the Highland Mining Company were reported to be 548,000 tons on January 1, 1888. The production expectation for the year was 2,150 ounces of gold per month or $35,000 in income.[78] Through the 1880s, ore was primarily mined from an open cut and underground workings accessed from the Highland Tramway. In 1890, a new two-compartment shaft, called the

Golden Prospect, was sunk on the Golden Prospect claim (see fig. 7.6). The two compartments were 5×9 feet and 6×9 feet, which allowed square-set timbers to be placed on timber trucks and pushed into the cage. The hoist consisted of a double Corliss engine with 20×60-inch cylinders. The flat reels were designed and fabricated to hold enough rope to permit hoisting from a depth of 2,000 feet. Ultimately, the Golden Prospect Shaft was sunk to a depth of 1,550 feet. The tramway level was situated on the 180-foot level, which connected the shaft with the 120-stamp Highland Mill. An Ingersoll-Sergeant air compressor, sized to power sixteen rock drills, was also purchased.[79]

Fig. 7.6. Golden Prospect (Highland) Shaft of the Highland Mining Company. Ed Brelsford, engineer, was killed May 10, 1901, as the result of a boiler explosion at the Golden Prospect Shaft. A sawmill was located at the Highland and was serviced by the Black Hills and Fort Pierre Railroad.—Photo courtesy of the Black Hills Mining Museum.

Deadwood Mining Company

After Fred and Moses Manuel, Alexander Engh, Henry Harney, and H. B. Young sold their interests in the Golden Terry claim to John W. Bailey and Thomas Durbin on January 29, 1877, Bailey bought out Durbin's interest.[80] Bailey then sold the north 1,000 feet of the Golden Terry claim and 50 feet of the south 500 feet of the claim to Peter Kimberly and Hannah Armes for $40,000 on July 20.[81] Kimberly and Armes initially called their mine the Golden Terry No. 2 Mine but later changed the name to the Deadwood Mine.

The north end of the Golden Terry claim became known as the Northerly Segregated Golden Terry, Mineral Survey No. 149, upon which the Deadwood Mine was located. The 6.90-acre Northerly Segregated Golden Terry claim extended from the north side of upper Bobtail Gulch to the south end of the Father DeSmet and Belcher claims located on the south side of Deadwood Gulch at Golden Gate.[82]

Kimberly and Armes sold the Northerly Segregated Golden Terry claim and the Deadwood Mine to Lloyd Tevis on November 6, 1877.[83] With Hearst having operating management of the mines owned or controlled by the Homestake capitalists, L. D. Kellogg was named superintendent of the Deadwood Mine.

On May 19, 1878, Hearst wrote another letter to Haggin, updating him on the status of the Deadwood Mine:

> *On my arrival here I found that Mr. Kellogg had run a tunnel some 160 feet under the mass of rock on the surface into the hill, and laterally with the vein, but below the quartz on the surface, and had no pay ore in said tunnel to my surprise. He had also run a cross-cut near the mouth of the tunnel or openings, and that was not satisfactory. So we changed the work and started a system of development that would either prove the mine good or bad in the shortest space of time. The time was so long, that I was much alarmed about the value of the mine, and so wrote you my fears, but as I telegraphed you, we have struck good ore in the west cross-cut. It still will, if it continues in No. 1 about the best average I have seen in the Hills, which will, if it continues a few feet farther, fix the value of the mine at a large figure.*[84]

After the Homestake capitalists formed the Deadwood Mining Company on October 4, 1878, Lloyd Tevis deeded the property to the company on October 31, 1878.[85] John T. Gilmer and O. J. Salisbury were major investors with Tevis in the Deadwood Mining Company.[86]

On February 17, 1879, a 60-stamp mill, called the Deadwood Mill, was completed. The mill had a capacity of 165 tons per day.[87] A tramway 400 feet long was constructed between the Deadwood Open Cut and the mill. Ore mined from the Deadwood Open Cut (see fig. 7.7) was shoveled into cars and hand-trammed to the mill. Later, upper and lower tunnels were driven northward from the face of the Open Cut. From the upper level, ore was mined and shoveled into raises that were driven from the lower level. Ore was transferred from the raises into railcars on the lower level.[88]

Fig. 7.7. The Deadwood Mine Open Cut and underground workings, 1902. The Deadwood Open Cut was situated on the south end of the Northerly Segregated Golden Terry claim. Eventually, the Father DeSmet Open Cut, located on the opposite side of the hill, merged with the Deadwood Open Cut. The bridge was part of a roadway that was established between Lead and Terraville.—Photo courtesy of the State Archives of the South Dakota State Historical Society.

Golden Terra Mining Company

The south 500 feet of the original Golden Terry claim was situated on the south side of upper Bobtail Gulch. John Bailey constructed a 20-stamp mill on the claim in March 1877.[89] In 1879, the mill was expanded to 30 stamps.[90] On October 16, Hearst was able to bond the property[91] through assignment of a purchase agreement Bailey had made with Moses Thompson for Bailey's nine-tenths interest in the south 500 feet. On January 23, 1878, as agent for the newly formed Golden Terra Mining Company, and upon satisfaction of the bond, Hearst instructed Bailey to deed the south 500 feet of the claim, excluding 50 feet, to the Golden Terra Company. The south 500 feet of the Golden Terry claim, on which the Golden Terra Mine was situated, was surveyed for patent as the Southerly Segregated Golden Terry claim, Survey No. 130, by J. D. McIntyre in September 1878.[92] Immediately, C. D. Porter was named superintendent of the Golden Terra Mine for the Golden Terra Mining Company.[93]

On May 19, 1878, Hearst wrote to Haggin, "Terra is moving along well, is opening up well, and is a good mine, but will go slowly as you are aware. That debt of $30,000 will bear heavily on it. I saw it was to $3 and so telegraphed you to sell it down to $2, thinking that if you did not do so, someone else would and buy it back at a lower figure. The mine is good at $2 to keep or to hold and as those smarties want to run it, give them a dose now and then, but don't lose much of your stock, at least below $2. But take it back at the lowest figures."[94]

On March 6, 1879, Hearst provided Haggin with an update on the status of the equipment for the new Golden Terra Mill. Once again, Hearst proposed to Haggin that their stock in the Deadwood Mining Company and the Golden Terra Mining Company be put on the market:

> The Terra machinery is all on the way from Sidney and I think will get in before the roads break up; when that mill gets going we will be able to make a splendid operation in actual results, so hurry up your plans and get entirely ready as soon as possible. The Corliss engine will be here in three or four days.
>
> I think it best to put the "Terra" and the "Deadwood" on the market with present mill capacity as now is a good time and we can make enough on them and have a sure thing; I think the prospects of a splendid mine in "Deadwood" are very flattering and perhaps will need another mill on the 'Deadwood' side, but do not want to wait.
>
> We will make a more thorough examination of the working of the "Highland," "Deadwood," and "Terra" when the "Deadwood" has made a cleanup showing the cost under our management, of reducing the ore both by our own and custom mills.
>
> Please write fully or telegraph what you think of my views in relation to all these matters; please have Mr. Tevis read this letter as I have received communication from him in relation to "Prince Oscar" and other matters here.[95]

After reading Hearst's telegram, Haggin telegraphed Hearst that he needed clarification about the request to sell their stock. Hearst responded on March 9 as follows:

> My telegram to you in relation to putting the "Deadwood" and "Terra" on the market meant simply this: That I have made up my mind to put them on the market with their present milling capacity—which is, on the "Deadwood" now, say 160 tons per day or thereabouts, Millet having reported to me (by guess) 180, but on investigation I find it 165, and the "Terra," when the new mill is completed will have 90 stamps, which I think is sufficient mill capacity for the mine—and on this I am determined to realize on both mines, to the extent at least of half our stock, or more, as we may conclude as to putting them on the market you can be the judge as to whether it will be better to keep the "Terra" where it is in San Francisco and by buying and selling it we can make more money out of it, and manage it better there, than to move it, and my advice is to try to experiment at any rate, because it will sympathize with the stock in New York anyhow. The machinery for the mill is all on the road and arriving rapidly and will be in running order towards the last of April, so you see you

will not have much time to arrange your plans. The "Deadwood" is now in full operation and the sooner you fix that on the New York market the better. I feel very sure that the "Deadwood" will develop into a mine that will require at least twice as many stamps as it has now, but that is too far in the future and I am not inclined to wait for it. I am willing for those who buy it, to take their chance and have the benefit of it, we are making enough, at what we will be able to realize at present milling capacity.
Let Tevis see this.[96]

The Golden Terra 60-stamp mill was completed on June 26, 1879 at a cost of $80,000. The mill was located just below the Deadwood Mill in Bobtail Gulch.[97] By August 1879, both the Deadwood and Golden Terra mills were operating superbly. Unfortunately, Haggin hadn't sold any of the stock of the two companies yet. Hearst was getting impatient. "Deadwood Mine is looking very well," Hearst wrote. "The mill is running nicely with every indication of a splendid cleanup this time. Terra mill and mine doing very well; late developments in the Terra look splendid. . . . I want you to hurry up the getting stocks on the New York board and get them in shape as I am very anxious to realize on at least one-half of our Deadwood stock; also a part of Terra as soon as it is best to do so."[98]

Hearst recognized the ore potential of the Golden Terra Mine and began negotiating to expand the Golden Terra Mining Company's property boundaries. A. S. Hinkley's interest in the Old Brig claim was purchased for $175 on August 16, 1879. On December 20, 1880, Hearst purchased A. B. Thomas's interest in the remainder of the claim for $300.[99] Hearst also wanted the Gopher and Golden Terra Extension claims, since they were very close to the Golden Terra Mine. Hearst attempted to purchase these two claims from the Fairview Mining Company, but was unsuccessful.[100] Taking another approach, Hearst and the Golden Terra Mining Company filed suit against J. C. Mahler and the other owners of the Gopher claim. On December 6, 1879, Judge Gideon Moody ruled in favor of the defendants, bringing a sigh of relief to a large number of shareholders who held forty thousand shares in the Gopher Mining Company.[101]

Deadwood-Terra Mining Company

J. B. Haggin, A. L. Hopkins, Richard P. Lounsbery, B. A. Haggin, A. B. Baulis, H. B. Parsons, and Philip W. Hohnes organized the Deadwood-Terra Mining Company as a New York corporation on August 23, 1880. The assets of the Deadwood Mining Company and the Golden Terra Mining Company were transferred to the new company.[102] Thomas McMaster, brother of Samuel McMaster, was named superintendent of the Deadwood-Terra Company. A short while later, general management of all the mines along the Homestake Belt that were owned or controlled by Haggin and the various capitalists, including the Deadwood-Terra, was assigned to Samuel McMaster.[103]

Under the new company, the Deadwood and Golden Terra mines were renamed the Deadwood-Terra Mine. Recognizing there was still significant ore potential in and around the mine, Hearst rather quickly convinced Haggin that the Deadwood-Terra Mining Company needed to expand the two mills and sink a new shaft to access the ledge of ore that was plunging to the south. The Deadwood Mill and the Golden Terra Mill were both expanded from 60 stamps to 80 stamps in 1881.

With the two mills already in place, Hearst and McMaster decided the Old Brig claim would be a good location for the new shaft. Ore could be hoisted up the shaft and trammed a short distance to either mill. The Terra Shaft, also called the Old Brig Shaft (see fig. 7.8), was subsequently collared on the Old Brig claim and sunk to the 300-foot level. The shaft was later deepened to the 800-foot level, its ultimate bottom. The location for the Old Brig Shaft worked out well for the Deadwood-Terra Mining Company and the Highland Mining Company. By 1898, a portion of the ore from the Highland Mine was being hoisted at the Old Brig Shaft and processed at the Deadwood-Terra Mill (formerly called the Golden Terra). Unfortunately on July 23, 1898, the hoisting works were destroyed by fire, resulting in a $50,000 loss. A new hoisting plant was completed in 1899 at a cost of $100,000.[104]

Fig. 7.8. The Old Brig (Terra) Shaft in Terraville. On August 23, 1880, the California capitalists and their business associates incorporated the Deadwood-Terra Mining Company in New York, consolidating the Deadwood Mining Company and the Golden Terra Mining Company. Cordwood was delivered to the steam boilers in the hoist house via chutes from the Black Hills and Fort Pierre Railroad (upper right).—Photo courtesy of the Black Hills Mining Museum.

On May 19, 1881, a large underground stope in the Deadwood-Terra Mine caved to the surface as it was being drawn down, killing Thomas Green, James Farley, and John Beakley. The stope was more than 400 feet long, 60 feet wide, and 60 feet high. Unlike other stopes in the Deadwood-Terra, this one was well timbered and had been mined eleven floors high. The force of the caving rock and timber created a tremendous concussion or blast of air. The carman, who was among the survivors, said he noticed that the timbers suddenly began twisting. In an instant, there was a sudden clap of thunder followed by a blast of air that blew him and his car 40 feet down the tramway. John Sullivan and Matthew Sligo were blown down a drift on the upper level and were badly injured. Six miners working in the south end of the mine were thrown down by the concussion created by the cave but managed to escape through the Highland Tunnel to the Homestake workings on the south side of the hill.

Samuel McMaster and a crew of miners worked their way into the south end of the caved area from the Highland Mine. The Deadwood-Terra superintendent and his crew worked their way into the caved area from the north end of the stope. Rescuers found that Thomas Green had been thrown some distance by the concussion of the cave and was dead. The bodies of James Farley and John Beakley were found amidst the tangle of rock and timbers. Nine other miners were trapped in the debris. Fortunately, the nine trapped miners sustained only minor injuries and they were able to slowly dig through 30 feet of rubble to one of the relief parties. A large funeral procession was held for the three miners and was reported by the *Deadwood Daily Pioneer-Times* to be "the largest affair ever seen in the Hills."[105]

By January 1, 1888, ore reserves for the Deadwood-Terra Mine were estimated to be 743,600 tons. The production expectation for the year was 3,000 ounces of gold per month or $50,000 in income.[106]

Father DeSmet Mining Company

The Father DeSmet claim was located on June 19, 1876, on the hillside south of Deadwood Creek near Golden Gate. The location was recorded on April 24, 1877.[107] After topsoil was removed, locators observed the vein of ore to be 60 feet wide. A small open cut was developed, whereupon it was discovered that the vein of ore had increased to a width of 154 feet.[108] The Father DeSmet Mining Company was subsequently organized around the Father DeSmet and Belcher mining claims.

To the east and adjacent to the Father DeSmet claim, ore was also discovered on the Golden Gate and Justice mining claims, which led to the creation of the Golden Gate Mining Company and the Justice Mining Company.[109] David E. Nichols, George D. Haven, and possibly others owned the Golden Gate Mine. Ore from the three mining companies was initially hauled to four common arrastras that were housed in a large building constructed by the owners of the Golden Gate Mining Company.[110] The arrastras had a combined capacity of eight tons per day.[111] The Elliott Lumber Company constructed a 20-stamp custom mill on the Mineral Point claim and commenced crushing DeSmet ore on a contract basis on August 24, 1877.[112] The mill was later expanded to 30 stamps and was devoted exclusively to ore from the Father DeSmet Mine. The mill and four arrastras had a combined capacity of 83 tons per day.

The Golden Gate Tunnel, also known as the Second Level, was driven into the hillside above the Elliott Mill. Toward the end of 1877, an inclined tramway was constructed between the portal of the Golden Gate Tunnel and the mill. The tramway featured an efficient double-track system. The weight of a loaded car being lowered to the mill on one track caused an empty car on the adjacent track to be hoisted back up the incline to the mine portal.[113]

Father DeSmet Consolidated Gold Mining Company

On December 8, 1877, California investor Augustus J. Bowie Jr., acting as a representative for other California investors, purchased the assets of the Father DeSmet Mining Company, Golden Gate Mining Company, and Justice Mining Company for $400,000 from David E. Nichols, George D. Haven, William Lardner Jr., H. S. Coleman, and John Flaherty. The deal included the Father DeSmet, Belcher, Golden Gate, and Justice mining claims, Placer Claim No. 17 above Discovery, the Arrastra House, blacksmith shop, barn, tramway, and the house and lot of Nichols and Haven.[114] Bowie then sold a four-fifths undivided interest in everything he had just purchased to the San Francisco resident Archibald Borland on January 12, 1878, but for only $160,000. Borland may have financed much of Bowie's purchase. August Hemme and L. R. Graves also acquired interests in the mine at about the same time.[115]

On January 23, 1878, the Father DeSmet Consolidated Gold Mining Company was incorporated. Augustus J. "Gus" Bowie Jr. was appointed superintendent at a salary of $1,500 per month plus living expenses.[116] Archibald Borland transferred his undivided four-fifths interest in the properties to the newly formed company on February 8, 1878, for consideration of "Certificates of Stock in said Incorporated Company hereafter to be issued."[117] Bowie deeded his one-fifths undivided interest in the properties to the Father De Smet Consolidated Gold Mining Company on April 2, 1879, for $1 and other consideration.[118]

Shortly after the Father DeSmet Consolidated Gold Mining Company was formed, Gus Bowie designed a new 80-stamp mill for the company, based on a design that he figured would be economical to construct and operate. While awaiting delivery of the machinery from the Union Iron Works in California, Bowie contracted with eight custom mills in the Central City area to crush

DeSmet ore. By July 1878, the custom mills had a total of 160 stamps crushing about 240 tons of ore per day. In the interim, a mill building with three main sections was constructed on the south side of Deadwood Creek. A new inclined tramway 280 feet long was constructed from the top of the mill to the portal of a new haulage drift located above the mill.[119] The new mill was commissioned in November 1878 at a cost of $100,755.73.[120] Unfortunately, a lack of water rendered the new mill inoperable. In the interim, ore from the DeSmet Mine continued to be crushed at the custom mills in the area.[121]

Anticipating there would likely be a shortage of water, the investors in the Father DeSmet Consolidated Gold Mining Company, along with others, formed the Wyoming and Dakota Water Company on January 25, 1878. After purchasing the Wyoming and Dakota Ditch, Flume, and Mining Company, which held water rights along the east fork of Spearfish Creek, the Wyoming and Dakota Water Company started digging a ditch from the upper reaches of Spearfish Creek in Spearfish Canyon.[122]

By spring, however, there was sufficient runoff in Deadwood Creek, and the new mill was able to operate at full capacity for a few months using water from the DeSmet, Bowie, and Palmer Ditch, which was associated with a water right on Deadwood Creek upstream from the mill.[123] When available, a limited amount of water was also purchased from the owners of the Foster Ditch, who had an earlier water right on Deadwood Creek upstream from the mill. Creek water for the mill was diverted into the ditches a few thousand feet upstream from the mill.[124]

Before the Spearfish Ditch was half completed, the Wyoming and Dakota Water Company went bankrupt and was sold at a Lawrence County Sheriff's sale to Samuel McMaster on April 23, 1881. McMaster was agent and superintendent for the Black Hills Canal and Water Company, which was owned by the same California capitalists who owned Homestake and other companies along the Homestake Belt.[125]

By early 1880, the Father DeSmet Mine was in a state of turmoil. J. C. McDonald replaced A. J. Bowie as superintendent in May[126] but only remained in the position until June 25 when L. L. Alexander replaced him.[127] Archie Borland resigned as director of the Father DeSmet Consolidated Company and the Wyoming and Dakota Water Company on August 12, 1880. Another board member, George D. Haven, resigned his position on November 12, 1880, at which time Charles H. Cook was elected to replace Haven.[128]

By December 1880, J. B. Haggin and Lloyd Tevis had purchased enough stock in the Father DeSmet Consolidated Gold Mining Company to have controlling interest in the company. Over the next several months, the Homestake capitalists were successful in taking over management of the company. A short time later, J. B. Haggin was elected president.

One of Haggin's first acts as president was to find a new superintendent for the DeSmet Mine.[129] He selected Harry M. Gregg to replace L. L. Alexander in

February. David Mason assumed Gregg's position at the Deadwood-Terra Mine. In one of his first reports, Gregg informed Haggin that the mine was a good one. "I find the DeSmet Mine to be a much better mine than common report had accredited it with being. The Golden Gate vein is the most easterly vein and is at present looking first rate, is about 30 feet wide, carries large quantities of iron pyrites (or white iron we call it here). We are working near the south end line and I thought at one time the vein would keep on the Golden Gate ground and run parallel with the east side line of the Deadwood and still not get onto the Deadwood ground. But now I think the vein will run into the Deadwood ground . . . thereby adding greatly to the value of the Deadwood-Terra Mine."[130] Based on Gregg's favorable assessment of the mine, the DeSmet Mill was enlarged from 80 stamps to 100 stamps, and a new 150-horsepower boiler was installed (see fig. 7.9).[131]

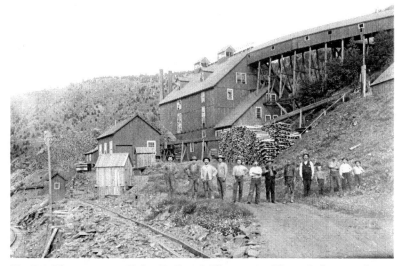

Fig. 7.9. The 100-stamp Father DeSmet Mill, ca. 1888. Through share purchases, J. B. Haggin and Lloyd Tevis gained a controlling interest in the Father DeSmet Consolidated Gold Mining Company by December 1880. The company was deeded to Homestake Mining Company in October 1901 at which time the mill was renamed the Mineral Point Mill. The mill was shut down in October 1918.—Photo courtesy of the Black Hills Mining Museum.

George Hearst, meanwhile, decided he had accomplished most everything that he had set out to accomplish for himself and his investors. Early in 1880, he decided it was time to return to his home and family in San Francisco. On January 21, it was reported that Hearst had decided to run for the governor of California.[132] After putting Samuel McMaster solely in charge of the Homestake interests in Lead, Hearst departed the Black Hills for good to reunite with his wife, Phoebe, and son, William Randolph Hearst.[133]

Under Haggin's direction, the board of directors for the Deadwood-Terra Mining Company and the Father DeSmet Consolidated Gold Mining Company voted to combine the operating management of both companies on May 1, 1886. George Webber, now superintendent of the DeSmet, was put in charge of both mines. The mines were quickly connected on the main tramway levels. Ore from the Deadwood-Terra Mine could now be hauled through an underground tunnel to the DeSmet Mill or hoisted to the surface at the Old Brig Shaft and processed in the Deadwood and Golden Terra mills. With the DeSmet Mill available, the Deadwood Mill was shut down in 1887. Its 80 stamps were added to the Golden Terra Mill in 1889.[134]

Caledonia Gold Mining Company

The Caledonia claim was located in 1877 by John B. Pearson and Frank Bryant. It was located north of the Homestake claim near the divide between the north fork of Gold Run Creek and Bobtail Gulch. Initial mining was done from an open cut and shallow underground workings. General John W. Gashwiler purchased the Caledonia Mine in 1878 for $65,000.[135]

Through Gashwiler's actions, the Caledonia Gold Mining Company was incorporated as a California corporation on September 2, 1878. The initial board of directors consisted of Sol Heydenfeldt, Henry D. Bacon, M. L. McDonald, Frank Locan, and J. W. Gashwiler, all who resided in San Francisco, California. The capital stock of the corporation was one hundred thousand shares valued at $10 million. At the time of incorporation, each of the directors was issued five shares, based on a par value of $100 per share.[136]

In April 1879, Louis Janin, who was the mining engineer Hearst and Haggin often contracted with, was asked by J. W. Gashwiler to perform an evaluation of the Caledonia properties. The Caledonia Gold Mining Company owned three claims, including the Caledonia, Queen of the Hills, and Grand Prize. Janin reported that three veins were cut by the Grand Prize Tunnel: the Discovery vein, the Blind Lode, and the Caledonia. The Discovery vein was 6-8 feet wide, the Blind Lode, 14 feet, and the Caledonia, 50 feet.

The veins were actively being developed by an open cut, a lower tunnel, and several small shafts and winzes. Janin found that the Discovery vein averaged $5 to $8 per ton, whereas the Caledonia vein averaged $7 per ton over a length of 700 feet. Based on his mine evaluation, Janin recommended to Gashwiler that a large mill of not less than 60 stamps be constructed at an estimated cost of $50,000 to $55,000. Janin figured the mill could process 3,000-4,000 tons of ore per month. Milling costs were estimated at $3 to $3.50 per ton.[137]

Based on Louis Janin's recommendations to J. W. Gashwiler, a 60-stamp mill, called the Caledonia Mill, was constructed in 1879-1880 at the head of Bobtail Gulch near the Deadwood and Golden Terra mills. George E. Webber

Jr. was named superintendent of the Caledonia Mine. He later accepted the job of general foreman of the Deadwood-Terra Mine. In 1883, he became superintendent of the Father DeSmet Mine.[138] Throughout 1881, most of the production from the Caledonia Mine was sourced from an open cut. A tramway tunnel was driven from the mill to a point below the open cut. An ore transfer raise was then driven upward from the tramway to the bottom of the cut. Ore mined from the open cut was transferred through an upper grizzly and a lower grizzly and hauled on the tramway to the Caledonia Mill.

Gashwiler himself was having financial difficulties. On July 8, 1881, the *New York Times* published an article entitled "Failure of a Large Mining Operator." The *Times* reported that "Gen. J. W. Gashwiler, the stock and mining operator, has filed a petition of insolvency. His liabilities amount to $520,000, and his assets consist of $213,000 in personal property and some real estate, the value of which is not stated." Some twelve different creditors were mentioned, with debts ranging from $10,000 to $200,000. Gashwiler attributed his failure to "the fall in the price of stocks and real estate, and the failure of some of his debtors."[139]

During a heavy rainstorm on July 20, 1881, a large volume of storm water caused a major slide in the Caledonia open cut, which covered all of the upper grizzly and part of the lower grizzly. Four men were killed, including James Roach, John Costello, Pat Hawkins, and L. H. Hamilton—the shift boss. William Gill, Daniel Cameron, and Andy Lawson, who were working at the lower grizzly, were pushed into the raise by the force of water and rock. Fortunately, they were able to crawl through some small openings between the boulders and were able to free themselves. They sustained only minor injuries.

The body of James Roach was found near the mouth of the tramway tunnel. Apparently, he had attempted to run from the mine but ended up being overcome by the onrush of water, mud, and rock. The three other men were completely buried at the upper grizzly. Their bodies were recovered individually over the next five days. The disaster closed the mine for the remainder of 1881.[140]

On the evening of October 5, 1886, disaster struck again at the Caledonia when a 50-pound box of giant powder exploded on the 300-foot level near the winze. Thirty-one miners were working on the level at the time. Four miners were killed as a result of the massive explosion, which was thought to have occurred when a spark from one man's pipe detonated the blasting caps that were stored near the powder. Five miners working some distance away were severely injured. The intensity of the explosion, which completely obliterated the shaft station, was heard throughout Terraville. Those killed included Phillip Wyman, Thomas L. Cheshire, Henry Rosevear, and John Pascoe. The largest funeral procession ever held in the Black Hills included the Lead Cornet Band, Central City Miners' Union, Odd Fellows, Knights of Pythias, mourners, and about one hundred buggies and wagons.[141]

Around 1887, Homestake Mining Company acquired operating control of the Caledonia Gold Mining Company. The Caledonia Shaft was deepened to 575 feet. The Caledonia Mill was expanded from 60 stamps to 80 stamps and recommissioned on October 1, 1889. An attempt was made to mine the large caved areas above the 400-foot level. A number of shrinkage stopes were mined as leading stopes from the 575-foot level to the 400-foot level. In October 1890, the pillars around the first shrinkage stopes were blasted, which caused the caved area above the 400-foot level to collapse unexpectedly into the newly mined stopes above the 575-foot level. In view of the unstable ground conditions in the mine, the mine and mill were shut down, and most all the employees were laid off.[142]

The Homestake Mergers

In 1893, things went from bad to worse at the Deadwood-Terra, Father DeSmet, and Caledonia mines. On April 18, fire broke out in a timber bulkhead area of the Mitchell stope on the 150-foot level of the Deadwood-Terra. Smoke and carbon monoxide filled the mine and spread quite rapidly to the Homestake and Father DeSmet mines on the 300-foot level. The three mines were quickly evacuated but had to be closed for several days until crews were able to isolate the fire by constructing bulkheads in the various drifts.

The Deadwood-Terra Mine and its mill were shut down because of unfavorable economics on August 30, 1893. The Father DeSmet Mine and the Caledonia Mine were also shut down late in the year for the same reason. For the first time since 1876, all of the mines and mills in the Terraville area were idle, and some six hundred people found themselves unemployed.

In 1894, Homestake's superintendent, Thomas J. Grier, announced that the mines could be reopened if the workers were willing to come back to work under a reduced wage scale. Grier proposed that those people who had previously been earning $3.50 per day would now receive $3, and those who had been earning $3 would receive $2.50 per day. Within a day, 275 people signed back on as employees. The Deadwood-Terra Mine and mill were restarted on August 14.[143]

At about the same time, Homestake developed a new mining plan for the Caledonia Mine, and it was reopened on a limited basis in 1894. A raise was driven from the 600-foot level of the Homestake Mine to a point below the bottom of the caved area in the Caledonia. Mining operations were then started below the cave. Ore was transported to the Old Abe Shaft, hoisted to the tramway level, and transferred to one of the three Homestake mills. Mining operations in the Caledonia ledge ultimately reached the 1,550-foot level at which point the ore was depleted. Between 1894 and the early 1930s, Homestake mined and milled approximately 5.4 million tons of ore from the Caledonia Ledge.[144]

Between 1893 and 1905, Homestake's board of directors approved the acquisition and merger of the various other companies that its principal owners already controlled. These companies included the Highland Mining Company, Deadwood-Terra Mining Company, Caledonia Mining Company, and Father DeSmet Consolidated Mining Company.

Homestake Mining Company acquired a one-year lease on all of the Highland Company's "mines, mining claims, mills, machinery, tramways . . . and all other property of every kind or nature" on July 27, 1899.[145] Through a judgment and decree, the Highland Mining Company was dissolved, and its assets were transferred to Homestake Mining Company on February 23, 1900.[146]

An agreement was reached between the Caledonia Gold Mining Company and Homestake Mining Company on January 22, 1901, whereby all of the assets of the Caledonia Company were transferred to Homestake for $500,000.[147] Similarly, on June 17, 1901, all of the assets of the Father DeSmet Consolidated Gold Mining Company were deeded to Homestake Mining Company for a consideration of $10.[148]

On February 28, 1902, the various mining claims and other assets of the Deadwood-Terra Mining Company were deeded to the Terra Incognita Mining Company for which Elliott M. West was president and William B. Cravath was secretary.[149] The board of trustees and shareholders for the Deadwood-Terra Mining Company subsequently voted to dissolve the company on March 19, 1902, and a Certificate of Dissolution was issued on April 25, 1902.[150] The Terra Incognita Mining Company deeded the Deadwood-Terra properties to Homestake Mining Company on April 9, 1902, for $200,000.[151]

Homestake now had legal title to most all of the mining claims and primary along the Homestake Belt. Other peripheral mining companies would be acquired later.

Chapter 8
Water Rights and Water Fights

Water Rights and Water Ditches

Besides needing abundant ore, owners and operators of the placer mines, lode mines, and mills in the Black Hills needed a fairly large and constant source of water to facilitate their mining and milling operations. Unfortunately, most of the richest lode claims in the Hills had little or no surface water. Consequently, the mills were commonly located on or very near the larger creeks and streams where a more steady supply of water was available to support crushing and milling operations. Ore was transported from the mine to the mill by wagon or rail, which proved to be very costly. Most of the mine and mill operators disposed of their mill tailings by simply sluicing them directly into the creeks.

Residents of townsites and mining camps faced similar difficulties in finding water suitable for domestic and firefighting purposes. Although creek water was available, it was unsuitable because of contamination from mining and milling operations located upstream. To complicate matters, townsite officials were forced to turn to the miners for water since the miners held most of the water rights. Competition and disputes over water and water rights rapidly became one of the biggest issues people faced in the development of the Black Hills.

The earliest water rights in the northern Black Hills were located on City, Deadwood, and Whitewood creeks in 1876 and 1877 by groups of miners who located placer and lode claims and mills along these particular drainages. The placer operations, lode mines, and stamp mills simply could not be worked economically without a constant supply of water. After all the available water was claimed along these three streams, the miners and other enterprising individuals expanded their search for water and subsequently located water rights on False Bottom Creek and Spearfish Creek.

The earliest water rights on Spearfish Creek and its tributaries were located by members of the Montana Expedition who established farms and ranches in the fertile lands of Spearfish Valley in 1876. Some of these individuals included Robert H. Evans, Henry Folsom, G. W. Reid, and J. B. Smith, who colocated the first water right on Spearfish Creek called the Evans-Tonn Ditch, in May 1876. The second water right on Spearfish Creek, called the Ramsdell Ditch, was located by Joseph Ramsdell, George B. Mann, Samuel J. Beck, William Seip, and Otto Uhlig in October 1876.[1]

Holders of the earliest water rights controlled virtually all available surface water from the main streams and tributaries in the northern Black Hills. Although other water rights continued to be filed over the next several decades, most were secondary in nature, which meant that holders of the secondary rights usually had insufficient water to conduct mining and milling operations. Holders of the primary water rights dug ditches and constructed flumes to divert creek water to their nearby mining claims and mill sites. Appendix A lists the principal water rights that were located in the northern Black Hills between 1876 and 1917, most of which were located for mining purposes.

A water right location typically listed the volume of water being claimed according to a unit of measure called the miner's inch. In most of the Black Hills mining districts, a miner's inch was defined as the amount of water that would flow through a 1-inch-square aperture in a 2-inch-thick plank, with a steady flow of water standing 6 inches above the top of the aperture. Except for a few adjudicated water rights, 50 miner's inches were equivalent to a flow of 1 cubic feet per second under South Dakota statutes.[2]

Beginning in 1878, several large water companies were organized to supply water to the mines, mills, and townsites on a commercial basis. Significant capital was required to purchase the various water rights, conduct topographic surveys, dig ditches, construct flumes, and divert water from the creeks to the end users. Because the earliest water rights were held by the miners, and in some cases, the farmers and ranchers of Spearfish Valley, the townsites and mining camps had to rely on the water companies to supply them with water for domestic and firefighting purposes. Demand for the limited amount of water was great. The mining companies, who controlled the water companies, made sure the mines and mills got their water as first priority. By the mid-1880s, water ditches were constructed from every major tributary in the northern Black Hills (see figs. 8.1 and 8.2). Most of the ditches and a majority of the water rights and ditches were acquired by Homestake Mining Company or its agents by 1890. A discussion of each of these companies, the water rights they held, and the ditches they constructed follows.

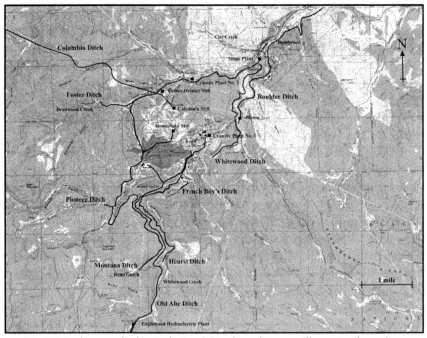

Fig. 8.1. Early water ditches in the Lead-Deadwood Area.—Illustration by author.

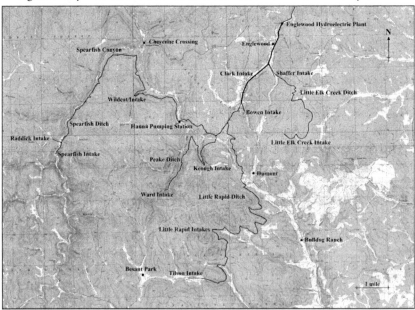

Fig. 8.2. The water ditches south of Lead. Beginning in 1906, water from the Spearfish, Peake, and Little Rapid ditches was piped into a high-pressure pipeline that supplied water at 210 pounds per square inch to drive the 400-kilowatt generator at the Englewood Hydroelectric Plant.—Illustration by author.

Wyoming and Dakota Ditch, Flume, and Mining Company and the Spearfish Ditch

On June 4, 1877, three different water rights were located on Spearfish Creek. W. L. Kuykendall, Moses Thompson, and James F. Berry located a water right, claiming all of the available water in the "main left fork of the east fork of Spearfish Creek," which amounted to about 800 miner's inches. The same three men located a separate water right claiming all of the available water in the "main right fork of the east fork of Spearfish Creek," which amounted to about 100 miner's inches. J. R. Whitehead, O. B. Thompson, and Alexander Benham located another water right, claiming all of the available water in the "east fork of Spearfish Creek," which amounted to about 500 miner's inches. These three water rights locations—located one-half mile above Raddick Gulch on main Spearfish Creek, in Raddick Gulch, and on the east fork of Spearfish Creek above Hanna, respectively—later became known as the Spearfish Water Right.[3]

The six men who discovered the Spearfish Water Right then decided to form a corporation named the Wyoming and Dakota Ditch, Flume, and Mining Company. The company was incorporated in the territory of Dakota on July 23, 1877, with all six men serving as directors.[4] The men subsequently deeded their respective interests in the three water rights to the corporation on August 13, 1877, for $84 per water right. At a meeting of the stockholders on November 10, it was decided to sell the company to Coll Deane of Oakland, California, for $7,000.

Under terms of a purchase agreement between the Wyoming and Dakota Ditch, Flume, and Mining Company and Deane, a new company was to be organized under the laws of California. The agreement called for constructing a flume from the east fork of Spearfish Creek to a point opposite of and south of Central City at an estimated cost of $60,000. The intended purpose of the ditch, as people would soon find out, was to provide a reliable source of water for the Father DeSmet Mill and the surrounding communities. The plan was to dissolve the existing company upon finalization of the sale. With approval from the board of directors of the Wyoming and Dakota, Ditch, Flume, and Mining Company on December 26, 1877, the purchase agreement was finalized and crews began constructing the flume.[5]

The *Black Hills Daily Pioneer* soon learned of the project and included the following article in the December 30 edition of the paper:

> The Wyoming and Dakota Flume Company are vigorously prosecuting the stupendous work of constructing a flume and ditch from Spearfish [Creek] to Deadwood and neighboring gulches, and hope to have the work completed by September next. The ditch when finished will be eighteen miles in length, three and a half feet deep and four feet wide, and will cost about $80,000. By it an abundant supply of water

will be secured for mining and all other uses and purposes. It will also be utilized for floating in timbers to supply the great demand that in a year's time will prevail throughout the district. The enterprise was projected, and is being prosecuted by California capitalists.[6]

Wyoming and Dakota Water Company

As Coll Deane was starting construction of the Spearfish Ditch, the new company, called the Wyoming and Dakota Water Company, was organized and incorporated under the laws of California on January 25, 1878. Five directors were named, two of whom included August Hemme, who was elected president, and Archibald Borland as vice president. Coll Deane, Moses Thompson, and William Littlebury Kuykendall were the other three directors. Kuykendall was named superintendent and agent for the company. Hemme was listed as a resident of San Francisco, California; Borland and Deane, of Oakland, California; and Kuykendall and Thompson, of Deadwood, Dakota Territory. Hemme, Deane, and Borland were listed as the initial holders of the capital stock in the company. Each held twenty thousand shares valued at $1 million, based on a par value of $50 per share.[7]

Some of the directors of the Wyoming and Dakota Water Company already had a financial or physical presence in the Black Hills. Archibald Borland and August Hemme were principal owners of the Father DeSmet Consolidated Gold Mining Company. William Littlebury Kuykendall was presiding judge over the Miners' Court trial when Jack McCall was acquitted by the jury for the murder of James Butler "Wild Bill" Hickok. The jury believed McCall's fabricated story that Wild Bill had killed McCall's brother.[8] Kuykendall was also elected probate judge for Lawrence County in November 1877.[9]

The Articles of Incorporation state that the purpose of the Wyoming and Dakota Water Company was to

> supply all Cities, Towns, and Counties of Wyoming Territory and Dakota Territory with fresh and pure water, by conducting and conveying the waters of the East fork of Big Spearfish Creek, or any other waters, into, and distributing the same by means of aqueducts and pipe through the streets and buildings of said Cities and Towns and other places, and selling the said water, to use said waters in conveying lumber of all descriptions by means of floating the same through canals and ditches, and generally to use said waters, and to acquire such real estate and other property as may be necessary for the purposes of the Company, and to grant, bargain, sell, and convey the same or any part or portion thereof.[10]

At a board of directors meeting held on December 6, 1878, a resolution was passed to revoke the power of attorney previously granted to W. L. Kuykendall. The board subsequently named A. J. "Gus" Bowie Jr. agent for the Wyoming and Dakota Water Company.[11] Bowie was also named superintendent of the Father DeSmet Consolidated Gold Mining Company at the time.

Shortly after its inception, the Wyoming and Dakota Water Company began procuring various water rights from others in an effort to source much-needed water for the Father DeSmet Mine. The three water rights on Spearfish Creek were already being developed through construction of the Spearfish Ditch from the headwaters of Spearfish Canyon. Certain other water rights were acquired in 1878 and 1880, which led to the construction of the DeSmet, Bowie, and Palmer Ditch and the Little Elk Creek Ditch.

DeSmet, Bowie, and Palmer Ditch

On November 19, 1878, Gus Bowie and Palmer Smith located a water right on Deadwood Creek approximately one mile above the confluence of Deadwood and Poorman Gulches. A separate water right of 250 miner's inches was located by J. A. Pierce and James Sapsley on October 18, 1878, about one-quarter mile above the Oro City Sawmill on Deadwood Creek "at the big bend." Water associated with these two water rights was directed into the DeSmet, Bowie, and Palmer Ditch, which became the primary source of water for the Father DeSmet Consolidated Gold Mining Company prior to 1881 when the company came under the control of J. B. Haggin and Lloyd Tevis.[12]

Despite having the rights to the DeSmet, Bowie, and Palmer Ditch, Archie Borland and Gus Bowie found themselves short of water for the Father DeSmet mine and mill. The water right associated with the Foster Ditch, an earlier right of 250 miner's inches held by Captain W. M. Foster, took most of the water available in upper Deadwood Creek. Recognizing the value of the Foster Ditch and its water rights to the DeSmet Company, George Hearst instructed Samuel McMaster to purchase the rights to the Foster Ditch, which he did in September 1878.[13]

Little Elk Creek Ditch

J. J. Buchanan, Edward Preble, and James J. Sutherland discovered the Little Elk Creek water right, consisting of 1,000 miner's inches on November 3, 1877. Acting as agent for the Wyoming and Dakota Water Company, A. J. Bowie was able to purchase this water right from various individuals in February 1880. Unfortunately, the water right was situated 2-3 miles southeast of Ten-Mile Station, later renamed Englewood, and two drainages away from the Father DeSmet Mine.[14]

Black Hills Canal and Water Company

Soon after the Homestake claim and the south portion of the Golden Terra claim were purchased by Homestake Mining Company, George Hearst recognized that he and his other California capitalists would have to begin making some strategic moves to acquire additional water rights in order to have sufficient water for their mines and mills along the Homestake Belt. Hearst's partners concurred, but decided that a separate company should be formed to manage the water issues. As a result, the Black Hills Canal and Water Company was formed under the laws of California on September 30, 1878. The stated purpose of the new company was to "acquire water rights . . . sell and supply water to mines, mills, towns, and cities and other consumers, for domestic, stock, agricultural, mining, mechanical, and manufacturing purposes."[15]

The first board of directors for the Black Hills Canal and Water Company included J. B. Haggin, George Hearst, William Willis, Joseph Clark, and A. E. Head, all of San Francisco, California. The capital stock was $1,000,000, which was divided into one hundred thousand shares with a par value of $10 per share. The five directors were the named subscribers to the stock, each holding five shares.[16]

On October 14, 1878, the board of directors of the Black Hills Canal and Water Company unanimously appointed Samuel McMaster as superintendent and agent for the company.[17] Hearst and Haggin were already in the process of buying all the water rights and ditches that they could, including the Pioneer Ditch, Montana Ditch, and other water rights on Whitewood Creek below Ten-Mile Station.

George Hearst kept close tabs on the actions of Archie Borland and the Wyoming and Dakota Water Company. In a letter dated March 6, 1879, to J. B. Haggin, Hearst wrote as follows:

> *I know that Bowlin [Borland] has not yet examined the proposed route of the Spearfish Ditch and I don't suppose that he has as yet determined what to do with the premises, but the prevailing opinion here is that the ditch can't be brought in any shape short of two hundred and fifty thousand dollars and to make winter water of it, it will cost as much more. Our present water rights, if they can be maintained, as I think they can, will supply at least a thousand stamps which will be sufficient for some years to come and perhaps for all time, but for the purpose of peace and harmony I will probably propose to sell to Bowlin [Borland] 1/5 of our stock in the water Co. at the neighborhood of cash.*
>
> *We can't very well overrate the value of our water rights and should the Spearfish water not come in, I believe our water to be worth as much as any other incorporation in the Hills as it requires so little attention*

and there is so little wear and tear that two men at the outside can attend to the whole thing the whole year around and consequently it is nearly all profit and everything else in the country must be exhausted before this water ceases to be a source of revenue as it is the only source of water for milling, mining, and drinking purposes for a large extent of the country in case I fail to pull off the Father DeSmet from interference in our water rights.[18]

The *Black Hills Daily Times* also kept track of progress on construction of the Spearfish Ditch. In the April 10 issue of the *Times*, it was reported that "work is progressing on the Spearfish ditch, and as soon as low freight rates prevail on the roads from the east, says the *Herald*, the necessary siphon material will be brought in from Chicago. It is thought that enough water can be brought in regularly to use as motive power in various localities, if required, and vast quantities of wood and possibly timber will be floated through the siphons."[19]

Between 1876 and 1879, agents of the Black Hills Canal and Water Company acquired various water rights. The primary ditches and associated water rights included the Pioneer Ditch, French Boys Ditch, Montana Ditch, and Foster Ditch.

Pioneer Ditch

On March 4, 1876, R. A. Webb and R. McLennan located a water right for 300 miner's inches on Whitetail Creek downstream from the mouth of Nevada Gulch. After completing a survey of the proposed route for the ditch to Lead City, the two men determined that the water right would have to be moved upstream to the confluence with Nevada Gulch. The men then constructed an intake dam and extended the Pioneer Ditch from Lead City to Nevada Gulch. Water was turned into the ditch in August 1876.[20]

Over the next several months, Webb and McLennan sold various undivided interests in the water right to John Belding, Ira Myers, W. H. Bull, James Karney, and H. W. Barber. Water from the Pioneer Ditch was initially conveyed to placer claims situated on the upper forks of Gold Run Creek. By late December 1877, J. B. Haggin had acquired most of the rights associated with the Pioneer Ditch. Under Haggin's control, the Pioneer Ditch supplied water to the Homestake 80-stamp mill, the White 30-stamp mill on the Homestake No. 2 claim, the Racine 20-stamp mill, and the Old Abe 20-stamp mill in October 1878. The charge for water was $2 per stamp per week, based on an allowance of one miner's inch of water per 5-stamp battery. After the Golden Star and Highland mills were completed in 1879 and 1880, the Pioneer Ditch and later, the Hearst Ditch supplied water to these mills.[21]

A small wooden-box flume about 4 inches square carried water from the Pioneer Ditch near the head of Poorman Gulch to the Deadwood and Golden Terra mills at the head of Bobtail Gulch in Terraville.[22] This section of flume was completed on November 10, 1878.[23] Haggin deeded his rights to the Pioneer Ditch to the Black Hills Canal and Water Company on June 7, 1897.[24]

French Boys Ditch

Henry Harney, Fred Manuel, Moses Manuel, and Alexander Engh located a water right of 400 miner's inches on May 10, 1876. The location was posted on Whitewood Creek approximately 700-1,000 feet above the confluence of Whitetail Creek and the mining camp of Pennington. The location notice was about 2½ miles above the intake for the Boulder Ditch. The four men then dug a ditch 1,000 feet long and constructed a water-powered arrastra on Whitewood Creek approximately 300 feet above the mouth of Whitetail Creek. Upon completing the work in September 1876, the four men commenced crushing ore from the Homestake claim in their arrastra.[25] Engh subsequently sold his one-quarter interest in the water right to Samuel McMaster on May 19, 1879. The Manuel brothers sold their one-half interest in the water right to McMaster on July 16, 1879. Harney did the same with his one-quarter interest on October 12 of that year.[26]

Montana Ditch

Two separate water rights locations were associated with the Montana Ditch. On June 16, 1877, Aron M. Bell and John M. Roberts located 600 miner's inches on the Reno Gulch fork of Whitewood Creek below Ten-Mile Station. The water right was located approximately 8 miles upstream from the head of the Boulder Ditch. Frank Abt, Frank Distelhurst, and John Gudts located 800 miner's inches from Whitewood Creek on January 16, 1878, at a location approximately 2 miles above the location filed by Bell and Roberts. Collectively, the owners of the two water rights formed the Montana Ditch Company and started digging the Montana Ditch to convey water to the head of the Pioneer Ditch on Whitetail Creek.[27]

After spending about $5,000 on construction of the Montana Ditch, the owners of the Montana Ditch Company sold their two water rights and the partially completed ditch to George Hearst and J. B. Haggin on July 17, 1878.[28] Immediately, Haggin and Hearst commenced completion of the Montana Ditch. Haggin and Hearst deeded the water rights associated with the Montana Ditch to the Black Hills Canal and Water Company on June 7, 1897.[29]

Foster Ditch

Captain W. M. Foster located the water right associated with the Foster Ditch on April 24, 1876. Foster claimed a water right of 250 miner's inches on Deadwood Creek, about one-quarter mile above the confluence of Deadwood and Poorman gulches. After developing the Foster Ditch primarily for the benefit of placer miners and custom mills along Deadwood Creek, Captain Foster sold a two-thirds interest in the water right to Wilson Ralston and Oliver Bidwell on July 13, 1877. Foster then sold his remaining one-third interest to George Moore on August 14, 1877. Ralston, Bidwell, and Moore then sold their interests to O. B. Thompson who, in turn, sold the Foster Ditch to William Frackelton on September 12, 1878. Two days later, on September 14, Frackelton sold the Ditch to Samuel McMaster.[30]

A. J. "Gus" Bowie, superintendent of the Father DeSmet Consolidated Gold Mining Company, evidently believed all along that the asking price for the Foster Ditch was too exorbitant and that he could sustain the DeSmet operations using groundwater from the mine. But since the water right associated with the Foster Ditch was superior to the rights associated with the DeSmet, Bowie, and Palmer Ditch, virtually no excess water was available for the DeSmet mine and mill from upper Deadwood Creek. Bowie probably figured no one else would want to pay the price for the Foster Ditch, particularly since the placer operations and custom mills along Deadwood Gulch were going out of business one by one. He thought he could simply wait and force a better price.

Samuel McMaster believed differently, as evidenced by his purchase of the Foster Ditch. Captain Foster was so elated that the DeSmet Company had missed out on the water that, after imbibing in a little liquor one day, Foster let Gus Bowie know exactly what he thought of the superintendent of the DeSmet Company. As it turned out, the amount of groundwater available on the DeSmet property played out, and Bowie didn't have sufficient water to run the new DeSmet Mill when it was commissioned in November 1878. Immediately, Bowie contacted George Atchison, president of the Boulder Ditch Company, and successfully negotiated a purchase of the Boulder Ditch in March 1879 on behalf of the Wyoming and Dakota Water Company.[31]

The Deadwood Water Question

On June 1, 1876, George Atchison located a water right consisting of 1,000 miner's inches on Whitewood Creek on Placer Claim No. 45, just below the mouth of Gold Run Creek. Atchison commenced to construct approximately 2 miles of ditch along the east side of Whitewood Creek to furnish water to placer claims along lower Deadwood Gulch and ultimately to placer claims in Boulder

Gulch [Boulder Canyon]. Toward the latter part of August, Atchison amended his location by posting a second notice on Whitewood Creek approximately three-quarters of a mile above the confluence with Gold Run Creek. Atchison then extended the ditch to the new location. The ditch became known as the Boulder Ditch.

Boulder Ditch

Water was diverted through the newly completed Boulder Ditch around June 7, 1877. The uncovered ditch was about 2-3 feet wide on the bottom, 3-4 feet wide on top, and 18 inches deep, providing a capacity of about 400 miner's inches. During the summer months of 1877, about 1 miner's inch of water was sold to the residents of Deadwood for bathing and livery purposes. Between the summer of 1877 and September 1878, an additional 150-200 miner's inches of water was sold to placer miners on Whitewood Creek as far downstream as Split Tail and Smith's gulches. At about the same time, Atchison sold three one-sixth interests in the Boulder Ditch water right to A. M. McKinney, Kirk G. Phillips, and William Hartley. He also sold a one-third interest to John Kane and retained a one-sixth interest for himself.

Meanwhile, upon purchasing the two water rights on upper Whitewood Creek and that portion of the Montana Ditch that was partially completed, the Homestake men started work in July 1878 to complete the Montana Ditch. Upon completion of the $25,000 project, which included a wooden flume 6 miles long, water was diverted into the head of the Montana Ditch at a point on upper Whitewood Creek just below Ten-Mile Station. Water flowed through the Montana Ditch and passed into the head of the Pioneer Ditch near the confluence of Nevada Gulch and Whitetail Gulch. The Pioneer Ditch conveyed water to the placer claims and mills that were situated along the forks of upper Gold Run Creek in Lead City.[32] The flume, as constructed, had a capacity of approximately 89 miner's inches.[33]

New French Boys (Old Abe) Ditch

On August 18, 1879, J. B. Haggin, George Hearst, and Samuel McMaster reposted the water right to the French Boys Ditch some 2-3 miles upstream from the original location on Whitewood Creek. The original location, 700-1,000 feet above Pennington, was simply too far downstream on Whitewood Creek and lower in elevation, compared to Lead City, to flow water by gravity to the mills in Lead City. This relocation of the water right, a questionable legal move by Hearst, Haggin, and McMaster, was done to gain elevation so that the water from Whitewood Creek would flow to the mills in Lead City and Terraville via the Pioneer Ditch.

In 1879, Hearst, Haggin, and McMaster completed the Old Abe Ditch. The head of this ditch was located approximately three-fourths of a mile below the head of the Montana Ditch on Whitewood Creek. The Old Abe Ditch was routed to intersect the lower section of the Montana Ditch. With this ditching arrangement, Hearst, Haggin, and Tevis were able to divert 600 miner's inches of water from upper Whitewood Creek. Unfortunately, for about six months of the year, the entire flow in Whitewood Creek only amounted to about 200 miner's inches. Under these conditions, there was virtually no water left in Whitewood Creek for Atchison and his partners for their Boulder Ditch. As a result, Atchison and his partners filed suit against George Hearst, J. B. Haggin, Samuel McMaster, and two unnamed defendants.[34]

Hearst now needed another legal opinion from J. B. Haggin. A telegram was sent, followed by a letter on May 19, 1878, as follows:

> *I telegraphed you in relation to water rights. The question is this, the Boulder Ditch takes out of Whitewood Creek a ditch at a given point that will run, say, 1,000 inches of water, but will not run water to our place on Gold Run as the place where they tap the creek is some 500 feet [lower].*
>
> *As you know our water comes from a point on Whitetail to Gold Run, Lead City, Homestake, etc. Now it is to our interest to own the water that comes here . . . . I find the Boulder Ditch Co. [also] sells the right to take out one hundred inches of water at a given point several miles higher up said Whitewood Creek, where said water will come to Gold Run which will materially interfere with us. This selection or claim to take the water out of said Whitewood Creek was made in January last, but during last summer some parties took up a right at a given point on Whitewood that would run the water to Gold Run, Lead City, etc. This location is not as old as the Boulder Ditch location, but is older than the second one [by the Boulder Ditch Company] . . . that they claim they can run the water to Gold Run, Lead City, etc. Now can the Boulder sell a water right and take it out in a younger location than the one spoken of above?[35]*

On March 10, 1879, as Atchison's lawsuit against Hearst et al. was underway, all of the interests in the Boulder Ditch and its associated water rights were purchased by Augustus J. Bowie, superintendent and agent for the Wyoming and Dakota Water Company. The *Black Hills Daily Times* reported on April 10 that "the Boulder Ditch has been purchased by this [Wyoming and Dakota Water] company, costing them, with a few necessary improvements, the sum of $50,000. This ditch takes water from Whitewood at a distance of three miles to the south of Central City. The company proposes to hydraulic certain claims below Deadwood first, but the hills all along the watershed between Deadwood and Whitewood will also be worked in due season."[36]

Bowie and Borland's purchase of the Boulder Ditch on behalf of the Wyoming and Dakota Water Company gave the two men significant bargaining power against George Hearst. The Boulder Ditch had an earlier right to water from Whitewood Creek, compared to the Montana Ditch and several other water rights Hearst had acquired along upper Whitewood Creek. Hearst needed water from Whitewood Creek for the benefit of the Homestake Company as well as the other companies along the Homestake Belt controlled by Hearst, Haggin, and Tevis.[37]

As a principal owner of the Father DeSmet Consolidated Gold Mining Company and the Wyoming and Dakota Water Company, Archie Borland may have rationalized the purchase of the Boulder Ditch as a way to avenge George Hearst for McMaster's purchase of the Foster Ditch the previous fall. The Foster Ditch, because of its earlier location, had primacy to the limited amount of water in upper Deadwood Creek over the DeSmet, Bowie, and Palmer Ditch. The Boulder Ditch, on the other hand, was located on Whitewood Creek, just above the mouth of Gold Run and couldn't supply water to the DeSmet Mill.

In a letter dated March 19, 1879, Hearst informed Haggin that "Archie Bowlan [Borland] is here yet and we have not as yet come to any understanding about the water arrangements. He is looking around, I think, trying to find some weak point for the purpose of attacking us to force a compromise to his advantage, but I think the more he looks around, the more he will be convinced that we are pretty well protected and he will be more satisfied to make the best arrangement he can with us in relation to water or otherwise, but if he acts differently we will try and clean him out and feel sure we can do so."[38]

Hearst was confident and defiant. On March 29, he provided Haggin with an update on the water rights issues, saying, "Everything here is looking well. Archie Bowlan [Borland] & Co. are doing everything they can both in the way of water and outside titles, I think for the purpose of getting a compromise out of us by which they can make something as he thinks he can play Gashwiler on us once more, but I have made my mind up to fight it out this time, no more compromises and feel sure we can take care of him and our interest. I think our property here is of more value than you realize and I have my doubts, whether it would not be better to keep the property as we now have it than to sell it for prices we can now get for it."[39]

Now that the Wyoming and Dakota Water Company owned the Boulder Ditch, A. J. Bowie and Archie Borland decided to join in the legal battles against George Hearst and company. A new lawsuit was filed to enjoin Hearst from polluting the Boulder Ditch with tailings from the Homestake mills in Lead City.[40] McMaster and Hearst responded to the charge, claiming they had the right to the water from Whitewood Creek as long as they returned the water to the creek above the intake dam for the Boulder Ditch.

To help mitigate the issue and bolster their position, Hearst and McMaster constructed settling dams on Gold Run Creek below the stamp mills. Hearst's attorneys took a more aggressive position, arguing that the Wyoming and Dakota Water Company should lose the water right associated with the Boulder Ditch on the basis that the Wyoming and Dakota Water Company wasn't showing any beneficial use of the water right.

Bowie responded to the accusation by acquiring a mining claim near Split Tail Gulch and assigning a man by the name of Pete McDonald to run a placer operation by using water from the Boulder Ditch. In addition, George Stokes, who was in charge of the Boulder Ditch for the Wyoming and Dakota Water Company, was instructed to find other uses for the water. He ended up selling some water to a laundry in Deadwood. Finally, Borland and Bowie came up with a better plan to show beneficial use for the Boulder Ditch water right: the Wyoming and Dakota Water Company would offer free water to the city of Deadwood.[41]

On May 23, 1879, Palmer Smith, acting superintendent of the Wyoming and Dakota Water Company, presented a written proposal to the Lawrence County Board of Commissioners to establish a water system within the city of Deadwood. Palmer proposed to furnish "clear and pure" water for "fire and all other public purposes" for a period of twenty years. The proposal called for the water company to construct reservoirs having a combined capacity of 750,000 gallons, bury water mains four inches and six inches in diameter from the reservoirs to and throughout the town, and erect fifteen fire hydrants at designated locations.

In return, the Wyoming and Dakota Water Company would receive compensation of one mill on each dollar of assessed value of taxable property from the city of Deadwood or the county of Lawrence. The proposal also stated that the water company would have the exclusive right to furnish water to the inhabitants of Deadwood at reasonable rates. The proposal sounded like a good one to the board of commissioners, so a city election was scheduled for June 3.[42]

In the election, the citizens of Deadwood accepted the water proposal with a 499-25 vote.[43] At a meeting of the county board of commissioners on June 5, representatives of the Black Hills Canal and Water Company appeared before the board to contest the election. The representatives said the commissioners were free to accept or reject the proposal but insisted that the terms of the proposal were such that the board of commissioners could not legally formulate a contract in conformity with the proposal. The representatives of the Black Hills Canal and Water Company claimed the "so-called election and vote were without authority, and not binding on anyone."[44]

The Black Hills Canal and Water Company then proposed to the board of commissioners that the company would construct a similar water system within Deadwood free of charge, supply Deadwood with water for firefighting purposes free of charge, and complete all work within ninety days. With the proposal, the Black Hills Canal and Water Company would have the exclusive right to supply domestic water to the residents at reasonable rates, and in the event of a disagreement between the water company and the board, the rates would be fixed by a separate commission.[45] After debating the situation, the board of commissioners decided on June 9, 1879, to solicit new proposals from interested parties and schedule a new election. The *Black Hills Pioneer* reported the resolution of the county commissioners on June 15 as follows:

> *Resolved, that the board of county commissioners submit to the legal voters of the town of Deadwood, on the 21st day of June, 1879, the question of whether it shall make a contract for water or not, and that the question to be submitted shall be "water contract" or "no water contract," and that a "short and accurate description" of all proposed contracts to be made with the board, shall be published in the Pioneer and Times newspapers during the time from now until the election, and that the voters shall designate on their respective ballots, which of the proposed contracts they will adopt, and that the proposed contract receiving the largest number of votes shall be the one to be made with the board. All the proposed contracts shall be submitted to the board on or before the 10th day of June 1879 at 4 o'clock p.m.*[46]

New proposals were received from the Wyoming and Dakota Water Company (WDWC), Black Hills Canal and Water Company (BHCWC), and Judge E. H. Barron, president of the City Creek Water Company (CCWC). The latter company was the entity that had been supplying water to the city of Deadwood. The proposal from the WDWC was amended to essentially match that presented by the BHCWC. Fifteen hydrants and water for the extinguishment of fires would be supplied free of charge; reservoirs would be not less than 150,000 or more than 750,000 gallons in size; domestic water rates could be fixed by a separate committee, if the parties could not otherwise agree; and the water system would be completed within four months of the date of having a signed contract. E. H. Barron proposed constructing a 1.5-million-gallon reservoir for fire and domestic purposes. Compensation of $5,000 would be made annually for the fire system.[47]

The Deadwood Water Question, as it was referred to, caused a great furor and much consternation within the community, with the citizens and even the newspaper editors taking sides on the issue. Al Merrick, manager of the *Deadwood Pioneer* sided with McMaster, who, in turn, provided financial support to the newspaper. Deadwood's postmaster, Dick Adams, became editor of the *Deadwood Pioneer* and went to work forming public opinion in support of the BHCWC. A. J. Bowie turned to Porter Warner for support from the *Evening Times*, but Warner showed reluctance. Bowie then purchased his own printing press and asked George W. Stokes, manager of the Boulder Ditch, to edit the new newspaper, which was called the *Evening Press*. Major Snyder and Gene Decker helped Stokes by writing in support of the WDWC.[48]

Amidst the furor, Judge E. H. Barron decided to withdraw his proposal on behalf of the CCWC. Samuel McMaster wrote his own letter on behalf of the BHCWC. McMaster assured the citizens of Deadwood that the water company had full intention in being able to supply the water. McMaster stated that "the company now has a ditch for conveying the waters of Whitetail and Nevada Creeks within about one mile of Deadwood with the exception of a portion of a tunnel being constructed near the Deadwood and Father DeSmet mines, which will be completed within a few days." McMaster went on to say that "with this ditch can easily be connected the Whitewood and Deadwood ditches if necessary." McMaster questioned how the WDWC could possibly furnish water to Deadwood, claiming the water "certainly cannot come from Spearfish [Creek] for years to come." He then requested that the citizens of Deadwood appoint a committee to examine the water supplies of the BHCWC and the means for delivering the water to Deadwood.[49]

The following letter by R. M. appeared in the June 13 issue of the *Black Hills Daily Pioneer*:

> *The people of South Deadwood are becoming enthusiastic, taking strong grounds on the water question. The meeting at the checkered front will be ordered by several prominent citizens who take the position that the Wyoming and Dakota Water Company has the best water, the largest supply, and at the greatest heighth, and are, therefore, entitled to the first consideration. The W. and D. Water Company made their proposition in good faith, it was accepted by the board of county commissioners in good faith, voted upon by the citizens in good faith, and carried in good faith, and the contract is rightfully theirs. Theirs is in the interest of the community and common fairness.*
>
> *The whole maneuvering of the Homestake Company and their honey-fugeling about the commissioners carries the stamp of bad faith on the face, and such is in keeping with that company's reputation. It is a bull-dozing, pumping, dangerous monopoly, grasping and selfish in the extreme, totally regardless of all rights or interests, life or liberty, but their own, which we propose to show up clearly at our meetings and through the press before the day of election.*
>
> *The Homestake Company is controlled by sly, artful, unscrupulous men whose moto is rule or ruin. Their game at present is to stand aloof, put on a show of sanctimonious indifference for the purpose of informing the W. & D. W. Co. and the public that they are simply acting in a true line of duty, that they have made a proposition to supply Deadwood with water, and if the people accepted it, all right, and if they rejected it, the*

company will be just as well satisfied. But this apparent indifference is a sham, a trick to deceive the people and lure them into inactivity, while the company, in the meantime, are energetically fortifying every position, carefully laying their plans so that on the arrival of Hearst to put on a few finishing touches at the eleventh hour, they can turn loose their dogs of war and with the free use of money sweep the field.[50]

The citizens and leaders of Deadwood responded quickly to McMaster's request for a committee to examine the BHCWC water. The *Black Hills Daily Pioneer* reported that "a committee of eight of our best citizens has been appointed to look into the matter of whether or not the Black Hills Canal and Water Company have what they claim." The committee included C. H. Wagner, H. B. Young, Henry Munday, Ben Holstein, J. A. Harding, Hugh McCaffrey, Nathan Coleman, J. A. Gaston, and W. Stillwell. In addition, the chief engineer of the Deadwood Fire Department appointed Dr. C. W. Meyer, Seth Bullock, H. B. Beaman, John H. Burns, and L. R. Graves to accompany the citizen's committee in the examination.[51]

Opponents to the proposal by the WDWC to supply water through the Boulder Ditch claimed that its water would not be pure and clear since the ditch passed below the Mount Moriah cemetery. The headline of one newspaper read, "How'll You Take Your Water, with Graveyard Seepage?"[52]

On the day of the election, the water committee's findings were published in the *Black Hills Daily Pioneer*: "We, the committee solicited by the Homestake company, Lead City, to ascertain if said company had water sufficient to supply Deadwood for fire and domestic purposes, would report that they visited the head of the Whitetail ditch, also Whitewood ditch, and if said company own and control the water flowing in said ditches, as represented by their agent, your committee believe they have an abundant supply of good water."[53]

The election was held on June 21 with the BHCWC prevailing over the Deadwood Water Question. The WDWC carried the Deadwood Precinct, but the BHCWC carried the South Deadwood Precinct by a wide margin. The *Black Hills Daily Pioneer* reported that the election was one of the most vehemently contested elections ever held in the Hills. The newspaper claimed that "every rounder and striker in the city had been provided with whiskey money the previous night, and they presented themselves yesterday morning in good, vigorous electioneering trim, with sufficient funds still remaining in their pantaloons to provide the elixir during the heat of the contest." The newspaper reported on the election as follows:

> The Homestake company had about one hundred men in abeyance which they intended to run in on the last quarter, but the strikers for the DeSmet company took possession of the polls, thus absolutely preventing

> all those enlightened American citizens from "exercising their God-given prerogatives." This was outrageous, for many of them had walked all the way from their homes on the Redwater to earn an honest dollar. But they were blocked out. The DeSmet men were too many. Sardines in a box were never packed like those DeSmet fellows were for ten feet on either side of the polls and for twenty feet back into the street. Many leading citizens were thus prevented from voting, and by resorting to this dodge the DeSmet folks were able to carry the election in the Deadwood precinct. They also had strikers down from Central for the purpose of bull-dozing the timid taxpayers of this city. But in South Deadwood, where every man was allowed to vote according to his own notion, the Homestake company won by a large majority.[54]

Immediately after the election, the BHCWC began work to supply water to Deadwood. Instead of supplying water from Whitetail Creek and Nevada Creek via the Pioneer Ditch, as promised, the company laid mains from City Creek and the reservoir on Forest Hill.[55] About the time the new water system was completed, fire broke out in the Empire Bakery on the morning of September 26, 1879.[56] Attempts were made to use water from the City Creek and Forest Hill system, Whitewood Creek, and the Boulder Ditch. By noon, however, most of Deadwood's business district had been destroyed.

The fire confirmed that the City Creek supply was woefully inadequate for firefighting, much less for domestic purposes. Wintertime conditions would result in even less water. Although the BHCWC had fulfilled its obligation to construct water mains, fire hydrants, and reservoirs with a combined capacity of 150,000 gallons, City Creek could only supply 3 miner's inches of water, which resulted in huge shortages.[57]

On September 29, just three days after the fire, well over 150 citizens and businessmen from the city of Deadwood signed a petition and forwarded the same to A. J. Bowie Jr., agent of the WDWC, asking for water from the Boulder Ditch. "Having become entirely satisfied that the citizens of Deadwood must look for future water supply for the protection of property in this town, to your company we now ask you to cause to be laid a six inch main along our principal street and to be caused to be connected therewith a suitable number of hydrants at the principal street corners, so that we may have the protection of the waters of Boulder Ditch; and we pledge ourselves to pay for such protection and the use of said water such reasonable rates as your company may charge."[58] No action was taken by the WDWC.

Two weeks after the Deadwood fire, the editor of the *Black Hills Daily Times* questioned why, prior to the election, the BHCWC led the citizens of Deadwood to believe that abundant water would be supplied from Whitewood Creek; but after winning the election, however, water was supplied from City Creek:

> But now, we wish to call the attention of our citizens to a fact that has, during the haste and excitement of the past two weeks, been entirely forgotten. It is this: That at the time the water proposals were before our citizens that no one expected, believed or proposed that the water would come or be furnished from City Creek. In the discussion then made in the public press and on the streets, City Creek water was entirely ignored and passed by. Why? For the very simple reason that this entire community believed that such a supply from that source was an impossibility, impracticable and out of the question. The advocates of the Black Hills Canal and Water Company knew that any claim to popular favor based upon the City Creek water would be disastrous to their proposal and they would be beaten ten to one.[59]

The Deadwood fire and inadequacy of the water system installed by the BHCWC also got the attention of the citizens and businessmen of Central City. On September 30, 1879, a general meeting of the residents of Central City was held in the Langrishe Theater in Deadwood to discuss a proposal by the WDWC to supply pipelines, hydrants, and water for firefighting purposes to Central City. T. C. Higby, E. P. Fowler, and Charles A. Girdler were empowered as trustees to enter into an agreement with the water company. In consideration of $1, the Central City Board of Trustees entered into an agreement with the WDWC on October 1, 1879.

The agreement specified that the WDWC would furnish and install pipe from the company's ditches above the Father DeSmet Mill to a point by Tyler's Livery Stable in the lower end of Central City. Five fire hydrants would also be supplied at different points along Main Street. The WDWC agreed that there would be no charge whatsoever for water used for firefighting purposes and that the company would maintain the water system within the city. The agreement stipulated that there would be no other franchise to supply water to the city and that the water company would have the exclusive right to extend service lines and provide domestic water to the residents "when deemed advisable."[60]

By mid-1880, the BHCWC was supplying 40 miner's inches of water to the Homestake 80-stamp mill and the Golden Star 120-stamp mill via the Old Abe, Montana, and Pioneer ditches at a cost of $2 per stamp per week.[61] Unfortunately, however, almost two years after the Deadwood fire, the city of Deadwood was still struggling with the inadequacy of water from City Creek. Again, attention was turned to the Boulder Ditch. "This water is diverted from the Whitewood Creek below the ditch which supplies the mills of Lead City," the Deadwood leaders observed, "and at present is useless for any purpose. Why cannot this water be turned into the mains?"[62]

After a long and expensive lawsuit in the district court between George Atchison et al. and George Hearst et al., Judge G. C. Moody issued a Conclusion of Law on July 23, 1880, and a judgment and decree on January 7, 1882. The judgment and decree affirmed the right of Hearst, Haggin, and McMaster to divert 500 miner's inches of water from Whitewood Creek at or above the head of the New French Boys Ditch without any obligation to return the water to Whitewood Creek above the head of the Boulder Ditch.

It was further adjudged and decreed that 89 miner's inches could be taken from upper Whitewood Creek relative to the two water rights that comprised the Montana Ditch, providing the water was returned to Whitewood Creek above the head of the Boulder Ditch. The Boulder Ditch was entitled to 500 miner's inches from Whitewood Creek, but the right was subordinate to the 500-inch and 89-inch water rights of Hearst et al. from upper Whitewood Creek.[63]

J. B. Haggin Maneuvers to Gain Control

The judgment and decree by Judge G. C. Moody in January 1882 was a good one for George Hearst and Homestake. However, it was a moot point since the BHCWC had acquired all of the assets of the WDWC in the interim. At the same time, J. B. Haggin had quietly gained control of the entire Father DeSmet Consolidated Gold Mining Company. The strategy and timing of the Homestake capitalists was, once again, impeccable. From its inception, the stockholders of the WDWC were nearly identical with those of the Father DeSmet Consolidated Gold Mining Company. Both companies were organized under the laws of the state of California, and both companies had offices at the same location in San Francisco. Archie Borland, August Hemme, L. R. Graves, and A. J. Bowie were major shareholders and directors for both companies. Theodore Widman was secretary for both companies. Borland held twenty thousand shares in the WDWC and forty-eight thousand of the one hundred thousand shares of the Father DeSmet Consolidated Gold Mining Company. With nearly identical stockholders, directors, and officers, the two companies worked in close harmony for the benefit of each other.[64]

The WDWC derived most of its funding from assessments on its own stock as well as large sums of money that were advanced by the Father DeSmet Company for the purpose of constructing ditches and canals that would benefit the mining company. In early August 1880, the directors of the Father DeSmet Company decided to discontinue advancing any more money to the WDWC. Moreover, the directors demanded a repayment of the balance due on previous advances, which amounted to about $90,000. At a meeting of the directors of the WDWC on August 12, 1880, Archie Borland and one other director tendered their resignations as directors of the company. Per Borland's recommendation and influence as a major stockholder, two replacement directors were elected at the meeting. The new board of directors then passed a resolution to convey all of the assets and liabilities of the WDWC to the Father DeSmet Company.

The latter company indicated it would then discharge all of the liabilities of the water company, which amounted to approximately $150,000.

On the same day, the directors of the Father DeSmet Company met. Following the resignation of Borland and the election of his successor, a resolution was unanimously passed to purchase the assets of the WDWC and arrange to discharge its liabilities. Unfortunately, the shareholders of the Father DeSmet Company later rejected the resolution of their directors to purchase the assets of the WDWC and assume its liabilities. As a result, the directors of the WDWC had to meet on August 20 to rescind their previous action. The same day, the directors of the Father DeSmet Company met and passed a corresponding resolution to annul their resolution of August 12 and reconvey the property to the WDWC.

On November 12, 1880, the directors of the WDWC met again. Following the resignation of George D. Haven as director and election of Charles H. Cook as successor, the board passed a resolution authorizing the WDWC to execute a promissory note in the amount of $90,787.03 to the Father DeSmet Company, which was the amount of indebtedness to the mining company. The same day, the directors of the DeSmet Company passed a resolution authorizing the sale, endorsement, and transfer of the note to Archie Borland. Having prior approval from the WDWC, Borland forwarded the money to the DeSmet Company and assumed the promissory note on November 15. It turns out that after Borland incurred the indebtedness, but before any money was advanced to the DeSmet Company, Borland sold all his interests in the WDWC and the Father DeSmet Consolidated Gold Mining Company to J. B. Haggin.

The directors of the WDWC met again on February 1, 1881. At this meeting, the directors passed a resolution that recognized the company's indebtedness to Archie Borland in the amount of $90,787.03, plus additional amounts for interest, attorney's fees, and overdrafts. The board approved a new promissory note to Borland in the amount of $109,049.76, payable within twenty-four days. The note was delivered to him on February 1, 1881. Borland then endorsed, transferred, and delivered the promissory note to Samuel McMaster. McMaster, in turn, filed suit against the WDWC in the district court in Lawrence County on March 17, 1881, to recover the amount due on the note.[65]

Meanwhile, by January 1881, J. B. Haggin and his partner, Lloyd Tevis, had purchased enough stock in the WDWC and the Father DeSmet Consolidated Gold Mining Company to have controlling interests in both companies. In February 1881, Haggin named Harrison M. Gregg superintendent of the DeSmet Company. The board of directors of the WDWC, in turn, passed a resolution on March 2 to replace L. L. Alexander with Harrison M. Gregg as superintendent and agent for the water company.[66]

The district court determined that the assets of the WDWC would be sold at a Sheriff's sale on August 18, 1881. On May 17, $11,219.13 was paid and credited against the note. An additional $27,455.90 was paid and credited on August 18 by Samuel McMaster, superintendent and agent for the BHCWC. McMaster was now owner of the three water rights and improvements that comprised the uncompleted Spearfish Ditch; DeSmet, Bowie, and Palmer Ditch and water rights on Deadwood Creek, which was supplying water to the Father DeSmet Mine; Central City water system and contract; Boulder Ditch and water right; three water rights on Elk Creek; and 10 acres of land that included the Spring Creek water right, which was supplying water to the city of Deadwood.[67]

After Haggin and his business associates gained control of the Father DeSmet Consolidated Gold Mining Company in January 1881, the Foster Ditch was used to supply water to the DeSmet Mill. McMaster deeded the Foster Ditch to J. B. Haggin and George Hearst on January 31, 1883. Hearst and Haggin deeded the water right to the BHCWC on February 17, 1905. After the Cyanide Plant No. 2 was shut down in December 1934, the Foster Ditch was used to supply water to the Slime Plant in Deadwood.[68]

In April 1881, Samuel and Richard Blackstone determined that construction of the Black Hills and Fort Pierre Railroad would destroy much of the lower part of the Montana Ditch, which had become part of the Old Abe Ditch. That portion of the Old Abe Ditch to be impacted by the railroad was rerouted a short distance to the east, and a 15×16-inch wooden flume was laid in the ditch.[69]

The BHCWC now had sufficient water to fulfill its water franchise obligations to the city of Deadwood. The head of the Boulder Ditch was moved upstream another three-quarters of a mile to a location called Indian Flats, about one-quarter mile downstream from the Savage Tunnel.[70] A new intake dam was constructed at that location. A new ditch and a pipeline were constructed along the west side of Whitewood Creek to McGovern Hill in Deadwood, where new water storage tanks were erected. The ditch was renamed the Whitewood Ditch. After 1906, when the Slime Plant was commissioned, the ditch was primarily used to supply water to the plant and became known as the Deadwood Mill Water Line.[71]

In August 1893, the city of Deadwood entered into an agreement with the BHCWC to construct 1,200 feet of 6-inch water main from the ditch along McGovern Hill to Charles Street and down Sherman Street. Additionally, new 4-inch pipes were installed along Harrison, Jefferson, and Jackson streets, as well as Lincoln Avenue. Several new fire hydrants were also installed. The cost of the improvements was estimated to be $6,200, which was paid by the City.[72] Having endured eighteen long years without adequate water for firefighting or domestic purposes, the city of Deadwood now had abundant water.

Other Acquisitions of the Black Hills Canal and Water Company

Several other important water rights, ditches, and pipelines were acquired by agents of the BHCWC. Some of the more significant of these included the Columbia Ditch, Hearst Ditch, Peake Ditch, and Little Rapid Ditch.

Columbia Water Company and Columbia Ditch

On December 21, 1877, John F. Moore located a water right of 100 miner's inches on upper False Bottom Creek. Moore then sold a one-half interest in the water right to William Lardner on December 19, 1878. Lardner named Elizabeth C. Lardner as an additional owner of his interest. The owners constructed an intake dam on upper False Bottom Creek approximately 1½ miles southwest of Maitland. From the intake dam, a ditch and flume were completed to the Sawpit and Hidden Treasure gulches for the purpose of furnishing water to the various mines and custom mills located around Central City. The ditch, which became known as the Moore and Lardner Ditch, was later extended to Terraville. Moore sold a 6/50 interest in his portion of the water right to Azby A. Chouteau on March 15, 1881.

On May 8, 1883, Moore, the two Lardners, and Chouteau entered into an agreement with the Caledonia Gold Mining Company to supply water from the ditch to the Caledonia 60-stamp mill for a period of three years, with an option to renew the contract for an additional three years. The owners of the water right and ditch subsequently deeded all of their interests to a new company called the Columbia Water Company on October 18, 1889. The Moore and Lardner Ditch was renamed the Columbia Ditch. A new agreement was then made on September 2, 1889, for the Columbia Water Company to supply water to the Caledonia Mill in Terraville for a period of five years.

On December 31, 1890, the BHCWC purchased 4,897 shares of stock in the Columbia Water Company from Azby Chouteau for $18,000. An additional 5,103 shares were purchased from William Lardner on January 6, 1891, for $35,000. The Columbia Ditch also supplied water to the Cyanide Plant No. 2 in Gayville after the plant was completed in 1902. The BHCWC deeded all of its interest in the Columbia Water Company to Homestake Mining Company on February 17, 1905. In November 1915, the water right was adjudicated to 21 miner's inches, based on historical usage, pursuant to litigation involving *Gus Keller et al. vs. Homestake Mining Company*.[73]

Hearst Ditch

J. P. Belding, Ira Myers, and George Atchison filed a location for 100 miner's inches from Whitewood Creek on January 1, 1878. The location was approximately 1 mile above Ten-Mile Station. Belding and Myers deeded their interest in the water right to George Hearst and J. B. Haggin on January 5, 1878.

Work commenced on construction of the Hearst Ditch in September 1880. The head of the ditch was located on Whitewood Creek below Ten-Mile Station and above what later became the tailings pile from the Wasp No. 2 Mill. The Hearst Ditch conveyed water to the Homestake mills in Lead City. Hearst and Haggin deeded their interest in the water right to the BHCWC on June 6, 1897.

In later years, the Hearst Ditch supplied water to the Kirk Power Plant. The ditch also supplied water to the municipal distribution system in East Lead via the Hearst Tunnel located near the South Lead Cemetery and South Mill Street.

In September 1881, Martin Sands located four separate water rights on Whitewood Creek. The locations totaled 110 miner's inches and were specific to Mineral Springs, Strawberry Springs, Whitewood Springs, and Flat Iron Springs. Most of these water rights were located below the head of the Hearst Ditch. The rights were deeded to J. B. Haggin on June 14, 1897. Haggin deeded the rights to the BHCWC on February 17, 1905.[74]

Peake Ditch

The Peake No. 1 and Peake No. 2 water rights were located in October 1889 by George Hearst and J. B. Haggin. The two water rights totaled 100 miner's inches.[75] The intakes were situated in Ward Draw and Keough Draw. Water flowed, by gravity, from the spring in Ward Draw through the Peake Ditch pipeline to the spring intake in Keough Draw. From this location, water flowed through the Peake Ditch pipeline through an inverted siphon to the Peake Weir House located at the top of Cyclone Hill, south of Lead Country Club.

Little Rapid Ditch

Four different water rights were associated with the Little Rapid Ditch. William Tilson claimed all the waters of the west branch of Little Rapid Creek on May 18, 1879. Tilson deeded the right to J. B. Haggin and George Hearst on October 8, 1885. Haggin and Hearst located Little Rapid Springs Water Rights No. 1 and No. 2 on August 18, 1885, and Little Rapid Springs No. 3 on August 27, 1885. These three locations totaled 130 miner's inches. The Little Rapid Water Rights were deeded to the BHCWC on June 7, 1877.[76]

The original Little Rapid Ditch, completed in 1885, consisted of an 8×10-inch wooden box flume, 61,000 feet long. In 1913 and 1914, the flume was replaced with a tile pipeline that included 10-inch—and 12-inch-diameter pipe. The intake for the Little Rapid Ditch was at Tillson Springs, about 1 mile east of Besant Park. The ditch passed the Harvey Ranch and continued to the Homestake and Little Rapid springs where additional water was fed into the pipe. The Little Rapid Ditch then passed by Irey Springs and Dumont. Water flow from the Little Rapid system was measured in the Little Rapid Weir House near the top of Cyclone Hill. Just north of the Little Rapid Weir House, the pipeline was routed into the high-pressure pipeline.

Now that the Homestake capitalists had all the water rights and ditches they needed, they decided that the Black Hills Canal and Water Company no longer served a useful purpose. The board members and shareholders of the company entered into an agreement to lease all its right, title, and interest in the water rights, ditches, flumes, water franchises, and other property to Homestake Mining Company on January 9, 1901.[77] The assets of the Black Hills Canal and Water Company were subsequently sold and conveyed to Homestake Mining Company on February 17, 1905, and the water company was dissolved.[78]

Homestake Water Collection System

Figure 8.3 illustrates the water collection system owned and operated by Homestake Mining Company as of March 1994. The system supplied domestic water to the Homestake mine and mill, city of Lead, Central City, and portions of Deadwood. Prior to about 1995, the active part of the historic system consisted of the those water rights, intakes, and pipelines associated with the headwaters of the main and east forks of Spearfish Creek, Elk Creek, Rapid Creek, and Whitewood Creek. All of these pipelines conveyed water, by gravity, to the city of Lead except for the pipeline associated with the Peake Ditch, which utilized the Hanna Pump Station to lift water to the Peake Weir House.

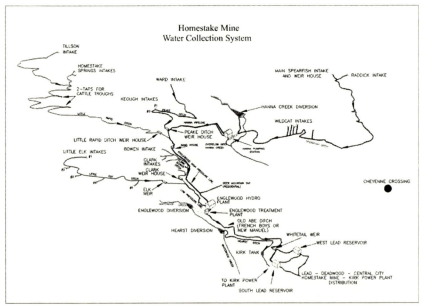

Fig. 8.3. The Homestake water collection system, 1994. This illustration depicts the water collection system that supplied water to the Homestake Mine as well as the city of Lead, Central City, and part of Deadwood. The entire system was deeded to the Lead-Deadwood Sanitary District, which now operates and maintains the active portions of the system.—Courtesy of Homestake Mining Company.

After Samuel McMaster purchased the Spearfish Ditch and its water rights at the Sheriff's sale in 1881 on behalf of the Black Hills Canal and Water Company, that ditch was completed. In about 1900, the Spearfish Ditch was retrofitted with 46,750 feet of 28-inch-diameter vitrified sewer pipe between the main Spearfish and Raddick intakes and the main sump at the Hanna Pump Station (see fig. 8.4). The pipeline passed through nine different tunnels, one of which was 249 feet long. At the Hanna Pump Station, the pipeline entered a concrete sump that contained an overflow weir. Any water in excess of what the three pumps could pump was returned to the east fork of Spearfish Creek through a 30-inch-diameter pipe.

Fig. 8.4. The Hanna Pump Station. Water from the main Spearfish Intake and Raddick Intake was conveyed through a 28-inch-diameter pipeline (the Spearfish Ditch) to the Hanna Pump Station. The water was then pumped to the top of Cyclone Hill where it flowed, by gravity, to the Englewood Hydroelectric Plant. From there, the water was conveyed, by gravity, through a pipeline in the Old Abe Ditch to Lead.—Photo courtesy of the Black Hills Mining Museum.

The Hanna Pump Station was originally equipped with three Reidler pumps, each having a capacity of 2,250 gallons per minute. Four wood-fired boilers rated at 200 horsepower each provided steam power for the pumps.[79] In 1909, the boilers were converted to a coal-fired system. A railroad spur was constructed from the Chicago, Burlington, and Quincy (CB&Q) line at Dumont to the Hanna Pump Station, which provided an efficient means for delivering coal to the boilers. To facilitate coal handling at the pumping station, miners were assigned to drive a drift into the hillside at the rear of the plant on the level of the fire-room floor. The miners then created a large coal bin by driving a raise from the drift and enlarging the raise to form the bin. A railroad bridge and dumping station were constructed over the bin to allow coal to be dumped directly into the bin. A small railcar was used to transfer coal from a chute at the bottom of the raise to the fire-room floor.[80]

The three Reidler pumps were started in November 1900. The pumps lifted the water 380 feet though a 16-inch-diameter wrought-iron pipe coupled to a 28-inch-diameter vitrified pipe 6,734 feet long. This pipeline passed through a tunnel 517 feet long and joined the Peake Ditch pipeline. The Peake Ditch consisted of a 16-inch-diameter vitrified pipe 23,100 feet long. This pipeline conveyed water from the Ward and Keough intakes to the Peake Ditch Weir House, located at the top of Cyclone Hill near Lead Country Club. At this location, the Peake Ditch pipeline was piped into the Spearfish Ditch pipeline from the Hanna Pump Station.

The pipeline of the Peake Ditch included a siphon called the Peake Siphon. The siphon consisted of 1,218 feet of 12-inch-diameter wrought-iron pipe laid in a ditch across the valley of upper Hanna Creek. The upstream end of the siphon on one side of the valley was a mere 12 feet higher than the downstream end of the pipeline at the weir house. Notwithstanding, the slight elevation difference provided sufficient energy to push the water up the side of the hill to the weir house. Water from the Spearfish Ditch and the Peake Ditch then entered a 30-inch-diameter vitrified pipe that extended 5,828 feet to a 12×32-foot concrete intake sump. This sump and weir, referred to as the Sand House, was located at the head of what was called the high-pressure pipeline.

Water from the Homestake and Tillson springs, located on some of the headwaters of the North Fork of Rapid Creek, was collected and directed through a pipeline in the Little Rapid Ditch. The flow of water in this pipeline was measured in the Little Rapid Ditch Weir House, also located just south of Lead Country Club. The pipeline in the Little Rapid Creek Ditch was piped into the 30-inch high-pressure pipeline some 2,757 feet upstream from the Sand House.[81]

Below the Sand House, the high-pressure pipeline consisted of 8,350 feet of 28-inch-diameter spiral-riveted steel pipe and 8,375 feet of 26-inch-diameter steel pipe laid in 1906. The static head or elevation differential across this pipeline was 485 feet, which provided water pressure of 210 pounds per square inch to drive the 400-kilowatt generator at the Englewood Hydroelectric Plant.[82] In 1979, the high-pressure pipeline was replaced with 17,000 feet of 24-inch-diameter steel pipe.[83]

Paralleling the high-pressure pipeline was a low-pressure line that collected all of the gravity-fed spring water from the Elk Creek, Schaffer, Bowen, and Clark springs. This water was used, in part, to drive the exciter generator at the Englewood Plant. The low-pressure pipeline consisted of 7,271 feet of 16-inch spiral-riveted steel pipe, 3,214 feet of 14-inch steel pipe, 2,375 feet of 12-inch steel pipe, 5,256 feet of 8-inch vitrified pipe, and 2,315 feet of 6-inch pipe.

From the Englewood Hydroelectric Plant, water from the high-pressure and low-pressure pipelines was directed into the Old Abe Ditch, which contained 19,445 feet of 30-inch-diameter vitrified pipeline between the Englewood Plant and the Whitetail Weir House. The Old Abe Ditch originally consisted of a 16×15-inch wooden flume 19,445 feet long. Years later, a new 18×18-inch wooden flume was constructed over the top of the original flume.

In 1906, the box flume was replaced with the 30-inch-diameter tile pipe and trestles were constructed in two locations to support the Old Abe pipeline. One trestle was 90 feet long. Another trestle, called the Reno Bridge (see fig. 8.5), was a lattice-truss trestle 300 feet long and 60 feet high. In 1929 or 1930, the tile pipe on the Reno Bridge was replaced with 30-inch-diameter wood stave pipe. The wood stave pipe on the bridge was replaced with new wood stave pipe in September 1970 and again in September 2008.[84]

Fig. 8.5. The Reno Bridge, 2007. The Reno Bridge was constructed in 1906 to support the 30-inch-diameter wood stave pipeline from the Englewood Hydroelectric Plant across Reno draw. The trestle is approximately 300 feet long and 90 feet high.—Photo courtesy of Todd Hubbard.

From the Whitetail Weir House, two 16-inch-diameter wrought-iron and steel pipelines were originally installed to convey water to the South Lead (Ellison) Reservoir. Later, two 12-inch-diameter pipelines were laid to convey a portion of the water from the Weir House to the West Lead Reservoir. The water filtration plant in Lead, constructed in the 1990s, receives water from the two 12-inch lines and supplies potable water to the West Lead Reservoir and an aboveground potable storage tank located adjacent to the South Lead (Ellison) Reservoir.

When the water collection system was completed in late 1906, the system consisted of 152,553 feet of collection pipe ranging in size from 6-30 inches in diameter. Four bridges carried the pipeline across valleys and draws. Twelve tunnels were driven to facilitate proper grade and routing of the various pipelines across and through mountainous terrain. The tunnels, which have a combined length of 3,428 feet, were drilled and blasted through ridges and outcrops of solid limestone to accommodate the Spearfish and Peake pipelines.[85] An extensive amount of preliminary survey work was required to determine initial routing and facilitate construction of the ditches and pipelines. All of the construction work was completed by hand and with the help of horses and mules.

In September 1999, the South Dakota Water Management Board validated Homestake's claims filed in 1993 to water rights associated with the Spearfish, Peake, Little Rapid, Clark-Bowen, and Little Rapid diversions. These diversions comprised the active portion of the Homestake water collection system. The validations include Vested Water Rights 1587-1, 1588-1, 1590-1, 1591-1, and 1594-1, respectively. Water associated with these rights was used for mining and milling operations at Lead, power generation at the Englewood Hydroelectric Plant, municipal purposes for Lead, Deadwood, Central City, and various domestic taps.

The validations authorized a "total volume beneficial use" not to exceed 5,800 acre feet of water annually averaged over a ten-year period as measured at the Whitetail Weir. The maximum combined diversion rate from all diversions was set at 13 cubic feet per minute, averaged over seventy-two hours. Homestake was permitted to divert not more than 2 cubic feet per minute on an instantaneous basis from the Little Rapid diversions, 8 cubic feet per minute from the Englewood diversion, 3.55 cubic feet per second from the two Peake diversions, and 5.89 cubic feet per minute from the Clark-Bowen diversion, subject to the combined maximum diversion rate.

The maximum combined diversion from the Peake diversion and the Spearfish diversion through the Hanna Pump Station was restricted to 14 cubic feet per minute on an instantaneous basis. The Spearfish diversion through the Hanna Pump Station could be used only if gravity diversions from the Peake, Little Rapid, and Clark-Bowen sources were not sufficient to meet the beneficial uses demand.[86]

Under Homestake's ownership and operating control, the water collection system provided water for Homestake's mining and milling needs, as well as municipal water for the communities of Lead, Central City, and Deadwood for more than 120 years. The vested water rights, water collection, and water distribution systems were deeded to the Lead Deadwood Sanitary District in phases between December 1993 and March 2003. Much of the system continues to be utilized today, except for the Hearst, Little Elk, and Schaffer ditches. The Hearst Ditch was abandoned in the 1970s, and the Little Elk and Schaffer ditches were abandoned in the early 1990s.[87]

Chapter 9
Early Mining and Haulage Methods

Mining by Open Cuts and Drawholes

The first hardrock ore mined by the Manuel brothers and others in 1876 was from shallow open cuts located high up on the hillsides above the north fork of Gold Run Creek. The same was true of most of the other mines along the Homestake Belt. As the miners continued to chase the quartz "leads" on their claims, the open cuts became larger and deeper. Two of the larger open cuts were the Mallory Open Cut on the Golden Star claim and the Homestake Open Cut on the Homestake claim.

Numerous adits were driven into the hillsides to follow the quartz stringers and ore. Many of these adits were driven from the faces of the open cuts. In other cases, adits were driven into the hillsides immediately below the open cuts. Inside these adits, cutouts were mined, and wooden ore chutes were constructed. From the chutes, raises were driven upward to intersect the bottoms of the various open cuts. Ore was mined from the sides and bottoms of the open cuts (see fig. 9.1), shoveled into railcars (see fig. 9.2), and dumped into the raises. A grizzly was located on a sublevel, a short distance above the chute, to control the size of ore that could pass through the raise to the chute (see fig. 9.3). The ore was drawn through the chutes at the bottoms of the raises and transferred into railcars (see fig. 9.4). A horse or mule was used to pull four to eight cars of ore through the adits to the Homestake and Golden Star mills (see fig. 9.5). Typically, two horse trains operated on a given level. After the mining progressed below the level of the gulches and mills, the ore had to be hoisted at one of the vertical shafts and transferred to the respective mill on the tramway level.[1]

Fig. 9.1. Double-jack drilling in an open cut. Solid bedrock was broken by drilling and blasting a series of short holes. Prior to the introduction of pneumatic drilling equipment, each hole was drilled by hand using progressively longer lengths of hand-drill steel that were driven with a sledge hammer. Single jacking was performed by a miner who held the drill steel in one hand and the hammer in the other. Double jacking, shown here, was a little more efficient. Each hole was subsequently charged with a blasting cap, black powder fuse, and either black powder or dynamite, referred to as "powder." A "round" of holes was detonated by lighting the black-powder fuse to each of the holes. The fuse detonated the blasting cap and powder in each hole.—Photo courtesy of Homestake Mining Company.

Fig. 9.2. Hand mucking in an open cut. Much of the early-day mining along the Homestake Belt was done in open cuts, including the Mallory, Homestake, Caledonia, Hercules, and Clara. Ore was drilled, blasted, and shoveled into 1-ton-capacity end-dump railcars. The cars were then trammed by hand to a drawhole raise, where the ore was dumped and transferred to the grizzly level.—Photo courtesy of the Black Hills Mining Museum.

Fig. 9.3. Breaking rock on a grizzly. Ore mined in the open cuts was shoveled into 1-ton end-dump railcars, hand-trammed and dumped into drawhole raises where the ore fell to the grizzly level. Here, a "grizzly man" is breaking the oversized rock and ensuring a smooth flow of ore to the chute, which was located on the next level below the grizzly level. Drawholes and grizzly levels were later used in a similar fashion as part of the process in recovering ore from crown pillars and caved areas. This miner is using a carbide lamp for illumination. Carbide lights were introduced into stopes on January, 16, 1912, according to personal notes of Thomas J. Grier. Candles were discontinued in the mine on March 1, 1912. In 1940, some 1,175 battery-powered cap lamps were purchased at a cost of $18,437.68 and carbide lights were phased out.—Photo courtesy of Black Hills Mining Museum.

Fig. 9.4. Drawing chute on the haulage level. This early-day photo shows a "chute drawer" using a bar to raise the control boards sufficiently to allow the ore to pass from the chute to the railcar. The ore car was then hand-trammed to the main drift where horses or mules were used to pull a train of cars to the mill, assuming the chute was on the tramway level. Otherwise, the ore cars had to be trammed to the shaft and hoisted on the cage to the tramway level. Note the candlestick stuck into the side of the chute for illumination.—Photo courtesy of the Black Hills Mining Museum.

Fig. 9.5. Hauling ore on the Highland Tramway.—Photo courtesy of Homestake Mining Company.

From July 1902 through June 1903, 1.279 million tons of ore were mined at the Homestake Mine. Some 20 percent of this total was derived from the open cuts and underground draw raises. This method of mining was very efficient and yielded low-cost gold production. Two miners working in one of the small open cuts could mine up to 200 tons per shift. This rock was handled by one person at the grizzly, one at the chute, and two driving the horse-drawn trains. This team of people averaged about 30 tons of ore per manshift, which was outstanding production. The cost for a crew of six to mine 200 tons per shift was $27, or $0.135 per ton of ore (see table 9.1).

Table 9.1 Cost of mining 200 tons of ore per shift using the Homestake drawhole method of mining

	No.	($/shift)
Miners @ $3.50/man/shift	2	$7.00
Grizzly Man @ $3.50/shift	1	$3.50
Chute Drawer @ $3.00/shift	1	$3.00
Drivers @ $3.00/man/shift	2	$6.00
Horses @ $0.90/horse/shift	2	$1.80
Blacksmith Labor	—	$0.50
Explosives @ $0.026/ton	=	$5.20
Total:	8	$27.00
		($0.135/ton ore)

Source: Bruce C. Yates, "Some Features of Mining Operations in the Homestake Mine, Lead, South Dakota," (Paper presented before the Black Hills Mining Men's Association, January 19, 1904), 25.

Open Square-Set Method

As the Homestake Mining Company was being organized in November 1877, George Hearst had already decided that the relatively wide, underground ore areas along the Homestake Belt would have to be supported with mine timbers as mining advanced horizontally and vertically through the ore. He wasted no time in introducing the square-set method of stoping, a method he successfully used when Hearst, Haggin, Tevis and Company owned the Ophir Mine on the Comstock Lode in Nevada.

Philip Deidesheimer developed the square-set system of stoping in 1860, and soon after he became superintendent of the Ophir Mine (see fig. 9.6). Prior to Deidesheimer's arrival at the Ophir, miners attempted to mine the rich ore using wooden props and rock pillars to support the backs of the stopes. As the stopes were mined wider, the miners found that the crude support methods were inadequate, and the stopes had to be abandoned because of uncontrollable cave-ins.

Using Deidesheimer's square-set method, a small sill cut was mined from an access drift previously driven through the ore. Upon completion of the sill cut, a sill floor consisting of a timber framework was laid out horizontally across the floor. The sill floor served to support the first square sets of timber, which, in turn, provided support for the first floor of timber sets and the stope below. As the rock was sequentially mined outward and upward, additional floors of square sets were installed until the designated limits of the stope were reached. The previous floors were filled with waste rock as the mining progressed upward, if such material was available.[2]

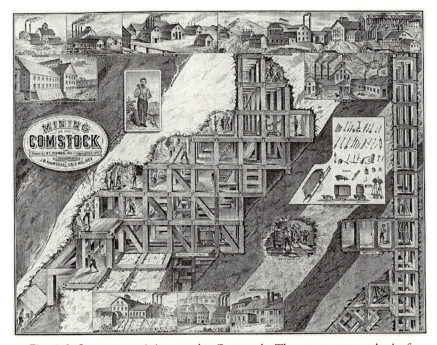

Fig. 9.6. Square-set mining on the Comstock. The square-set method of stoping was developed by Philip Deidesheimer in 1860 at the Ophir Mine on the Comstock Lode. George Hearst, who was an owner in the Ophir at the time, wasted no time in introducing the method at the Homestake Mine in 1878.—Library of Congress (LC-DIG-pga-01999).

At the Homestake Mine, the bottom tier of square-set cubes erected on the sill floor was known as the sill-floor sets. These sets were 9 feet tall to accommodate the chutes. Above the sill floor, all subsequent floor sets were 8 feet 5½ inches high. A sill-floor set occupied 324 cubic feet, which equated to 32 tons of solid rock. An upper floor set contained 304.5 cubic feet, which was equivalent to 30 tons of solid rock. Figure 9.7 shows a timber stope on the Pierce Ledge on the 200-foot level of the Homestake Mine.

Fig. 9.7. A timber stope in the Pierce Ledge, 200-foot level. This photo, by W. B. Perkins, shows square-set mining in a stope on the Pierce Ledge on the 200-foot level. Note the sill-floor sets and chutes for loading ore cars. The small railcar was used for transporting mine timbers and other supplies into the mine. Failure to backfill open, square-set stopes such as this one ultimately led to the subsidence problem that forced abandonment of Lead's original business district.—Photo courtesy of the Black Hills Mining Museum.

Prior to about 1900, the mining system used between the 500-foot level and the surface simply consisted of an array of open cuts, drawhole raises, and square-set stopes. The sill-floor excavations for the square-set stopes typically extended from hanging wall to footwall and sometimes for the entire length of the ore area. The sill-floor timbers were then placed over the entire sill cut. As the timber stopes were mined upward, the timber sets were either left open or partially filled with waste rock, if such rock was available. Often after the first timber stopes had been mined upward several floors, the entire block of ground over the area of the initial sill cut became quite unstable. In many cases, the entire area caved, despite attempts to install solid timber cribs under the various rock pillars on the sill level (see fig. 9.8). Attempts were subsequently made to recover ore in the caved areas by using the drawhole method.

Fig. 9.8. A solid timber crib. Prior to about 1900, solid timber cribs such as this one sometimes had to be installed to provide ground support in large sill cuts that became unstable after one or more stopes were mined upward from the common sill. As mining advanced below the 500-foot level, the practice of cutting large sills in advance of stoping was mostly discontinued.—Photo courtesy of Homestake Mining Company.

Four square sets of nine posts were constructed to support the drawhole raises that were used to tap ore in the caved areas. This configuration was called a "nine-post raise." The nine-post raise was usually driven in the footwall to the elevation of the crown pillar, sill pillar, or caved area. In the same fashion as with drawhole raises driven to intersect the open cuts, a grizzly was incorporated in the nine-post raise to control oversize rock. A miner was positioned at the grizzly whose job it was to manually break oversize rock, so it would pass through the openings between the grizzly bars. A ladder way, called a "manway," was also carried in the nine-post raise, which provided access to the grizzly level and the raise above. If the raise was being used to tap a caved area, the miners would initiate the cave by barring and blasting the caved material. If the draw raise was used to recover a solid pillar, the nine-post raise was used as a manway, from which a new square-set stope was started in the pillar.[3]

Over a fifteen-year period ending in the 1930s, more than 10 million tons of ore were recovered from caved areas using the drawhole method (see fig. 9.9). The caved material consisted of a mixture of caved ore, square-set timbers, wall rock, and waste fill. Although mill grades were low, mine management contended the method was profitable. Very labor-intensive sorting methods were employed at the chutes to separate undesirable materials from the ore. Any waste rock that could be identified was culled out and dumped into completed stopes as backfill. Those mine timbers not damaged were reused in other square-set stopes. Thousands of cords of nonsalvageable wood were hoisted to the surface and made available as firewood to the residents of Lead.[4]

Fig. 9.9. The drawhole method of mining. The drawhole method, in conjunction with open cuts, was a very efficient method of mining from 1876 to about 1900. Afterward, use of the drawhole method was moderately successful in tapping ore from caved areas where large sill cuts had been made.—Photo courtesy of Homestake Mining Company.

Homestake System of Stoping

A somewhat random order for sequencing the mining of the various stopes and pillars was practiced above the 500-foot level. About the time the 600-foot level was being developed shortly after 1900, W. S. O'Brien, general mine foreman, and Bruce C. Yates, mine engineer, started experimenting with a more ordered system of stopes and pillars. This system was intended to reduce the massive caving that was occurring as a result of the practice of creating large sill cuts from which randomly placed stopes and pillars were established. With the new system, stopes were mined transverse to the strike of the ledge of ore. The primary stopes were mined 60 feet wide along strike and from hanging wall to footwall. Rock pillars of ore that were 60 feet long separated the primary stopes.

The mining plan was initiated by driving a main crosscut from the Star Shaft across the entire ledge of ore. From this east-west crosscut, access drifts were driven in both directions along strike in the footwall. Transverse pillars, 60 feet wide along strike, were established on each side of the main crosscut from hanging wall to footwall. Beyond each pillar was a 60-foot stope, 60-foot pillar, and so on, until the full extent of ore was reached in both directions on each mine level. Years later, the width of the pillars was reduced from 60 feet to 42 feet, then 40 feet. Each primary stope was designated as No. 1 Stope North, No. 2 Stope North, No. 1 Stope South, No. 2 Stope South, etc., relative to the Star Shaft. The pillars were similarly designated and numbered as No. 1 Pillar North, No. 1 Pillar South, etc.

Initially, all of the primary stopes were mined using the square-set method. After a primary stope was completed and filled with waste rock, the adjacent pillars were also mined using the square-set method and filled with waste rock. This method, called the Homestake System of Stoping, proved successful in reducing the massive caving that had been plaguing the mine above the 500-foot level. The one disadvantage of the new Homestake System involved the large amount of timber that was being consumed. Mine timbers had to be sourced from areas as far away as lower Elk Creek and Nemo.[5]

Moreover, Homestake now had to comply with the Forest Reserve Act and the Organic Administration Act. The Forest Reserve Act of 1891 allowed the president of the United States to designate forest reserves within the public domain. The Black Hills Reserve, created on February 22, 1897, and the Organic Administration Act, which was passed later that year, regulated the harvest of trees on federal land, including the unpatented mining claims held by Homestake.[6]

As a solution, O'Brien and Yates decided to mine the primary stopes without timber, except for the sill-floor sets. These stopes were called "bull pens" or shrinkage stopes. The crown pillar above a shrinkage stope and the pillars adjacent to the stope were mined as square-set "timber" stopes.

The sill cuts for the shrinkage stopes were mined to the full width of the limb of ore. Each 60-foot-wide stope was supported with eleven lines of sill timber. After the horizontal sills were laid, two lines of track were laid on sills over the length of the transverse stope. One or more additional track lines were laid perpendicular to the two main carways. Sill-floor posts and caps were erected over the double-track system. The track lines were protected by placing double lagging over the posts and installing lagging or lacing slabs behind the posts along the sides of each track line.

Mining commenced in an upward direction by drilling and blasting successive "breast cuts" across the stope until the perimeter limits were reached on all sides of the stope. Upon completion of this cut, the sill-floor sets were completely filled with broken ore, except for the carways and the cribbed manways that were carried upward along the walls of the stope. The miners repeated the cycle by drilling and blasting another series of horizontal breast cuts across the entire stope.

Concurrently, shovelers removed side lagging and shoveled enough broken ore from along the main carways and perpendicular track lines to provide the miners with sufficient overhead room from which to drill and blast successive breast cuts. Years later, the two timbered carways were replaced with a single 7×7-foot crosscut driven on the centerline of the transverse stope from footwall to hanging wall. From this crosscut, drifts were driven at 90 degrees to the crosscut on 30-foot intervals. Each of these drifts served as shoveling points from which relief rock and final cleanup were performed for the shrinkage stopes.

Two or three sets of wood manways were carried along the walls of the stope to provide access and ventilation between the bottom sill and the working faces of the stope. Above the 1,100-foot level, where mine levels were 100 feet apart, the shrinkage stopes were mined upward to a height of approximately 80-85 feet. Below the 1,100-foot level, where mine levels were 150 feet apart, the shrinkage stopes were mined to a height of approximately 125 feet.

At this elevation in the shrinkage stope, one or more raises were driven through the crown pillar to the level above. The broken ore was then drawn out of the completed stope by the shovelers (see fig. 9.10). Whenever possible, as one end of the stope was emptied of ore, waste rock was dumped into the appropriate raise from the level above. The waste rock helped provide support for one end of the stope while the remaining ore was removed from the shoveling points at the other end of the stope. After all of the ore was removed and the stope was filled with waste rock, the remaining crown pillar of ore was mined to the sill floor of the next level above using the square-set method. To minimize caving problems, the highly-stressed crown pillar was usually sequentially mined using small 25×60-foot square-set stopes.

Fig. 9.10. Hand mucking in a shrinkage stope, 1922. Another photo, by Lease, shows five different track spurs that have been extended to the toe of the rock pile in a shrinkage stope. The shovelers are hand mucking rock into 1-ton end-dump cars. Note the person drilling blockholes with a pneumatic drill.—Photo courtesy of Homestake Mining Company.

Chutes were not utilized under the shrinkage stopes in the early years because oversized ore caused hang-ups in the chutes. Initially, management personnel believed that the stoping operation was more efficient if the miners drilled and blasted the oversize at the shoveling (mucking) points across the sill floor. In 1917, the thinking changed and the shrinkage method was again modified.

After the entire sill cut and first lift were mined, a timberline with multiple chutes was installed on the sill floor next to each pillar adjacent to the primary stope. A pyramid of waste rock was then established between the two timberlines in the sill cut by dumping the waste through one or more raises from the mine level immediately above the stope (see fig. 9.11). Mining then commenced upward. Broken ore was allowed to fall over the timberlines and pyramid of waste rock.

Once it was leveled, the broken ore in the stope served as a temporary work platform for the miners. Another series of breast cuts was systematically drilled and blasted across the back of the stope (see fig. 9.12). Blockholing of oversize rock was completed on the rock pile within the stope. Only a sufficient amount of "relief rock" was drawn through the chutes to maintain the top of the rock pile at the proper working height for the benefit of the miners. Successive "lifts" or slices of ore were subsequently mined from the stope back until it was advanced to the bottom of the crown pillar immediately below the next level.

Fig. 9.11. Development of a shrinkage stope. Two timberlines have been constructed (one is outside the picture) and a pyramid of waste rock has been placed between the two timberlines from the level above.—Photo courtesy of Homestake Mining Company.

Fig. 9.12. Drilling a breast cut from the top of the rock pile in a shrinkage stope, ca. 1920.—Photo courtesy of Homestake Mining Company.

The pyramid of waste rock between the two timberlines enabled all of the ore to be drawn out of the stope through chutes located in the two timberlines. By incorporating chutes and relying on gravity to move the broken ore through the chutes, the new stoping arrangement largely eliminated the need for hand shoveling throughout the life of the shrinkage stope.[7]

Prior to the early 1880s, drilling in the mine was done by hand. Single-jack drilling was accomplished by one miner, who held the drill steel in one hand and struck the end of the steel with a hammer held in his other hand. After each blow, the miner rotated the drill steel about one-half turn. Double jacking required two miners (see fig. 9.13). One miner held and rotated the drill steel while the other struck it with a heavy sledgehammer. By the early 1880s, hand-drilling methods were replaced with a pneumatic drill mounted on a bar and column. After about 1900, two different types of drills were used including the Cleveland 5D and the Ingersoll-Rand No. 248. Stoper drills were used for drilling raise rounds. These included the Ingersoll-Rand CCW11 and the Cleveland No. 44SW stoper drill. A Cleveland No. 44TW drill was used for blockholing work.[8]

Fig. 9.13. Old and new way of drilling. The two miners on the left are blockholing by "double jacking." Prior to the introduction of pneumatic drills in about 1883, all drilling was done by hand in stopes and drifts.—Photo courtesy of Homestake Mining Company.

Surface and Underground Haulage

Homestake's first means of surface haulage made use of horses and mules to pull short trains of ore cars out of the mine and across the tramway to the Homestake Mill. In 1879, as a test, George Hearst decided to purchase a steam-powered locomotive to haul ore on the tramway to the 80-stamp Homestake Mill. Hearst envisioned that the locomotive could also be used to transfer supplies and materials very efficiently on the surface.

On August 26, 1879, Homestake took receipt of its first steam-powered locomotive and commenced hauling rock on the surface tramway (see fig. 9.14). The locomotive was named the J. B. Haggin after the company's president and treasurer. The J. B. Haggin, No. 4669, was purchased at a cost of $2,250 from the Baldwin Locomotive Works in Philadelphia. The locomotive was transported by rail to Bismarck, Dakota Territory. From Bismarck, it was hauled to Lead City by bull team. The total freight cost was $385.20.[9] An additional $16,030.71 was spent to purchase thirty ore cars and install 1,500 feet of 22-inch gauge track between the mine and the Homestake and Golden Star mills. Between August 26, 1879, and August 1, 1880, the J. B. Haggin hauled 165,422 tons of ore to the mills at an operating expense of only $0.0544 per ton. The locomotive weighed 10,000 pounds and could haul 30 tons of ore.[10]

Fig. 9.14. The J. B. Haggin, Homestake's first steam locomotive, purchased in 1879.—Photo courtesy of the Black Hills Mining Museum.

In 1880, a second steam-powered locomotive identical to the J. B. Haggin was purchased by the Highland Mining Company for use on its tramway between the mine and the Highland Mill. The Highland's tramway locomotive was named the W. R. Hearst after George Hearst's son, William Randolph. This locomotive, No. 5125, was also purchased from the Baldwin Locomotive Works at a cost of $2,650. Freight charges were additional, which amounted to $63.10 between Philadelphia and St. Paul and $339.45 from St. Paul to Lead City.[11]

Homestake purchased a third steam locomotive in July 1889 for use on the tramway between the Deadwood-Terra Mine and the Pocahontas Mill (see fig. 9.15). The locomotive No. 10090, was named the I. C. Stump after a Homestake director. The locomotive was purchased from the Baldwin Locomotive Works for $2,121 plus $217.30 for freight.[12]

Fig. 9.15. The I. C. Stump steam locomotive. This photo, dated October 11, 1906, shows the I. C. Stump steam locomotive at Terraville, along with Dick Pancoast, engineer, and Fred Symons, brakeman.—Photo courtesy of Black Hills Mining Museum.

In the meantime, horses and mules continued to serve as the primary means of hauling ore and transporting supplies underground (see fig. 9.16). Moving the animals through the shaft required special harnesses and rigging (see figs. 9.17 and 9.18). In 1889, there were seven horses and fifteen mules in use underground at the Homestake Mine.[13] By 1903, after acquiring the other primary "belt" mines, Homestake's complement of horses and mules increased to ninety animals underground.[14] Although both types of animals were used, horses were found to adapt better to the work.[15]

Fig. 9.16. Old Smokey pulling a train of 1-ton ore cars on the 400-foot level of the Homestake Mine.—Photo courtesy of the Black Hills Mining Museum.

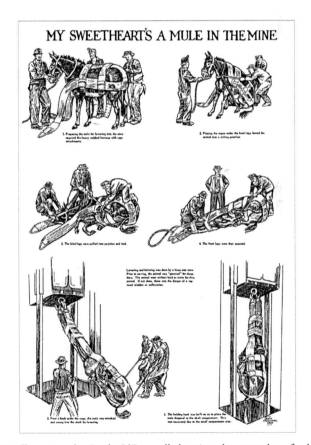

Fig. 9.17. Illustration by Buck O'Donnell showing the procedure for lowering horses and mules down a mine shaft.—Photo courtesy of Black Hills Mining Museum.

Fig. 9.18. Lowering a horse down the shaft to the 900-foot level.—Photo courtesy of Homestake Mining Company.

On May 26, 1902, the *Lead Daily Call* reported that "a good looking, vigorous colt was born to one of the work mares this morning on the 300-foot level of the Homestake. This is the first colt ever foaled underground in this section of the country."[16] Named Teddy after President Theodore Roosevelt, the mule was brought to the surface and became everyone's favorite at the July 4 celebration in Lead in 1902 (see fig. 9.19).[17]

Fig. 9.19. Teddy the mule. Teddy was born on the 300-foot level of the mine on May 26, 1902. He was brought to the surface and became everyone's favorite during the July 4 celebration that same year.—Photo courtesy of Black Hills Mining Museum.

The following year, the *Lead Daily Call* reported on August 17, 1903, that a horse named Maggie had been brought to the surface to participate in the town's Labor Day parade. Maggie worked in the mine for fifteen years and was brought to the surface only one other time for a few days in 1893 because of a mine fire. The reporter quipped, "In her 15 years of service she had hardly lost a shift and has pulled more ore and waste than any other animal in the mine."[18]

In 1900, the company began investigating the possibility of replacing its steam-powered tramway locomotives with new ones powered by compressed air. The *Lead Daily Call* reported on the test that Homestake had just approved.

> Preparations are now well underway by the Homestake for the introduction of engines that will have for their power compressed air instead of steam. The preparations now are only of an experimental nature, and if the new engines prove satisfactory the little steam locomotives that the residents of Lead have known for the past twenty odd years will be a thing of the past, and in their place will be the compressed air engines. To put in the new proposition will entail a cost of twenty-five or thirty thousand dollars. It will likely be about three months before the experiment will be made, as there is necessary considerable preparations to be made. The main purpose of using compressed air is to lessen the risk of fires, and to do away with the smoke that the present engines create underground.[19]

On April 19, 1901, it was reported that the new three-tank, compressed-air locomotive had arrived in Piedmont and would be delivered to Lead in a few days. To accommodate "refueling," Homestake placed air-receiver tanks at the Ellison Shaft and in the stamp mills. The three-tank air locomotive, assigned Homestake No. 1, was 19 feet 8 inches long and weighed 28,000 pounds.[20] The locomotive was capable of hauling 100 tons of ore per train. By comparison, the steam locomotive could only haul 32 tons of ore per train. By June 1903, based on success of the test, all of the steam locomotives were removed from service on the mill tramways (see figs. 9.20 and 9.21).[21]

Fig. 9.20. Compressed-air locomotive No. 1 and a train of ore cars on the Ellison Tramway. These men are posing with the three-tank, compressed-air-powered locomotive and a train of cars that are about to be pushed across the Ellison Tramway over Gold Run Gulch to the Golden Star Mill. The Ellison Trestle was decommissioned on March 25, 1934, because of the surface subsidence problem. The T. J. Grier residence is shown in the upper center.—Photo courtesy of Evelyn (Schnitzel) Murdy.

Fig. 9.21. The three-track train crossing in Lead, South Dakota. This photo, by D. S. Cole, was taken in about 1901, shortly after Homestake purchased its first compressed-air locomotive to begin replacing the steam locomotives on the mill tramways. This three-track crossing was located just west of the First Finnish Evangelical Lutheran Church, shown in the center of the photo. The J. B. Haggin steam locomotive, commissioned in 1879, is pushing sixteen cars of ore from the Ellison Shaft to the Golden Star Mill. The new three-tank air locomotive is also shown on the Ellison Tramway. The Fremont, Elkhorn, and Missouri Valley train with engine number 208 is on the middle bridge. From the depot in Lead, the Fremont, Elkhorn, and Missouri Valley tracks traversed along present-day Grandview Drive to Gayville and on to the depot in Deadwood. The Deadwood Central Train with engine number 497 is on the lower level. These tracks traversed through lower Gold Run Gulch to Pluma and Deadwood.—Photo courtesy of the Black Hills Mining Museum.

Based on the initial success of the air-powered locomotive on the tramways, Homestake took receipt of its first 18-inch-gauge air locomotive, assigned No. 2, on December 28, 1903, from the H. K. Porter Company in Pittsburgh, Pennsylvania, for use underground. This locomotive was much smaller than the tramway locomotive. The underground locomotive was 11 feet 3 inches long, 3 feet 4 inches wide, weighed 10,000 pounds, and operated on 18-inch-gauge track. The second underground locomotive, assigned No. 3, was received in November 1904. After a short and successful testing period, five additional 10,000-pound air locomotives, numbered 4-8, were received on May 1, 1906 (see fig. 9.22). No. 9, a single-tank locomotive for the tramway, was received in 1907.[22]

Fig. 9.22. Compressed-air locomotives from the H. K. Porter Company. This photo, dated May 1, 1906, shows compressed-air locomotive numbers 4, 5, 6, 7, and 8 that were purchased from the H. K. Porter Company in Pittsburgh, Pennsylvania. The cost was about $1,990 each.—Photo courtesy of the Black Hills Mining Museum.

On February 3, 1908, Homestake received four additional locomotives for use underground. One of the units, assigned No. 13, was a new design for use in smaller drifts underground. This unit was only 8 feet 10 inches long, 3 feet wide, and weighed 7,000 pounds. Eventually, Homestake operated ten of the smaller locomotives underground.[23]

The air locomotives worked well underground. The *Deadwood Daily Pioneer-Times* reported that the introduction of the air locomotive "marked the passing of the horse as motive power." By November 26, 1908, there were only seventeen horses left in the mine and a number of those were brought to the surface within days. Two air locomotives were already operating satisfactorily on the surface, and a third unit was on order.[24] Seventeen air locomotives were in use underground.[25]

Homestake ended up purchasing a total of thirty-six air locomotives between April 19, 1901, and May 7, 1931 (see table 9.2). The air locomotives were numbered 1-35, except for No. 1A, which was purchased on January 31,

1928. No. 1 was the only three-tank locomotive and was used on the surface tramway. Nos. 1A, 9, and 20 were also 22-inch-gauge locomotives used on the surface tramways but were of single-tank construction. The other thirty-two locomotives were operated on 18-inch-gauge track.

All of the locomotives were powered by compressed air at 900-1,000 pounds per square inch tank pressure (see figs. 9.23 and 9.24). Galvanized, high-pressure, compressed-air pipelines were installed in the main headers of each level from the surface through the 4,100-foot level. In 1981, six air locomotives were still being operated on the 800-, 1,100-, 2,600-, 3,650-, and 4,100-foot levels. The last of the air locomotives were removed from service in 1983.[26]

Table 9.2 Compressed-air locomotives used at the Homestake Mine

Loco No.	Track Gauge (in.)	Main Tank Pressure (p.s.i.)	Aux. Tank Pressure (p.s.i.)	Main Tank Capacity (cu. ft.)	Avg. No. of Cars per Trip	Loco Weight (lb.)
1	22	900	170	165	31	28,000
1A	22	900	250	137	31	25,000
9, 20	22	900	170	70	11	17,000
8, 25, 26	18	900	145	28.5	25	10,000
13-19, 21-23	18	900	145	20.5	12	7,000
2-7, 10-12, 24, 27-35	18	900	145	40.0	25	10,000

Source: Data from Locomotive Repair Office Records, Homestake Mining Company.

Fig. 9.23. A compressed-air locomotive and cars at the ore dump. Air-powered locomotives were first used on the surface in 1901 and underground in early 1904. The last ones were removed from service in 1983.—Photo courtesy of Homestake Mining Company.

Fig. 9.24. Charging an air locomotive. This photo, taken in 1958, shows Bernard Reausaw charging his air-powered locomotive with 900-1,000 p.s.i. compressed air. The "air motor" had to be "filled" many times each shift, depending on the haulage duty. Recharging was accomplished in a few minutes.—Photo courtesy of Homestake Mining Company.

Battery locomotives were introduced into the mine in 1934, when two 6-ton units were purchased. Four 8-ton units were added shortly thereafter. The 18-inch gauge, 6-ton and 8-ton battery locomotives became the primary means for transporting ore, waste, supplies, and workers for the remaining life of the mine. As part of the deep-level development project initiated in the 1970s, all levels below the 6,800-foot level were converted to 30-inch-gauge track. Track systems on the surface and the main transfer level on the 4,850-foot level were fitted with three rails to accommodate the 18- and 30-inch-gauge equipment.[27]

With the introduction of the battery locomotives, the standard mine car was converted from a 1-ton end-dump car to a Granby-style car with a nominal capacity of 3 tons. The Granby car incorporated a small side-mounted "fifth wheel." As the locomotive passed by the "camelback" at the dumping station, the protruding fifth wheel of each Granby car traveled over the camelback ramp (see fig. 9.25), which raised the hinged box of the car and allowed the payload of broken rock to fall into the ore or waste raise. This arrangement resulted in a huge improvement in safety and efficiency for the mine because the motorman or a helper no longer had to dump the cars manually. With the Granby-style car, an entire train could be dumped in less than one minute.

Fig. 9.25. A typical "camelback" ore dump. As the motorman approached the ore dump, he or she activated a switch, which caused the hinged dump screen to open. As the fifth wheel of each Granby car in the train traveled over the camelback, shown on the right, the door of the car opened, allowing the ore to fall out of the car and into the raise. A ten-car train, containing about 30 tons of ore, could be dumped in about one minute.—Photo courtesy of Homestake Mining Company.

On the 4,850-foot level, which was the main underground ore-transfer level between the primary winzes and the surface shafts, a tandem, electric trolley unit was used to pull a twenty-car train of 8-ton-capacity bottom-dump cars. The trolley transferred ore from chutes at No. 4 Winze, No. 6 Winze, Main Ledge, and 9 Ledge to the Yates and Ross shafts. In 1951, the air locomotive on the surface tramway was replaced with a 10-ton electric trolley for transferring ore from the Yates, Ross, and Ellison shafts to the South Mill.[28]

Chapter 10
Arrastras, Stamp Mills, and Amalgamation

The Arrastra and Amalgamation

Because of oxidation, Precambrian ore mined from the upper reaches of mines along the Homestake Belt was found to generally be "free-milling" ore. This meant that most of the gold could be recovered simply by pulverizing the ore to liberate the finely disseminated particles of gold and amalgamating them with mercury. Many of the earliest and smallest mining and milling operations along the Homestake Belt used an arrastra to crush and pulverize the ore and amalgamate the gold. By about 1878, most of the belt mines abandoned their arrastras in favor of the stamp mill, which was much more efficient and productive in recovering gold by crushing and amalgamation.

A typical arrastra consisted of a circular dugout that was lined with durable stones (see fig. 10.1). A vertical, rotating spindle was mounted in the center of the arrastra from which multiple horizontal arms were affixed. Large sturdy dragstones of up to 300 pounds each were fastened to the horizontal arms with chains. A waterwheel, horse, or mule was used to rotate the arms, causing the dragstones to crush and pulverize the charge of ore spread across the floor of the arrastra.

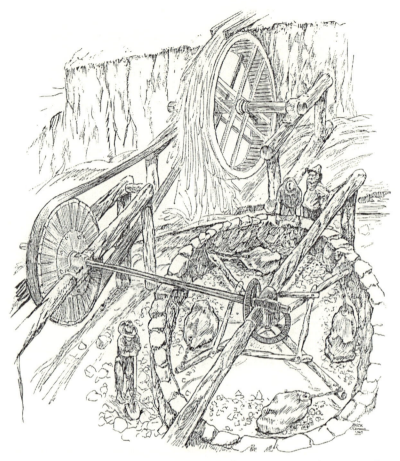

Fig. 10.1. Illustration by Buck O'Donnell of a water-powered arrastra. This illustration depicts the type of arrastra Fred and Moses Manuel, Henry C. Harney, and Alexander Engh constructed in August 1876. Energy from the waterwheel was transferred to the driveshaft and pinion gears, which caused the four arms to rotate. The dragstones crushed the ore as it was shoveled or fed into the arrastra.—Courtesy of the Black Hills Mining Museum.

As the ore was being crushed and pulverized in the arrastra, water was added to control pulp density. After a few hours of grinding, mercury was added to amalgamate the small particles of gold as they were liberated from the host rock. Mercury, also known as quicksilver, has an affinity for gold, silver, copper, and certain other metals. The mercury readily combined with small flakes of gold and silver to form an amalgam. The heavy amalgam became trapped in the spaces between the paving stones. After approximately twelve hours of grinding, additional water was added to the arrastra to thin the pulp. Plugs were then pulled, or a gate opened, and the mostly barren tailings were carefully drained and washed from the arrastra. The amalgam was collected, strained through a chamois, and heated in a retort. During retorting, the mercury was driven off as a vapor and recovered by distillation. The remaining gold and silver "sponge" was sold to an assayer or refiner who melted the sponge and poured the dore' into buttons or bars. The average arrastra crushed and pulverized an average of 170 tons of ore annually.[1]

The first ore from the Homestake claim was processed in an arrastra constructed by Fred and Moses Manuel, Hank Harney, and Alexander Engh on Whitewood Creek about 300 feet above the mouth of Whitetail Creek.[2] Over the winter of 1876-1877, the four men recovered $5,000 worth of gold from their arrastra.[3] By late August 1877, Moses Manuel abandoned the arrastra and constructed a 10-stamp mill on the Homestake claim. Manuel also contracted with owners of other custom stamp mills in the Lead area to have additional ore processed, thereby increasing the overall rate of production from the Homestake Mine.[4] This arrangement worked well for the Manuels and their partners but was woefully inadequate for George Hearst and his investors.

The Stamp Mill and Amalgamation

The stamp mill was much more efficient for crushing and pulverizing the ore compared to an arrastra. Depending on the hardness of the ore, most small custom stamp mills could crush 2-3 tons of ore per stamp per day. A typical stamp mill consisted of one to five stem-and-shoe assemblies mounted in a wooden timber frame (see fig. 10.2). This array of stamps, called a battery, was used to pulverize the ore in a cast-iron box called a mortar. A 10-stamp mill usually consisted of two 5-stamp batteries.

Each stamp assembly consisted of a wrought-iron or steel stem, tappet, head or boss, and shoe (see fig. 10.3) The stem was approximately 3 inches in diameter and 10-14 feet long. Attached to the bottom end of the stem was a cast-iron "head" or "boss" approximately 8 inches in diameter and from 15-24 inches long. The stem and boss assembly alone weighed 200-400 pounds. Attached to the boss was a replaceable shoe that pulverized the ore as it passed over a stationary die. The upper part of the stem included a moveable collar called a tappet. It regulated the stroke of the stem assembly. The tappet was periodically adjusted to compensate for wear on the shoe. Each stem was matched with a cam that was keyed to a rotating camshaft. As the camshaft rotated, the curved surface of each cam caused the corresponding tappet and stem to be lifted and rotated. The stamp would then fall as the tappet fell off the end of the cam. The varied positions of the cams on the shaft caused the stamp assemblies to rise under power and fall, by gravity, in a staggered fashion.[5]

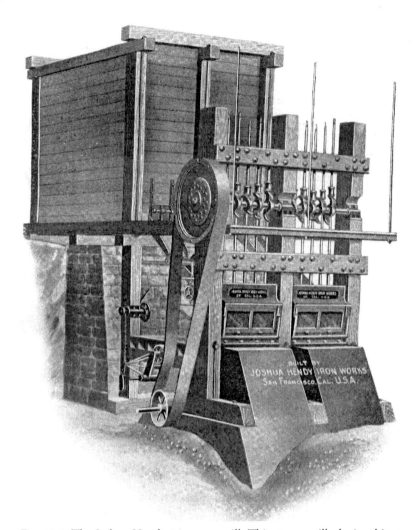

Fig. 10.2. The Joshua Hendy 10-stamp mill. This stamp mill, depicted in a Joshua Hendy Iron Works brochure in 1911, was very similar to the stamp mills supplied by the Fraser & Chalmers Company of Chicago and the Union Iron Works of San Francisco that were used in the various mills along the Homestake Belt. Most of the Homestake mills were designed and constructed around pairs of 5-stamp batteries.—Courtesy of MS Book and Mineral Company.

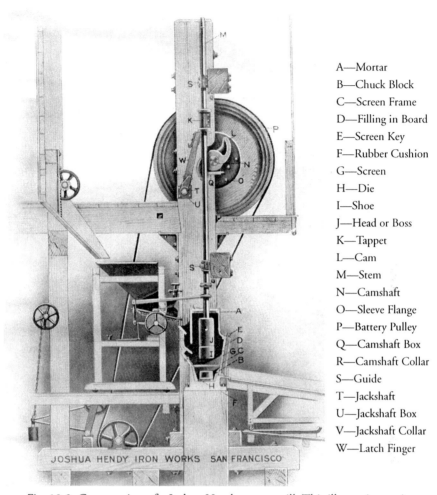

Fig. 10.3. Cross section of a Joshua Hendy stamp mill. This illustration, as it appeared in a technical brochure by the Joshua Hendy Iron Works in 1911, depicts a cutaway side view of a typical stamp assembly.—Courtesy of MS Book and Mineral Company.

Run-of-mine ore was usually fed through stationary crushers located near the hoisting works or in the upper part of the mill building. After the ore was crushed to 1½-1¾ inches, it was transferred to a bin located above the stamp batteries. From the bin, ore was fed into the mortars of the stamp batteries. Water was added, which helped move the ore over the stationary dies in the mortar. Each time a stamp fell, a portion of the ore was crushed. Mercury was added to the mortar at periodic intervals throughout the shift to amalgamate the gold as it was liberated from the host rock by the pulverizing action of the stamps. A significant percentage of the gold was periodically recovered as amalgam on copperplates inside the mortar.

After the ore was sufficiently pulverized, as determined by the size of the openings in the discharge screen, the "pulp," as it was called, was continually washed through the screen, located at the front of the mortar. The pulp passed over mercury-coated copperplates on the apron, which was located outside of the mortar. Additional gold was captured on the apron plates. The plates on the apron were approximately 10-12 feet long and abutted the mortar 4 inches below the discharge opening. The table was sloped 2 inches per linear foot, which caused the pulp to flow slowly enough to facilitate amalgamation of much of the remaining gold in the pulp. Below the apron plate was the sluice plate, which trapped much of any gold remaining in the pulp. The amalgam was then collected and placed in a "retort," where the gold and silver were separated from the mercury as a dore' using a process called retorting (see fig. 10.4). The mercury vapor was distilled in the retort and recovered as part of the process.[6]

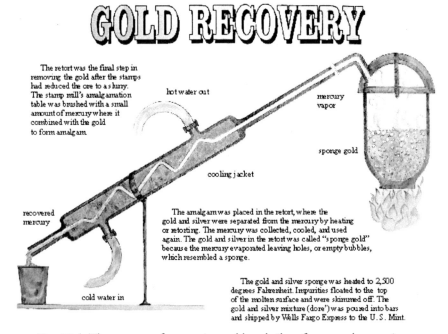

Fig. 10.4. The process of recovering gold and silver from amalgam using a mercury retort.—Courtesy of Hangtown's Gold Bug Park and Mine, City of Placerville, California.

The Large Stamp Mills along the Homestake Belt

Excluding the small 10-stamp mill of the Columbus Gold Mining Company in Sawpit Gulch, there were seven large stamp mills operating along the Homestake Belt by December 1880 (see table 10.1). As originally constructed, the seven mills had a combined capacity of 580 stamps. Homestake Mining

Company owned two of the mills, the Homestake and Golden Star (see fig. 10.5). Three of the other mines and mills were managed by Homestake but were technically owned by other companies controlled by the various directors and capitalists of Homestake. These other mills included the Highland Mill (see fig. 10.5), owned by the Highland Mining Company, the Deadwood Mill (see fig. 10.6), owned by the Deadwood Mining Company, and the Golden Terra Mill (see fig. 10.6), owned by the Golden Terra Mining Company. The latter two companies were consolidated as the Deadwood-Terra Mining Company on August 23, 1880.[7]

Table 10.1 The large stamp mills along the Homestake Belt

Name of Mill	Date Commissioned	(stamps) Initial/ Additional	(stamps) Capacity	Date HMC Acquired	Date Shut Down
Homestake	Jul 12, 1878	80	80	Owned	Nov. 18, 1922
	Sep. 1894	20	100	—	—
	Apr. 1897	100	200	—	—
	1915	20	220	—	—
Golden Star	Sep. 1, 1879	120	120	Owned	Mar. 1923
	May 1890	40	160	—	—
	Aug. 1895	40	200	—	—
Highland (Amicus)	mid-1880	120	120	Feb. 23, 1900	Dec. 1934
	1895	20	140	—	—
	Jul 1904	100	240	—	—
Deadwood	Feb. 17, 1879	60	60	Apr. 9, 1902	1887
	1881	20	80	—	—
Golden Terra (Pocahontas)	Jun 26, 1879	60	60	Apr. 9, 1902	Dec. 1934
	1881	20	80	—	—
	1889	80	160	—	—
Caledonia (Monroe)	1879	60	60	Jan. 22, 1901	Sep. 1925
	Oct. 1, 1889	20	80	—	—
	Before 1900	20	100	—	—

Father	Nov. 1878	80	—	Jun 17, 1901	Oct. 1918
DeSmet (Mineral Point)	1881	20	100	—	—
Tungsten Mill	Early 1916	5	5	Owned	Jun 1918
South Mill	Sep. 1922	120	120	Owned	1953

Sources: Data from Lounsbery and Haggin, Financial Agents, *Report of Homestake Mining Company from January 1, 1878 to September 1, 1880*; Joel Waterland, *The Spawn & the Mother Lode* (Rapid City: Grelind PhotoGraphics & Typesetters, 1987); Mildred Fielder, *The Treasure of Homestake Gold* (Aberdeen: North Plains Press, 1970); Homestake Mining Company, *Sharp Bits* 19, no. 2 (Summer 1968); Homestake Mining Company, "Mill Report for 1916."

Fig. 10.5. The Homestake, Golden Star, and Highland mills. This photo, taken by the Detroit Publishing Company in about 1900, shows (from left to right in the center) the Homestake Mill, Golden Star Mill, and Highland (Amicus) Mill. Above the Highland Mill are the Blacksmith Shop (left), Foundry (right), and Old Abe Hoisting Works (top). The Highland Mine is shown in the upper left directly above the Golden Star Shaft. The superintendent's residence is shown to the left of the Ellison Tramway.—Library of Congress (LC-DIG-ppmsca-18063).

Fig. 10.6. The Deadwood, Golden Terra, and Caledonia mills in Terraville. This Locke and McBride photo, taken in the mid-1880s, shows a portion of Terraville along with the Deadwood, Golden Terra, and Caledonia mills. Wood chutes extend to the Terra (Old Brig) hoist house, Deadwood Mill, and Golden Terra Mill. The Terraville Tunnel was located in Shoemaker Gulch, which extends to the left above the Caledonia Mill.—Photo courtesy of the Black Hills Mining Museum.

   1—Black Hills and Fort Pierre Railroad and Wood Chutes
   2—Old Brig Shaft and Hoist House
   3—Deadwood Mill
   4—Golden Terra Mill
   5—Hearst Mercantile Store
   6—Timekeeper's Office
   7—Caledonia Mill

Homestake Mill

Immediately after acquiring the Homestake claim, George Hearst realized he needed a much larger stamp mill to meet the needs of his business plan for the Homestake Mine. An 80-stamp mill was soon approved and ordered in November 1877 from the Union Iron Works in San Francisco. A Mr. Gilson from California was appointed superintendent of construction for Homestake 80-stamp mill. Several other people from the Union Iron Works helped erect the 80-stamp mill, which was considered to be the largest stamp mill in the world. These people included Meno Voight, blacksmith, and Samuel Wray, Smith

Russell, and Mr. McGregor as millwrights. Al Corwin, John Russell, and Mr. Higgins traveled from Nevada and Utah to assist in the project.

Upon arriving in Cheyenne, Wyoming Territory, the men found they would have to lay over for three days before a stage could take them to the Black Hills. Knowing that hostile Indians might be encountered along the Cheyenne-Deadwood route, or in the Black Hills, the men decided to go to the outskirts of Cheyenne and indulge in some target practice. Upon returning to town, the men were met by an officer who promptly arrested them and took them to see Judge Slaughter. The judge had little sympathy for the men's story and fined them for practically all the cash they had in their possession.[8]

The men from Union Iron Works reached Deadwood on February 21, 1878. Here, they were met by William A. Farish, superintendent of the Homestake Mine. From Deadwood, the men traveled to Central City, up Bobtail Gulch, through Terraville, over the hill, and down north Gold Run Gulch to Lead City. About the only mining that had been completed to date was Gwinn's Tunnel on the Homestake No. 2 claim and a shaft 60 feet deep, called Young's Hole, on the Homestake claim. Homestake's Incline Shaft, located near where the Golden Star Shaft was later sunk, was in the process of being developed.

The mill constructors found that the only mill nearby was the 30-stamp mill that had been built by A. G. Smith and commissioned as the M. L. White Mill in December 1877. The mill was primarily used to crush ore from the Homestake No. 2 claim. Gilmer and Salisbury, part of the Homestake syndicate, subsequently purchased the Homestake No. 2 claim in March 1878. Immediately thereafter, George Hearst and Samuel McMaster were assigned responsibility for operational management of the Homestake No. 2 claim. According to the men from the Union Iron Works, every afternoon Hearst and McMaster would arrive at the 30-stamp mill and work alongside the mill crew. Hearst pounded out samples in a hand mortar, and McMaster panned the samples. Mr. Swartout was appointed foreman of the White Mill in June 1878.

Some 100 tons of ore were sorted from the grading operations for the 80-stamp Homestake Mill. This ore was milled at John Costello's custom mill. On February 26, carpenters started to raise the building for the 80-stamp mill. The building was enclosed, and shingles were on the roof by March 8. That evening, a huge snowstorm developed. Three days later, the men found there was 4½ feet of snow on the ground, which hampered delivery and setting of the mill machinery.[9]

Conditions failed to improve. In two of his regular updates to J. B. Haggin in May 1878, George Hearst articulated the difficulty he and his crews were having in attempting to construct the world's largest stamp mill amidst the springtime conditions in the Black Hills:

> *I have delayed writing for the reason that things had not sufficiently matured in our matters and various interests. First, things were going to bad as fast as possible for anything and more than you can imagine. But things are getting along in very good shape in and about Homestake, and the mine is looking first rate and is perhaps a more valuable property than any of us had supposed. The mill is getting along slow partly on account of the very bad weather and condition of the roads. But my opinion is the sooner you get rid of Mr. Gilson the more money you will make and save (this is private) as he must now go on and complete the mill, but when completed and sunk, the amount of money that it will cost will make even your hair curl. I think him a good deal of a humbug.*
>
> *We are having one of the worst storms of the season; it has been snowing and raining continuously for four days and nights and the snow is from one to six feet deep according to locality and mud lay in the traveled roads two feet. You will perceive that all kinds of movements have stopped, save very difficult walking. I came up from Deadwood today and to get here we had to lead or drive our horses through snowdrifts from four feet to eight feet of snow. It was laughable to see the horses floundering and rolling through the snow; it seemed ridiculous, it being the 19th of May. It is a very warm snow and will disappear as soon as the sun appears, which I hope and think, will be soon. If the snow had not melted to a great extent as it fell, God only knows how deep it would have been.*[10]

Homestake's 80-stamp mill, commissioned on July 12, 1878, was completed at a cost of $164,500. The mill was powered by a 190-horsepower steam engine that was 20½ inches in diameter and 40 inches long.[11] Steam was supplied to the engine at 90 pounds per square inch from two wood-fired boilers, each of which was 54 inches in diameter and 16 feet long and contained forty-six tubes 3½ inches in diameter. The engine shaft formed a one-line shaft for the entire mill and was located at the battery sill. This arrangement minimized construction costs but had the disadvantage of having to shut the entire mill down if anything went wrong with the line shaft. Shortly after the mill was completed, the line shaft broke and had to be replaced with a stronger shaft.[12]

Golden Star Mill

On December 1, 1878, a new 120-stamp mill was ordered to supplement the 80-stamp mill that had been operating since July 12. A contract was made with the firm of Walter & Lind to construct the new Golden Star Mill for $28,000. However, additional land was needed for the new mill. Numerous houses and lots had already been purchased in December and January south and east of the Homestake Mill to accommodate the new mill.[13] Unfortunately, the land being graded for the new mill site also involved properties that had not been purchased since they were part of the pending litigation involving the Giant claim and the Old Abe Mining Company. One of these properties involved a shaft that was being sunk by Dan Reagan and Mike Hines on the Pride of the West claim owned by A. Cohen and A. L. London.[14]

On the morning of January 17, 1879, the graders preparing the site for the new Golden Star Mill were instructed by Angus McMasters to tear down the shanty over the Grand Prize Shaft and fill the shaft with waste rock. Cohen and London, who were at the shaft, found they were outnumbered. Around noon, Cohen and London appeared at the Homestake office with Deputy Sheriff James Lynch. Lynch presented Samuel McMaster with a cease and desist order to prevent the Homestake graders from filling the Grand Prize Shaft. In the meantime, J. Lewis, Andy Johnson, and other employees of Cohen and London regained possession of the shaft and proceeded to cover the shaft with boards and planks.

That afternoon, Samuel McMaster appeared at the site and saw that the shaft had been covered but was not filled. He started to remove the boards himself, but Lewis pulled him away. McMaster grabbed one of the boards and made a threatening gesture toward Lewis. Lewis, in turn, drew his pistol and warned McMaster to go away and keep away or he might get hurt. McMaster then had warrants issued against A. Cohen, P. Cohen, Andy Johnson, A. L. London, J. Lewis, and John Dowe for malicious trespass.

McMaster thought this would alleviate the situation and allow the graders to resume work. However, he soon discovered that several men were still protecting the shaft. William Travis, John Smith, Clark and Lee, employees of Homestake, armed themselves and took a position in an old cabin near the Lincoln Shaft. Angus McMasters, who was standing near the front door of the cabin, hollered at the men guarding the shaft. "Stand aside," he said, "if you don't want to be shot." Before the men could respond, a shot rang out from the cabin, and Aleck Frankenberg, one of the men guarding the shaft, staggered and fell. Two more shots were fired in rapid succession. The men inside the cabin then made a hasty exit through a window in the rear of the cabin.

Immediately, the men guarding the shaft and a few other bystanders stormed the cabin. A shotgun found in the cabin was later traced to Samuel McMaster as its owner. Dr. Bacon and Dr. White treated Frankenberg, who sustained facial wounds to his chin and right temple. Warrants were issued against Samuel McMaster, Angus McMasters, and William Travis, and a search was instituted for them and the two other men. Within an hour after the shooting, volunteers erected a new shafthouse in place of the one that had been torn down. Threats of lynching the men who shot Frankenberg were "loud and deep."[15] There was also talk of ordering Sam McMaster and Angus McMasters out of the city on the basis that "our citizens are aroused and have stood the insolence and abuse of these companies long enough."[16]

A. M. Willard, who coauthored *The Black Hills Trails* with Jesse Brown in 1924, was employed as a treasure guard for the State, Express and Transportation Company at the time of the shooting. As a treasure guard, Willard also held the appointment of deputy sheriff and could assist the sheriff whenever required. On the night of the shooting, Willard was in Deadwood waiting for his next stage run to Fort Pierre. The next morning he went down to the sheriff's office and offered his services to Sheriff John Manning. While the two were visiting, Sheriff Manning was summoned to the jail at once. Willard recounted the following in *The Black Hills Trails*:

> *I went up with the sheriff. We found the jailer, Black Jack Manning, a relative of the sheriff's, walking the floor very much excited. When I asked him what the trouble was, he told a very strange story. It seems that a notorious character by the name of John Flaherty had come to the jail in the middle of the night and, according to Black Jack's story, Flaherty offered Jack a large roll of bills said to contain five thousand dollars if he would open the jail door and let the gunmen go. I said, "Well, Jack, did you take the money?" He was very indignant to think that I would ask him such a question. I said, "Jack, you have lost a very fine chance," and he replied, "Would you have taken that money and let those murderers get away?" I said, "No, but if you had taken that money and locked Flaherty up you could turn in the money to the court and thereby made a fine case against this gang, and besides, you would have made a very fine name for yourself." Jack never thought of that. He was a good, square boy, but was not up to all the tricks going around at that time. The sheriff had been a very interested listener all during the conversation and said that he believed he did not know all that was going on inside of his office but would look into things a little closer from now on and asked me to cooperate with him.*[17]

A few days later, twenty-two-year-old Aleck Frankenberg died. A coroner's inquest concluded that the "deceased came to his death from exhaustion, caused by excessive hemorrhage from gunshot wounds received at the hands of an agent or agents employed by the California Company, Homestake No. 1, or Giant Consolidated, or their superintendent."[18] Frankenberg's funeral was the largest that had ever taken place in Lead at the time.[19] Within a few days of the shooting, Angus McMasters, John Clark, Lee Smith, and William Travis were arrested and put in jail.[20] On January 29, the *Black Hills Daily Pioneer* followed up on the story: "It is reported that strenuous efforts are being made to have Angus McMasters released on bail. Apropos, it is wondered why Sheriff Manning refused to take and serve a warrant for the arrest of Samuel McMaster last Friday night. This is merely a quest after information."[21]

A trial by jury lasted several days in Judge Gideon C. Moody's courtroom. Unfortunately, the witnesses to the alleged murder were nowhere to be found. Hearst, meanwhile, had been in San Francisco. After arriving back in the Black Hills, he wrote to Haggin on March 6 describing how all of the civil suits and the "shooting scrape" were making it "pretty lively around the courthouse:"

> *I arrived here in good health, had a rather cold trip, but the weather has been very good since my arrival and is now getting quite warm. I found all of our troubles at fever heat, and most of them at a culminating point, the shooting scrape combined with the various civil suits, made it pretty lively around the Court House.*
>
> *We have all the boys out of jail except three, and they are there for murder and will be tried this term of court. Their trial was to have come off this week but the most important witnesses could not be found, suppose they have gone to Salmon River, the case is now set for next Wednesday. I hope all will be present and the case go on, as, the way it is now, it consumes time and sets us back in our civil business, which is not small, as you will readily see by the number of protests, pro and con.*
>
> *As to the lawyers, I think we have made no mistakes in retaining Thomas for the reason that he is a son-in-law of Barnes, presiding Justice of the Territorial Supreme Court, and he (Barnes) got the President to appoint Moody (our District Judge) who has the appearance of being a very stern and upright Judge and although Mc[Master] does not like Thomas, still I think he presents his views in a clear and able manner and it would have been impossible to have gotten our papers in proper shape unless we had him or some other such man.*[22]

On March 19, 1879, the jury found the three men "not guilty." They were acquitted and set free. Later that day Hearst wrote another letter to Haggin informing him of the results of the trial: "The parties who were indicted for killing the man Frankenburg, in the mill site trouble, were today acquitted, but the hostility of the judge seemed to be very great at the acquittal, as he took the jury very much to task and dismissed them from the court. What other trouble may rise from it, we don't know, at any event we will try to make the best of it. All the parties arrested in connection with the affair are now out except one, who was acquitted by the present grand jury, but is held over under bonds for the next grand jury."[23]

The *Black Hills Pioneer* provided a detailed accounting of the outcome of the trial and Judge Moody's remarks on March 20:

> *The jury, in the case of the Territory vs. Clark, Travis, and Smith, having returned into court yesterday morning with a verdict of not guilty, the judge expressed extreme astonishment at the result, and animadverted severely upon the conduct of the causes which produced such a result, saying in substance, that the facts proven upon the trial were so manifestly plain, pointing to the guilt of the defendants, that it seemed to him no man could possibly doubt their guilt. The judge further said that if men could be shot down like dogs in broad daylight, the perpetrators of the act, arrested almost upon the spot, witnesses spirited away, the evidences of the criminal act concealed, and all supplemented by evident perjury and subornation of perjury, and the criminals promptly acquitted by a jury, then a jury trial was simply an unmitigated farce, and that he had rather entrust his rights, if such was to be the cause of the administration of the case, to a half dozen Piegan Indians; for then he would be assured that at least some of his natural rights would be respected.*
>
> *He also said there was one possible consolation about all this, and that was that the actors in this scheme to defeat the ends of justice had, by their conduct in this case, placed themselves wholly in the power of unscrupulous men, and that if at any time they failed to respond to whatever demands were made upon them by the tools they had used, those tools could, with perfect impunity, speak the truth and bring them to the bar of justice for these acts, and that here, publicly and openly, he invited every one of those men that had been used in this case, at any time when their demands were not responded to, either now, or months, or years hereafter, to come forward, speak out, and bring these men to justice.*
>
> *After some further remarks, not at all complimentary to such a mode of procedure, he discharged the jury from the case and from further attendance upon the court as jurors, saying that he had no further use for them.*[24]

In his book, *The Black Hills Trails*, A. M. Willard reported Judge Moody's comments slightly differently. According to Willard, Judge Moody scorned the jury, saying, "I would sooner have a jury of Pagan Indians try cases in this court than men of your kind, you can always secure money now from the Homestake Company as they have bought you outright." The judge then turned to Sheriff Manning and said, "Mr. Sheriff, if you ever bring any of these men into my court again to serve as jurors, I will commit you for contempt of court. They are a disgrace to the country." Later, according to

A. M. Willard, the bailiff was arrested and charged with being implicated and was said to have handled the money that reached the jury. Nothing was ever proven, Willard wrote, "although it was a well-known fact that several of the jurors had plenty of money after the trial was over." The jury became known locally as the "Pagan Jury."[25]

In February 1879, the Corliss engine for the 120-stamp Golden Star Mill was shipped from Providence, Rhode Island, to Bismarck on the Northern Pacific Railroad. In Bismarck the 89,000 pounds of machinery were transferred to the Northwestern Stage and Transportation Company for final transport to Lead City using freight wagons.[26] At about the same time, the mill itself was received from the Fraser & Chalmers Company of Chicago. The *Black Hills Daily Times* announced, "Another big mill for the Homestake Mine. Two freight trains, aggregating about seventy-five wagons and belonging to the Northwestern Transportation Company, are expected to be in tomorrow or the next day, and choak up the gulch from here to Lead City. The trains are loaded principally with machinery for the second great Homestake mill, located at Lead."[27]

The 120-stamp Golden Star Mill, located on the Lincoln claim, was commissioned on September 1, 1879, at a cost of $251,500, including property acquisitions. A 90×156-foot building housed the stamp mills and other machinery. A tramway entered the end of the building 54 feet above the floor.[28] A 300-horsepower Harris-Corliss engine with a 26-inch-diameter bore and 60-inch stroke powered the Fraser and Chalmers Mill. Four boilers, each 56 inches in diameter and 16 feet long, supplied steam to the engine.

Highland Mill

By spring 1879, a decision was made to construct a new 120-stamp mill for the Highland Mining Company. R. D. Millet was selected to design the mill. Millet had recently designed and installed the stamp mills for the Deadwood Mining Company and the Golden Terra Mining Company using equipment from the Fraser & Chalmers Company. Hearst and McMaster were pleased with those two mills. On July 31, 1879, after negotiating the price, McMaster contracted with Fraser and Chalmers to furnish the machinery for the new 120-stamp mill for a lump sum price of $51,000.[29]

A 114×92-foot mill building was constructed to house the 120 stamps along with a 56×76-foot addition to house the Corliss engine and four steam boilers (see figs. 10.7 and 10.8). The mill became known as the Highland Mill and commenced crushing in mid-1880. After the 120-stamp Highland Mill was commissioned, the White 30-stamp mill was decommissioned.[30]

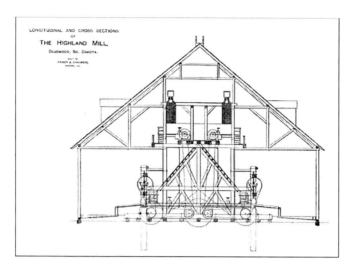

Fig. 10.7. Cross section of the Highland Mill by Fraser and Chalmers showing the ore bins and drives for the two opposing lines of stamp batteries.—Source: Data from H. O. Hofman, "Gold-Mining in the Black Hills," *American Institute of Mining and Metallurgical Engineers Transactions* 17, (February 1889).

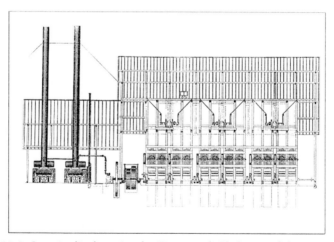

Fig. 10.8. Longitudinal section by Fraser and Chalmers of the 120-stamp Highland Mill. The section shows the four boilers, tramway, ore bins, and twelve 5-stamp batteries. The other twelve 5-stamp batteries were situated immediately behind this line of stamps, as shown in the accompanying cross section.—Source: Data from H. O. Hofman, "Gold-Mining in the Black Hills," *American Institute of Mining and Metallurgical Engineers Transactions* 17, (February 1889).

The mill machinery, supplied by Fraser and Chalmers, consisted of a Corliss engine fitted with a 26×60-inch cylinder, a 20,000-pound flywheel 20 feet in diameter, and two pulleys 12 feet in diameter designed to drive 30-inch-wide belts. Four tubular steam boilers 54 inches in diameter and 16 feet long were also supplied. The boilers were designed to operate in pairs of two. The boiler stacks were 42 inches in diameter and 70 feet high.

Ore was fed onto six 4×11-foot grizzlies that were fitted with iron bars spaced two inches apart. Oversize rock from the grizzlies was fed through six 15×9-inch Blake rock crushers. The 120-stamp mill was configured with twenty-four batteries of five stamps each. The inside front of each mortar was fitted with a copper amalgamation plate ⅛ inch thick and 7 inches wide. The twenty-four amalgamation tables were fitted with a copperplate ⅛ inch thick, 55½ inches wide, and 10 feet long. At the end of each pair of tables was a sluice fitted with a copperplate ⅛ inch thick, 6 inches wide, and 10 feet 6 inches long. The sluices from every four sets of batteries fed a single sluice fitted with a copperplate ⅛ inch thick, 18 inches wide, and 12 feet long. Each of these six sluices fed a long sluice fitted with copperplates ⅛ inch thick, 14 inches wide, and 10 feet long.[31]

Operation, Standardization, and Optimization

Between January 1878 and September 1880, water for the Homestake mills was purchased from the Black Hills Canal and Water Company at the rate of $2 per stamp per week, which equated to about eleven cents per ton of ore crushed. Forty miner's inches of water (approximately 0.8 cubic feet per second) were required to operate the Homestake and Golden Star mills.[32] By 1887, the cost of water for the three mills in Lead had increased to $3.50 to $4 per stamp per week.[33]

Unfortunately, the stamp mills had an insatiable appetite for wood as well as ore. Large amounts of cordwood were needed to fuel the boilers that provided steam for the engines in the mills and hoist houses. The Black Hills and Fort Pierre Railroad delivered wood to the various Homestake mills and hoisting works from April through December each year. The Road, as it was called, was closed during the adverse winter months, which meant that a large inventory of cordwood and mine timbers had to be available at each of the mines and mills prior to the onset of winter. Rail spurs were constructed above the mills and hoist houses to facilitate unloading and delivery of cordwood using specially constructed chutes.

The chutes were 700-1,500 feet long, 25 inches wide, and 12 inches deep. The bottoms and lower 9 inches of the sides were lined with ¼-inch-thick iron plates. During the summer months, a small stream of water was run through the chutes to provide lubrication and cooling. A discharge "nozzle" was affixed to the bottom end of each chute, which directed the fast-moving cordwood to the desired part of the woodpile.[34]

Between May 1, 1878, and September 1, 1880, the cost of cordwood delivered to the Homestake and Golden Star mills averaged $4.75 per cord, which equated to $0.27 per ton of ore milled. For the first twenty-eight months of operation, the two boilers in Homestake's 80-stamp mill consumed an average of 10.7 cords of wood per day. At the 120-stamp Golden Star Mill, the four boilers required 17.6 cords of wood per day during the first twelve months of operation and about fourteen cords of wood per day thereafter as operating efficiencies improved. Cordwood alone accounted for 25 percent of total operating costs of these two mills. Approximately eighty-three cords of wood were required to fuel all seven of the large mills along the Homestake Belt each day throughout the mid- to late-1880s.[35]

Despite high fuel costs, the new mining and milling designs worked well. Ore was removed from the mines, hauled to the respective mill, and dumped onto a grizzly. The grizzly consisted of an array of wrought-iron bars, 1 inch wide and 2-4 inches deep. The bars were spaced 1½-2 inches apart. Each grizzly was 3-4½ feet wide, 10-14 feet long, and positioned at an angle of about 40 degrees, depending on the mill. As the ore was dumped onto the grizzly, the minus 2-inch ore passed through the bars and fell into the mill bin. The plus 2-inch ore rolled down the grizzly and onto the crusher floor. The oversize ore was then hand shoveled into the rock breakers.

As early as 1881, after Homestake had acquired operating control of six of the seven large stamp mills along the Homestake Belt, the mills were standardized as much as possible. The Homestake Mill consisted of machinery from the Union Iron Works of San Francisco. The Fraser & Chalmers Company of Chicago supplied the machinery in the Golden Star, Highland, Deadwood, and Golden Terra mills. Homestake also modified the Father DeSmet Mill, which was originally designed by Superintendent Augustus J. Bowie, after the Homestake capitalists gained controlling interest in the Father DeSmet Consolidated Gold Mining Company in 1881. The Caledonia Mill was the only mill that was significantly different from the others, partly because of the hardness of the ore, which required a greater stamp drop.[36]

Eli Whitney Blake of New Haven, Connecticut, designed the first jaw crusher in 1854, based on the need to provide crushed stone for road construction projects.[37] Blake's crusher was designed with a stationary jaw and a moveable jaw. The closing action of the moveable jaw caused the stone to be crushed to the desired size. The California gold miners quickly adapted the Blake crusher in their mining operations to crush ore. By 1879, the Blake Rock Crusher Company had sold over five hundred crushers to mines and quarries, including the Homestake Mine.[38]

Sixteen Blake jaw crushers were purchased by Homestake Mining Company and the Highland Mining Company. The first four were installed at the new Homestake Mill in 1878. The Golden Star and Highland mills were constructed with six each in 1879 and 1880, respectively. A No. 5 crusher was used, which reduced the run-of-mine ore to minus 1¾ inches. One crusher was used for every 20 stamps. The three stamp mills in Lead were configured with two lines of stamp batteries that were situated directly below the crushed ore bin. The bottom of the crusher bin was constructed to form of an inverted V, which facilitated the transfer of ore into the "feeders." The feeders regulated the flow of ore into the various stamp mortars. Two types of feeders were commonly used, including the Hendy Challenge feeder and the Tullock Automatic feeder. One feeder was required for each stamp mortar.[39]

When Homestake expanded the Caledonia Mill from 60 stamps to 80 stamps in 1889, a single No. 6 Gates crusher was purchased to replace the three No. 5 Blake crushers.[40] The Gates crusher, which incorporated one of the first gyratory design concepts, was designed and patented by Philetus W. Gates in 1881.[41]

Based on the success of the Gates crusher at the Caledonia Mill, Homestake quickly replaced all of its other Blake crushers with the more efficient Gates No. 6 crusher. The Gates crushers were installed at the respective hoisting works instead of the mills. Three Gates crushers were installed at the Homestake Mine, two at the Highland Mine, and two at the Deadwood-Terra Mine.[42] By 1912, nineteen Gates crushers were in operation at the shafts, with six at the Ellison, four at the Golden Prospect, four at the Old Abe, three at the Old Brig, and two at the Golden Gate.[43]

The original mortar used in the Homestake mills consisted of a one-piece casting that weighed 5,400 pounds. It was 54½ inches high and 54¾ inches long. The feed opening was 24 inches long, 4½ inches wide, and 7 inches deep.[44] After years of experience using stamps made by the Union Iron Works and Fraser and Chalmers, Homestake modified the mortar in the early 1890s by narrowing it through the area of the inside lip. This change increased the flow rate of pulp through the mortar yet held the material long enough to facilitate effective amalgamation inside the mortar. The modified mortar weighed 7,300 pounds. Its overall dimensions increased slightly to 58¼ inches high, 56¾ inches long, and 28¼ inches deep.[45]

The Homestake 850-pound stamp assemblies used throughout the 1880s consisted of a wrought-iron stem 14 feet long and 3½ inches in diameter that weighed 340 pounds, cast-iron head 18 inches long by 9 inches in diameter at the top and 8 inches in diameter at the bottom that weighed 240 pounds, cast-iron shoe 8¼ inches in diameter by eight inches high that weighed 140 pounds with its shank, and a cast-iron tappet 12 inches long by 9¼ inches in diameter that weighed 130 pounds.[46] When Homestake modified the mortar in the early 1890s, the overall weight of the stamp assembly was increased to 900 pounds. At the three mills in Lead, the drop was increased to 10½ inches, with each stamp falling eighty-eight times per minute. This arrangement produced an extremely fine pulp with approximately 80 percent passing through a 100-mesh screen.[47]

The preferred order of drop for the stamps in a 5-stamp battery at Homestake's mills was 1-3-5-2-4. The camshaft was configured so that the center stamp dropped last in the sequence. Although other variations in the order of drop yielded less wear on the mortar, the 1-3-5-2-4 arrangement was found to maximize production.[48] After much use, a die wore to a convex surface, whereas a shoe became slightly concave at its base. A die was changed at nominal three-month intervals after it had worn to within one inch of the foot plate.[49]

Prior to the development of Charles W. Merrill's cyanide process at mines in Montana, California, and Arizona between 1893 and 1897 and implementation of the process at Homestake's Cyanide Plant No. 1 in 1901, gold was recovered almost exclusively by amalgamation in the stamp mills. The majority of the gold recovered was collected on one or two amalgamated copperplates inside each mortar.

A second set of amalgamated copperplates, collectively referred to as the apron plate, was mounted on an inclined table outside the mortar. The purpose of the apron plate was to collect any particles of gold, free mercury, or amalgam not collected or retained on the inside plate. As the ore became sufficiently pulverized to pass through the screen, the motion of the stamps and flow of water caused the fine pulp to flow over the amalgamated apron plate. A splash board was installed across the head of the apron plate to reduce the disturbance on the plate as the pulp was discharged from the mortar. In all of the Homestake mills, except for the Deadwood and Golden Terra, the apron plate was ⅛ inch thick, 55½ inches wide, and 10 feet long. In the other two mills, the plates were 12 feet long. The apron plates in the Caledonia Mill were 8 feet long. The apron plates in the Homestake mills were fastened with screws to a wooden table that sloped 2 inches per foot. The last 4 feet of the 16-foot-long table was reduced to 4 feet wide and contained two overlapping blankets that served to trap the heavy sands. Every half hour, the blankets were cleaned.

A replacement apron plate was prepared by scouring it with sandpaper and emery cloth. The surface of the copperplate was then dressed with a coat of potassium cyanide. After two days, mercury was sprinkled on the cyanide coat and rubbed into the plate with a moist cloth and tailings. The amalgamated plate was then installed on the table. For the next 2-4 weeks, a higher-than-normal amount of mercury was added to the mortar at regular intervals to normalize the apron plate. Additional potassium cyanide was added to the mortar if the apron plate became discolored from copper salts.[50]

Mercury was added to the mortars at the rate of about 10 pounds 14 ounces per day at the Deadwood Terra Mill in 1895. The Homestake Mill averaged 12 pounds 2 ounces per day; the Golden Star Mill, 24 pounds 3 ounces per day; and the Highland Mill, 16 pounds 5 ounces per day. For the year ending June

1, 1894, the Homestake Mill consumed 2,034 pounds of mercury, while the Golden Star consumed 3,440 pounds. Approximately 22 percent of the mercury consumed was not recovered.[51]

The pulp from the apron plate and blankets flowed into a mercury trap located at the foot of each table. The gold, amalgam, and mercury not collected on the mortar plate or apron plate were mostly captured in the mercury trap. The pulp from each battery then passed through a sluice box and through a secondary mercury trap located outside the mill. In the Homestake 80-stamp mill, 80 ounces of amalgam and 144 ounces of mercury were recovered in the inside mercury trap each month. An additional 10-12 ounces of amalgam and 40 ounces of mercury were recovered monthly in the outside mercury traps.

The mercury trap at the foot of each table consisted of a 14×17-inch wooden box, 24 inches deep. The box contained three wrought-iron baffle plates and a bottom copperplate. The pulp flowed under the first baffle plate, over the middle plate, and under the third plate. The sluice box, which received pulp from the inside trap, consisted of a wooden trough, or launder, with a copperplate on the inside bottom. The sluice box sections were 8-10 feet long, 18 inches wide, and had a drop of about 1 inch per foot. The outside trap consisted of a wooden box 48 inches long, 14 inches wide, and 48 inches deep. This trap also had three inside partitions. Prior to construction of Cyanide Plant No. 1, the tailings from each outside trap were discharged into the two settling dams situated on Gold Run Creek.[52]

Through the 1880s, each of the seven stamp mills along the Homestake Belt required twelve to twenty-five total employees who worked two twelve-hour shifts per day. A foreman was assigned responsibility for two or three mills and was paid a wage of $6.50 per hour. Depending on the size of the mill, a typical crew on each shift included a millwright, an engineman, a fireman, a head amalgamator, one or more amalgamators, one or more crushermen, an oiler, one or more feeders, and possibly a laborer. A millwright at either the Homestake Mill or Golden Star Mill was paid $4.50 per hour; an engineman, $3.50; a fireman, $3; a head amalgamator, $4; and an amalgamator, $3.50. The rest were paid $3 per hour, except for the laborer, who was paid $2.50 per hour.

At the start of each day shift, the mill crew was responsible for shutting down the mill and collecting amalgam from the apron plates (see fig. 10.9). This process consisted of shutting off the water supply to the batteries, hanging the stamps, and removing the splash board. The loose sands were washed off the apron plates. Amalgam was loosened from the plates using heavy whisk brushes and rubber scrapers, swept to the head of the apron, and brushed into an iron receiver. Lastly, the plates were brightened by brushing them with tailings moistened with potassium cyanide. The plates were then redressed with mercury and smoothed with a paintbrush (see fig. 10.10). Four hours were required to clean and redress the twenty-four apron plates at the Golden Star Mill.

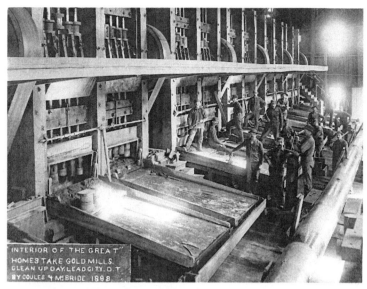

Fig. 10.9. Cleanup day at the Homestake 80-stamp mill, 1888. This photo by Coules and McBride shows members of the mill crew scraping and brushing the amalgam off the apron plates in the Homestake Mill.—Photo courtesy of Homestake Mining Company.

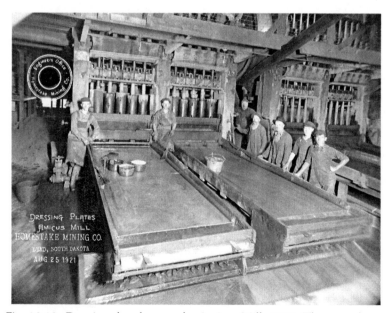

Fig. 10.10. Dressing the plates at the Amicus Mill, 1921. The crew shown here has just finished redressing the apron plates with mercury at the Amicus Mill.—Homestake Collection, Adams Museum, Deadwood.

The amalgam was placed in a Wedgewood mortar and diluted with fresh quicksilver (mercury) and water. The amalgam was worked by hand to free any impurities, which were carefully washed off the surface of the amalgam. Once the amalgam was brightened through cleaning, it was passed through a strainer to remove excess quicksilver. The remaining amalgam paste was then placed in a piece of linen and squeezed by hand to remove excess mercury. The amalgam balls were stored until a large batch could be retorted (see fig. 10.11).

Fig. 10.11. Balls of gold amalgam. Amalgam removed from the mortar plates and apron plates of the stamp mill batteries was filtered through linen cloth or canvas bags and squeezed into balls to remove excess mercury and facilitate retorting.—Photo courtesy of Homestake Mining Company.

Twice each month, the mills were shut down to remove amalgam from the copperplate inside the mortar. This was accomplished by removing the splash board, curtains, screen frame, and chuck blocks. The curtains, screens, and chuck blocks were washed and placed on the aprons where the amalgam was removed by scraping. The mortar plate was also carefully scraped to remove the amalgam. After all the amalgam was removed, quicksilver was applied to the mortar plate to redress the plate and soften any residual amalgam. The redressed plate was scoured with tailings to brighten the dressing. As a final step, the quicksilver was smoothed with a paintbrush. Any required maintenance or replacement of parts on the stamp assembly was concurrently completed.

The amalgam collected from the apron plates, chuck blocks, screens, mortar plates, sluices, mercury traps, old shoes, and dies was placed in a large cleanup pan 5 feet in diameter. After adding water and 600-700 pounds of quicksilver, the pan was rotated at 30 revolutions per minute for approximately three hours to remove any impurities. The amalgam was then placed in a strainer to remove the liquid mercury. Any remaining liquid mercury was removed by placing the amalgam paste in canvas bags full of water and squeezing the bags. The resulting balls of hard amalgam contained approximately 38 percent gold.

Gold and silver were separated from the mercury through retorting and melting. Bulb retorts and cylindrical retorts made of cast iron were used in the 1880s to accomplish this work. The cylindrical retort was 12 inches in diameter, 36 inches long, and held 1,000 pounds of amalgam. However, a charge of 500 pounds was common and required six hours and one-quarter cord of wood to complete the process. The resulting dore' was melted into bars.[53]

In about 1901, the composition, number, and arrangement of the apron plates were modified to improve gold recoveries outside the mortar (see tables 10.2 and 10.3). Two plates of plain copper, each of which was 54 inches wide and 12 feet long, were placed in line with the flow of pulp from each pair of 5-stamp batteries. Below these two plates, three silver-plated copperplates were placed transverse to the flow. These three plates were electroplated with two ounces of fine silver per square foot of plate.[54] A third row of silver-plated copperplates was also placed in line with the flow of pulp. A final set of silver-plated copperplates was installed at the plate houses, where the final step of gold recovery by amalgamation took place. The electroplating was performed at the assay office where one of Homestake's first electrical motors was used to drive a generator to facilitate the process.[55]

Table 10.2 Plate arrangement and approximate distribution of recovery at the Amicus Mill[1]

Row No.	Type	No. of Plates	Plate Area (sf)	Amalgam (%)	Values (%)
Mortar Plate	Copper	2	4	34.3	46.5
1	Copper	2	108	46.4	43.0
2	Silver-plated	3	162	10.6	5.7
3	Silver-plated	3	162	5.6	3.1
4	Silver-plated	3.5	189	3.1	1.7
		—	625	100.0	100.0

(1) Based on two, 5-stamp batteries crushing 42 tons per day.

Source: Data from Allan J. Clark and W. J. Sharwood, "The Metallurgy of the Homestake Ore," *American Institute of Mining and Metallurgical Engineers* 22, (1912): 77.

Table 10.3 Distribution of gold recovery in the Homestake stamp mills (May through August, 1910)

Amalgam From:	Retorts Bullion Percentage	Mill Recovery Percentage of Total Value	Fineness of Refined Bullion		Ratio Silver to Gold
			Gold	Silver	
Batteries	43.8	49.4	814.5	175.5	0.215
First Row Plates	28.1	37.5	797.5	180.5	0.226
Silvered Plates					
Second Row	23.8	4.4	573.0	418.0	0.730
Third Row	25.7	2.3	564.5	425.5	0.755
Fourth Row	24.4	0.9	538.5	451.5	0.839
Cleanup	20.7	4.0	779.0	186.0	0.237
Traps	22.4	1.5	766.0	215.0	0.280
All Sources:	32.5	100.0	780.0	205.5	0.263

Source: Data from Allan J. Clark and W. J. Sharwood, "The Metallurgy of the Homestake Ore," *American Institute of Mining and Metallurgical Engineers* 22, (1912): 81.

An average of one-quarter ounce of mercury was required per square foot to "set" the silver-plated copperplates before commencing amalgamation.[56] In 1902, as a result of the additional rows of silver-plated copperplates, overall gold recovery by amalgamation increased by approximately $250,000, which was approximately 5.6 percent of total revenue.[57]

Over a four-month period in 1910, the overall gold recovery by amalgamation in the Pocahontas and Monroe stamp mills at Terraville was 72 percent of the estimated ore value. At the three stamp mills in Lead, gold recovery was reported to be 74 percent.[58] During the first several decades of operation, Homestake attempted little or no sampling in the stamp mills to determine the actual head grade of the ore. In general, the ore that was milled was not weighed or assayed. A sampler, who collected samples of ore from the various stopes where the ore was being mined and delivered to the respective mills, made the estimate of ore value. On average, the sampler pulverized and panned fifty to fifty-five samples per day. The estimated ore value was based on the amount of visible gold found in the pan; no attempt was made to assay the remaining concentrates. As a result, reported recoveries from the stamp mills were probably overstated.[59]

The Mills under Homestake's Ownership

By December 1880, the Homestake capitalists had acquired enough stock in the Father DeSmet Consolidated Gold Mining Company to have controlling interest in the company and operational control of the mine and mill. The Homestake capitalists acquired controlling interest in the Caledonia Gold Mining Company in about 1886. Homestake now had management and operating control of all the major mines and mills along the Homestake Belt.[60]

In 1887, the Deadwood-Terra Mining Company shut down the Deadwood Mill. Its 80 stamps were moved to the Golden Terra Mill in 1889. The remodeled 160-stamp mill became known as the Deadwood-Terra Mill.[61]

Between 1893 and 1905, the boards and shareholders of the Highland, Caledonia, Father DeSmet, Deadwood-Terra, and Homestake companies, all of which were controlled by the Homestake capitalists, voted to merge their respective companies into Homestake Mining Company. The Highland Mill was subsequently renamed the Amicus, the Caledonia Mill was renamed the Monroe, the Father DeSmet Mill was renamed the Mineral Point, and the Deadwood-Terra Mill was renamed the Pocahontas.

Under a lease agreement, Homestake began crediting production from the Highland Mine to the Homestake Mine in May 1898.[62] Assets of the Highland Mining Company were deeded to Homestake Mining Company on February 23, 1900.[63] The Amicus Mill was considered by Homestake to be the best of the older stamp mills. In July 1904, an additional 100 stamps were added to the mill, boosting its capacity to 240 stamps.[64] When the mill was remodeled in 1915 and 1916, the weight of each stamp assembly was increased to 1,100 pounds.[65] In 1924, the tube mills were moved from the Regrind Plant to the Amicus Mill to reduce the pulp size.[66] After the South Mill was enlarged in 1934, the Amicus Mill was decommissioned in December.

With the expansion of the Amicus Mill in 1904, the six operating stamp mills had a combined capacity of one thousand stamps. In 1915, an additional twenty stamps were added to the Homestake Mill, bringing its capacity to 220 stamps. Total milling capacity from the six large mills owned by Homestake was now at 1,020 stamps. The Tungsten Mill also added five stamps. With these expansions, milled production increased to a new high of 1.678 million tons of ore in 1917 (see table 10.4).[67]

Table 10.4 Mill production in tons through the Homestake stamp mills (1917-1920)

	(220 stamps) Homestake	(200 stamps) Golden Star	(240 stamps) Amicus	(160 stamps) Pocahontas	(100 stamps) Monroe	(100 stamps) Mineral Point	(5 stamps) Tungsten Mill	(1,025 stamps) Total
1917	361,450	313,144	403,986	241,102	178,882	171,978	7,081	1,677,623
1918	348,300	254,157	399,798	285,604	180,206	153,937	6,628	1,628,630
1919	259,564	218,635	325,706	228,933	154,984	0	4,766	1,192,588
1920	241,375	274,167	395,638	242,783	120,981	0	0	1,274,944

Source: Homestake Mining Company, "Metallurgical Summary, 1917-1920."

Homestake shut down the Mineral Point Mill in October 1918, which reduced overall capacity to 920 stamps from the six large mills. Upon commissioning the 120-stamp South Mill in September 1922, the Homestake Mine reached its peak capacity of 1,040 stamps, but only for about two months. The South Mill was intended to replace the Homestake Mill and the Golden Star Mill, both of which needed to be moved because of subsidence issues and expansion of the mine. The latter two mills were subsequently shut down in November 1922 and March 1923, respectively.[68] After the Monroe Mill was shut down in September 1925, total milling capacity was reduced to 520 stamps housed in the Amicus, Pocahontas, and South mills.

Retorting and Refining

Every ten days, amalgam from the stamp mills was taken to the assay office (see fig. 10.12) where the amalgam was weighed and placed in cast-iron trays. The trays were placed in one or more of five horizontal, cylindrical retorts, each of which held 7,000 ounces of amalgam (see fig. 10.13). After retorting was completed, the crude bullion was weighed, divided, and remelted in graphite crucibles in coke-fired furnaces. Each crucible contained four pounds of pulverized borax crystals. Upon completion of melting, the bullion was cast into bars. A small amount of water was poured on top of each bar to cool the slag, which was then pulled or scraped off. Using tongs, each bar was immersed in a bath of very dilute nitric acid and then in cold water. The cooled bars were scoured with wet sand, rinsed, and dried. Each bar was sampled by drilling small holes in opposite corners of the bar and assaying the bullion samples. Finally, the bars were individually weighed, stamped, bagged, and sold, completing the gold-recovery process using stamp mills, amalgamation, retorting, and refining prior to 1901 (see figs. 10.14 and 10.15).[69]

Fig. 10.12. The Homestake Assay Office, 1896. The Homestake Assay Office was located on the south side of East Main Street across from the Central Steam Plant. Amalgam from the stamp mills was processed in five cylindrical retorts.—Photo courtesy of the Black Hills Mining Museum.

Fig. 10.13. Amalgam retorts in the old assay office. Each of the retorts held 7,000 ounces of amalgam.—Photo courtesy of Evelyn (Schnitzel) Murdy.

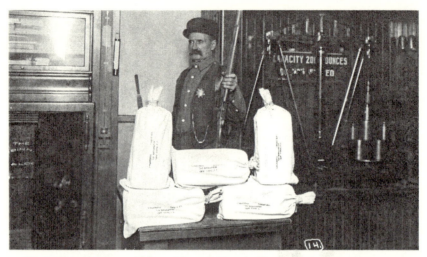

Fig. 10.14. Sacking gold bars, Harry J. Teer, bullion guard.—Photo courtesy of Evelyn (Schnitzel) Murdy.

Fig. 10.15. Wells Fargo Express Company guards transferring $250,000 of gold bullion from the Homestake Mine, 1890.—Library of Congress (LC-DIG-ppmsc-02598).

Chapter 11
The Need for Cordwood, Mine Timber, and a Railroad

Justification for the Railroad

For the first several decades of operation, the Homestake Mine consumed a tremendous amount of wood for lumber, mine timbers, and cordwood. Virtually, all of the hoists, stamp mills, and sawmills were powered by steam engines and boilers that required cordwood for fuel. Lumber was needed throughout the surface and underground areas. Pine logs were sawn and framed into various types of mine timbers that were used to frame the shafts and provide ground support in the square-set stopes.

One of Homestake's first sawmills was located in Strawberry Gulch. The mill was supervised by James Van Dyke.[1] On October 12, 1879, the sawmill, including an inventory of 60,000 board feet of lumber and timber, was destroyed by fire. The fire was suspected by company officials to be "the work of an incendiary." A new sawmill was constructed in Murphy Gulch in the Yellow Creek area. This mill had a capacity of 9,000 board feet per day.[2]

By mid-1879, Homestake began constructing a 22-inch-gauge railroad to haul ore and deliver cordwood, mine timbers, and other supplies to the hoisting works and stamp mills in Lead City. On August 26, 1879, the J. B. Haggin locomotive was commissioned and began its primary work task of hauling ore to the Homestake Mill. The wood-fired, steam locomotive was a huge success.

As Homestake began to acquire operating control over some of the other mines located along the Homestake Belt, George Hearst and Samuel McMaster realized that vast amounts of additional pine were going to be needed for timber for the square-set stopes, ties for underground railways, and cordwood for the boilers in the numerous hoist houses and mills. Unfortunately, most of the hillsides around Lead and Deadwood had already been clear-cut by Homestake, various other mine owners, and the townspeople. Hearst and McMaster found themselves having to pay $4.75 per cord for cordwood, $25 per thousand board feet for lumber, and $12-$18 per thousand board feet for mine timbers delivered to Lead City. The two men concluded that a railroad was needed to transport wood or coal from new areas outside of Lead.[3]

During March of 1879, Hearst wrote a series of letters to J. B. Haggin describing the coal fields on the edge of the Black Hills and how it would be more cost-effective to fuel their boilers with coal instead of wood. Hearst laid the groundwork for justification of a railroad to haul the coal as follows:

> *I am informed and believe that at a distance of twenty-five miles or thirty miles from here there exists large bodies of coal; in fact, we are using it in the blacksmith shop and find it to give good satisfaction. I understand that eastern parties are thinking of building a railroad from these coal banks to Deadwood and this place, and figure the cost at about $300,000 for the grading and ties laid down; the iron can be had on time by mortgaging the road bed. I intend looking into the matter carefully, as fuel will soon be a serious question with us and this coal is good and can be mined for, not to exceed $1.50 per ton, and delivered here for a like sum, and as one ton of it is equal to about two cords of wood. It would certainly be much cheaper than wood even now and I calculate that the consumption would reach about two thousand tons daily. Should I find all this, upon examination to be fact, I think we would be making a grave mistake to allow the control of it to fall into hands other than our own and I may write to you to send Short out to make a survey of it, at all events I would like to have your opinion on the subject at your earliest convenience.*[4]

Apparently, Haggin wasn't convinced that a railroad would pay back their investment. Hearst responded, clarifying that at this juncture, he was only committing to having a survey done to get an idea of the cost and feasibility for a railroad. "In relation to the railroad," he wrote, "I have no doubt about persons interests in roads that do not pay, are perfectly willing to get out of them, but those who happen to have paying enterprises of that kind, are not so anxious to let them go; however, I am not building a Railroad but am only having a survey made, which will cost less than $2,000 with a view of finding out the probable cost of building one and of securing a right-of-way for it, because someday a Railroad will be built."[5]

Hearst got his survey. After it was completed, he was even more enthusiastic about the feasibility of a railroad. In one of his informational updates to Haggin dated March 19, Hearst added more justification for his plan: "Write me what your think of our railroad scheme here that we think a great deal about and now think can be done much cheaper than I first wrote you as I am offered by parties here to build and equip it for $12,000 per mile."[6]

By March 23, Hearst hadn't heard anything back from Haggin, so he sent another letter: "Please answer, as you receive my letters and telegrams either by letter or telegram as I am getting very impatient.... Have you seen Short since he returned from surveying the railroad? You seem to be afraid of that road, but I tell you, within five years after completion it will pay 25% of its cost construction."[7]

With two Homestake mills on line and new shafts being sunk, Hearst and McMaster needed cordwood and mine timber. The same held true for the Highland, Deadwood, and Golden Terra companies that were controlled by the Homestake capitalists and managed by Hearst and McMaster. On September 30, 1880, McMaster placed a notice in the *Black Hills Daily Times* requesting bids for 40,000 cords of wood and 1 million board feet of mine timber. The notice specified that the cordwood had to be split, cut into 4-foot lengths, and delivered to the yards of the respective hoisting works and mills, "in such quantities as will keep not less than one month's supply on each yard." Mine timbers, 10 inches square and 12 inches square, were to be supplied in 12-foot and 18-foot lengths. Hewed lagging was to be supplied in 6-inch and 8-inch thicknesses.[8] A one-year contract was awarded.

By April 1881, J. B. Haggin and Lloyd Tevis had approved Hearst's plan to construct a narrow-gauge railroad from Lead City. The new 36-inch-gauge railroad would serve to transport mine timber and cordwood from forested lands south and east of Lead. The "J. B. Haggin" steam locomotive was already proving to be a success in hauling ore on 22-inch-gauge tramway tracks from the Homestake Mine to the Homestake and Golden Star mills. In a like fashion, the "W. R. Hearst" locomotive was performing quite satisfactorily hauling ore for the Highland Mining Company to the Highland Mill.

McMaster decided that construction of the railroad would be completed on a contract basis. On April 24, McMaster put a notice in the *Black Hills Daily Times* soliciting bids for grading 12½ miles of the "Homestake and Custer" railroad from Lead City to the upper reaches of Elk Creek between what would later become the logging camps of Woodville and Brownsville.

> April 24. Notice is hereby given that the undersigned will receive bids until the evening of the 27th day of April, A.D. 1881, to grade twelve and one-half miles of the Homestake and Custer railroad, or separate bids to grade from one to three miles of the same. The profiles and specifications for the first three miles of the road are subject to inspection to all persons proposing to bid at the office of the engineer of the Homestake Mine in Lead City. The right is reserved to reject any and all bids. Dated April 22, 1881. Samuel McMaster.[9]

On April 29, the work was awarded to S. G. Burn, president of the Lead City Miners' Union. McMaster put Richard Blackstone in charge of supervising the railroad's construction. By August, one mile of track had been laid. The name of the railroad was soon changed to the Homestake Railroad and later, the Black Hills Railroad Company.[10] On July 10, 1882, the railroad company was renamed the Black Hills & Fort Pierre Railroad Company by its new directors.[11]

The first locomotive placed on the Black Hills and Fort Pierre Railroad was named the George Hearst (see figs. 11.1 and 11.2). It was purchased from the H. K. Porter Company in Pittsburgh, Pennsylvania, for $7,840 plus freight charges of $1,619.40. The locomotive was shipped on October 20, 1881,[12] and arrived in Pierre on November 5, 1881. From there, the locomotive was shipped by ox team and arrived in Lead on November 29, 1881.[13]

Fig. 11.1. Installing the cow catcher on the George Hearst locomotive. The first locomotive for the Black Hills and Fort Pierre Railroad was named the George Hearst and was commissioned in Lead, South Dakota, on December 8, 1881.—Photo courtesy of the Black Hills Mining Museum.

Fig. 11.2. The train crew of the George Hearst locomotive. The George Hearst was renumbered as No. 494 by the Burlington and Missouri River Railroad after it purchased the Black Hills & Fort Pierre Railroad Company in 1901. The locomotive was removed from service and dismantled in November 1902, after twenty-one years of service.—Photo courtesy of the State Archives of the South Dakota State Historical Society.

On December 3, the *Black Hills Daily Times* reported, "The locomotive for the Homestake Railroad is here, but as yet the iron horse has not yet come out of his stall or stable. The critter has not been placed on his legs, neither has his water trough or apparatus for fuel consumption been entirely fixed yet. In a few days the machine will be placed on the tracks, and then the first railroad in the Black Hills will be in operation."[14] The first trainload of cordwood was delivered to Lead City in December 1881 (see fig. 11.3). Hearst's vision for a railroad was now a reality.

Fig. 11.3. The George Hearst delivers the first load of wood on the Black Hills and Fort Pierre Railroad, December 1881.—Photo courtesy of the State Archives of the South Dakota State Historical Society.

In 1882, another one-year contract for cordwood and timber was awarded, this time to Fred Evans of Hot Springs. Evans placed advertisements in most all the southern Hills newspapers seeking five hundred men, on a subcontract basis, to cut 40,000 cords of wood for $1.25 per cord. Evans was responsible for cutting and delivering the cordwood to various points along the newly established Black Hills and Fort Pierre Railroad.[15]

The second locomotive for the Black Hills and Fort Pierre Railroad was named the Louis T. Haggin after J. B. Haggin's son. This locomotive was also purchased from the H. K. Porter Company in Pittsburgh at a cost of $9,800 plus $1,150.16 for freight. Between 1883 and 1901, five additional locomotives were purchased for the company.

Along the Route

George Hearst and Samuel McMaster originally envisioned the railroad would be constructed to the coal fields around Cambria, Wyoming Territory.

After a survey was completed, the plan was abandoned. In a similar fashion, the board of directors for the Black Hills and Fort Pierre envisioned the railroad might be extended to Fort Pierre, Dakota Territory. There, the railroad could receive freight from steamboats on the Missouri River and the Chicago and Northwestern's subsidiary, the Dakota Central Railway, in Pierre. The plan would also result in the creation of an efficient means of transportation between Fort Pierre and the Black Hills. Unfortunately, this plan was also abandoned because of the large capital requirement and problems procuring a right-of-way across portions of the Great Sioux Reservation. Instead, attention was directed toward completing the road to the vast timber lands south and east of Lead City. Ultimately, the railroad was extended through Elk Creek Canyon to a terminus named Piedmont (see fig. 11.4).[16]

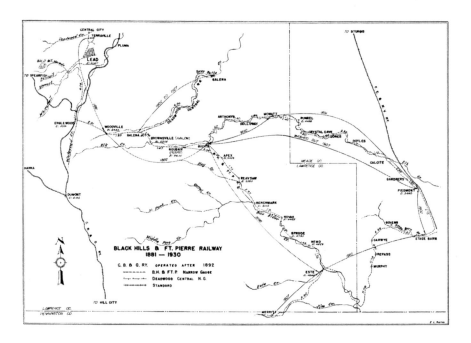

Fig. 11.4. Map of the Black Hills and Fort Pierre Railroad, 1881-1930. In 1889 and 1890, the Black Hills and Fort Pierre Railroad was extended through Elk Creek Canyon to a new terminus that Richard Blackstone platted and named Piedmont. After the rail line through Elk Creek Canyon was washed out in 1907, the line was abandoned and replaced with a new connection from Buck's Landing to Piedmont through Stagebarn Canyon.—Fielder Collection, Devereaux Library Archives, South Dakota School of Mines and Technology.

Lead City to Terraville

From the head of Poorman Gulch in Lead, a series of spurs were constructed to deliver cordwood, mine timbers, and other supplies to the various mills and hoisting works of Homestake and the other companies owned or controlled by the Homestake capitalists. One such spur extended around the north fork of Gold Run Gulch to the Golden Prospect (Highland) Shaft. Another spur was constructed to unloading stations above the Golden Star and Old Abe shafts as well as the Homestake, Golden Star, and Highland mills in Lead. Yet another spur extended to an unloading station above the Old Brig Shaft. Other spurs extended above the Deadwood and Golden Terra mills and above the Father DeSmet Mill. In the mid-1890s, a spur was constructed to South Lead to facilitate delivery of cordwood, mine timbers, and coal to the Ellison Shaft and later, cordwood and coal to Cyanide Plant No. 1.

Lead City to Reno Gulch

From Lead City, the Black Hills and Fort Pierre Railroad was constructed past the West Lead Cemetery and the Homestake Powder House, a structure that was constructed a short time later. A bridge was constructed across the mouth of Nevada Gulch. The route continued to Whitetail Siding, then made a horseshoe curve and climbed to Summit Siding near the top of Whitetail Summit.[17] From there, the rail line proceeded to Reno Gulch, where Reno Siding and Homestake's Reno Powder Magazines were constructed in 1919.[18]

The grade between Whitetail Siding and Summit Siding was extremely steep and considered to be one of the most treacherous stretches on the entire railroad. To make matters worse, the horseshoe curve and bridge were located at the bottom of the grade where the tracks crossed Whitetail Gulch and continued to Lead. Two railroad men were killed on December 30, 1927, when the brakes froze and their runaway train jumped the track on the horseshoe curve at the bottom of Whitetail Summit.[19]

Attempted Train Robbery at Reno Gulch

Reno Gulch was the scene of an attempted train robbery on October 12, 1888. The twelfth day of each month was customarily payday for the various people working from the various logging camps and sawmills situated along the Black Hills and Fort Pierre Railroad. Alex McKenzie, manager of the Hearst Mercantile, was responsible for delivering the $12,000 payroll to camps along the railroad each month. Like clockwork, McKenzie boarded the train on the twelfth of each month, armed with his valise and sawed-off shotgun, and delivered the payroll.[20]

It was a well-known fact among the logging and mining camps that the pay train dutifully made its run on the twelfth of each month and that McKenzie was normally the only extra person assigned to guard the payroll. Various scenarios as to how the train might someday be held up were lightheartedly bantered about on a regular basis by "tin horn gamblers" at the Gem Theatre in Deadwood.[21]

After all the years of conjecturing and speculating, a genuine attempt to rob the train was finally made. In fact, as it was later learned, the same gang of robbers had attempted to hold up the train on two previous occasions but failed in their efforts. Ideas for their first robbery attempt were initially made at a dance the men were attending in Sturgis. The plans were finalized a few days later at the John Telford Ranch on Alkali Creek east of Sturgis. The plan was to hold up the train on August 12 at the big trestle on Bear Butte Creek near Brownsville. The robbery attempt was made, but it had to be scrapped when one or more members of the gang failed to show up with extra guns.

The gang's second attempt was made on September 12 at the Woodville Camp. This time the gang consisted of John Wilson, Napoleon "Jack" Doherty, "Spud" Murphy, and one other person. The four men arrived at Woodville the night before, cut the wires on either side of the station, and hid in the bushes bright and early the next morning. Unfortunately, for the robbers, the train was late because of a picnic being held for the Homestake machinists and mechanics. The robbers got tired of waiting and left. However, the train crew figured out what was happening when they found they had to repair the communication line.[22]

The gang's third attempt began on the afternoon of October 11 when John Telford departed his ranch in his buggy and headed toward the Lawrenson icehouse near Pluma. In his wagon, he had a Winchester rifle, a shotgun, wrench, axe, bar, and lantern. Wilson, Doherty, and Murphy left the ranch a short time later on horseback, armed with revolvers, and met up with Telford at the icehouse at about ten o'clock that evening. Together the four men proceeded to Reno Gulch. At daylight, Wilson and Doherty removed one set of fishplates and spikes, spread the rails, and respiked them about 4 inches wider than normal. The men then built a fire and waited for the pay train to come.

That same morning, the twelfth, Alex McKenzie's rheumatism was acting up so William A. Remer, assistant payroll secretary, made the trip instead. Remer was with John B. Commiskey, engineer, Reece Morgan, fireman, Charles Lavier, brakeman, and Charles H. Crist, conductor (see fig. 11.5). Also on the train were Richard Blackstone, railroad superintendent, and H. P. Anderson, both of whom were seated on some freight on a flatcar. H. P. Swindler, section boss, Dell Fockler, Ben Bell, and E. J. Zeljadt, the section crew, rode on the last flatcar. Remer sat in the cab of the engine with his double-barrel shotgun, extra gun, and three extra shells. Blackstone was armed with his Sharps rifle.

Normally, the train bypassed the Reno Siding and continued on toward Brownsville at full speed. Luckily, this day, as the train approached Reno Siding,

Commiskey slowed the train to let the section crew off. Suddenly, Commiskey felt the wheels of the engine jolting across the ties where the robbers had removed the fishplates and spread the rails. Instinctively, Commiskey reversed the engine and brought the train to a stop.

Fig. 11.5. Part of the Black Hills and Fort Pierre Railroad train crew who were on board when the Homestake payroll train was held up on October 11, 1888. Back row, left to right: W. W. Sweeney, section boss, Reese Morgan, fireman, John B. Commiskey, engineer, Richard Blackstone, railroad superintendent, and Ben Bell, conductor. Front row, left to right: Tillis Hirbour, Charles Lavier, brakeman, Charles Crist, conductor, and Dell Fockler, engineer.—Photo courtesy of Jerry Bryant.

The moment the train came to a stop, the masked men stood up from behind some bushes and yelled, "Hands up!" The men commenced shooting at the people in the cab of the engine. One shot from Wilson's rifle passed between William Remer and Reece Morgan, struck the boiler head, and broke the glass in the fireman's window. Morgan stepped off the train with one hand up and walked slowly toward the trestle. However, Remer took aim at Wilson and Doherty and fired a shot. Doherty, in turn, took aim and fired at Blackstone and Anderson. Blackstone's rifle jammed. Remer then loaded the extra gun and gave it to Commiskey.

Blackstone shouted to the others that he could see some horses tied up in the trees. Murphy then appeared, jumped on one of the horses, and rode away. Remer, in the meantime, crawled back into the cab of the engine. Using a broom handle, he blew the train whistle erratically in an attempt to get some help. The crew on the other engine at Woodville heard the strange whistles and proceeded to the scene. After searching the area, W. W. Sweeney, fireman on the Woodville engine, found Wilson lying on the ground badly hurt with one wound below his left eye and another on his side. Wilson begged to be killed to end his pain and misery. Wilson then lied and said his other two partners were named Clark and Jones. He said that Clark (Murphy) had made a getaway and that Jones (Doherty) was also shot in the face and was still in the area.

Zeljadt, meanwhile, ran all the way back to Lead City to get help. There, he reported that Commiskey and Remer had been killed and that the payroll had been stolen. Superintendent Grier and a deputy sheriff immediately formed a posse, which struck out in search of the robbers. Commiskey got the train back on the track and returned to Lead City with Wilson and the deputy sheriff. Remer continued to Brownsville with the payroll. There, the men were relieved to learn that none of the train crew had been killed. As Remer was taking the valise out of the toolbox, the crowd could see that the bag still contained its contents. The crowd proceeded to let out a yell that "could have been heard on Dead Man's Hill." From that day forward, Remer became known as Buck Shot Bill.[23]

"Spud" Murphy, whose real name was Alfred G. Nickerson, and John Telford were arrested in Rapid City a few days later. A posse, which included Sam Blackstone and others, traveled to Salt Wells in Wyoming based on a report that "a man who had been shot in the face" was seen there. In December, Murphy and Telford were tried and sentenced to fifteen years in the state penitentiary. Homestake put up a reward of $1,000 for Doherty, dead or alive.

On February 6, 1889, Doherty was arrested in Douglas, Wyoming, by John T. Williams, deputy sheriff of Converse County. Samuel Blackstone then went to Douglas to identify Doherty. After doing so, Blackstone and Deputy Sheriff Williams then proceeded to return to Deadwood with Doherty. En route, the men encountered some of Doherty's friends, who made an attempt to free him. Blackstone and Williams stood the men off and delivered Doherty to Sheriff Knight in Deadwood. The reward was paid to Sheriff Williams. Supposedly, Williams split the reward with a gambler in Douglas who had revealed the location of Doherty.

John Wilson, while recuperating in jail, escaped on February 15, 1889. Three days later, he was captured in Boulder Canyon and taken to the jail where Doherty was imprisoned. On the night of March 12, Doherty, Wilson, and another prisoner named Stewart escaped from jail, stole three horses, and made a getaway. Wilson and Doherty went to Rossland, British Columbia, and sought asylum there. Later, it was reported that Wilson was killed in Whitman County, Washington, D. C. while resisting arrest. Telford and Murphy were released from prison in February 1891 based on a legal technicality. Telford moved to Salt Lake City and became a contractor. Murphy assumed his legal name of Nickerson and became a respected citizen and superintendent of a Sunday School in Nebraska.[24]

Reno Gulch to Buck's Landing

From Reno Gulch, the Black Hills and Fort Pierre Railroad continued to West Siding, located near the Englewood Hydroelectric Plant that was constructed in 1906. From West Siding, a spur was constructed to Ten-Mile Station (Englewood). East Siding was located on the east side of the bridge that spanned Whitewood Creek. From East Siding, the rail line continued to Horseshoe Grove and Woodville. Horseshoe Grove was located about two-thirds of the way between the bridge at Whitewood Creek and Woodville.

Although both sites were popular stops for people to have picnic outings, Woodville was the more popular site because of a small lake there. Both sites represented logging camps and timber landings that were active until the mid-1880s. When the need for cordwood and mine timber increased and initial logging areas were depleted, the railroad and logging camps pushed eastward. The camp of Hagginsville or Hagerstown was located between Woodville and Brownsville near Colorado Siding. Fred Evans, who had the main timber contract with Homestake, based his operations in Hagginsville. Jack Leeper, John Moyer, and B. Johnson operated boardinghouses in Hagginsville for as many as one hundred woodsmen and railroad men.[25]

The grade between Woodville and Brownsville was steep and treacherous as described by Jim Bullard, who worked on the Black Hills and Fort Pierre Railroad:

> *The biggest thrill I ever had on my 49 years and three months on the shining steel was on the morning of Jul 12, 1902, going down the mountain from Woodville on the old Black Hills & Fort Pierre narrow gauge railroad. I was the only brakeman . . . big, good natured John Cooksey was the engineer, Hy Wamsley was fireman, and the conductor was a long, tall, freckle faced fellow from Box Butte County, Tom Brigg . . . . After tipping the mountain I commenced setting hand brakes, but could not get much action out of those old brakes on the flat cars. We did not have many cars with air. When big John Cooksey the engineer commenced whistling for brakes I knew we were really running away.*
>
> *When that little mountain hog, the 488, went around the curve she went down the dump, cars piled up behind her and yours truly went up in the air, came down and landed right between two big boulders where there was a nice bed of soft dirt and sand that had been whipped in there by the wind. Some landing! Cooksey and Hy Wamsley, the fireman, went clear of the engine as she turned over, and they both landed in the top of a tree without a bruise. We had a paymaster for the Homestake Company in the caboose and an old lady passenger with five dozen eggs going to*

Nemo to visit her son. The paymaster had a Winchester and $5,000 in silver and gold to pay off employees around Nemo and other stations along the line. They had a late payday. It took five long days and nights to get that 488 back on the rails.

This narrow gauge railroading was apart from any other kind. It was hard, cruel, [and] dangerous. Many died, were crippled for life or just could not take it and walked off the job . . . . Runaways were the thing the narrow gauge man dreaded. Once they got a good start they were gone and you went with them. The only way to prevent them was to club 'em tight with the good old brakie's friend, the brake club, and when you heard that long sharp blast from the hogger you knew it was time to tie 'em down just a little harder or you were gone.[26]

At Galena Junction, a spur line was constructed to Robinson's Wood Camp near the mining and logging camp of Galena. Years later, the spur line was operated by the Deadwood Central Railroad.[27] Less than a mile beyond Galena Junction was Avalon, which soon became part of Brownsville. Brownsville was named after David Brown, a contractor who supplied horses, sleds, skids, tools, and supervision around the logging camp. At its peak, Brownsville was home for about six hundred people, most of whom were associated with the lumber and timber business.[28]

After the Black Hills and Fort Pierre Railroad reached Brownsville in 1882, the population of the camp increased even more for a short time. John Monheim was named postmaster of Brownsville on July 21, 1882.[29] The Hearst Mercantile Company operated a large store in Brownsville for a few years, which was managed by Alex McKenzie. After he moved to Lead in 1886 to manage the Brick Store, which was also owned by the Hearst Mercantile Company, C. H. Gilbert assumed management of the store in Brownsville. A few years later, the store at Brownsville was closed and was moved to Elk Creek Station, close to where Homestake had established a new camp and sawmill.[30]

From Brownsville, the railroad followed Elk Creek to Perry, originally called Lewisville. Lewisville was established for people employed at the Carter Mine and the Uncle Sam Mine, both of which were discovered in 1878. The name was soon changed to Perry when Gifford Parker, manager of the Uncle Sam Mine, applied to have a post office established.[31]

A spur line of the Black Hills and Fort Pierre Railroad was constructed to the Uncle Sam Mine (see fig. 11.6) in 1885. After the owners of the mine purchased the 60-stamp Esmeralda mill in Blacktail Gulch, the mill was dismantled and hauled to the Uncle Sam Mine across the Black Hills and Fort Pierre Railroad. The Uncle Sam Mine became part of the Cloverleaf Mining Company when it was organized in 1898. Pierre Wibaux, a Frenchman with cattle and banking interests in Montana, was named president of the Cloverleaf. With Wibaux and the Cloverleaf infusing new life into Perry, the settlement was renamed Roubaix after Wibaux's hometown in France. A large, two-story schoolhouse was constructed in Roubaix shortly after 1900.[32]

Fig. 11.6. The Uncle Sam Mine at the town of Roubaix. This winter photo shows the Uncle Sam Mine, which operated from 1878-1937. According to a Bureau of Mines Bulletin by Paul T. Allsman, the mine recovered 43,885 ounces of gold and 290 ounces of silver from 220,931 tons of ore over this period. The Black Hills and Fort Pierre Railroad can be seen immediately behind the row of vegetation in the center of the photo. The railroad ceased operation on January 13, 1930, at which time the rail line was salvaged.—Fred Mosley collection.

Less than 1 mile from Perry was Portuguese Siding, located just west of Buck's Rock. Buck's Rock was a large rock outcrop named after Charles Buck, a German immigrant who lived in the area. Next came Buck's Landing, located at Buck's Rock. Construction of the railroad was temporarily stopped at this location in 1882.[33]

The Black Hills & Fort Pierre Railroad Company was a profitable operation for the Homestake capitalists and other investors. For the eight-month period ending June 30, 1889, gross receipts amounted to $62,860.46, mostly from freight service. Only $724 was derived as revenue from passenger service. Expenses totaled $34,876.80. Directors in 1889 included George Hearst, J. B. Haggin, T. J. Grier, D. K. Dickinson, and G. C. Moody. Hearst was president of the company.[34]

Buck's Landing to Piedmont via Elk Creek Canyon

Based on the financial success of the Black Hills & Fort Pierre Railroad Company and the continuing need for cordwood and mine timber, a decision was made to extend the railroad to Homestake's new sawmill near Mowatt. Proceeding east from Buck's, the rail line was extended to a settlement that later became known as the Junction. Several homes were located at the junction, including one owned by Fred Sawyer, a timberman who cut track ties and sold them to the various railroad companies in the Black Hills. In more recent years, the families of Ariel Larson, Harold Larson, and Charles Larson owned homes at the junction.[35] In 1889, the railroad reached Mowatt, located 1.45 miles east of the Elk Creek camp.[36]

The railroad was completed through Elk Creek Canyon to Piedmont on October 18, 1890.[37] At this terminus, mining equipment and supplies could be transferred from the Fremont, Elkhorn, and Missouri Valley Railroad, which was completed between Rapid City and Whitewood in 1887. At the same time, additional company-owned and contract logging camps and sawmills were established along the rail line at Anthony's, Elk Creek Station, Mowatt's, Haven's, Runkle's, Jones', Doyle's, Miller's Platform and Piedmont.

Anthony's was named after Anthony Weier, a German miner who moved to Elk Creek from Crook City after reports were made of rich silver strikes in Galena. His housekeeper was Sarah "Aunt Sally" Campbell, a black woman who accompanied the Custer Expedition through the Black Hills in 1874. Aunt Sally came to Crook City from Bismarck. At Crook City, she cooked for miners and cared for sick people, one of whom was Anthony Weier. After Anthony recovered from his typhoid fever, the two moved into a cabin on Elk Creek. A large wye was constructed at Anthony's to facilitate switching of cars and engines.[38]

About 1½ miles downstream from Anthony's was Elk Creek Station, previously called Holloway. Established in June 1884, the Elk Creek camp grew rapidly.[39] By 1889, hundreds of people were directly or indirectly associated with construction of the Black Hills and Fort Pierre Railroad and operation of Homestake's new logging camp and sawmill, located another 1½ miles downstream. The Hearst Mercantile Company constructed a large store in the Elk Creek camp. Before the store items were moved into the new 40×80-foot building, a large dance was held for some three hundred of Homestake's employees and contractors working in the area.[40] By October 1888, Elk Creek had grown sufficiently to justify a train depot, post office, saloon, and several stores.[41]

Forest fires and floods plagued the railroad and residents living along Elk Creek. In 1886, a forest fire began near Custer Peak. The fire quickly spread 15 miles to Roubaix and the Elk Creek camp. Fortunately, the Black Hills and Fort Pierre train crew came to the rescue, picking up residents along the way. In some cases, the train had to speed to safety over bridges already in flames. The fire destroyed twenty-three bridges in Elk Creek Canyon alone.[42]

In October 1889, another wildfire broke out near the Elk Creek Station. This time, a large complement of firefighters was successful in containing the blaze before it had a chance to race out of control. Still another wildfire ravaged the Elk Creek area in the fall of 1893. As before, the train passed through the camp sounding its whistle, picking up women and children, and shuttling them to safety. More than once, the fire raced ahead of the train, forcing the engineer to speed through a wall of flames. Eric Pietila and members of his family narrowly escaped by lying in the creek until the fire passed by. Most everything in the Elk Creek camp burned except the train depot and the Homestake store, both of which were miraculously saved. The store at Runkle was also destroyed as were about 10 miles of track and numerous bridges. Fortunately, the Black Hills & Fort Pierre Company decided to rebuild the line, and the work was quickly completed.[43]

Mowatt's was located about 1½ miles east of Elk Creek Station. The camp was named after Jack Mowatt, who became general manager of Homestake's sawmills in 1887. In February 1887, Mowatt moved one of the Homestake sawmills to Greenwood. Beyond Mowatt were the camps of Havens and Runkle. Runkle, famous for its apple trees, was named after George Runkle after he purchased one of Homestake's sawmills and established a business there in 1892. In 1895, a boiler explosion occurred at the Runkle Sawmill, killing the fireman and head sawyer and injuring three other men. Pieces of the boiler works were found some 300 feet away.

Some 2½ miles southeast of Runkle was Crystal Cave. Starting in about 1894, Crystal Cave became a favorite daytime excursion for people living in Lead and Deadwood. One could purchase a roundtrip train ticket in Lead for $2, depart town at 8:00 a.m., tour the cave and Knife Blade Rock, and be home again by 7:00 p.m. Lunch and a tour of the cave cost twenty-five cents.[44]

Southeast of Crystal Cave along the rail line in Elk Creek Canyon were the camps of Jones', the Quarry, Doyle's, Calcite, and Miller's Platform. William E. Jones owned the Jones Sawmill, located 2.45 miles down the canyon from Crystal Cave. The Doyle camp was named after William D. Doyle and was located about 1.14 miles east of the Jones Sawmill. Because the Doyle Sawmill was larger than the Jones Sawmill, the camp was also a little larger and had two sidings on the railroad.

At its peak, some eighty-five people were employed as loggers, sawyers, and quarry men at Doyle's. The Doyle camp was large enough to have its own post office, a school, saloon, store, boardinghouse, and numerous residences. As was the case with many of the area sawmills, the Doyle Sawmill supplied lumber, mine timbers, and rail ties to Homestake on a contract basis.[45] William Doyle was also manager of the Black Hills Quarry Company. The quarry was located between the Jones Sawmill and the Doyle Sawmill. Sandstone mined from this quarry was used in the construction of the Miners' Union Hall in Lead.

Calcite was located 4.34 miles northwest of Piedmont on the Black Hills and Fort Pierre Railroad. The camp was home to seven Homestake employees who mined limestone at the quarry and produced lime in a limekiln located on site. The quarry was located high on the ridgetop above Calcite. The first lime plant was constructed in about 1900. Two sets of rails extended up and down the steep hillside between the quarry and the lime plant. A steam hoist positioned on the ridgetop was used to lower carloads of limestone from the quarry to the kiln on one set of the double-track system. The weight of the loaded cars helped raise the empty cars on the adjacent track.[46]

Limestone mined from the quarry was placed in the kiln, which was fired using 4-foot lengths of pine cordwood. The kiln was fabricated with four furnace doors through which the cordwood slabs were loaded. Some seventy hours were required to process one batch of lime. Lime was loaded into railcars on the Black Hills and Fort Pierre Railroad and transported to Lead, where the lime was used in Homestake's cyanide plants and Slime Plant in Deadwood. A few years later, Homestake added a second limekiln to the plant at Calcite, enabling near-continuous production of lime (see fig. 11.7).[47]

Fig. 11.7. Homestake's limekilns and quarry at Calcite. The lime plant, located about 4 miles northwest of Piedmont in Elk Creek Canyon, operated from about 1904 to 1931. The quarry was located on the ridgetop. Note the crude trackline above and to the right of the lime plant on which carloads of limestone rock were lowered down the hill to the lime plant. The rail line for the Black Hills and Fort Pierre Railroad is in the foreground.—Homestake Collection, Adams Museum, Deadwood.

Evan Evans was Homestake's resident superintendent for many years at the Calcite operation. Evans lived in a house just east of the lime plant while the six other employees lived in a separate bunkhouse nearby. After the flood of 1907 destroyed the rail line and bridges in Elk Creek Canyon, the lime had to be loaded into wagons and taken to the Fremont, Elkhorn, and Missouri Valley Railroad at Piedmont. By that time, the Elkhorn provided service to Deadwood. A spur allowed lime to be delivered directly to Homestake's Slime Plant in Deadwood. Upon completion of the Black Hills and Fort Pierre line through Stagebarn Canyon in 1910, lime shipments were resumed on the railroad. The operations at Calcite continued until 1930 at which time they were shut down concurrent with abandonment of the Black Hills and Fort Pierre Railroad. Over its operating life, the Homestake lime plant at Calcite produced more than 70,000 tons of lime.

On October 15, 1933, the Civilian Conservation Corps (CCC) established a camp at Calcite and constructed a new fifty-man barracks and other buildings. The bunkhouse formerly used by Homestake was converted to a hospital. The CCC camp was abandoned in October 1935.[48] A stone monument marks the location today.

Approximately, 1 mile southeast of Calcite was Miller's Platform. The camp consisted of a lumber and timber landing along the railroad. The camp was named after Newel Miller, who nearly lost his life while grading the replacement rail line in Stagebarn Canyon. The last train stop along the rail line in Elk Creek Canyon was Gardner's. The camp was named after Captain "Cap" C. V. Gardner, who owned the Spring Valley Ranch in the area at the time.[49]

The Town of Piedmont

By 1890, work was nearing completion on the Black Hills and Fort Pierre Railroad. The terminus was planned to be located approximately 1½ miles northwest of the town of Sacora along the Fremont, Elkhorn, and Missouri Valley Railroad. The location for the Black Hills and Fort Pierre terminus was selected because of the suitability and availability of land and water, compared to Sacora, located nearby. For the same reason, Homestake decided to create a new townsite at the planned terminus to accommodate the railroad workers and support a new freight-transfer station.

Homestake purchased a large tract of land in Spring Valley from Herbert S. Hall on April 11, 1890, for the purpose of constructing a townsite. The purchase included the west half of section 10, except for 5 acres on which C. V. Gardner lived. Under the direction of Thomas J. Grier, Homestake superintendent, the townsite of Piedmont was platted by Richard B. Blackstone, superintendent of the Black Hills and Fort Pierre Railroad. Piedmont means "at the base or foot of the mountain."

The Homestake capitalists of George Hearst, J. B. Haggin, and others formed the Piedmont Townsite Company for the purpose of establishing the townsite and selling city lots to employees of the railroad and prospective business owners in Piedmont. Nine city blocks, each measuring 300 feet square, were laid out adjacent to the Fremont, Elkhorn, and Missouri Valley Railroad. A strip of land 275 feet wide between the Elkhorn right-of-way and First Street in Piedmont was reserved for use by the Black Hills and Fort Pierre Railroad. A month or so later, seven additional city blocks were platted into the townsite.

The *Rapid City Black Hills Weekly* reported on the pending lot sales in its newspaper on May 23: "By telephone last evening from Piedmont it is learned that Messrs. Grier and Whitney and McPherson, cashier of the First National Bank of Deadwood, were in that prospective city yesterday. While there they arranged for a sale of lots to take place at ten o'clock on May 22. McPherson yesterday purchased lots on which he and the people he represents propose at once to erect a bank building."[50]

In the same edition, the *Rapid City Black Hills Weekly* reported on the progress of completing the rail line through Elk Creek Canyon to Piedmont: "For several days past an immense amount of iron and other material for the Black Hills and Fort Pierre Railroad has passed through Rapid City, and there is now on hand at Piedmont very near enough rails to iron the entire route from that place to where the road now ends. Yesterday, three extra freight trains pulled through the city all loaded with iron and all bound for Piedmont. Grading on the road is progressing rapidly and tracklaying will shortly be commenced on the main line. One or two sidetracks have already been built and others will be soon."[51]

Just north of town, a "turn-around" wye was installed on the Black Hills and Fort Pierre tracks, and a new section house was built at that location. The Black Hills and Fort Pierre Railroad constructed its own freight buildings and a transfer station near the Elkhorn Depot. This allowed freight and equipment to be unloaded and transferred from standard-gauge cars on the Elkhorn tracks to narrow-gauge cars on the Black Hills and Fort Pierre tracks. The newly constructed station, located between the two sets of tracks and just east of the Piedmont Hotel, soon became a shared depot. W. E. Royce was the first station agent at Piedmont for both railroads. In no time, Piedmont had its own post office, the Western Bank and Trust Company, a Methodist church, two section houses, saloon, pool hall, blacksmith shop, livery stable, and a school.

Elnora, Carrie, and Henry Leroy built the Piedmont Hotel in 1896. Homestake may have persuaded the Leroys to construct the hotel, based on a $220 mortgage issued to them from Thomas J. Grier of the Piedmont Townsite Company. J. B. Haggin, G. Hearst, and T. J. Grier held the mortgage until April 17, 1902, at which time clear title was given to the Leroy owners. Passengers who arrived on the Elkhorn Railroad stayed overnight at the hotel. The following morning, they boarded the Black Hills and Fort Pierre train and continued on to Lead City.[52]

After the Hearst Mercantile Company completed construction of one of its Brick Stores in Piedmont in March 1891 (see fig. 11.8), the post office was moved to the store. Tom D. Murrin was named manager of the store and postmaster for the post office. The bricks for the store were manufactured just north of Piedmont and hauled to the building site by Newel Miller. The Brick Store sold just about anything and everything, including farm implements. The Hearst Mercantile Company sold the store to Robert R. Engle on October 24, 1924.[53]

Fig. 11.8. The former Hearst Mercantile Company store in Piedmont, 2008. The Brick Store was constructed at the corner of Main and Second streets in Piedmont in 1890-91. After the store was completed, some thirty couples from Lead attended a ball in Piedmont in June to celebrate completion of the new store.—Photo by author.

On June 6, 1890, the *Rapid City Black Hills Weekly* announced that "a magnificent 25-ton locomotive arrived at Piedmont last Thursday and was set up on Friday and Saturday. On Sunday tracklaying began, and yesterday sufficient was down to test the new engine. It was found perfect. Graders are pushing work surprisingly and no doubt exists that the extension will be completed by the first of August. Very little building is in progress in Piedmont. The Socorro [Sacora] depot has not yet been moved to the new town as reported a few days ago. A number of flat cars were taken by wagon from Lead City to the Piedmont end, for use on the construction."[54]

The engine was No. 4 for the Black Hills and Fort Pierre Railroad, arriving on a flatcar from the H. K. Porter Company, Pittsburgh, on the Fremont, Elkhorn, and Missouri Valley Railroad.[55] The new Black Hills and Fort Pierre line through Elk Creek Canyon to Piedmont was completed in August, just in time to put the new locomotive to work.

Residents of the Black Hills wasted no time in taking one of the first scheduled excursions through Elk Creek Canyon. Unfortunately, two accidents occurred during the first month of operation. On September 9, just southeast of Runkle, a train bound for Lead City collided with an excursion train that was taking a ball team from Lead City to Piedmont. Several people were injured in the collision, and both engines were damaged. A few days later, two people were killed, and several others were injured northwest of Runkle when the vibration of an excursion train caused a tree to fall onto an open car full of people. Notwithstanding, the route proved to be very popular among Black Hills citizens.[56]

The new train service was also a boon for Lead merchants. On just one day in October 1890, some twenty-one freight cars of merchandise arrived in Lead for merchants. Twelve additional cars of freight were expected later that day. Lead businessmen had never had to deal with so much freight at one time.[57]

The passenger train ride between Piedmont and Lead City soon became a favorite pastime among Black Hills residents, who found the ride exhilarating and the views breathtaking (see figs. 11.9 and 11.10).

Fig. 11.9. The Black Hills and Fort Pierre narrow-gauge train at Giant Bluff in Elk Creek Canyon, ca. 1890.—Library of Congress (LC-DIG-ppmsc-02545).

No. 38

BLACK HILLS & FT. PIERRE R.R.

TIME TABLE.

No. 8. GOING SOUTH	STATIONS.	No. 7. GOING NORTH
4.53 P. M.	Lv.........LEAD.........Ar.	11 56 A. M.
	1.90 Miles.	
5.01 "	.....Whitetail Crossing.....	11.48 "
	1.05 Miles.	
5.06 "	......Whitetail Summit......	11.43 "
	3.34 Miles.	
5.16 "	.........Englewood..........	11.33 "
	2.33 Miles.	
5.26 "	.........Woodville..........	11.23 "
	1.40 Miles.	
5.33 "	........Colorado Siding......	11.16 "
	2.20 Miles.	
5.42 "	.......Galena Junction......	11.06 "
	1.12 Miles.	
5.48 "	........Brownsville.........	10.59 "
	1.52 Miles.	
5.55 "	...........Perry............	10.51 "
	.86 Miles.	
5.57 "	......Portuguese Siding.....	10.47 "
	3.14 Miles.	
6.14 "	.........Anthoneys..........	10.31 "
	1.60 Miles.	
6.18 "	.........Elk Creek..........	10.26 "
	1.45 Miles.	
6.24 "	..........Mowatt's..........	10 20 "
	2.86 Miles.	
6.35 "	..........Runkles...........	10.09 "
	2.46 Miles.	
6.46 "	........Crystal Cave........	9.58 "
	2.42 Miles.	
6.57 "	...........Jones............	9.47 "
	.43 Miles.	
6.59 "	..........Quarry............	9.45 "
	.95 Miles.	
7.02 "	..........Doyles............	9.42 "
	1.91 Miles.	
7 07 "	.........Millers............	9.37 "
	3.44 Miles.	
7.18 P. M.	Ar.......PIEDMONT.......Lv.	9.26 A. M.

TAKES EFFECT
DECEMBER 20, 1896, 12 M. NOON.
RICHARD BLACKSTONE, SUPERINTENDENT.

Through Tickets ◆

To all parts of the

WORLD

AT

LOWEST RATES.

✦✦✦✦

Agency for the following

STEAMSHIP LINES

American,   Hamburg American,
White Star,   Anchor,
Red Star,   Dominion,
Cunard,   Union,
North German Lloyd,   Netherland,
French Line,   Hansa,
Thingvalla,   Baltic,
and Beaver.

Close connections at Piedmont with F., E. & M. V. trains for all points.

The Scenery on this Line is acknowledged by old travelers to equal, if not surpass, in beauty the famous Rocky Mountain scenery of the Denver & Rio Grande.

Insist on having Tickets to and from Lead City via this route.

All freight for Lead City and Galena, S. D., should be routed "Care B. H. & Ft. P." at Piedmont (Junction F., E. & M. V.), or at Englewood (Junction B. & M. R. R.)

For any information as to Rates, Routes, etc., call on or address

R. F. TACKABURY, GENERAL AGENT,
J. W. CURRAN, CITY TICKET AGENT,
LEAD CITY, SO. DAKOTA.

Fig. 11.10. Timetable of the Black Hills and Fort Pierre Railroad, 1896. The timetable lists two of the one-way trips that could be taken between Lead City and Piedmont. A one-way trip took approximately two and one-half hours to complete.—Courtesy of the Black Hills Mining Museum.

The following news article from the *Rapid City Journal* on September 11, 1890, provided an excellent visualization of the trip:

A Romantically Picturesque Route to Travel

The completing of the Black Hills and Fort Pierre Railroad from Piedmont to Lead City has added a great attraction to travel in the Black Hills country. The company, during the short time the road has been in operation has enjoyed good business, almost everyone having business in the upper country, taking the Piedmont route. The trip is one well worth taking, and when the attractiveness is better known to the traveling public the company will find its present accommodations sadly inadequate. Leaving the Elkhorn road at Piedmont, the traveler boards a Pullman palace car, and for a mile or so is taken whirling along the base of the mountain, when suddenly, without an intimation, the flying cars are enclosed between Elk Creek Canyon's walls of rock. Winding and twisting the road runs, every rod revealing to the traveler grander sights and more beautiful specimens of nature's architecture, hugging, for a few hundred yards, the sides of a stupendous cliff whose walls tower skyward until the lofty pines which adorn its summit have shrunken to pigmy proportions. The train makes a chute across the canyon passing beneath an overhanging ledge of rock and curving around a jutting point, a narrow part of the canyon is reached.

On either side are reared precipitous walls of solid rock and there seems to be scarce room enough for the cars to find passage through. The traveler involuntarily withdraws from close proximity to the window, so near do the fleeting walls appear to be, but smiles at his fears, a moment later. Widening out the canyon has been changed by nature into a beautiful park, the green foliage of pines contrasting with that of the deciduous trees which autumn's touch has changed to flaming red or crimson with bits of yellow. Narrowing again bottling cliffs rise high above the car tops, new beauties revealing themselves at every rod.

Conversation is neglected, at every window is a face; the platforms, despite the fact that flying smoke and falling cinders render them uncomfortable, are filled with people eager to drink in the beauties of the scene of which but a fleeting glimpse is given as the cars rush on. The ones passed are forgotten in those that follow, for the route from Piedmont to Lead City is an ever changing panorama of magnificent scenery, and the traveler never wearies. His eyes find pleasure in the beautiful views, the wild and rugged scenery, his mind is occupied with inspired thought and o'er him steals a reverential awe of that great power whose hands have built up such giant piles, painted with such delicate touch the falling leaves of autumn. Nature builds better and more beautiful than man, and really seems to have used her power to paint and build with lavish hand in the country through which the Black Hills and Fort Pierre railroad takes the way.

The route traveled is one pleasing succession of surprises from the moment Piedmont is left and Lead City is reached. Lead City with its mines and mills will also be found an interesting place in which to spend a day or so. Good hotel accommodations can be had and the tourist will find plenty that is new and novel to hold him in interest while he tarries there. Deadwood, Terraville, and Central have their attractions, too, and taken together a more pleasant trip or one possessing greater interest cannot be taken. The accommodations on the narrow gauge are first class in every respect, the road bed is smooth and well kept, and every precaution is being taken by the company to avoid the possibility of accidents. When

> *its beauties are known the Piedmont route will be the best traveled route in the Black Hills, picturesque and romantic, grand and majestic the scenery along it cannot be surpassed anywhere in the world.*[58]

In 1902, the Black Hills and Fort Pierre Railroad extended its tracks some 2.79 miles to a point just above the mouth of Stagebarn Canyon to facilitate Homestake's harvest of timber holdings in that area.[59] William D. Linscott had a sawmill there, having moved it from Tilford in 1900. Unfortunately, the flood of 1907 completely destroyed the Linscott Sawmill. After the flood, W. Holstein established his sawmill close to the former location of Linscott's sawmill.[60]

Buck's Landing to Este

A separate branch of the Black Hills and Fort Pierre was completed between Buck's Landing and Este on October 1, 1898. Several mining and logging camps were located along this section, including Apex, Reausaw, Benchmark, Greenwood, Novak, Spruce, and Nemo. Located about 3 miles southeast of Buck's Landing, Apex was so named because it was located on the divide between Elk Creek and Hay Creek. Apex consisted of a small logging camp and railroad siding where cars were switched, and lumber and timber were received from area logging crews. The next site, called Reausaw, was located 1.77 miles southeast of Apex and was named after the Fred Reausaw family who homesteaded on Hay Creek. Today, the railroad embankment forms the dam for Reausaw Lake.[61]

The Black Hills & Fort Pierre Company maintained a depot at Benchmark, located at the confluence of Hay Creek and Boxelder Creek. Many of the residents of Benchmark worked for the nearby Lucky Strike Mining Company, formed in 1902. Remnants of the Lucky Strike mine and mill can be seen from the Benchmark Road approximately 1 mile west of Nemo Road.[62]

By the first quarter of 1898, the Black Hills & Fort Pierre Railroad Company had advanced toward the mining camps of Greenwood and Novak. On April 5, 1898, Charles Flormann and his wife Fredricka granted the company the right-of-way deed at the cost of $5. This allowed the construction of the railroad through their property in Greenwood. The deed also included "the use and occupancy of the boardinghouse, stable, office, and the church situated upon said premises at Greenwood during the construction of said road, or until October 1, 1898."[63]

The Greenwood Camp (see fig. 11.11) was founded in about 1884 as Laflin. Robert Flormann was named the first postmaster of Laflin on August 11, 1884.[64] Near Greenwood were several investment mines, many of which were considered swindles.[65] The Greenwood Mine, Smoots Mine, and Safe Investment Mine (see fig. 11.12) were the primary mines located near Greenwood.

Fig. 11.11. The mining camp of Greenwood. This photo of Greenwood shows the old livery stable on the left and the boardinghouse on the right. A church, barn, and blacksmith shop were located behind the boardinghouse. Charles Flormann's residence is the second house, from front to back, in the middle of the photo. A store is shown in the upper left.—Photo courtesy of Colette (Flormann) Bonstead.

Fig. 11.12. The Safe Investment Mine near Greenwood. A 60-stamp mill was completed in 1907 but closed by year's end because of lack of ore.—Waterland Collection, Devereaux Library Archives, South Dakota School of Mines and Technology.

By 1900, some two hundred residents resided in Greenwood. The camp had a church where school was also held, post office, store, livery stable, blacksmith shop, and numerous homes. A stage line that operated between Rapid City and Deadwood had formerly used the livery stable.[66] Charles Flormann was listed as "clerk of [School] District and enumerator" in the School Census dated June 1, 1887.[67] After the mines in the area fizzled and the area's lumber and timber operations relocated to Nemo, the post office was moved to Novak on April 19, 1905, with Flormann continuing as postmaster.[68]

Novak was located about one-half mile southeast of Greenwood. Originally, the camp was called Slabtown. After Greenwood lost a large percentage of its population, the Safe Investment Mining Company arranged to have the post office moved to Slabtown, and the camp was renamed Novak.[69]

The First Federal Timber Sale

In 1898, the Black Hills and Fort Pierre Railroad was extended from Buck's Landing to Este to facilitate Homestake's harvest of the healthy stands of timber in the Nemo and Este areas. Under the Mining Law of 1872, miners were entitled to use the timber on an unpatented mining claim for the mineral development of that particular mining claim. Congress enhanced the ability of western mining companies to obtain lumber by passing the Free Timber Act in 1878. This act authorized miners to take for mining purposes "any timber or trees growing or being on the public [domain], subject to such rules and regulations as the Secretary of the Interior may prescribe for the protection of the timber." The rules and regulations prohibited the cutting of any tree less than 8 inches in diameter and required that brush and treetops be disposed of in a manner that would prevent the spread of forest fires.

In 1894, the United States filed suit against Homestake for illegally cutting 6,828,160 trees on public lands. The suit alleged that Homestake was staking mining claims in the Black Hills and cutting trees on those claims solely for the benefit of the Homestake Mine in Lead. The plaintiff sought damages, including interest, of $688,804. Investigators for the General Land Office estimated the market value of the lumber to be $2-$3 million. After an attempt by the plaintiff to have the trial moved to Sioux Falls failed, the case was heard in Deadwood. Four years later, Judge John Carland found Homestake guilty of the charge and fined the company $75,000.[70]

In 1896, Secretary of the Interior Hoke Smith asked the National Academy of Sciences to develop a national forest policy and recommend possible forest reserve sites that might be managed by the U.S government. A National Forestry Commission was formed to carry out the request. Members of the National Forestry Commission visited the Black Hills and other public lands throughout the West as part of the evaluation process. Regarding its investigation of the

Black Hills, the commission reported, "It is evident that without government protection these forests, so far as their productive capacity is concerned, will disappear at the end of a few years, and . . . their destruction will entail serious injury and loss to the agricultural and mining population of western North and South Dakota."[71]

After reviewing the recommendations of the commission, President Grover Cleveland acted to create thirteen forest reserves on February 22, 1897. One of these reserves was the Black Hills Forest Reserve, comprising 957,680 acres. The forest reserves continued to be managed by the General Land Office. Facing pressure to abolish the reserve system altogether, Congress quickly passed the Organic Administration Act in June 1897. The act provided assurance to companies and consumers that the newly created forest reserves would continue to supply timber for their future needs.

On November 3, 1897, Gifford Pinchot, working for secretary of the Interior Cornelius N. Bliss, presented Homestake Mining Company with a proposed timber harvest plan in an effort to win the support of Homestake and end the practice of staking unpatented mining claims merely for the timber they contained. Mine Superintendent T. J. Grier and Attorney Gideon C. Moody informed Pinchot they supported the concept of the forest reserve plan and indicated they would help fund a mechanism for the government to sell timber. Homestake submitted an application on April 8, 1898, for the right to harvest timber in accordance with Pinchot's plan.

In anticipation of having its application approved, Homestake established a logging camp about 2 miles south of Nemo and constructed a circular-rig sawmill at the camp. The camp was named Este after the Este family who lived nearby. Over the next eighteen months, the General Land Office studied the application and visited the proposed harvest site. In October 1899, Homestake resubmitted its bid of $1 per thousand board feet for all of the timber that could be harvested within the proposed cutting area. Upon acceptance of Homestake's bid on November 5, 1899, the first regulated timber harvest and sale on federal forest reserve lands, called Timber Case No. 1, was started.[72]

Homestake's timber cutters used crosscut saws and axes to fall the trees. Sawlogs were hauled by horse and wagon to the sawmill at Este. Tie hackers and timber hewers used broadhead axes and spuds to peel the pine bark from the sawlogs and hew the track ties. Cutters were paid by the cord and hewers by the lineal foot. Mine timbers and bridge timbers were sawed using the new circular-rig mill. The sawed mine timber proved to be much more satisfactory than the hand-hewn timber.[73]

In 1907, a spur line of the Black Hills and Fort Pierre was completed between the Este camp and Merritt to facilitate completion of the timber sale (see fig. 11.13). Over the duration of the timber sale, which was completed in 1908, some 15 million board feet of lumber and 5,100 cords of cordwood were harvested. The circular-rig sawmill at Este continued to be used until 1913, at which time a new band saw head rig was built at Nemo, and the Este camp was closed.[74]

Fig. 11.13. Fireman Bob Leeper with Black Hills and Fort Pierre engine No. 536 at Homestake's lumber and timber camp in Merritt.—Photo courtesy of the State Archives of the South Dakota State Historical Society.

The Burlington Group Assumes Control

The Fremont, Elkhorn, and Missouri Valley standard-gauge railroad reached Deadwood on December 29, 1890, from Rapid City via Piedmont and Whitewood. A month later, the Grand Island and Wyoming Central branch of the Burlington and Missouri River railroad also reached Deadwood from Hill City via Ten-Mile Station, Pennington, and Pluma. With these additions, railroad service now connected the northern Black Hills with the rest of the world. The railroads brought forth another boom in mining and construction and all of the related support and spin-off businesses.[75] With arrival of the railroad, Ten-Mile Station was renamed Englewood, as described by the *Black Hills Weekly Times*:

> At the dinner table the Times men learned that Ten Mile Ranch has been christened Englewood; that it will be a regular station, at which an agent will be placed and at which water tanks are to be erected. Pennington will be known in railroad circles as Kirk.[76]

On July 27, 1901, the *New York Times* announced that the Chicago, Burlington and Quincy Railway Company had purchased the Black Hills & Fort Pierre Railroad Company from Homestake Mining Company for $1,091,037.40.[77] Chicago, Burlington and Quincy's subsidiary, the Burlington and Missouri River Railroad, assumed operating management of the Black Hills and Fort Pierre Railroad. The Chicago, Burlington and Quincy had already established its Burlington and Missouri River line into the Black Hills with standard-gauge branches from Englewood to Spearfish, Hill City to Keystone, and Edgemont to Deadwood. At about the same time the change in ownership and management of the Black Hills and Fort Pierre Railroad was taking place, a 6-stall roundhouse was completed in Lead (see fig. 11.14).

Fig. 11.14. Engine No. 488 on the turntable in front of the Black Hills and Fort Pierre Roundhouse in Lead. The roundhouse was constructed in 1901 to service the narrow-gauge locomotives of the Black Hills and Fort Pierre Railroad. The turntable served to align the engines with the proper service stall or turn the locomotives around.—Photo courtesy of Gary Richards.

To accommodate use of standard-gauge cars of the Burlington and Missouri River, a third rail was laid from the standard-gauge track at Fantail Junction to Lead. With this arrangement, both narrow-gauge cars and standard-gauge cars could be used in the same train. At the same time, the Burlington and Missouri River Railroad renumbered the Black Hills and Fort Pierre engines from 488 to 494, the latter being the No. 1, "George Hearst" locomotive (see table 11.1). In addition to the seven narrow-gauge locomotives, the Burlington and Missouri River Railroad acquired eight boxcars, two "way" cars, two gondola cars, sixty-four flatcars, one tank car, two reefers, and two passenger cars.

Table 11.1 Narrow-gauge locomotives of the Black Hills & Fort Pierre Railroad Company

Loco No.	Builder	Shop No.	Cost	Date	Type	B&MR No.[1]	CB&Q No.[2]
1	H. K. Porter	457	$7,840.00	Oct. 20, 1881	2-6-0	494	N/A
2	H. K. Porter	493	$9,800.00	Jun 22, 1882	2-6-0	493	530
3	Baldwin	6691	$7,410.31	1883	2-8-0	492	536
4	H. K. Porter	1145	$6,675.00	May 1890	2-6-0	491	531
5	Baldwin	11769	$6,645.00	Apr. 1891	2-6-0	490	532
6	Baldwin	17612	$7,595.00	1900	2-6-0	489	534
7	Baldwin	18888	$7,595.00	1901	2-6-0	488	533

(1) Burlington and Missouri River Railroad.
(2) Chicago, Burlington, and Quincy Railway.

Sources: Data from Mildred Fielder, *Railroads of the Black Hills* (Seattle: Superior Publishing Company, 1964); T. J. Grier, personal notes, Black Hills Mining Museum.

Toward the latter part of 1904, the Chicago, Burlington and Quincy Railroad Company assumed direct management of its subsidiary lines within the Black Hills. The names of the subsidiary companies, including the Burlington and Missouri River, Grand Island and Wyoming Central, Black Hills and Fort Pierre, and Deadwood Central were dropped. Once again, the narrow-gauge locomotives operating on the former Black Hills and Fort Pierre lines were renumbered. The George Hearst had already been removed from service and dismantled in November 1902. The other six locomotives were renumbered as follows: No. 2, 530; No. 3, 536; No. 4, 531; No. 5, 532; No. 6, 534; and No. 7, 533. Over the next several years, all but numbers 533 and 534 were removed from service.[78]

Flood of 1907 Forces Abandonment of the Line through Elk Creek Canyon

Periodic flooding wreaked havoc on the Black Hills and Fort Pierre Railroad. In 1894, a flash flood washed out numerous bridges and miles of roadbed in Elk Creek Canyon between Elk Creek Station and Crystal Cave. The Elk Creek line was rebuilt. However, the final blow came on June 12, 1907, when a cloud burst and flash flood ravaged Stagebarn Canyon and Elk Creek Canyon. The flood washed out the majority of the track and bridges between Buck's and Calcite in

Elk Creek Canyon. One 6-mile stretch of Elk Creek Canyon alone contained sixty-three bridges, most of which were damaged or destroyed by the flood.

With the southern branch of the railroad already at Este, a decision was made to abandon the northern line from Buck's to Calcite through Elk Creek Canyon and extend the line from Stagebarn Station to Este through Stagebarn Canyon. An agreement was reached between Homestake Mining Company and the Burlington whereby Homestake would procure the 10.84-mile right-of-way and pay for the construction labor and materials if the Burlington crews would complete the work. Newel Miller of Piedmont subcontracted to complete most of the new route through Stagebarn Canyon.[79] The 10.84-mile link between Este and the Stagebarn station was completed on August 16, 1910, and Homestake was able to resume getting its lime and other freight directly across the former Black Hills and Fort Pierre narrow-gauge line.[80]

Concurrent with rerouting the Black Hills and Fort Pierre Railroad from Elk Creek Canyon to Stagebarn Canyon, several new logging camps were established along the railroad between Este and the Stagebarn camp to access Homestake's large timber holdings in the area. Such camps included Goiens, Carwye, Repass, and Murphy. The camp at Goiens was located near the north fork of Stagebarn Canyon approximately 1.8 miles west of the Stagebarn camp. In about 1911, Newel Miller constructed numerous trails from the Goiens logging camp to the various timber holdings of Homestake. Horse teams were used to transport sawlogs to the railcar sidings located between Goiens and Carwye in Stagebarn Canyon.

One of the larger logging camps was Carwye, located at the head of Stagebarn Canyon. Carwye was named for the wye that was constructed to switch the engine and cars around to facilitate travel up and down the steep grades in Stagebarn Canyon.[81] Carwye supplied cordwood to Homestake's lime plant at Calcite.[82] Repass was located about midway between Carwye and Murphy. A large barn on the homestead ranch of Robert E. Repass provided shelter for the horses used in the logging operations based at the Repass camp.

The last two locomotives to be operated on the narrow-gauge line between Lead and Piedmont were CB&Q Nos. 533 and 534, the youngest two of the seven originally purchased.[83] The entire Black Hills and Fort Pierre Railroad line was officially abandoned on January 13, 1930, as outlined in a letter dated August 28, 1956, from the Public Utilities Commission:

> The narrow gauge railroad line extending from Englewood to Calcite, a distance of 41.86 miles and known as the Black Hills & Fort Pierre Railroad, was abandoned, effective as of January 13, 1930, by order of the Interstate Commerce Commission in its Finance Docket No. 7910. The South Dakota Railroad Commission, in its order of January 6, 1930 (Docket 5878), authorized the abandonment of all

stations on the line, vis: Woodville, Galena Junction, Avalon, Roubaix, Apex, Reausaw, Bench Mark, Novak, Spruce, Nemo, Este, Murphy, Repass, Carr Y, Goiens, Stagebarn, Piedmont, Gardners and Calcite. The application for abandonment was filed jointly by the Black Hills & Fort Pierre Railroad Company and the Chicago, Burlington & Quincy Railroad Company, on October 25, 1929.[84]

The final run from Piedmont to Lead across the Black Hills and Fort Pierre line was made on March 20, 1930. J. C. Lang was conductor, Frank Carter, engineer, William Elrod, fireman, and Tom Gorman, brakeman. By May, the rails and ties had been pulled, and the Black Hills and Fort Pierre Railroad was finished. From that point forward, timber was hauled by truck from the Nemo Sawmill to Lead. The other standard-gauge lines of the Chicago, Burlington, and Quincy Railroad continued to service Homestake until the late 1970s.[85]

The Larger Homestake Sawmills

Based on the success of the first federal timber sale and sawmill at the Este camp, Homestake purchased various placer claims and constructed a new sawmill at Nemo in 1913. The Este operation was subsequently shut down. Like most of the larger mining and logging camps, Nemo had its own post office and school. A Hearst Mercantile store, similar to the one in Lead, was also contracted to provide consumer goods for Homestake's employees living in Nemo (see fig. 11.15). Gabe Fredrickson managed the store until 1926 at which time Pat Deen took over. As with the other stores of the Hearst Mercantile Company, Homestake employees were granted open charge accounts at the Nemo store. Most any food or household item could usually be obtained at the Hearst Mercantile Store.[86]

Fig. 11.15. The Hearst Mercantile Store in Nemo, 1921.—Photo courtesy of the Black Hills Mining Museum.

Horses provided the means for skidding and hauling logs at the Homestake logging camps (see fig. 11.16) until 1923 when crawler-tractors and four-wheeled Schutler log wagons were introduced. A few years later, larger crawler-tractors and eight-wheeled Schutler log wagons were introduced at the Nemo operation. The Schutler wagons were used to transport logs several miles to landings along the Black Hills and Fort Pierre Railroad. Wood was loaded onto flatcars and delivered to the Nemo Sawmill. Here, the wood was sawed, framed, and reloaded on cars and delivered to Lead (see fig. 11.17).

Around 1930, the Linn Tractor and Trailer (see fig. 11.18) replaced the Schutler log wagon.[87] Two of the Linn trucks were assigned to the Nemo operation with John "Happy" Larson and Albin Erickson as operators. The Linn truck could travel about 10 miles per hour empty and 1-2 miles per hour loaded. Timber haulage across the narrow-gauge Black Hills and Fort Pierre Railroad to Lead was abandoned in 1929 after rubber-tired Mack flatbed trucks were introduced. The Mack trucks were replaced with tractor-trailer trucks in 1938.[88]

Fig. 11.16. A Homestake logging crew, ca. 1888-89. The man sitting on the log wagon is Ed Owens. The two men on the cook's left side are Dave and Morgan Theophilus. According to a "Census of Manufactures, Lumber and Timber Products" report filed with the U.S. Department of Commerce for 1909, Homestake employed eighty-five people in its logging operations and seventy-three people in its sawmills at the time.—Photo courtesy of the Black Hills Mining Museum.

Fig. 11.17. A load of wood on the Black Hills and Fort Pierre Railroad at Nemo.—Photo from Keith Shostrom collection.

Fig. 11.18. The Linn Tractor-Trailer, introduced into the Homestake logging operations in about 1930. John "Happy" Larson is the driver.—Photo courtesy of Homestake Mining Company.

The sawmill at Nemo operated two shifts per day, six days per week.[89] Three 250-horsepower Scotch marine boilers supplied steam power for the mill. Dutch ovens were used to burn the large volumes of sawdust. One 350-horsepower Corliss engine and three smaller engines supplied power to the various small motors and generator that provided illumination at the camp. Logs were dumped into a hot pond heated by exhaust steam from the mill boilers. The hot pond served to reduce the frost level in the logs to facilitate sawing in the wintertime.

The Nemo Sawmill consisted of a 10-inch-band head rig with a capacity of about 5,000 board feet per hour. The rig also contained 7-foot-diameter wheels, live rolls, gang edger, slasher, trimmer, and main cutoff saw. The timbers were squared at the head rig and dropped onto the roll case, which carried them to the cutoff saw. The cutoff saw sawed the timbers to the proper lengths for mine use. Timbers varied in size from 10-14 inches square and from 5 feet 6 inches in length to 9 feet 8 inches in length. The latter size was used for sill-floor posts in the mines. Additionally, the cutoff saw produced mine wedges, wedge blocks, shingle blocks, and mine lagging. Mine timbers were conveyed to the framing mill for special cuts.

Lumber from the head rig passed through the gang edger, over the trimmer, and to the sorting chains. From here, the lumber was loaded onto lumber buggies and delivered to the pile yard for drying or seasoning. After seasoning, the lumber was moved to the planing mill for finishing. The plant also produced lath, shingles, and mouldings. Operation of the Nemo Sawmill was discontinued on July 1, 1935, five years after the railroad discontinued operations.[90]

A second new sawmill was constructed at Moskee, Wyoming, in 1923 as part of the Golden Gate Mine and Timber Company (see fig. 11.19).[91] A third sawmill, Camp 5, was established in 1934 approximately 1 mile north of Vanocker Canyon (see fig. 11.20). Homestake's three sawmills produced some 11.410 million board feet of lumber and timber in 1934 of which 9.631 million board feet was used at the mine. The remainder was either sold commercially or used at the camps.[92] The Camp 5 operation was shut down on December 31, 1937.[93]

Fig. 11.19. The Moskee Sawmill and forest fire of August 9, 1936. This photo, by Black Hills Studios, shows the forest fire that threatened Homestake's Moskee Sawmill. The Moskee Sawmill was shut down in 1940 after Homestake's Spearfish Sawmill was commissioned.—Photo courtesy of Johnny Sumners and the Black Hills Mining Museum.

Fig. 11.20. Homestake's sawmill at Camp 5, which was operated from 1934 through 1937.—Photo courtesy of Jerry Bryant.

The Moskee Sawmill had a capacity of 3,000 board feet per hour initially. The equipment consisted of a rotary head rig, live rolls, gang edger, cutoff saw, trim saw, lath mill, and planning mill. The boiler plant initially consisted of one Scotch marine boiler. The plant was powered by one 75-horsepower engine and one 175-horsepower engine. A small engine and generator were used for camp lighting. Houses, wood fuel, and water were provided at no charge to employees at the Moskee and Nemo camps.[94]

In November 1940, a new sawmill was completed at Spearfish (see fig. 11.21), and the Moskee sawmill was permanently shut down. By 1954, the Spearfish Sawmill was producing approximately 17 million board feet of products annually. Logs were sourced from Homestake and National Forest lands in the Bear Lodge Mountains and the Black Hills of South Dakota and Wyoming. Homestake's timber holdings in 1954 amounted to more than 60,000 acres of land.[95]

Fig. 11.21. Homestake's Spearfish Sawmill. The Spearfish Sawmill was commissioned in November 1940 at a cost of approximately $500,000. After a fire that destroyed the sawmill on February 19, 1980, the property was sold to Pope and Talbot Inc. in 1981.—Photo courtesy of the Black Hills Mining Museum.

At the Spearfish Sawmill, tree-length logs were dumped into a receiving pond to rid the logs of dirt and other foreign materials. From here, the logs were conveyed to a large circular "buck-up saw" where they were cut to proper length and subsequently dropped into a "hot pond." The logs were then conveyed to a head-rig saw and rough cut into various sizes of mine timber and lagging. From here, plank material was sent to a double-end trim saw. Slab material was sent to a horizontal-band "resaw" and sawed into lath and other small products. A planer was used to make such products as tongue-and-groove lacing slabs for the mine, ship lap, and flooring.[96] To balance production needs and allow Homestake's Spearfish Sawmill to focus on needs of the mine, additional mill-run lumber was outsourced to other sawmills in the 1950s and 1960s. In 1960, more than 7 million board feet of finished lumber was purchased from the Neiman, McLaughlin, and Nicholson sawmills.[97]

On August 1, 1972, the assets of the lumber and timber department of the Homestake Mine, including the Spearfish Sawmill, were transferred to the Homestake Forest Products Company, a wholly owned subsidiary of Homestake Mining Company.[98] Tragedy struck on February 19, 1980, when a fire destroyed the entire sawmill section. The bin sorter, planing mill, office, and kilns were not damaged.[99] Fortunately, the Homestake Mine was no longer dependent on mine timber based on decisions to use rock bolts and cablebolts instead of support timber and utilize open-stoping methods in lieu of square-set methods. For these reasons, the Spearfish Sawmill property was sold to Pope and Talbot in February 1981, which effectively brought a close to another chapter of Homestake's rich history.

Chapter 12
The Lead Townsite Development

Occupants of the Townsite versus the Mineral Claimants

The history of gold discoveries in the area of Lead dates back to January 7, 1876, when William Gay, Alfred H. Gay, M. J. Ingaldsby, E. D. Haggard, D. Muckler, and John B. Pearson staked the Giant Lode (No.115) and the Gold Run Lode (No. 116) near the north fork of Gold Run Creek.[1] In February 1876, Thomas E. Carey crossed over the divide from Deadwood Creek and discovered placer gold farther down Gold Run Creek, a tributary of Whitewood Creek. Shortly after Carey's discovery, the camps of Washington and Lead City were established (see figs. 12.1 and 12.2).

Fig. 12.1. Lead City and Washington, 1876-77. This view, looking south toward Mill and Bleeker streets, is one of the earlier photos of Lead City and Washington taken in late 1876 or early 1877. Lead City is the settlement between the north fork of Gold Run Gulch, shown in the foreground, and the main fork of Gold Run, shown in the upper center. In the *Gold Belt Cities: The City of Mills*, reference is made to a 1905 souvenir edition of the *Lead Daily Call*, where it is mentioned that the building in the left center marked with an "x" above

its roof is the first store of Peter A. Gushurst, a pioneer Lead merchant. Much of South Lead was later established in the area to the upper left. The Washington camp is that area northeast of the north fork of Gold Run, partially shown in the lower left.—Photo courtesy of Fielder Collection, Devereaux Library Archives, South Dakota School of Mines and Technology.

Fig. 12.2. North Mill Street, Lead, 1878. The view in this photo is to the north, down Mill Street. Small open cuts can be seen on the side of Clara Hill, where work has been done on the Homestake, Golden Star, Giant, Old Abe, and Clara mining claims, among others. The taller building in the upper left (marked with an "x") is the Holvey Building, constructed in 1877 at the northwest corner of Main and Mill streets.—Photo courtesy of Black Hills Mining Museum.

W. P. Raddick, one of the first prospectors in Gold Run Gulch, provided the following historical sketch in 1906, which was reprinted in the *Lead Daily Call* on August 5, 1926:

> *The ground selected for the site of the new town was between the north and south forks of Gold Run and south of the Homestake and Golden Star mines, this being the only suitable spot upon which to start the foundation of the new city. Foremost in this movement were James D. Coffin, Frank Judkins, William Quiggley, and Robert Jones, better known as "Smoky Jones."*
>
> *Next in order was the selection of a name for the new city and after little discussion the name of "Lead City" was decided upon, the name*

being suggested by the lead or lode mines in and around it. On the tenth day of July, 1876, some brush was cut and lines run and lots staked off. This work was continued on the eleventh, on which day the work was completed. In the afternoon of that day, the founding party assembled at the camp of "Smoky" or Robert Jones, which was situated under the spruce and pine trees on ground which is now East Main Street, and between the old First National bank building and the Northwestern Depot.

The personnel of the founding party assembled in Jones' camp on that July afternoon is as follows: William Quiggley, Frank Judkins, E. Q. Kipplinger, David Snyder, Jasper King, Samuel Parker, James D. Coffin, Robert Waldschmidt, William Baldwin, Jack Daly, Charles Jones, Michael Reddy, John King, W. P. Raddick, Charles DuSette, Robert or 'Smoky' Jones, and John Hensey.[2]

At the miner's meeting, called to order by Charles DuSette, Charles Jones was elected secretary and James D. Coffin, recorder. It was decided that no other officials would be elected, nor would any kind of city government be established, since the country was still a part of the Great Sioux Reservation and the government would not have any legal standing. Notwithstanding, the organizers decided that the local law of the miners would govern over any disputes.

The miners decided that city lots would be 50×100 feet and the fee for acquiring and recording a lot would be $1.25.[3] A person was limited to one lot and had sixty days in which to build a residence. A house was required to be at least 10×14 feet and 6 feet high. By July 27, approximately twenty houses were under construction.[4] Later, in the absence of any other laws or ordinances, many of the miners elected to construct a cabin on one-half of their lot and sell the other half. This resulted in numerous lots being only 25 feet wide.[5]

That afternoon, the miners laid out the north-south streets of Mill, Bleeker, Gold, Wall, Siever, and Galena. East-west streets parallel to the south fork of Gold Run Gulch included Pine, Main, and Saunders. Pine Street was located one block north of Main, and Saunders was one block south. According to Raddick, a few weeks later, a different party laid out the mining camp of Washington.[6]

At the same miner's meeting, John King, Fred Manuel, and Robert Jones were elected trustees, with King acting as chairman. One of their first charges was to identify a route for a wagon road between Lead City and Deadwood.[7] Gold Run Gulch, like Deadwood Gulch, was virtually impassable throughout most of 1876 because of the large amount of fallen timber and underbrush that covered the gulch.

A horse trail was established across the ridgetops between Washington and Deadwood, and it became the first practical route between the mining camps of Lead City and Deadwood.[8] The horse trail passed through Washington and over McGovern Hill to Deadwood. Another horse trail extended from Lead City over the hill to Bobtail Gulch and down that gulch to Deadwood Gulch. A wagon road was badly needed. Two possible routes were identified—one via Poorman Gulch and Deadwood Gulch, the other through Gold Run Gulch to Whitewood Creek. The miners, who preferred the latter route, decided that the Deadwood businessmen should assist in constructing one or both routes.[9]

A few days later, the editors of the *Pioneer News* ran the following editorial:

> We have received a communication from Lead City, the only town here purposely built, you might say, for quartz miners, which deserves the perusal and careful attention of our citizens. Would it not be well to acquiesce in the sentiments of the Lead City correspondent and assist in the construction of a wagon road by which the two cities can be connected? It must be conceded that Gold Run is as rich in placer diggings as the richest and that a great many valuable loads have been discovered and are at the present time being actively worked, and the more the placers and lodes of the section are worked, particularly the more thorough the development of the lodes, the better they prove themselves to be. Taking all the facts into consideration regarding the prospective importance of Lead City, we would suggest that the citizens take this matter in hand and give it their earliest attention, thereby showing their appreciation of the progressive steps already being taken by our Lead City neighbors.[10]

By July 1881, a toll road was constructed from the camp of Cleveland on the south side of Deadwood to Lead City via Whitewood Creek and Gold Run Gulch. The toll road was called the Cleveland Toll Road.[11]

A majority of the quartz miners of Lead City convened again on August 2, 1876, for the purpose of organizing a quartz-mining district. The Gordon Quartz Mineral District was established. E. M. Tower was elected president and James Coffin, recorder. The miners commenced drafting laws and boundaries for the new district.[12]

Over the winter of 1876 and spring of 1877, J. D. McIntyre, a U.S. deputy surveyor, laid out the Denver addition to Lead. The Denver addition was bounded on the east by Galena Street and on the west by Poorman Gulch Road and Ruby Basin Road. In 1878, McIntyre helped prepare a townsite patent for the residents of Lead City. McIntyre informed the residents that for a $25 filing fee, a lot owner would receive a warranty deed to their lot once the townsite patent was issued. McIntyre's proposal met with strong opposition from the holders of the mining claims, however, and the application process was stopped.[13]

Washington was also laid out in July 1876. The town is shown on a later plat called the "Plat of the Town Site of Lead City," which was prepared by George S. Hopkins on May 11, 1885. Washington consisted of six blocks situated east of Spring Street, south of Old Abe Street, and north of the Gold Run Toll Road, which is now East Main Street along Highway 85. The town of Washington included Spring Street, Fox Street, Grantz Street, and Main Street, now called Washington Street. Most of the Washington camp was located within 1,000 feet of Spring Street.

On January 17, 1881, a plat was prepared by William L. Smith and submitted to Leonard Gordon, county probate judge, as part of a new application for a townsite patent for Lead City. The plat represented 138.28 acres. However, a dispute quickly arose when the holders of the unpatented mining claims, mostly Homestake Mining Company, objected to the application, claiming the mineral claimants were entitled to patents on the lands, not the holders of town lots. Meanwhile, Homestake and other holders of mining claims continued to file for and receive patents to many of their claims.[14]

In October 1888, James Milliken, owner of the Savage Tunnel and mining claims covering parts of South Lead, requested that Judge Moody conduct a hearing to determine land ownership. Judge Moody met with attorneys representing Homestake and the residents, but nothing substantive resulted from the hearing.[15]

The citizens of Lead City held a municipal election in August 1890 to decide on incorporation of the city. The resulting vote was 471-9 in favor of incorporation as the city of Lead. A new townsite application was submitted and approved, based on the plat drawn up by George S. Hopkins, dated May 11, 1885. The name of the city was shortened to Lead, in accordance with the decision to incorporate. Cyrus H. Enos was elected mayor. Four wards were decided upon and were represented by Charles Barclay, Ernest May, P. A. Gushurst, D. J. O'Donnell, John K. Searle, Frank Abt, Thomas Conners, and Michael Cain. Joseph B. Moore was appointed city attorney; Thomas Neary, chief of police; and Henry Schmitz, city auditor.[16]

Locators of the mining claims subsequently filed an appeal with the commissioner of the General Land Office in Washington, D. C., protesting the townsite patent that had been issued by local authorities. The General Land Office rejected the appeal on the basis that no substantial mineral values were present on the mining claims and that the land was more valuable for townsite purposes. The mining locators made a further appeal to the secretary of the Interior, the agency of the government that administrated the Mining Law of 1872. During the appeal process, the parties decided to negotiate a settlement.[17] Attorney G. C. Moody represented Homestake. Attorneys J. P. Wilson and Thomas H. Breen represented the townsite claimants. The plat that had been prepared by Hopkins in 1885 was submitted to Adoniram J. Plowman, county probate judge and successor to Judge Leonard Gordon. Judge Plowman certified the Hopkins Map as correct on October 29, 1891.

On February 5, 1892, a settlement agreement was reached between "Occupants of the Lead City Townsite" and "Mineral Claimants, J. B. Haggin and Homestake Mining Company." Haggin and Homestake, as holders of the mining claims, were given the right to continue procuring patents to their mining claims. The agreement specified that Homestake would "convey by quit claim the right to the surface occupancy of all such lots and blocks and parts of lots and pieces of ground situated upon said claims as are now occupied by persons residing or claiming property in said Lead City Townsite." The agreement allowed Homestake "to continue to possess the right to mine underneath said surface, and to use it for all such underground mining purposes by running drifts, tunnels, stopes, excavations of ore or other mining operations, including the sinking of shafts from the surface."

Pursuant to the agreement, if any part of the mining operations would ever "come to the surface or so near the surface as to endanger the surface occupancy," Homestake would provide the surface occupant with a sixty-day notice. Within ninety days of such notice, the surface occupant was required to remove all buildings and structures. If the surface occupant failed to remove the structures, Homestake could remove such structures or regard them as abandoned and not be liable for damages. The agreement named Cyrus H. Enos, Peter A. Gushurst, and Ernest May trustees for the Occupants of the Lead Townsite. The signed Townsite Agreement was filed for record on March 18, 1892.[18]

The quitclaim deed was issued to the trustees on March 1, 1892. Homestake Mining Company, James B. Haggin, and James Milliken, owners of the mining claims, conveyed "the right to the surface occupancy" of all the lots and blocks described in the deed and shown on the Hopkins Map. The terms and conditions of the Townsite Agreement were restated in the deed. The deed specifically stated that "it is not intended to dedicate to the public any of the streets or alleys in said townsite area, except the surface use thereof, subject to the same rights and reservations as herein set forth, relating to the occupied lots and parcels of ground."[19]

The lots and blocks to which the right to surface occupancy was deeded are described in the quitclaim deed and are shown on the Hopkins Map. The land areas were generally bounded by Summit Street on the south; thence north on Barclay to Hill Street, thence west on Hill to Sawyer Street; thence north on Sawyer to Main Street; thence northwest to include Blocks 7, 10, and 25 north of Pine Street and west of the Star Mill; thence west to include the block known as Homestake Boulevard along the west side of Gold Street; thence south to the south side of Railroad Avenue; thence west on Railroad Avenue to Siever Street; thence south on Siever to Main Street; thence west on Main Street to Stone Street; thence south on Stone to Addie Street; thence east on Addie to Gold Street; thence south on Gold to Summit Street; thence east on Summit to Barclay (see fig. 12.3). The deed was filed for record on January 5, 1893.[20]

Fig. 12.3. Lead City, 1889. Most of Lead's businesses and residences shown in this photo are located within the area covered under the Townsite Agreement of 1892. Pursuant to the agreement, occupants of town lots were given the right to occupy the surface. Homestake retained mineral rights and was allowed to continue mining below the surface.—Photo courtesy of the State Archives of the South Dakota State Historical Society.

1—Black Hills and Fort Pierre Railroad
2—Homestake (Vertical) Shaft (1878-1890s)
3—Golden Star Shaft (1881)
4—Homestake Hotel (W. Side of N. Mill Street; May 1878)
5—T. J. Grier Residence and Offices
6—Homestake (a.k.a. Old Eighty) (July 12, 1878, through November 18, 1922)
7—Old Abe Shaft (B&M No. 1; January 1879, through December 27, 1934)
8—Golden Star 120-Stamp (September 1, 1879, to March 1923)
9—Highland (Amicus) 120-Stamp (mid-1880 to December 31, 1934)
10—Hearst Mercantile "Brick Store"
11—World's Fair Restaurant (E. Side N. Mill St.)
12—First National Bank (NE Corner, Main and Mill)
13—Miners' Union Hall (NW Corner, Main and Bleeker)
14—Dickinson & Cornes Drug Store
15—Masonic Hall (SE Corner, Main and Bleeker streets)
16—Catholic Church

Subsequent to the Townsite Agreement and quitclaim deed from Homestake, Edwin W. Mitchell, city engineer, remapped and updated the entire city on a plat dated June 25, 1895. Mitchell renumbered most of the lots and blocks on his plat. On December 27, 1898, the U.S. Land Office issued a townsite patent to Judge Leonard Gordon, trustee for the occupants of the Lead City Townsite, but only for tracts A, B, C, D, and E within the original townsite boundary. The five tracts totaled 16.661 acres, as shown on a new plat drawn by Frank Morris, U.S. surveyor general for South Dakota (see fig. 12.4).

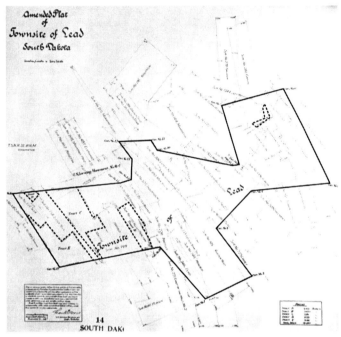

Fig. 12.4. The original Townsite of Lead, South Dakota per Frank Morris's plat, December 27, 1898. The U.S. Land Office issued the trustee of Lead City a townsite patent for tracts A, B, C, D, and E (shown in dashed outlines), totaling 16.661 acres, on December 27, 1898, as shown on this plat by Frank Morris, U.S. surveyor general for South Dakota. Owners of lots within these five tracts were given fee simple or absolute title to their lots, including mineral rights. Owners of lots outside the five tracts remained subject to the 1892 Townsite Agreement, which gave them the right to occupy the surface of their respective lots for as long as the surface wasn't needed for mining purposes.—Photo courtesy of Fielder Collection, Devereaux Library Archives, South Dakota School of Mines and Technology.

Except for the 0.43-acre Tract A, which was located east of the Homestake stamp mills, the four other tracts were roughly bounded by Gold Street on the east, Grand Avenue on the west, Addie Street on the south, and Main Street or

Railroad Avenue on the north. The five tracts were excluded from the mineral patents that were granted to the holders of the mining claims. Most of the other lots that were situated outside the five tracts were on the mining claims held by Homestake and the few other mining companies. The townsite patent for tracts A through E was filed for record on February 1, 1899.[21]

On August 7, 1900, the original townsite boundary was expanded, and J. P. Crick, city engineer, prepared a new townsite plat. The five tracts within the original townsite boundary are depicted on Crick's plat. In the 1980s and 1990s "surface occupants" of the town lots subject to the 1892 Townsite Agreement were given the opportunity by Homestake to obtain title to the surface of their town lots. Even today, one or more of the Hopkins, Mitchell, Crick, Stoner, or other city plats are referenced in the legal descriptions for lots and blocks with the city of Lead, depending on the location.

The Gushurst and Manuel Connection

Peter Albert "Al" Gushurst arrived in Deadwood on June 1, 1876, via Fort Laramie and Custer. Almost immediately, Gushurst started making good money by purchasing remaining supplies at huge discounts from prospectors leaving the Black Hills. He would then resell the supplies to the prospectors who were invading the Hills in search of their fortunes. By August, Gushurst completed construction of the Big Horn Store on Main Street in Deadwood (see fig. 12.5). The store was named after the Wolf Mountain Stampede, which was a mini gold rush where prospectors from the Black Hills rushed to the Big Horn Mountains of Wyoming Territory to partake in a rich gold find that turned out to be a fabrication.

Fig. 12.5. Peter A. Gushurst's Big Horn Store in Deadwood, Dakota Territory, Stanley J. Morrow, 1876.—Library of Congress (LC-DIG-Stereo-1s00438).

Gushurst had no more opened the Big Horn Store for business when several placer miners appeared and informed him that his store was situated on their mining claim. Not wanting to get involved in a legal battle over his rights as a lot owner, Gushurst settled with the miners for $175, plus his rifle. A few days later, another group of miners from Lead City suggested to Gushurst that he should move his business to Lead City. Based on their need for a store in Lead, the miners enticed Gushurst by offering to move his stock of goods to Lead free of charge. To make the invitation even more appealing, the miners offered Gushurst an existing cabin to move into. He subsequently accepted the offer, surmising that business might eventually be better in Lead City, anyhow. Gushurst's intuition told him the placers in Deadwood Gulch would soon play out and that the lode mines in Lead City would draw significantly more people to Lead. Gushurst then sold the Big Horn Store to Jake Goldberg and began preparing to move to Lead.[22]

The miners helped Gushurst make the move to Lead City on August 10, 1876. Almost immediately, he opened for business in a cabin located between the forks of Gold Run Gulch. In 1877, Gushurst built a new store at the northeast corner of Main and Gold streets. This store was used until 1898 at which time Gushurst had the building dismantled to make room for an all-new brick building he constructed a short time later (see fig. 12.6).

Fig. 12.6. Dismantling the Gushurst store in Lead, 1899. This photo was taken on May 17, 1899, by M. F. Brennan of the *Pioneer* Group. The occasion involves Peter A. Gushurst, a pioneer grocer in Lead, whose crew is in the process of dismantling his store at the northeast corner of Gold and Main streets to make room for a new brick building that Gushurst will soon construct. Gushurst is the man seated in the front row with the dog. —Photo courtesy of the Black Hills Mining Museum.

Gushurst was instrumental in helping lay out a third settlement, named Golden, in the Lead City area. Golden was located east of Washington and extended down the north flank of Gold Run Gulch. Eventually, both camps were merged into Lead City.

Gushurst married Josephine A. "Josie" Aikey on May 26, 1878, in Lead City. She was born November 8, 1861, in Kelso, Sibley County, Minnesota. Josie was a niece to Fred and Moses Manuel. It was early July 1877 when Fred Manuel and Hank Harney took a break from their mining claims and made a trip to Chicago.[23] On their way to Chicago, the men stopped to visit Fred's relatives in Minnesota. While there, Fred took his niece, Josie, to Chicago to see the sights. The trip must have been an eye opener for her, considering the rawness of the West and the fact that she was only fifteen years old.

Upon returning to Minnesota, Josie begged her parents and Uncle Fred to let her come to Lead City with Fred. Josie said she could be Fred's housekeeper for a while. Reluctantly, Fred agreed and they made their way to Lead City in the fall of 1877. After the Manuel brothers and their partners sold the Homestake claim to the California capitalists, the Manuels decided to move away. By this time, Josie was having such a good time in Lead City that she convinced Uncle Fred to allow her to stay in Lead. Fred then made arrangements for Josie to stay with a family named the Watsons, who were good friends of the Manuels.

Al Gushurst and Josie Aikey first met at a dance during the July 4 festivities in Lead in 1877. The dance was the first formal dance ever held in Lead City[24] and was held in the attic of a building built by Paul Jentges at the northwest corner of Main and North Mill streets. The building housed the Holvey Drug Store a short time later.[25] Aikey and six other women, who constituted the entire female population of Lead City at the time, attended the dance.[26]

Since the Watsons' residence on Main Street was only a half block away from Al's store, Josie came to see Al in his store quite often. In May 1878, the two were married. The marriage was the first one in Lead City. When the time came for Al to present the bride's ring at the wedding, the best man couldn't find the ring and neither could Al. After a few moments, Josie could see that Al was very embarrassed. Josie then reached into the pocket of her wedding dress and produced a gold ring she had removed from her finger prior to the ceremony. This ring was a gift she received from Uncle Fred months before and was made from some of the first gold he recovered from the Homestake Mine. Al graciously took the ring and placed it on her finger, and the wedding ceremony was completed.[27]

First Murder in Lead City

At a town meeting held in the spring of 1877, the residents of Lead City decided to reelect its board of trustees. Their newly assigned duties involved tending to civil matters and arbitrating disputes. Peter Albert Gushurst, Timothy

O'Leary, and Major Ledwich were chosen as trustees. One of the first issues the trustees faced involved some of the businessmen and residents who found they could not find a suitable lot for their cabin or business. A few of these people, such as John McTeague, decided to appropriate a portion of the streets on which to build. McTeague chose part of Bleeker Street for his cabin. He went so far as to mark out the boundaries of his lot with flour and declared that anyone violating the boundary "would have need for the coroner."[28]

Other people followed McTeague's example and marked out their own lots in the middle of Bleeker and Pine streets with logs. The rest of the citizens demanded that the board of trustees take action against the squatters, claiming that if Lead was going to be functional, it needed to have streets that were passable. Gushurst was asked to take action, beginning with McTeague. Ignoring McTeague's threats that he would be shot if he crossed the flour line, Gushurst crossed the line and tried to convince McTeague that he should abandon his cabin before the citizens stormed the place with guns and tore the place down. McTeague rebutted that he would "kill everyone in this end of the United States" rather than abandon his cabin site.

After the board of trustees caused a warrant to be issued against McTeague, Sheriff Seth Bullock took him into custody. That night, McTeague's cabin was mysteriously blown to pieces with dynamite, and by daybreak, Bleeker Street was again passable. Meanwhile, "Shotgun" Layther, a friend of McTeague, attempted to jump the lot of Ben Wadsworth on Pine Street. An altercation resulted between the two men, and several shots were fired.

The following day, February 25, 1878, McTeague, who had just been released from jail, decided he was going to straighten the matter out with Wadsworth. McTeague came to Wadsworth's lot armed with a revolver. Wadsworth stood firm on his lot with his double-barreled shotgun. He warned McTeague three times to stay off his lot, but McTeague paid little attention. Wadsworth pulled the trigger on one barrel, and McTeague fell, mortally wounded. Wadsworth was arrested for murder but was later acquitted by Judge Hall at a preliminary hearing in Deadwood.[29]

Schools

The first public school was held in 1877 in a small log house on north Bleeker Street. Julia B. Snyder was the first teacher with fourteen students. Frank Abt, Henry Hill, and J. C. Booth served on the first school board, which was organized under the district township.[30] In June 1878, the first formal school district was established in Lead. Henry Hill was elected clerk, Thomas Pryor, treasurer, and Frank Abt, director. Thirty-two students were enrolled in public school during the summer months that year. Classes were held in an upstairs room of Belliveau's store, under the direction of Professor Dean.

In September 1878, school was held in a house on Pine Street opposite the old hose house. Professor Wheeler and Julia Snyder were teachers for the fall term. In 1879 and 1880, classes were moved to a house on Bleeker Street where Professor Darling and Ms. L. Chapman taught the students. The former Catholic Sister's Hospital on Wall Street was transformed into the first public schoolhouse in 1881 (see fig. 12.7). J. S. Thompson was principal, and E. J. Bishop, Anna Graham, and Ms. Burnham were teachers.[31] Lead's other school was located in the Washington area.[32]

Fig. 12.7. The Lead City District School on Wall Street. School was held in this building from 1881 until about 1914, when it was razed to accommodate the new Central Campus building.—Photo courtesy of the Black Hills Mining Museum.

The Lead Independent School District No. 6 was formed on April 26, 1895. On August 23, ground was broken for a high school along the north side of Addie Street between Wall and Siever streets. The high school building on Wall Street was completed early in 1896 at a cost of $35,000 (see fig. 12.8). The building served as Lead's High School through the graduating class of 1940. After that, the building was used for some of the elementary classes until 1981, when it was demolished. A new high school was completed at 320 South Main Street in 1940.[33] On July 1, 1971, the Lead and Deadwood Independent School Districts were combined. The first combined high school graduation was held in May 1972, and 192 students were graduated.[34]

Fig. 12.8. The school buildings of Lead's Central Campus. This photo shows (from left to right) the Assembly Hall constructed in 1903, Central Elementary School (1914), and Lead High School (1896). The three buildings were located north of Addie Street between Wall and Siever streets.—Photo courtesy of the Black Hills Mining Museum.

In 1900, after hearing of the Lead Womens' Club's interest in establishing a kindergarten for four- and five-year-old children, Phoebe Hearst offered to provide funding to support what became known as the Hearst Free Kindergarten. Hearst had funded other such kindergartens in California and Montana as early as 1883. Phoebe personally helped dedicate the new kindergarten, located in the Christ Episcopal Church on Main Street, on May 27, 1901. In 1934, the Hearst Free Kindergarten was merged into the Lead Public School System. However, because the kindergarten was still being held in the Episcopal Church, Homestake assumed the cost of the rent for that space. The Hearst estate paid for a new heating plant, linoleum, and paint for that part of the church used for the kindergarten.[35]

As the population of Lead boomed so did the need for additional school buildings. By 1905, Lead was the second largest city in South Dakota. Approximately 1,900 children were enrolled in the Lead School District in June 1908, compared to 1,766 the previous year.[36] With Lead's population approaching 10,000 before World War I,[37] new schools were continually being constructed. These included the South Lead School (built in 1900) (see fig. 12.9), West Lead School (1901), Assembly Hall on the north side of the Central Campus (1903), a high school annex (1926), Washington School (1937), and the new high school (1940).[38]

Fig. 12.9. The South Lead School. The South Lead School was constructed in 1900 to accommodate the population growth in Lead. Morning and afternoon kindergarten classes were held in the room on the left. First-grade class was held in the room on the right. The school was located on South Mill Street across the road from the South Lead Cemetery and below the road to the Ross Shaft.—Photo courtesy of the Black Hills Mining Museum.

A Few Other Early Lead Businesses

Lead was home to many drugstores over the years. Each store marketed its own "cures" (see fig. 12.10). Dr. Daniel K. Dickinson's brother, W. R. Dickinson, also known as Billy Dick, operated Dickinson's Drug Store on the corner of Main Street and Bleeker Street. After the two Dickinson brothers left for California in 1904, W. R. Dickinson sold his drugstore to Charles L. Stillman and Albert Stillman.[39]

Fig. 12.10. Medicine bottles of Lead's early-day druggists. W. R. Dickinson, brother of D. K. Dickinson, had his own drugstore and marketed Homestake Sarsaparilla (center) and Homestake Cough Syrup (right center) for a short time.—Photo by author.

Other Lead druggists included Alexander Lawie, Knowles and Ingalls,[40] and A. Ottman. Ottman operated the City Drug Store. J. R. Tapster operated the Red Drug Store at the corner of Main and Mill streets. L. P. Jenkins and J. F. Fisk also had drugstores near Main and Mill streets. The Brown Pharmacy was located at the northwest corner of Main and North Bleeker streets. The Eagle Drug Store was located at 14 North Bleeker on the corner of Pine.

Many of the miners were notoriously thirsty. Fortunately, for them, Lead businessmen offered an array of saloons where, it was said, "all the mining was done" (see fig. 12.11). Fidele "Fred" Vercellino operated the Cristoforo Colombo Saloon from 1901 to 1914. Cotton's Bar was located at the corner of Main and Mill streets. The Welcome Saloon was located on Main Street between Mill and Bleeker streets. The list goes on.[41]

WHERE ALL THE MINING WAS DONE

Fig. 12.11. A pencil sketch by Buck O'Donnell entitled, *Where All the Mining Was Done*.—Courtesy of the Black Hills Mining Museum.

Lead had at least two brothels, including the Green Front and Annie's Place. Annie's Place was located on North Bleeker Street and was owned by Annie Woods. One of the popular songs of the World War I era was originally written as a love note to a girl who worked at Annie's Place. George A. Norton, who had fallen in love with Maybelle, wrote the love note on the back of an envelope. Ernie Burnett later put the words to music. The song, called "My Melancholy Baby," was published in 1912.[42]

The first hotels in Lead City included the Miners' Hotel, constructed by James Long in June 1877,[43] and Kate Graham's hotel, which was located at the corner of North Wall and Main streets.[44] Several other hotels were constructed over the next few years, including the Martin Hotel, the Springer House, owned by Ed Springer, the Abt Hotel, owned by Frank Abt, and the Homestake Hotel (see fig. 12.12).[45]

Fig. 12.12. The Homestake Hotel, July 4, 1910. The original Homestake Hotel was constructed in May 1878. After a fire damaged the hotel on March 3, 1903, it was rebuilt, and the brick addition was completed.—Photo courtesy of the Black Hills Mining Museum.

The first custom assay office in Lead City was located in Kate Graham's Hotel on the northwest corner of Main and Wall streets. S. R. Smith had the first undertaking business. Jack Daly had the first blacksmith shop, located on the northeast corner of Main and Mill streets. Ernest May, who was an owner in the Wheeler brothers' placer claim in Deadwood Gulch, opened a store in Lead in 1877. May was an investor in various mining companies in the northern Black Hills and was one of the first men to introduce a chlorination plant for treating ores of the Black Hills. May also served one term as mayor of Lead.

The Catholic Sisters of Charity had the first hospital in Lead in 1879. Later that year, Homestake established its first hospital. A branch of the First National Bank of Deadwood was established in Lead in 1878. It was known as the Bank of Lead and had John Ainley as its first manager. The bank was later reorganized as the First National Bank of Lead. E. M. Rainwater sold the first liquor in Lead in 1876. Thomas P. Jones had the first butcher shop, and John Howe had the first bakery. Both were located on Mill Street.[46]

Like Deadwood, the business district of Lead was decimated by fire, but twenty-one years later. At about 2:00 a.m. on March 8, 1900, flames appeared between the Searle butcher shop and an adjacent building owned by Charles Dalkenberger. Chris Nelson, a night clerk at the Campbell House, first spotted the fire and quickly sounded the alarm. The fire quickly raced to the northeast and burned the Miners' Union, the Faust building, and all of the buildings in the block bounded by Main, Mill, Bleeker, and Pine streets. Using explosives, miners blasted several buildings to create a firebreak and prevent the fire from reaching the Homestake mills and the Hearst Mercantile. Over the next few years, the business district was rebuilt in the same area. One of the new buildings included the Smead Hotel, an immaculate five-story brick-and-stone building constructed by Walter E. Smead at the corner of Pine and Mill streets (see fig. 12.13).[47]

Fig. 12.13. The Smead Hotel, constructed at the northwest corner of Pine and North Mill streets in 1901. The area is occupied by the Open Cut today.—Photo courtesy of the Black Hills Mining Museum.

Churches

The first formal church services in Lead City were held in 1877 by a group of Catholic people who gathered for worship in a large log cabin situated near the 80-stamp Homestake Mill. In 1878, a new Catholic church was built on North Gold Street with Father Bernard Mackin, priest. Around 1894, the church was razed, and a new one was constructed on Siever Street next to the parochial school the Sisters of St. Benedict had opened in 1892. This school operated until 1966 at which time students were transferred to the public schools. The church burned in 1900 but was rebuilt by Father Redmond in 1902.[48]

The Congregational Society Church was organized next on August 27, 1878. The congregation moved into a newly constructed building on the corner of Wall Street and Julius Street in 1892. In 1911, the congregation merged with the United Presbyterian Church.

The Reverend W. D. Phifer organized the first Methodist society in Lead on November 15, 1880. Their first services were held in various locations, including the Opera House, log schoolhouse, and Miners' Union Hall. On February 3, 1881, a charter was granted to the congregation and a new building was built. This building, dedicated on August 11, 1881, served the congregation until April 18, 1909, when a larger building was dedicated. After selling the latter building in 1932, another new building was constructed at 111 South Main Street in 1933 and dedicated on January 28, 1934. The building was expanded in 1955-56.

The Reverend E. K. Lessell, a seminary student from Connecticut, held the first Episcopal services in Lead in 1878. On August 16, 1887, construction was started on a new building at the corner of Wall and Addie streets. This building was dedicated as Christ Church on March 4, 1888. The congregation outgrew this building and constructed a new one on Main Street a few years later. In 1900, the Phoebe Hearst Free Kindergarten was started in the basement of the church. The church and rectory were moved to 631 West Main Street in 1920 to facilitate relocation of Homestake's Hearst Mercantile Store from North Mill Street.[49]

Reverend Henrik Tanner was called to Lead in May 1889 to organize a Finnish parish. The Finnish Evangelical Lutheran Church of the Black Hills was organized in May 1889 with sixty-six charter members from Lead, Terraville, Brownsville, and Belle Fourche. Initially, services in Lead were held in the Lead Temperance Society Hall. However, the congregation soon decided to discontinue meeting there since the hall was also being rented out for "all kinds of worldly pleasures." In July 1889, Lot No. 11 was purchased on East Main Street for the purpose of constructing a church building. John Niemi and Matt Saari completed the new building on March 13, 1891, at a cost of about $1,300. The church was dedicated on May 24, 1891.

On May 30, 1890, the four-church parish separated into independent congregations under Reverend Tanner. The Lead congregation incorporated on July 28, 1890, under the name of First Finnish Evangelical Lutheran Church of

Lead City. It joined the Suomi Synod on February 18, 1891. In 1903, Charles Alaniva, a twenty-four-year-old Homestake miner, painted a wall-sized mural of Christ's Ascension. The painting was given to the church as a gift during Easter service in 1907. Beginning February 13, 1949, a decision was made to conduct services in English only and to record church business in English.

Reverend Jack Hill accepted a call on January 1, 1959, to serve a parish consisting of the Lutheran Church in Lead and Cave Hills Lutheran Church in Buffalo. In 1961, construction was started on a new building next to the old one on East Main Street. The new building was dedicated on February 4, 1962, as First Lutheran Church. The former building was donated to the Lead Womens' Club and was moved to the Sinking Gardens, which was a park located between Main Street and the Open Cut. In 1992, as part of Homestake's expansion of its Open Cut surface mine, the building was moved to 100 East Main Street where it was converted to JL's Gift Shop. The wall mural of Christ's Ascension still adorns one inside wall of the building.

Immanuel Lutheran Church in Whitewood joined the Lead and Buffalo Finnish Lutheran churches in September 1967. The three-church parish became known as the Prairie and Pines Parish. In 1992, it became necessary to move the newer First Lutheran Church building to facilitate Homestake's expansion of the Open Cut. First Lutheran Church and Bethel Lutheran Church voted to merge in 1992 and became known as Shepherd of the Hills Lutheran Church at 825 West Main Street.[50]

The First Baptist Church was organized in Lead on September 3, 1891. Worship services were held at two different locations on Main Street. The sites later became Texaco Service Station and Grant Hamilton's Hardware Store. On January 13, 1935, a weeklong dedication was held to commemorate the new building at 611 Railroad Avenue.

The Scandinavian Lutheran Church was organized on January 15, 1895. Worship services in Norwegian were initially held at the First Finnish Evangelical Lutheran Church. On September 11, 1896, the Scandinavian church purchased a building at the corner of Wall and Addie streets from the Episcopal Church. Under Reverend H. J. Peterson, the Scandinavian Lutheran Church was renamed Bethel Lutheran Church on January 4, 1918. In 1939, a new building was constructed to the west on Addie Street. With the congregation still growing, a larger building was constructed at 825 West Main Street in 1955. The building was expanded in 1963 to provide space for Sunday school classrooms and offices.[51] Bethel Lutheran merged with First Lutheran as Shepherd of the Hills Lutheran Church in 1992.

The First United Presbyterian Church held its first services on November 18, 1894, in the hose house on Mill Street in Lead. In about 1911, the church merged with the Lead Congregational Society Church. The combined congregation moved to 12 Baltimore Street in 1939 as First United Presbyterian Church.

Other historic Lead churches include the Christian Science Society of Lead, organized in 1899, the Assembly of God Church, organized in 1935, and the Berean Baptist Church, organized on April 4, 1956.[52]

Celebrations

Parades, special events, and ball games became favorite forms of entertainment for residents of the Black Hills for many years, especially on Labor Day and July 4 in Lead. The 1916 July 4 celebration was no exception, and it proved to be particularly exciting for residents of the Black Hills. Planners for the Lead celebration were made aware of a professional tightrope walker named Ivy Baldwin. Baldwin proposed stretching a high wire across the Open Cut, which at that time was about 450 feet wide from peak to peak and some 500 feet deep. Baldwin claimed he could safely perform the act using nothing but the high wire and his balancing pole. A deal was made. Baldwin would perform his feat at 11:00 a.m. on July 4.

Following a rousing round of midmorning fireworks in the Open Cut, the time came for Baldwin to do the impossible. Baldwin started across the wire step by step, all the while manipulating the pole to maintain his balance (see fig. 12.14). Every so often, the wind added to the challenge for Baldwin and the suspense of the crowd. Undaunted, Baldwin paused and calmly waited for each gust of wind to subside. Some people couldn't bear to watch, knowing that a fall was inevitable. At about the midpoint, which was also the highest point above the bottom of the Open Cut, Baldwin sat cross-legged on the wire. From his unique vantage point, Baldwin rested and immersed himself in the breathtaking views of the Open Cut, the crowd, and South Lead. After about five minutes, he got up and made his way to the other side of the Open Cut, completing the entire crossing in about thirty-five minutes.

Fig. 12.14. Ivy Baldwin's tightrope walk over the Open Cut on July 4, 1916.—Homestake Collection, Adams Museum, Deadwood.

After a parade on Lead's Main Street, the area residents moved to the ballpark to partake in the afternoon festivities, which included a traditional baseball game. Ivy Baldwin was also present, preparing a hot-air balloon for his next feat. After the game was over, it was Baldwin's turn to perform again. He climbed into the basket of the balloon, ascended to a height of about 1,000 feet and jumped out (see fig. 12.15). After a few moments, he deployed a small chute to slow the rest of his descent. Over the next several minutes, Baldwin performed a series of daring, acrobatic acts from a trapeze bar. Once again, Baldwin safely reached the ground under the roar and applause of a very excited and appreciative crowd.[53]

Fig. 12.15. Ivy Baldwin's parachute jump from a hot-air balloon over the city (Mountain Top) ballpark on July 4, 1916.—Photo courtesy of the Phoebe Apperson Hearst Library.

Another of Lead's famous celebrations was held on August 6-7, 1926, to commemorate a half century of mining operations in the Black Hills and the Homestake Mine. Much of the festivities for the Golden Jubilee, as it was called, were held in Poorman Gulch where Homestake crews had transformed the gulch into a pioneer mining town patterned after the early-day mining camps, including Lead and Deadwood. An entire array of log structures was created, or recreated, including the Gem Theater, the Green Front, saloons, stores, cabins, and more. Demonstrations of the various aspects of placer mining, quartz mining, and stamp milling were also presented. A gold rush parade was held on August 6, followed by a historical pageant on August 7 that featured more than four hundred participants.[54]

Lead's Labor Day celebration was always another favorite of area residents. The first such celebration was organized by the miners and held on September 3, 1888. Annual celebrations in the earlier years included a picnic, dances, games, and speeches held at some special out-of-town locations such as Woodville. Area residents were shuttled to the festivities on the Black Hills and Fort Pierre Railroad.

After the Homestake Aid Association was formed, its members primarily took charge of organizing the annual event.[55] A typical celebration consisted of a town parade (see fig. 12.16) and various events at the Mountain Top ballpark, including the championship baseball game between two Homestake teams. The children were entertained with their own games, events, and rides, which included the "cage" and the "mine train" (see figs. 12.17 and 12.18). The last such Labor Day celebration was held in 1965.

Fig. 12.16. Lead's Labor Day Parade, 1961. The buildings shown were later razed to make way for the last addition to the east end of the Homestake Hospital.—Photo courtesy of Homestake Mining Company.

Fig. 12.17. The "cage ride" at Lead's Labor Day Celebration.—Photo courtesy of Homestake Mining Company.

Fig. 12.18. The "mine train" ride at a typical Labor Day Celebration in Lead.—Photo courtesy of the Black Hills Mining Museum.

Chapter 13
Sister Cities Near the Homestake Belt

Mining Brings Mining Camps

The initial gold rush of 1875-1876 resulted in the establishment of numerous mining camps throughout the Black Hills. Mining camps were established whenever and wherever a seemingly rich discovery was made. Most of the camps were located on creeks and tributaries at or very close to the location where the original gold discovery was made. Placer miners found they needed to live close enough to their diggings to guard against trespassers and claim jumpers. The creeks also provided water for their livestock and helped satisfy the domestic needs of the miners. Partnerships of miners often chose to work their claims day and night with hopes of being able to retire rich and return home to their families. This, too, required living quarters close to the diggings for oversight and good communications.

The earliest mining camps consisted of clusters of canvas tents scattered along the creek banks. Whenever a miner or small group of miners thought they had a good-enough claim that might last for several months, they replaced their canvas tent with a wood shanty or log cabin, particularly during the winter months. As the richest placers started to play out in 1877, more of the miners turned their attention to finding and developing the hardrock lode mines. The lode mines offered more of a promise of longevity in an area; hence, the more formalized mining camps and townsites were located around the centroids of the lode mines. In many instances, the lode mines were discovered on the flanks of the valleys not too distant from the placers.

Gayville was the first mining camp to be established near the Homestake Belt (see fig. 13.1). Shortly thereafter, many other mining camps were established in upper Deadwood Gulch as the placer miners and lode miners moved up the gulch and its tributaries in search of the mother lode. Gayville was located across Deadwood Creek from Blacktail, which was established a short time later at the mouth of Blacktail Gulch.

Lancaster City was located in Blacktail Gulch near the Gustin Mine. Lincoln was located near the head of Bobtail Gulch. South Bend was located in Deadwood Gulch just above the mouth of Bobtail Gulch. Terraville was situated in Bobtail Gulch just below Lincoln. Above South Bend was Central City and Golden Gate. Anchor City was located at the confluence of Deadwood Gulch and Poorman Gulch. Above Anchor City, toward Lead, were the small camps of Oro City and Poorman Gulch, both of which were situated in Poorman Gulch. At the head of Deadwood Creek was Silver City, also known as Pimlico.[1]

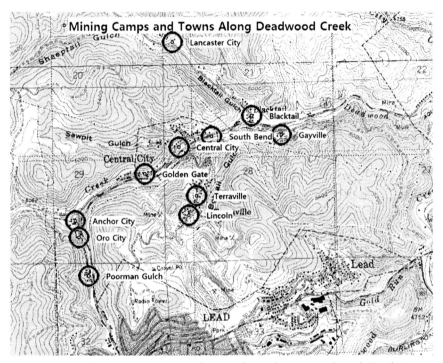

Fig. 13.1. The mining camps and towns located near Deadwood Creek in upper Deadwood Gulch. The city of Lead was established along the north and south forks of Gold Run Creek. The camp of Silver City, or Pimlico, is not shown but was located at the head of Deadwood Creek approximately 2 miles west of Anchor City.—Illustration by author.

By mid-1877 the supporting and entertainment businesses had established a presence throughout the Black Hills, and the overall population began to increase very rapidly. In May 1877, the population of the Black Hills was estimated to be about 15,250 people. Deadwood and its southerly suburb, Whitewood, had 5,500 residents; Lead City, 1,500; Gayville and Central City, 3,000; Crook City, 400; Centennial City, 100; Rapid City, 250; Bear Butte Gulch, 500; and the Central Hills, 2,500. An additional 3,000 people migrated among the camps.[2]

Gayville

William Gay and his brother Alfred established the first mining camp in the northern Black Hills in November 1875. The camp, known as Gayville, was located along Deadwood Creek immediately below the mouth of Blacktail Gulch (see fig. 13.2).

Fig. 13.2. "Bird's-eye view of Gayville, 1877" by Stanley J. Morrow. The Hildebrand Mill is shown on the left.—Photo courtesy of the W. H. Over Museum.

Gayville, sometimes called Troy, grew rapidly in 1877. In May, the *Black Hills Daily Times* described the camp as follows: "It has six hotels, eight stores of general merchandise, two clothiers, one hardware, three bakeries, two meat markets, one assay laboratory, two drugstores, two shoe houses, one brewery, two wagon and blacksmith shops, three express and transportation companies, and one news depot. . . . From the present outlook there is a bright future for the growth and prosperity of Troy." The reporter for the *Daily Times* reported in September that "the main street of the town was completely blocked during a portion of our visit there . . . mule trains, ox trains, express wagons, hacks and buggies were locked in and mixed up in great confusion. It was a perfect jam, and a fearful amount of profanity was breathed upon the air before the entanglement was straightened out."[3]

During its peak years, residents of Gayville had 250 homes, 30 businesses, and 1 assay office.[4] C. C. Fyler operated a tollgate on the Deadwood-Central City Toll Road just east of the present-day Lawrence County Maintenance Shop. A monument exists today at the site of the former tollgate. The tollgate had other operators in later years, including Robert Brown, T. Brennan of the Western

Specialty Company, and S. A. Crist, who managed the county's Poor Farm after it was constructed on the north side of Deadwood Creek in the 1890s.[5]

One of the first murders in the northern Black Hills occurred on July 9, 1876, when John R. Carty and Jerry McCarty killed Jack Hinch. Carty and McCarty were playing cards with Bill Trainor at a saloon in Gayville. Hinch, who was watching the card game, became convinced that his friend Trainor was being swindled. Hinch decided to call it a night and retired to his place of lodging at Turner & Wilson's Saloon. Carty and McCarty arrived later at the saloon, awakened Hinch, and persuaded him to go have breakfast with them. As Hinch was getting out of bed, McCarty fired two shots at Hinch and Carty attacked him with a knife. Hinch died the following morning.

McCarty and Carty hastily departed town but were followed by a large posse of Hinch's friends. The two traveled to Fort Laramie at which point they separated. When the posse reached Fort Laramie, members of the party convinced I. C. Davis, a U.S. marshal, to pursue Carty and McCarty. Davis apprehended Carty but wasn't able to catch up with McCarty. Davis returned Carty to the Black Hills. Upon reaching the outskirts of Deadwood, Davis wrapped Carty up in a blanket and instructed him to lie on the bed of the wagon so he would be hidden as they passed through Deadwood. By the time Davis reached Gayville, Carty had nearly succumbed from heat and lack of oxygen and had to be revived.

The next day, August 1, a miners' court was held, and Carty was tried for murder. Bill Trainor and his friends were present and attempted to bring a speedy end to the court and have Carty hanged. However, Marshal Davis saw what the crowd was up to and took action to see that Carty was given a fair trial. From atop a barrel, Davis addressed the excited crowd, "Boys, I have brought this man from Fort Laramie, through a country swarming with Indians, in order that you might try him for his life. When I took him, I gave him his choice to be taken to Yankton and tried by the courts, or to come back to the Hills to be tried by the miners. He chose to come here, and by god, that he shall have. Try him and if you find him guilty of murder, hang him and I will help you pull the rope. But until he has had a fair trial, the man or men who touch a hair of his head will first walk over my dead body."[6]

With this said, the crowd settled down and began making arrangements for a trial. After casting lots to decide who would preside as judge, a man by the name of O. H. Simonton, who had just arrived in the Black Hills from Fort Pierre and served as justice of the peace in Chicago, was elected. A jury of twelve men was selected from a panel of forty. The jury included E. B. Parker, Ed Durham, J. H. Balf, John Kane, G. Shugardt, George Heinrich, A. C. Lobdell, C. W. Shule, John W. Gill, S. M. Moon, and George Atchison, and a man named Curley B. Chapline was selected to be prosecutor. Mills and Hollis were selected to defend Hinch. R. B. Hughes, reporter for the *Black Hills Weekly Pioneer*, and Preacher Henry Weston Smith were also present for the trial.

Later that same day, a trial was held. After deliberating until ten o'clock that evening, the jury reached a verdict of "Guilty of assault and battery." The jury explained that although Carty was an accessory to the murder, he was not the one who inflicted the fatal injuries on Hinch. Carty was released and escorted out of the Black Hills.[7]

A man by the name of Tom Carter came to Gayville from Pittsburgh in 1877. Carter fabricated a pulverizer for the Golden Terra Mine and operated a business called the Gayville Foundry. Lester Forbes built a mill in the Gayville area under the guidance of "spirits." However, just as the mill was nearing completion, the spirits apparently stopped communicating with Forbes and he left town. The mill was never operated and was torn down in 1905.

J. B. LeBeau came to the Black Hills in 1876 from St. Louis, where he was employed in the jewelry business. After arriving in the Black Hills, LeBeau started a gold-jewelry manufacturing business in Gayville. He later relocated the business to Central City. During a town fire on April 28, 1888, LeBeau lost about $3,000 in cash that he had hidden under a rug.

In 1879, Gayville had about thirty other businesses, including a bakery, planing mill, Plummer's grocery store, Gayville brewery, Anderson's brewery, Pierce's grain warehouse, Hanley's saloon, and L. Miller and Company, a business that sold doors, sashes, siding, flooring, moulding, and glass. Homestake's Cyanide Plant No. 2 was constructed at Gayville in 1902. The plant was located on the south side of Deadwood Gulch, directly opposite of the mouth of Blacktail Gulch.

Gayville sustained major fires in August 1877 and May 1886. The latter fire started when Mrs. William Henley lit a fire in her stove and went outside to feed her chickens. The house caught fire and soon engulfed several other buildings, including the schoolhouse. Fortunately, no one was hurt and the building was saved.

In 1901, Gayville was home to some eighty-three men who were assigned to construct the Fremont, Elkhorn, and Missouri River Railroad from Gayville to the Washington section of Lead.[8]

William Gay, one of the originators of Gayville, was involved in an altercation himself in his mining camp. Gay befriended a good-looking girl from the Badlands of Deadwood, and she moved in with him. Not far from Gay lived a family who had a son named Lloyd Forbes. Forbes grew fond of Gay's live-in girlfriend. One day, Forbes sent the girl an anonymous letter asking her to meet him at a certain place by moonlight. Upon reading the letter, the girl laughed and showed it to Gay. Gay became furious and tracked down the postal carrier, General James Fields, forcing him to divulge the name of the person who sent the letter. Gay then went to Forbes' house, confronted him, and accidentally shot Forbes behind his left ear. Forbes died a day or two later.

Gay surrendered to the local authorities. At his trial, Gay was found guilty, but he spent only a short time in the penitentiary. After he was released, he went to Montana, got into an altercation, and shot another man. During the subsequent investigation and trial, some ten thousand people signed and presented a petition attesting to Gay's upstanding stature. The petitioners requested his release. However, the supporting efforts failed to achieve the desired results, and William Gay was hanged for murder.[9]

Blacktail, Lancaster City, and South Bend

Blacktail was located at the mouth of Blacktail Gulch. The Lardner Party located the first placer claim on upper Deadwood Creek on November 9, 1875. This claim was situated just below the mouth of Blacktail Gulch in the general area of what later became Gayville and Blacktail. On the flanks of Blacktail Gulch were several mines, including the Wooley-Peacho, Jupiter, Pinney, Gustin, Minerva, Deadbroke, Esmeralda, and others. The operators of these mines primarily pursued the basal cement ores of the Deadwood Formation.

John Hildebrand's 20-stamp mill was a custom mill situated on the northeast side of the confluence of Blacktail Gulch and Deadwood Gulch. By about 1896, a new mill was constructed at about the same location. It became known as the Gibbs, Cook, and Parker Mill or Ben Cook's Mill (see fig. 13.3). The Baltimore and Deadwood Mill was located on the northwest side of the confluence of the two gulches. It later became the Portland Mill in about 1902 and the Columbus Consolidated Mill shortly thereafter. The Esmeralda, Deadbroke, and Minerva mills were located a short distance up Blacktail Gulch.[10]

Fig. 13.3. View of Gayville showing the Gibbs, Cook, and Parker Mill at far left, Cyanide Plant No. 2 (upper center), and the Columbus Mill (formerly the Baltimore and Deadwood Mill) at lower right.—Photo courtesy of the Black Hills Mining Museum.

One of the businesses in Blacktail was the Bowman and Flannery Saloon, which also housed Old Buck's barbershop and a billiard hall. John Delauncy ran the Franklin Saloon. Other businesses included Mrs. O'Hara's Capital Boarding House, Washburn's grocery store, and Louie Misel's slaughterhouse. Blacktail also had its own schoolhouse. Some of the teachers included Gertrude Lang, Nettie Allen, and Esther Pendo. In later years, the structure was transformed into a home owned by Charles Martin.[11]

Notwithstanding the difficulties it presented, Blacktail Gulch was one of the routes prospectors used to get in and out of Deadwood Gulch on horseback. Deadwood Gulch itself was initially a quagmire of beaver dams and deadwood. This is probably the reason the members of the Custer Expedition and Newton-Jenney Expedition chose not to visit much of the northern Black Hills—the area simply wasn't accessible with a horse or mule, much less a wagon.

Nimrod Lancaster, one of the owners of the Gustin Mine, laid out Lancaster City on September 15, 1877. The town was located in Blacktail Gulch, about one-half mile north of Deadwood Gulch. Some two hundred lots, measuring 25×100 feet each, were laid out. Lancaster envisioned a long life for the town, based on the two 10-stamp mills that had already been established, the mines, and "plentiful" water in Blacktail Creek.[12]

South Bend was home to several early-day mills, including the Tual, Casey, and Webber Amalgam Mill, Whitney Amalgam Mill, Sheldon Edwards Amalgam Mill, and Cassel and McLaughlin Amalgam Mill. All were custom stamp mills. The camp had numerous saloons, including James Wilson's, the Nugget, Graham's saloon, and Charlie Utter's saloon. Utter was a friend of Wild Bill Hickok. Other businesses included a grocery store owned by the Phillip brothers, a blacksmith shop run by Brum, the Occidental Hotel, operated by P. S. Tetrault, and a wagon shop. Hi Yum was a Chinese businessman who operated an opium house, which was raided in 1879. Yum refused to pay the fine, however, and was sentenced to ten days in jail. Other Chinese people operated a laundry service business and a store that catered to the Chinese residents.[13]

Central City

Above South Bend in upper Deadwood Gulch was Central City. On January 20, 1877, a community meeting was held to merge several of the small but nearly contiguous camps into an organized townsite. William Lardner, one of the pioneers of upper Deadwood Gulch, was appointed chairman. A. H. Loudon was elected secretary, and George Williams was elected recorder. A townsite committee was elected to lay out the town. This committee consisted of Edward McKay, Frank S. Bryant, and George Williams. Upon the recommendation of I. V. Skidmore of Central City, Colorado, the committee decided to name the town Central City. South Bend and Golden Gate were included within the legal limits of Central City.[14]

In 1876-77, numerous mines and mills were situated in and around Central City. The Father DeSmet mine and mill was situated on the south side of Deadwood Creek. The Hidden Treasure Mine was located in Hidden Treasure Gulch north of Deadwood Creek. Numerous other mills were located right in town (see figs. 13.4 and 13.5). By mid-1877, Central City boasted a population of around three thousand people. The mouths of Sawpit Gulch and Hidden Treasure Gulch and the area along upper Deadwood Gulch were soon lined with saloons, gambling houses, bordellos, homes, and businesses (see fig. 13.6). The Shoofly Saloon was one of the first saloons to open for business.

1—Father DeSmet Mill
2—Jones and Pinney Mill
3—Union Mill
4—Cunningham and Dorrington Mill
5—Bogle Mill
6—Thompson Mill
7—Pearson Mill
8—Fairview Mill
9—Hidden Treasure Mill

Fig. 13.4. Pollock and Boyden photo of Central City and custom stamp mills, July 1880. Mill locations are identified in Willard Larson's *Early Mills in the Black Hills* (Isanti, Minnesota: privately printed, 2001).—Photo courtesy of the Black Hills Mining Museum.

Fig. 13.5. View of Central City from above the Father DeSmet Mill. The photo also shows the Hidden Treasure Mine (upper left), Wooley-Peacho Mine (upper right), Great Eastern (formerly the Cunningham and Dorrington) Mill (lower left), and the Bogle Mill (right center). This photo was probably taken shortly after the flood of May 11, 1883, since many of the other stamp mills were gone.—Photo courtesy of the Black Hills Mining Museum.

Fig. 13.6. Central City near the mouth of Hidden Treasure Gulch. The large building on the right is the Central City Miners' Union. The train depot is shown in the foreground.—Homestake Collection, Adams Museum, Deadwood.

In the fall of 1877, one of the first schools in the Black Hills was opened in Central City under the direction of schoolmaster Dolph Edwards.[15] By September, the town had a post office, telegraph office, and the first brick building in the Black Hills. In any given week in 1877, as many as fifteen different businesses were in some phase of construction.[16]

Central City had as many as five different newspapers in its heyday. James S. Bartholomew published the *Herald* on a daily basis in 1877 using the first steam press in the Black Hills. The paper went out of business in 1881. Charles Collins, the Sioux City newspaperman and longtime promoter of the Black Hills, published the *Central City Champion* on a weekly basis in 1877-78. The *Central City Enterprise* was published daily in 1881-82. The *Black Hills Index* was started in 1882 but went out of business within a year. In 1885, the Black Hills Miners' Union began publishing the *Black Hills Register*, which lasted until about 1907.[17]

In 1877, Judge David B. Ogden, who lived in Anchor City, conducted the first church services at Golden Gate. The first worship services in Central City were held in the school building in 1878. Later, services were held in the Opera House. The Methodists established a church in 1877. The church was formally organized in November 1878 when Reverend James Williams was sent to the Black Hills from the Northwest Iowa Conference. A congregational society was established in 1877. Soon thereafter, the group constructed their first church building and separate parsonage under Reverend B. F. Mills.[18]

Reverend John Lonergan started Catholic worship services in Central City in 1877. In 1878, Father B. Mackin established Catholic churches in Lead and Central City. Three priests were ministering to the people of the Black Hills by 1881. One of these was Reverend P. J. Colovin, who served in Central City. Beginning in 1915, the chaplains of St. Joseph's Hospital in Deadwood included Central City in its parish. Bishop John J. Lawler of Lead assigned Reverend John Novak as resident pastor at Central City in 1916. From 1919 to 1921, Reverend J. J. O' Reilly served the parish at Central City. He also ministered at St. Thomas at Terry, St. Michael at Trojan, and St. Joseph at Terraville. O'Reilly was succeeded by Reverend P. J. Kelly, who served from 1922 to 1924; Reverend J. J. Lynch from 1925 to his death on February 27, 1929; and Reverend William J. Boat, who served for nine additional years.[19]

Henry Rosenkranz, his father-in-law, Sam Kaiser, and their partner, Dan Warner established the Rosenkranz Brewery in 1878. Rosenkranz was born in Germany on October 14, 1846. He immigrated with his parents to New York in 1855. At the age of sixteen, Henry landed a job at a brewery in Buffalo, New York. He then became a foreman at a brewery in Niagara Falls four years later. In 1868, he moved to Kansas City, Missouri, where he went to work at another brewery. From there, Rosenkranz went to Helena, Montana, where he worked in the mines at Unionville. After four months, he found other employment around Helena, working at a distillery, on a government survey crew, and in a brewery.

Upon learning of the Black Hills gold rush, he came to the Black Hills, arriving in Deadwood on August 10, 1876. In the spring of 1877, Rosenkranz opened a retail liquor store in Central City. That September, he and several other men went on a prospecting trip to the western Black Hills. While on that trip, the party encountered a band of hostile Indians who ended up killing all of their horses and a member of the party named Thomas Carr. Rosenkranz returned to Central City and started the Rosenkranz Brewery with Kaiser and Warner. After a fire ravaged Deadwood in 1879, much of Rosenkranz's competition was wiped out, and his business flourished. Rosenkranz bought out Warner's interest in the brewery in 1880.[20]

Rosenkranz then purchased the Phoenix Iron Works in 1882 and started a new brewery that included a bottling plant. He bottled his beer under the brand name of Gold Nugget Beer. By 1884, the brewery was turning out twenty-five barrels of beer per day. The 22×50-foot malt house was three stories high. The brewery also had an 18×40-foot engine house that housed a 12-horsepower engine, a 20×30-foot bottling house, an 18×30-foot patent icehouse, and a 14×40-foot storage cellar. The Rosenkranz Brewery employed six men and had two teams of horses for the delivery of beer to the various saloons in the surrounding communities.

Because of the prohibition law passed by the South Dakota legislature in 1889, Rosenkranz was forced to close the brewery. He resorted to selling ice and coal for a short while and later opened a saloon in Central City with a partner, Frank Ourth. Rosenkranz leased the brewery to a man named Faulkner and the O'Connor brothers, who began brewing "near beer" under the business name of Black Hills Brewing and Malting Company.

In 1895, the shareholders of the Minneapolis Brewing Company purchased the brewery from the O'Connor brothers. After five years of operation, the company was restructured again in 1900 at which time the name of the brewery was shortened to the Black Hills Brewing Company (see fig. 13.7). The officers of the new company included Sigmund F. Wiedenbeck, president, Armin Neubert, vice president, Gustav J. Heinrich, secretary, and Hubert B. Schlichting, treasurer and manager. All of these men were also officers or managers of the Minneapolis Brewing Company.

Fig. 13.7. The Black Hills Brewing Company. In about 1890, Henry Rosenkranz sold his brewery to a Mr. Faulkner and the O'Connor brothers, who resumed operations under the name of the Black Hills Brewing and Malting Company. The business changed ownership again in 1895 and became known as the Black Hills Brewing Company. Highway 14A now occupies most of the area that contained the brewery. The Open Cut and the Father DeSmet Mill are visible to the right of the brewery.—Photo courtesy of Black Hills Mining Museum.

As part of the business reorganization, the brewery was reconfigured to include a patented design of Armin Neubert and new bottling works (see fig. 13.8). By 1905, plant capacity was one thousand barrels per month. The company ran an advertisement in the *Deadwood Daily Pioneer-Times* in 1914 claiming that "Gold Nugget beer has answered the call for a strictly high-grade, high class old fashioned beer, and this beer has the distinction of being brewed in one of the most sanitary breweries from the choicest hops and barley grown. It has been properly aged before being placed on sale, and is not charged with carbonated gas. Our sole aim has been to see how good and appetizing we could make our beer, and today our books show hundreds of satisfied customers. Order a case today through any of our distributors—the Black Hills Brewing Company."[21]

Fig. 13.8. The bottling works of the Black Hills Brewing Company.—Photo courtesy of the Black Hills Mining Museum.

In 1917, the South Dakota legislature again succumbed to the temperance leaders and enacted a prohibition making it illegal to produce or sell alcohol. The legislation struck a blow to operators of the Black Hills Brewery, who were forced to reorganize as Black Hills Products. A new product was introduced, called Cherry Blossom, and was initially bottled as a soft drink. After experiencing low sales volumes of this product, Manager Schlichting began adding a few hops making it a near beer and more. This modification soon drew the attention of the law authorities, who subsequently ordered the brewery to dump its existing inventory. Schlichting responded by giving away free drinks and free cases of beer. The remaining kegs of beer were dumped into Deadwood Creek, causing the creek to flow with a head.

Schlichting then resigned and moved to California. Armin Neubert, vice president, took over management of the business. A new bottled product was introduced called Gold Nugget Ginger Ale along with another near beer called Byro. As before, the state officials forced the company to stop producing Byro. A short time later, the company was dissolved on October 14, 1928.[22]

On May 11, 1883, a warning was sent out from the Ten-Mile Ranch advising residents of Central City, Gayville, and Deadwood to be on the lookout for a flood. Hard, steady rains and warm weather had taken a toll on a recent, heavy snow. The waters of Whitewood Creek and Deadwood Creek rose at an alarming rate. Residents living close to these two streams were soon forced to evacuate their homes. Central City was the first to be hit. Several stamp mills along the valley bottom were destroyed or badly damaged. A boardinghouse owned by Pat Early was swept away, as was the house owned by Henry Rosenkranz. The icehouse at the brewery was destroyed, which forced the sale of warm beer throughout the ensuing summer months. Using giant powder [dynamite], miners proceeded to blow up eight to ten buildings along Deadwood Creek in an effort to clear blockages and open up a larger channel way for the raging floodwaters.

The floodwaters were relentless for days. By the time the creek returned to its normal state, it became obvious that thousands of tons of sediment had been washed away, exposing the bedrock all along Deadwood Creek. The few placer claims that had been active were soon finished up and abandoned. Ten levels of the Deadwood-Terra Mine were either flooded or heavily damaged. The Deadwood-Gayville Toll Road was completely destroyed. The Occidental Hotel was also destroyed, as were ten cabins located nearby.

On the morning of April 25, 1888, disaster struck again in Central City. This time it was fire. The fire started in Belliveau and Jensen's restaurant at about 5:30 a.m. By the time the fire was discovered, the entire building was engulfed in flames. Unfortunately, the restaurant was located in the middle of the business district. When the fire crews attempted to activate the water mains, they discovered the mains were dry. Very rapidly, the fire spread from one building to another. Once again, the miners resorted to the use of giant powder in an effort to create a firebreak. Fire departments from Lead and Deadwood were called to the scene. The focus was now aimed at saving the surrounding towns, not Central City. By the time the fire was put out, some 140 residences and businesses had been destroyed (see fig. 13.9). Fortunately, no human lives were lost.[23]

Fig. 13.9. Central City, shortly after the fire of April 25, 1888. The structure on Deadwood Creek in the left center of the photo is the former Thompson Mill. Above and to the left of it at the toe of the hill are the Bogle and Jones & Pinney mills, respectively.—Photo courtesy of the Black Hills Mining Museum.

Immediately, the town started rebuilding with lumber furnished by Homestake Mining Company. Unfortunately, Central City never was able to regain the momentum it enjoyed between 1876 and May of 1888. The communities of Central City, Golden Gate, and Terraville pooled resources to establish a bigger and better fire crew. The C and G Hose Company No.1 organized on April 25, 1889, the first anniversary of the great fire. The hose company won the state championship for the 200-yard "hub-and-hub" race, posting a time of 21:25 seconds.[24]

Lincoln and Terraville

Lincoln and Terraville were situated at the head of Bobtail Gulch (see fig. 13.10). The camps were established in 1876 after prospectors began looking for the source of the placer gold found in lower Bobtail Gulch and upper Deadwood Gulch. Their search resulted in John B. Pearson and his party staking the Giant, Gold Run, and Old Abe claims on January 7, 1876. In April, Alexander Engh, Henry Harney, and Fred and Moses Manuel staked the Golden Terry lode claim on the divide between Deadwood Creek and Gold Run Creek.

Fig. 13.10. The town of Terraville. This photo shows the Deadwood-Terra (Pocahontas) Mill on the left and the Caledonia (Monroe) Mill on the right. The Methodist Church is shown in the upper center.—Photo courtesy of the Black Hills Mining Museum.

Within about two years, Lincoln was merged into Terraville. Hundreds of miners resided in Terraville while working for companies such as the Highland, Deadwood-Terra, Caledonia, and Father DeSmet. In April 1880, a new school building was erected, along with a private hospital operated by Dr. J. O. Gunsolly. By 1886, the hospital became one of Homestake's hospitals with W. R. Dickinson and Dr. Robbins as physicians. Homestake opened one of its Hearst Mercantile stores in Terraville in 1887.

Because many of the miners were single men, numerous boardinghouses and hotels soon became thriving businesses in Terraville. Such names included the Lincoln House, Newton's Boarding House, Miners' Home Hotel, Emmet House, and the Caledonia House. Jack Gray ran the Caledonia House, located in Shoemaker Gulch. Gray hired several girls to help run the Caledonia House in 1881 but ended up firing them because they flirted too much with the men who lived there. Gray then hired several boys, but they didn't work out either, so he hired most all black people to help out. Mrs. Gray fired them the first day. The Grays finally hired some "educated gentlemen, pious fellows" who made everyone happy. In later years, the Caledonia House was home to the James R. Harris family.[25]

The northeast end of the Terraville Tunnel (see fig. 13.11) was located close to the Caledonia House. The tunnel extended through the hill between Lead and Terraville. Homestake permitted miners and other members of the general public to use the tunnel to walk between the two communities. However, a sign posted on the portals of the tunnel warned that it was forbidden to take baby carriages through the tunnel. Pedestrians traveling through the tunnel in the earlier years always had to be on the lookout for an oncoming ore train. When a train was encountered, the pedestrians had to hug the wall until the train passed by.

In the 1950s and 1960s, the children of Lead made use of the Terraville Tunnel as a shortcut to the ice-skating rink at Blacktail. The rink was housed in Homestake's inactive Cyanide Plant No. 2, which ceased treating ore in 1934. A long, steep stairway with 280 steps extended up and down the hillside between Terraville and Central City for the "convenience" of the skaters and area residents.[26]

Fig. 13.11. The Terraville Tunnel. Homestake Mining Company allowed its employees and residents to travel through the Terraville Tunnel between Lead and Terraville until the late 1960s. In the early days, when the tunnel was used for haulage of ore, a pedestrian had to watch for the ore train, hence the sign prohibiting baby carriages in the tunnel.—Homestake Collection, Adams Museum, Deadwood.

The Methodists became active in Terraville in 1879. The congregation organized the United Methodist Church and dedicated a new building on November 2, 1889. The names of most all the children who ever attended Sunday school at the church were recorded on the sloped ceiling in the attic. Some forty-nine different ministers served at the church, including Roberts (1889), F. G. Boylan, J. W. Lucas (1909-10), W. J. Davison (1911-14), ending with the Rev. Dwayne Knight. The building was razed in 1982, along with the rest of Terraville, to facilitate Homestake's Open Cut surface mine that commenced production in fall 1983.[27]

The *Deadwood Daily Pioneer-Times* reported on a "first" on May 8, 1908, when it was noted that Richard Blackstone had driven his new automobile through Terraville the previous day. The *Pioneer-Times* stated the event "attracted considerable attention as it was the first auto ever seen in Terraville."

Golden Gate, Anchor City, and Poorman Gulch

Golden Gate is situated immediately above the original site of Central City. Today, the two communities are seamless. Four arrastras were constructed in Golden Gate in 1877. The arrastras crushed ore from the original Father DeSmet, Belcher, Golden Gate, and Mineral Point claims. By 1879, the mines were consolidated into the Father DeSmet Consolidated Gold Mining Company.[28] Some of the first businesses in Golden Gate included the Conley House, Golden Gate Saloon, Dawson Hotel, and the DeSmet Saloon.

In 1887, a haulage tunnel, called the Golden Gate Tunnel, was driven between the Old Abe Shaft and the Father DeSmet Mill, located across from Golden Gate. The tunnel was driven by Homestake Mining Company to access ore above the 300-foot level of the Homestake Mine. Eventually, the employees of Homestake living in Golden Gate and Central City used the Golden Gate Tunnel as a shortcut to get to and from work. This arrangement worked especially well in the wintertime. Upon reaching the Old Abe Shaft, the employees would catch a cage, travel to the surface, and report for work in Lead. Those who weren't employed in the shops reported for work, returned underground, and completed their shift of work.[29]

Anchor City and Poorman Gulch were home to many of the miners who worked at the Hidden Fortune Mining Company, Hoodlebug, Nutless Mine, Grantz Gold Mining and Milling Company, and others. Silver City, known as Pimlico between 1887 and about 1894, was located near the head of Deadwood Creek along the Fremont, Elkhorn, and Missouri Valley Railroad, about two miles upstream from Anchor City. Residents of Silver City worked for mines close by, such as the Trinity, Pennsylvania, Big Four, Capital, and upper Hidden Treasure mines.[30]

Today the mining camps of Terraville and Lincoln are nonexistent. The camps of Gayville, Blacktail, South Bend, Golden Gate, Anchor City, and Oro City form the fringes of Central City along Highway 14A.

Chapter 14
Cyanide Charlie's Contribution

The Advent of Cyanidation

As early as 1888, it was widely recognized in mining circles that Homestake was not recovering significant gold values from its stamp mill tailings. The problem related to the sulfides, or sulphurets, as they were called, which amounted to 3-5 percent of the tailings. H. O. Hofman suggested classifying the tailings to separate the coarse sands, which he said would contain the heaviest concentrates. The coarse sands could be pulverized in a wet rolls or a Chilean mill, followed by amalgamation and barrel chlorination. Hofman predicted the cost of concentrating and treating the tailings would probably not exceed $1 per ton of concentrates.[1]

T. A. Rickard commented in 1897 that unfriendly critics were claiming that "haste and waste" best described the methods of Homestake Mining Company. The critics suggested that Homestake's metallurgical methods were crude and extravagant and that the amount of gold and sulfides being discharged was "totally inadequate."[2]

As mining progressed a little deeper in the Homestake Mine in the 1880s and 1890s, Homestake recognized that gold recoveries were slowly diminishing. Assays showed the gold value in the tails from Homestake's stamp mills was averaging $2.25 per ton.[3] In 1894, in an effort to improve overall recovery, the narrow sluice plates on the apron were replaced with copperplates electroplated with silver. The temperature of the water was maintained at about 50 degrees, which helped keep other minerals from fouling the surface of the plates. The silver-plated copperplates tended to capture more of the fine gold and added about $0.30 per ton to the overall gold recovery.[4]

Subsequent investigations revealed that recoveries were being unfavorably impacted by an increasing amount of sulfide minerals, primarily pyrite, arsenopyrite, and pyrrhotite. Much of the gold was encapsulated by the sulfides. This observation indicated that the ore was becoming slightly "refractory," and treatment processes other than amalgamation would likely be required to improve total recovery. Moreover, the loss of mercury and amalgam using the amalgamation method contributed to the gold-recovery problem and cost of milling.

In an effort to recover more of the gold in the stamp mill tailings, Samuel McMaster constructed a Blanket House in 1881. Strips of Brussels carpet were placed on inclined tables. The tailings passed over the carpet. Every four hours, the carpet was cleaned and the gold-bearing sulfide material was saved. By 1885, some 1,200 tons of concentrate had been saved with an estimated value of about $40,000. Although the concentrates averaged $33 per ton, there was no technology available to recover the gold, except by smelting. Unfortunately, the nearest smelters were in Omaha and Denver, and the cost of shipping and smelting made the concentrates uneconomic. McMaster decided to stockpile the concentrates.

Shortly after the Deadwood and Delaware Smelter was constructed in Deadwood in 1891, Homestake decommissioned the Blanket House and began shipping the stockpiled concentrates to the Deadwood smelter. A net smelter return equaled to 50 percent of the assayed value of the concentrates, less shipping and smelting charges of $9 per ton, yielded only a small profit to Homestake. As a result, the smelting option for treating stamp mill tailings was discontinued. Still intrigued by the idea of smelting, McMaster and Grier purchased a Perfection Concentrator from the Colorado Iron Works in 1892 to concentrate the mill tailings. The results were about equal to that achieved in the Blanket House.[5]

The Black Hills Gold and Silver Extraction Mining and Milling Company constructed a 75-ton-per-day cyanidation plant in Deadwood in 1893 (see fig. 14.1). The plant, known as the Rossiter Plant after Bryon and Mike Rossiter, commenced operating in December 1893 using the patented MacArthur-Forrest cyanide process to treat ore from the company's mine in Sheeptail Gulch. Later, as part of an experiment, the plant was utilized to determine the viability of treating Homestake's low-grade mill concentrates using the MacArthur-Forrest cyanide process. The tests were unsuccessful.[6]

Fig. 14.1. The Rossiter Cyanide Plant in Deadwood. The Rossitor Plant, named after Bryon and Mike Rossiter, was located toward the lower end of Deadwood and upstream from the Slag Pile. It was the first plant in the Black Hills to use cyanidation for gold recovery. The plant was commissioned in December 1893 using the patented MacArthur-Forrest cyanide process to treat gold ore from the company's mine in Sheeptail Gulch.—Waterland Collection, Devereaux Library Archives, South Dakota School of Mines and Technology.

In 1894, Homestake contracted with Professor Englehardt in Denver to determine if it might be feasible to roast the sulfide concentrates and treat them with bromine. A bulk test using 30 tons of concentrates resulted in 97 percent recovery of the assayed gold value. Englehardt estimated that the treatment cost using his bromine process would be about eighty cents per ton of concentrates. Homestake personnel were impressed with the test results.[7]

Concurrently, the Buxton Mining Company was conducting large-scale bromine tests on its ore at the Buxton Mill in Fantail Gulch near Lead. Based on mixed results reported by the Buxton people, Homestake metallurgist Allen J. Clark decided not to pursue further testing using Processor Englehardt's bromine process.[8]

A short time later, two other men, one of whom was John Williams, built a small concentrating house in Pluma in which they installed four two-compartment Hartz jigs. This arrangement proved satisfactory on bulk samples of tailings from the Homestake stamp mills, yielding several tons of concentrates valued at $8 per ton daily. Based on the results of this work, Grier decided to construct his own jig house across from the assay office. This concentrating plant was commissioned in May 1895.

Grier's new jig house consisted of twelve Hartz jigs that produced 50 tons of concentrates daily at an average grade of $6-$8 per ton. Using techniques based on gravity, the jigs proved successful in separating and concentrating the heavier, gold-bearing sulfides from the lighter sulfides. The concentrates were delivered to the smelter in Deadwood for additional testing to determine if smelting was now economic. Because the Deadwood and Delaware Smelter needed pyrites for treatment of ores from other mines anyhow, an agreement was soon reached. Homestake resumed sending its concentrated mill tailings to the Deadwood and Delaware Smelter. The jig house continued to operate until the Cyanide Plant No. 1 was commissioned.[9]

During the year ending June 1, 1894, the Homestake mills produced 915,010 pounds of sulfide concentrates assaying $5-$8 per ton. The sulfides consisted primarily of pyrite, arsenopyrite, and pyrrhotite. At the Deadwood and Delaware Smelter, the concentrates were converted to an iron matte for a fee of one-half the assay value of the concentrates. The iron matte was sent to the Omaha and Grant Smelting and Refining Company in Omaha for final smelting and refining.[10]

In the fall of 1897 B. B. Thayer, a consulting engineer for Homestake in San Francisco, California, called upon Charles Washington Merrill to discuss Merrill's cyanide process for possible application at the Homestake Mine in South Dakota. Thayer was aware that Merrill had successfully implemented the Merrill cyanide process at the Standard Consolidated Mine at Bodie, California, in 1893-1894, at the Harqua Hala Mine in Arizona in 1894-1895, and at Montana Mining Company's Drumlummon Mine in Marysville, Montana, in 1895-1897. After two or three days of conversation, Thayer recommended to Homestake's corporate personnel that Merrill visit the mine to investigate using his cyanide process to recover gold from the stamp mill tailings. The recommendation was accepted.[11]

In January 1898, Merrill made his first visit to Lead. William A. Remer, former paymaster for Homestake and Lawrence County sheriff from 1892 to 1896[12] recorded the following in his diary: "January 14, 1898: at 1:30 walked to Lead—at HS office—talked with Smead about Grier being down on me—knows nothing of it—met Mr. C. S. Merrill of San Francisco—cyanide man."[13]

After making an initial assessment of the Homestake mills and tailings, Merrill indicated to Homestake that they were probably doing better at the Deadwood and Delaware Smelter than could be achieved using the Merrill cyanide process. Notwithstanding, Merrill agreed to perform some laboratory work on the tailings. Years later, in reflecting on his first site visit and initial comments to Homestake, Merrill wrote, "However, while there, I was impressed with the tremendous total values in the immense tonnage of their very low grade tailings which were being run to waste, and took samples thereof back with me to San Francisco and began work on them, hoping to recover some of these waste values by treatment with the Cyanide Process."[14]

Merrill's metallurgical testing in his San Francisco laboratory produced favorable results. After convincing himself that his cyanide process would work at the Homestake Mine, he returned to Lead and met with Superintendent Thomas J. Grier and Edward H. Clark, manager of the Hearst Estate. Merrill proposed that Homestake construct a pilot plant to facilitate metallurgical testing on a slightly larger scale. Merrill explained how a cyanide solution would be used to dissolve the gold contained in the sand-sized fraction of the stamp mill tailings. Zinc dust would be used to precipitate the gold from the cyanide solution. The gold would then be recovered by cupellation using lead.

Grier and Clark wasted no time in consenting to Merrill's proposal. Merrill also received approval to bring W. J. Sharwood to the project. Sharwood was a chemical engineer who worked with Merrill on the Montana Mining Company project. Allan J. Clark, a metallurgist Homestake had hired in 1897, was also assigned to the project.

During the spring of 1898, a pilot plant with a capacity of 60 tons per day was erected just south of the Amicus Mill on ground later occupied by the south end of the Central Steam Plant. The plant included jig classifiers to separate the sand fraction from the finer slime fraction. Only the sand fraction would be economic to treat, based on initial results of testing in the laboratory. Six vats, used to leach the gold from the sands, were constructed.

After initial testing, Merrill discovered that the sands needed to be aerated to allow oxidation of the sulfides to enhance dissolution of the gold and overall recovery. This was accomplished by forcing low-pressure air through the false bottom of the sand vats. Otherwise, the pilot plant was working well. By late 1899, Thomas Grier became convinced that it was much more economic to treat the sand tailings using the Merrill cyanide process than by smelting.[15] "Consequently,"

Merrill wrote, "the company determined to erect a large plant to treat the sand portion of the tailings from the mill located at Lead, which proved successful and materially increased the total recovery and profits from the treatment of the ore."[16]

Grier entered into the first five-year contract with Charles Merrill in 1899. Merrill was tasked to direct and oversee construction and operation of the new cyanide plant. His compensation was solely based on a percentage of the gold recovered using the Merrill cyanide process. Along with W. J. Sharwood and Allan J. Clark, Merrill brought other assistants to Lead, including Charles C. Broadwater, Louis D. Mills, and Frank H. Ricker. Broadwater was a graduate of the Royal School of Mines; Mills, a graduate of Stanford University in mining and metallurgy; and Ricker, a mechanical-engineering graduate from Iowa State College. Merrill also brought his wife, Clara, to live with him in Lead for the duration of the project.[17]

Cyanide Plant No. 1

In April 1901, Cyanide Plant No. 1 was constructed on the south side of Gold Run Creek in east Lead to begin treating sands from the Homestake, Golden Star, and Amicus mills (see figs. 14.2 and 14.3). Gravity was used to transport the stamp mill pulp from the three mills to a new building, called the Cone House, located just below the three mills (see fig. 14.4). This building housed twelve gravity-settling cones 7 feet in diameter that thickened the stamp mill product called the "pulp." A portion of the water from the Cone House was returned to the stamp mills. Thickened sands and heavier slimes from the Cone House were piped directly to secondary gravity-settling cones in the cyanide plant.

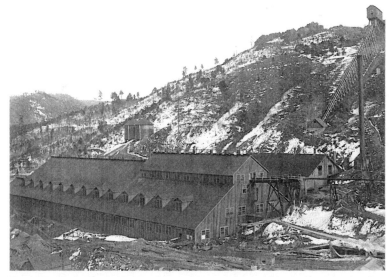

Fig. 14.2. Cyanide Plant No. 1 in Lead. Commissioned in April 1901, the plant had an initial capacity of 1,200 tons per day.—Photo courtesy of Homestake Mining Company.

Fig. 14.3. W. B. Perkins photo of the Cyanide Plant No. 1 after its first major upgrade.—Photo courtesy of the Black Hills Mining Museum.

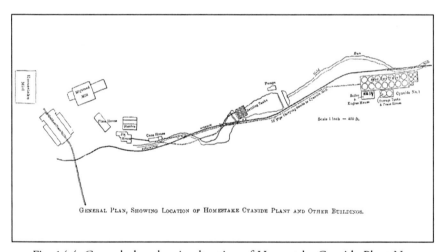

Fig. 14.4. General plan showing location of Homestake Cyanide Plant No. 1 and other facilities, 1903. This plan illustrates how the stamp mill product from the Homestake, Golden Star, and Amicus mills was delivered to the Cyanide Plant No. 1. Water and fine slimes were held in the settling tanks and dams. Once settled, a portion of the water was pumped to the Highland water tank and reused in the mills.—Source: C. W. Merrill, "The Metallurgy of the Homestake Ore," *American Institute of Mining & Metallurgical Engineers Transactions* 34 (October 1903).

Prior to construction of the Slime Plant in 1906, the finest slimes obtained from the gravity cones in the Cone House were discarded and discharged to Gold Run Creek. A series of slime-settling tanks were installed between the settling dams to clarify water before a portion was pumped to the Highland tank.[18]

At the Cyanide Plant No. 1, coarse sands were passed through six additional settling cones. After further classification in these cones, the stamp mill pulp was separated into nonleachable slimes of minus-200 mesh (i.e., two hundred openings per lineal inch of screen), which represented about 40 percent of the entire pulp. The slimes were discharged to Gold Run Creek. The other 60 percent of the pulp consisted of direct-leachable coarse sands of which about 40 percent were plus-100 mesh.[19]

The Cyanide Plant No. 1 originally housed fourteen redwood leaching vats arranged in two lines of seven. Each vat was 44 feet in diameter, 9 feet deep, and held approximately 610 dry tons of sand. The vats were filled with sand using two Butters and Mein distributors, each of which traveled along a track positioned over its respective line of vats (see fig. 14.5).[20]

Fig. 14.5. Butters and Mein distributor filling a redwood sand-leaching vat. The Cyanide Plant No. 1 eventually housed twenty-three sand-leaching vats, each of which was 44 feet in diameter.—Photo courtesy of Homestake Mining Company.

The bottom of each vat contained a bottom filter consisting of coconut matting covered with 8-ounce canvas. The filter was placed over a series of 2×4-inch pine strips of wood set on edge and positioned 2 inches apart on the tank bottom. The bottom edge of each strip contained 2×6-inch notches spaced 12 inches apart. The bottom of each vat was fitted with four side gates hinged and fastened by swing bolts. A center discharge opening in the tank bottom was controlled by a threaded rod and handwheel that extended through a vertical column to a height just above the top of the tank staves. The top of the column also supported the Butters and Mein distributor.

Two lines of overhead pipelines were used to convey the strong and weak cyanide solutions to each tank. Water was supplied in separate pipes. Below the filter were four valves connected to the air supply and the weak and strong solution pumps. Some 8½ hours were required to fill a vat with classified sands. After draining water from the vat, air was injected upward from the bottom of the tank to aerate the sands. Leaching with a strong sodium cyanide solution followed the aeration phase. The vat was then drained and aerated again. A second strong cyanide solution was then applied, followed by drainage and aeration. The sands were then subjected to a third leaching cycle using a weak sodium cyanide solution. Finally, the sands were rinsed and sluiced from the vat, completing a nominal 154-hour vat cycle (see table 14.1). Capacity of the plant was initially designed at 1,200 tons per day.[21]

Table 14.1 Typical sand treatment cycle (1930)

	Hours	Minutes
Filling	8	20
First draining	15	40
First aeration	14	—
Strong-solution leaching	16	—
Second draining	12	—
Second aeration	11	—
Strong-solution leaching	14	—
Third draining	10	—
Third aeration	10	—
Weak-solution leaching	12	—
Washing	29	30
Sluicing	<u>1</u>	<u>30</u>
Total:	154	—

Source: A. W. Allen, "The Homestake Enterprise," *Engineering and Mining Journal* 132, no. 7 (October 12, 1931).

In his annual report for 1901, Grier reported, "The Company's 1,200 ton cyanide works, in course of erection when the last annual report was made, started on the tailings from the mills, after numerous vexation delays, very late in April of this year. The first clean-up at the end of May is most gratifying, exceeds anticipations and is profitable in a handsome degree." In fact, some $40,000 of gold was recovered for each of the next several months from the sand fraction of the stamp mill pulp.[22] The annual gold yield at the Cyanide Plant No. 1 amounted to about $327,000 for the eight-month period in 1901 and $581,000 for 1902.[23] Merrill's cyanide process was working.

Total treatment costs at the Cyanide Plant No. 1 amounted to $0.353 per ton in 1902. The average value of the sand fraction was $1.65 per ton.[24] In 1904, six additional leaching vats were added, making a total of twenty vats. The additional tanks permitted greater sand-retention time in all the leaching tanks, enhancing dissolution and overall recovery of gold without sacrificing plant throughput.[25] By 1912, plant throughput was averaging 1,600 tons of sand per day.[26] Subsequent expansions increased the number of leaching vats in the No. 1 Plant to twenty-three.

Success at Cyanide Plant No. 1 didn't come without much effort on the part of Merrill and his staff. During the first few years of his contract, Merrill was known to have consistently worked 14-18 hours each day at the plant, seven days per week. Most everyone in town knew Charlie's wife, Clara, since she actively participated in the various social functions within the community. Mrs. Charlotte C. Clark once commented, "I never got to know Mr. Merrill very well because he was always at the mine, it seemed. But Mrs. Merrill was a very friendly, social person, and I saw her often, knew her well, and liked her very much."[27]

Years later, some of Homestake's old-timers also recounted how very few people had even met Charlie, since he was rarely seen anywhere except at the Cyanide Plant No. 1. Merrill soon became known as Cyanide Charlie, since, according to one old-timer, he "talked, ate, drank, lived and dreamed about nothing but his cyanide process."[28] Frank Ricker, a member of Merrill's staff, could not understand how Merrill could work until ten or eleven o'clock at night and report back to work at two or three o'clock in the morning.

In *The Merrill Story*, Ricker recollected, "I was on the night shift a long while, and I well recall how frequently Charlie Merrill would leave the plant late—maybe ten or eleven at night—only to come back at two or three in the morning to make sure everything was going all right. I never did understand how he could keep going with so little sleep and rest, but he did, and showed no signs of strain—he was always kindly and gracious, even in correcting us." Notwithstanding, there was a "method in his madness," Ricker said.[29]

Ricker commented how Charlie Merrill was "constantly observing, constantly testing, constantly experimenting—both to overcome, or modify existing difficulties to improve his equipment, processes and methods." Ricker concluded that "as a result of Charlie's experiences with Cyanide No. 1, when he built Cyanide No. 2 he was able to effect improvements which cut the operating costs from about 21 cents a ton to about 17 cents a ton, and when you are dealing with large tonnages, that amount of saving is quite an item."[30]

Cyanide Plant No. 2

Based on the operating success of Cyanide No. 1, a decision was made in 1901 to construct a second sand treatment plant on the south side of Deadwood Creek at Gayville. This plant, called Cyanide Plant No. 2, was located about one-half mile east of the Father DeSmet Mill (see fig. 14.6). The Cyanide Plant No. 2 was designed to treat the sand fraction of tailings from the Mineral Point (formerly Father DeSmet), Pocahontas (Deadwood-Terra), and Monroe (Caledonia) stamp mills. Construction of the 60×360-foot building started in the fall of 1901. Efficient labor and a relatively mild winter allowed the 600-ton-per-day plant to be commissioned in 1902.[31]

Fig. 14.6. Cyanide Plant No. 2 at Gayville, 1908. From 1902 through December 31, 1934, Cyanide Plant No. 2 was used for cyanidation of the sand fraction of ore pulp from the Mineral Point, Pocahontas, and Monroe stamp mills. Later, the building housed a community ice-skating rink until 1965, when the building was demolished. In 2006, Homestake constructed a water treatment plant at the site to treat seepage from the Open Cut waste rock facilities.—Homestake Collection, Adams Museum, Deadwood.

The Cyanide Plant No. 2 contained five redwood leaching vats, each of which was 54 feet in diameter and initially 10½ feet deep. A few years after construction, sheet-iron strips were bolted to the top staves of each vat, which increased the tank depth to 13 feet and vat capacity to 1,250 tons. Some 36-40 hours were required to fill a vat at the Cyanide Plant No. 2.[32]

In 1903, with two new cyanide plants operating quite satisfactorily at 1,800 tons per day, T. J. Grier decided to add 100 stamps to the Amicus Mill. This expansion, completed in July 1904, increased the capacity of the Amicus to 240 stamps. The six large stamp mills now housed 1,000 stamps. The three mills in Lead had 640 stamps; the two mills in Terraville, 260 stamps; and the Father DeSmet Mill, 100 stamps.[33] The 1,000 stamps were now capable of crushing and pulverizing an average of about 3,600 tons of ore per day. For the fiscal year ending in May 1904, milled production increased to 1,299,057 tons, yielding approximately $4.776 million in bullion.[34]

Renovations were also undertaken at the Golden Star Mill starting in 1903. The stamp assemblies were retrofitted with 900-pound stamps. The drop of the stamps was increased to 10½ inches, resulting in eighty-eight blows per stamp per minute. Mortars were narrowed to 12 inches wide at the lip. A No. 8 screen (about 35 mesh) was placed in each mortar, with the bottom of the screen positioned 10 inches above the top of the dies. This combination produced a very fine pulp of which 80 percent passed a 100-mesh screen. Water was fed into the 5-stamp mortar at a rate of approximately 20 gallons per minute or 4 gallons per stamp.[35] However, by 1907, water flow was increased to 7.8 gallons per stamp per minute.[36]

With 1,000 stamps in operation, Homestake's six mills milled approximately 3,500-3,900 tons of ore each day from 1903 through 1905.[37] Of this amount, the sand-sized fraction, consisting of approximately 60 percent of the product from the stamp mills, was delivered to the two cyanide plants. The fine slime product was uneconomic and was wasted. The 640 stamps from the three mills in Lead supplied the Cyanide Plant No. 1 with approximately 1,200-1,400 tons of classified sand product per day. The 360 stamps of the three mills in the Terraville area supplied the Cyanide Plant No. 2 with approximately 800-900 tons of sand product daily.[38]

Amalgamation continued as the primary means of recovering gold at the three stamp mills in Lead. After amalgamation, the stamp mill pulp was transported by gravity to the Cyanide Plant No. 1 in East Lead. There, the tailings, or pulp, were classified in hydraulic cone classifiers to separate the coarser sand fraction from the finer slime fraction. In a similar fashion, the pulp from the three stamp mills in Terraville and Central City was delivered to the Cyanide Plant No. 2 in Gayville. Gold was removed from the sands in the two sand plants using Merrill's Cyanide Process. Merrill's next assignment was to figure out a way to recover the gold from the slime fraction that was merely being discharged with the spent sands into Gold Run Creek and Deadwood Creek.[39]

Slime Plant

In 1903, Merrill began devoting more of his time to determine how the gold values might be recovered from the minus 200-mesh (slime) product of the stamp mill tailings. This finely pulverized portion of the pulp, called the slime fraction, contained gold values of approximately $0.80 to $1.10 per ton, based on a gold price of $20.67 per ounce.[40]

Merrill's description of the problem and solution was as follows:

> Therefore, we conducted experiments looking toward profitable treatment of the slimes, a muddy portion of the tailings, which were being separated from the sands because they were too expensive to treat. The experiment on this slime gave hopeful indications, so we determined that the two flows of slime pulps, one from the Lead mills and the other from the mills in the second valley, should be brought together on a hill above the confluence in Deadwood.
>
> In the treatment of all the tailings, a problem was the presence of reducing material in the tailings, which, unless oxidized, prevented the dissolution of gold by the cyanide solution. To accomplish this oxidation, it was necessary to aerate both the sands and slimes. This was not too difficult to accomplish with the sands, but owing to the very fine communation of the slimes, they presented a problem.
>
> The first step in my proposed process for treatment of the slimes was to dewater them, and this was accomplished by filtration in very large filter presses, and then aerating the filter press cakes by air pressure. After aeration the cake was leached with cyanide solution thus dissolving the gold, and then precipitated by my zinc dust process. The key to the successful process was my invention whereby the treated filter cake could be removed by sluicing the filter press. This method supplanted the costly hand removal of the treated cake as practiced in Australia, the cost of which would have been completely prohibitive at Homestake. Furthermore, my method permitted the construction of 26-ton presses, whereas the largest press previously built had a capacity of only four tons—a most important economic advantage.
>
> In view of this, I conferred with Grier and told him I wanted to build a small experimental filter press with a mechanism for sluicing out the slime cakes after treatment. He agreed, and after successful experiments with this small sluicing press, I told him I would like to build a larger press as the next step. When I suggested a press having a capacity of ten tons, he immediately asked if I couldn't build a larger press. To this I replied, "Yes, I certainly can and would like to, but it is going to cost $10,000 just for the ten-ton press." I mention this remark of Grier's because it demonstrated the breadth of his vision and his whole attitude—proposing a larger press instead of the $10,000 press, just for an experiment! The ten-ton press proved successful and consequently a very large plant was erected, containing a large number of filter presses, each having a capacity of 26 tons per charge. The plant was located above the confluence of the valleys mentioned above.

> Both the sand and slime plants broke the world records for the low cost of the treatment of sands and slimes, and, after over fifty years, the advances in other metallurgical details have not altered the methods we used at Homestake, or operated more cheaply.[41]

In 1904, George Moore, B. C. Cook, and D. C. Boley began assisting Merrill in designing the Slime Plant. The preferred site for the plant was in Deadwood, just south of the confluence of Deadwood Creek and Whitewood Creek (see fig. 14.7). This location allowed the slimes from Cyanide Plant No. 1 and Cyanide Plant No. 2 to be transported, by gravity, through separate pipelines to the new slime-treatment plant (see figs. 14.8 and 14.9).

The slimes were accumulated, or "thickened," at each sand plant in a thickener tank previously designed by John V. N. Dorr. Dorr successfully tested and proved his slime thickener at the Lundberg, Dorr, and Wilson Mill in Terry. The Dorr slimes thickener consisted of a cylindrical tank, which contained agitator rakes attached to a vertical driveshaft. The shaft was suspended from a drive mechanism located on a supporting structure at the top of the tank. Operation of the thickener caused the slimes to settle to the bottom of the tank where they were constantly agitated. The clear solution then flowed out the top of the tank. Upon completion of the pipelines in 1905 and the Slime Plant in 1906, the slime fraction of pulp from the six stamp mills could now be treated economically.

Fig. 14.7. The Slime Plant in Deadwood. The Slime Plant was designed by Charles W. Merrill to recover gold from the slime fraction of ores classified at the Cyanide Plant No. 1 in Lead and the Cyanide Plant No. 2 in Gayville. The Slime Plant operated from 1906 until 1973 at which time the plant was replaced by the Carbon-in-Pulp Plant in Lead.—Photo courtesy of the Black Hills Mining Museum.

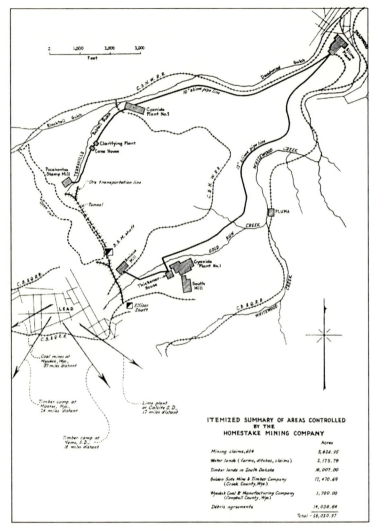

Fig. 14.8. Plan view of Homestake's stamp mills, cyanide plants, and slime-treatment plant, ca. 1931. After the ore was crushed and amalgamated in the stamp mills, tailings were transported, by gravity, to Cyanide Plant No. 1 and Cyanide Plant No. 2. At these two plants, the tailings were classified. The sand fraction was treated at the cyanide plants using the Merrill cyanide process. The slime fraction was delivered through separate pipelines to the Slime Plant in Deadwood where the slimes were treated for gold recovery in Merrill filter presses.—Source: A. W. Allen, "The Homestake Enterprise," *Engineering and Mining Journal* 132, no. 7 (October 12, 1931): 289. Reproduced by permission.

Fig. 14.9. The slime pipeline between the Cyanide Plant No. 2 in Gayville and the Slime Plant in Deadwood, 2001.—Photo courtesy of Homestake Mining Company.

The main building of the Slime Plant, measuring 65×270 feet, was initially equipped with twenty-four Merrill filter presses, each of which contained ninety frames covered with two layers of No. 10 cotton-duck canvas. Each filter press was charged with 24-26 tons of the finely pulverized ore called "slime." Each press had a capacity of about 70 tons per day, giving the entire plant a capacity of 1,700 tons per day.[42] By 1912, four additional filter presses were installed.[43] Ultimately, thirty-one filter presses were housed in the Slime Plant.[44]

The Slime Plant consisted of three main buildings and facilities located on four benches. Each bench was 24-30 feet apart, vertically (see fig. 14.10). The building on the uppermost bench housed two sludge-storage tanks that received the incoming pulp from the sand treatment plants in Lead and Gayville. The building also housed one wash-water tank. Each of these three tanks was 26 feet in diameter, 24 feet deep, and constructed of California redwood. Directly above the sludge tanks was the equipment for feeding lime to the pulp.

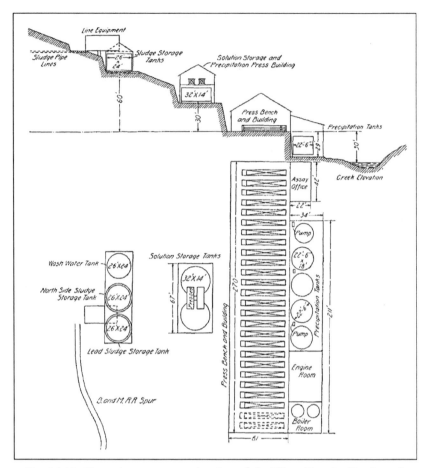

Fig. 14.10. General arrangement drawing of the Slime Plant in Deadwood, 1907.—Source: Mark Ehle, "The Homestake Slime Plant," *Mines and Minerals* (March 1907): 3.

Thirty feet below the Sludge Tank Building was another bench on which the Solution Storage and Precipitation Press Building was located. This building housed two redwood tanks 32 feet in diameter by 14 feet deep. One tank was used for weak cyanide solution and one for strong solution. The upper floor of this building contained two precipitation filter presses.

Thirty feet below the Solution Storage Building was the large Press Building that housed two floors, stair-stepped 29 feet apart, vertically. The upper floor housed the twenty-four large filter presses (see fig. 14.11). The lower floor contained two precipitation or weak-solution redwood tanks 22½ feet in diameter by 18 feet deep, two standardizing or strong-solution tanks the same size as the weak-solution tanks, a wastewater storage tank, solution pumps, a compressor room, and a boiler room. An assay office was located in a separate building. Immediately below the lowest floor was Whitewood Creek. A 100,000-gallon sluicing water tank was located on the hillside 150 feet vertically above the press floor. This tank received water from the Whitewood Ditch, Foster Ditch, and settling dam in Lead, and it supplied sluicing water for the filter presses and water for the wash-water tank.

Fig. 14.11. Filter presses in the Slime Plant. The Slime Plant eventually housed thirty-one filter presses that were used to dewater, aerate, and leach the minus 200-mesh slime fraction of ore from the cyanide plants in Lead and Gayville.—Photo courtesy of the Black Hills Mining Museum.

From the bottom of the two sludge-storage tanks, slime was transferred to a constant pressure tank, where lime was added to neutralize the slime. The slime was then directed through a feed pipeline that extended through a pipe gallery to the Press Building. Inside this building, the header pipe was routed over the midpoints of the filter presses in the Press Building. Between each pair of presses, lateral 6-inch pipes directed the slime to 4-inch branch pipes that directed the slime to the upper ends of each pair of presses.

Each of the twenty-four filter presses consisted of a supporting frame 46 feet long. The frame supported ninety-one flush plates and ninety-two distance frames suspended from side rails. Each 4×6-foot plate contained a circular opening 2½ inches in diameter in each corner, an upper-center hole 4 inches in diameter, and a lower-center hole 6 inches in diameter. Each press was fitted with three types of frames referred to as "standard," "top feed and bottom discharge," and "top-gate." A No. 8 canvas cloth covered the faces of each plate. Two long thrust bolts bound the plates and frames of the press units firmly together against the front standard.

The ends of each press were charged with pulp under a pressure of 35 pounds per square inch through the upper 4-inch holes. The pulp passed through the cored holes of each frame compartment. Water from the pulp flowed through the canvas-bounding faces and along the corrugations of each plate to the lower-corner holes. The water was collected in the wastewater tank and pumped up the hill to the sluicing water tank. The charge of "filter cake" in each press was then subjected to an eight-hour cycle that required one hour for filling, six hours for oxidation, leaching, and washing, and one hour for sluicing.[45] By 1931, the slime-treatment cycle was increased to ten hours.[46] During the oxidation phase, the filter cakes were aerated with very low-pressure compressed air. The cakes were then leached with cyanide solution to dissolve the gold. After a final rinsing, the barren cakes were sluiced out of the filter presses and discharged into Whitewood Creek.

The "pregnant" or gold-bearing solution from the filter presses was collected in weak-solution precipitation tanks on the lowest floor. At that location, zinc dust was added at the pumps, and the pregnant solution was pumped to two precipitation presses in the Solution Storage Building. The zinc preferentially replaced the gold in the cyanide complex, causing the gold to precipitate. The gold-bearing precipitate was subsequently captured in the two precipitation presses. The precipitate was then transported to the refinery in Lead, where the gold and silver were recovered.[47]

At the refinery in Lead, precipitate was treated with acid and placed in a large steam drier. The precipitate was thoroughly blended with litharge, borax, silica, and powdered coke. A solution of lead acetate was sprinkled on the precipitate. The latter was then briquetted under a pressure of 4,000-6,000 pounds per square inch and fused in a cupel furnace. After a sufficient cooling period, the glass slag was broken from the outer surface of the crude bullion. The lead was then removed by cupellation. The remaining crude bullion was subsequently cast into dore' bars 975-985 fine.[48]

In 1908, after ten years of highly successful work in Lead, Merrill's second five-year contract came to an end. Merrill's work to implement the Merrill cyanide process at Homestake led to the construction of Cyanide Plant No. 1, Cyanide Plant No. 2, the Regrind Plant, and the Slime Plant. Homestake would reap the benefits of Merrill's work over the remaining life of the mine. Merrill himself was compensated rather well, as indicated in William A. Remer's personal diary: "June 25, 1907: long visit with Grier [in] front [of the] Brick Store—told me Merrill's contracts were all for five years—does not have to stay only two on Slime Plant—gets 15 percent of net returns on Cyanide Plant and 7½ percent on Slime Plant—has had nearly $300,000 so far."[49]

Merrill and his wife, Clara, returned to San Francisco with their three children, all of whom were born in Lead. One of Merrill's final recommendations to Superintendent Grier was to construct a regrind plant to reduce the size of the sands, thereby improving liberation and dissolution of the fine gold that was tightly encapsulated in the sulfide minerals.[50]

Regrind Plant

Acting on Merrill's recommendation, Grier immediately constructed a regrind plant that was completed in 1908. The plant initially housed four tube mills, including a 5×14-foot Denver Engineering Mill, two 5×18-foot Allis Chalmers Mills, and one Hardinge 6×6-foot mill (see fig. 14.12). The four mills used pebbles as a grinding medium and had a combined capacity of 462 tons per day.[51] Two additional tube mills were added a few years later. In 1924, the tube mills were moved from the Regrind Plant to the Amicus Mill.[52]

Fig. 14.12. Tube mills in the original Regrind Plant. Six tube mills like this one were used to achieve fine grinding in the Regrind Plant. After the ore was crushed and amalgamated in the stamp mills in Lead, the ore was transferred to the Regrind Plant by flume. Chert pebbles were used in the tube mills as a grinding medium. After undergoing fine grinding in the tube mills, the "pulp" was slurried to the Cyanide Plant No. 1 where the pulp was classified. The sand fraction was processed at Cyanide Plant No.1. The slime fraction was slurried to the Slime Plant in Deadwood for treatment.—Photo courtesy of Evelyn (Schnitzel) Murdy.

Before Charles W. Merrill arrived in Lead, total gold recovery at the Homestake Mine was averaging 70 to 75 percent—all from the stamp mills and plate houses.[53] As a result of Merrill's work, two sand plants, a slime plant, and a regrind plant were added, boosting overall recovery to 92-94 percent.

In 1909, Homestake's 1,000 stamps crushed about 125,000 tons of ore per month. Approximately, 57 percent of the pulp was treated at the two sand plants and 41 percent at the Slime Plant. The remaining 2 percent of the slimes were wasted.[54] Using the Merrill cyanide process and Merrill filter presses, gold recovery increased to 93.7 percent in 1909 of which 72.4 percent was recovered from amalgamation in the mills, 0.7 percent in the regrind plants, 13.0 percent in the cyanide sand plants, and 7.6 percent in the Slime Plant.[55] Figure 14.13 illustrates the amalgamation, cyanidation, and refining processes in place at Homestake in 1915—the same year the stamp milling capacity was increased to 1,020 stamps.

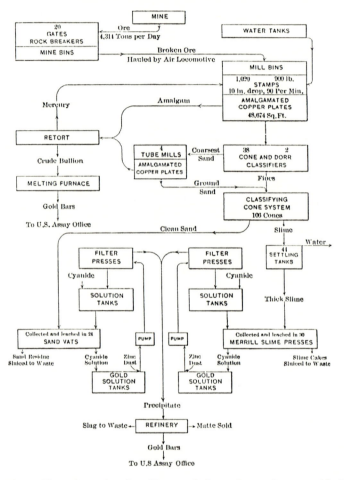

Fig. 14.13. Flow sheet showing Homestake's amalgamation, cyanidation, and refining processes, 1915.—Source: Allan J. Clark, "Notes on Homestake Metallurgy," *American Institute of Mining and Metallurgical Engineers Transactions* 52 (September 1915).

Homestake's Tungsten Mill

Homestake acquired the properties of the Columbus Consolidated Gold Mining Company, which included the properties of the Hidden Fortune Mining Company, in 1912. Over the next few years, a tungsten-bearing ore known as wolframite was found to exist in the former Harrison Mine west of the Open

Cut. In 1916, Homestake constructed a 5-stamp mill in the former Regrind Plant and began treating the wolframite ore for recovery of tungsten.

Ore was hand-selected and hauled to the mill in wagons. After being pulverized in the stamp mill, the ore was concentrated on a Wilfley No. 6 Concentrating Table. Concentrates were dried, sacked, and shipped. Tailings were delivered to the Cyanide Plant No. 1 for recovery of gold. During the first year of operation, some $229,915 worth of tungsten was recovered and sold for the World War I war effort. The Tungsten Mill was shut down in June 1918.[56]

South Mill

In 1919, an experimental ball mill was installed in the Pocahontas Mill to test its effectiveness in the reduction of Homestake ore. At about the same time, Bruce C. Yates, superintendent of the mine, recognized that a new mill was needed to replace the Homestake, Golden Star, and Amicus mills, whose locations were in jeopardy because of continuing surface subsidence. A decision to construct a new mill, called the South Mill, was made in 1920. Grading for the site was started in January 1921 on the south side of Gold Run Gulch in east Lead.[57]

R. G. Wayland and Allan J. Clark designed the South Mill. Clark was Homestake's chief metallurgist, who had almost twenty-five years of service with the company at the time. The South Mill was commissioned in September 1922 (see figs. 14.14 and 14.15). The new mill housed twenty-four 5-stamp batteries that were arranged in two rows of 60 stamps each. The rows were referred to as the North Battery and the South Battery (see fig. 14.16).

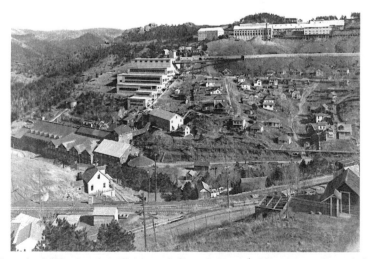

Fig. 14.14. The South Mill (upper left) and Cyanide Plant No. 1 (lower left). The building to the lower right of the South Mill is the South Mill Thickener House. Note the Fremont, Elkhorn, and Missouri Valley Railroad tracks on Grandview Drive.—Photo courtesy of Homestake Mining Company.

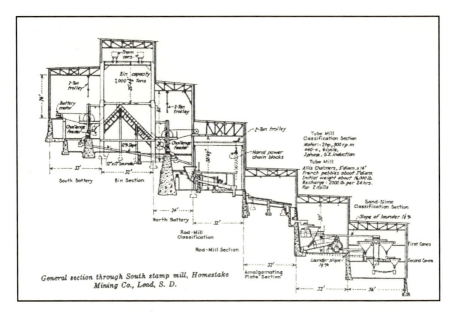

Fig. 14.15. General section through the South Mill. When commissioned in 1922, the South Mill contained 120 stamps that were arranged as North Battery and South Battery under a 7,200-ton-capacity ore bin. After the sand product from the stamp mills was dewatered, the pulp was directed to six 5×10-foot Allis Chalmers rod mills and two 5×14-foot Allis Chalmers tube mills.—Source: Edward Hodges Robie, "Milling Practice at the Homestake Gold Mine," *Engineering and Mining Journal* 122, no. 15 (October 9, 1926): 566. Reproduced by permission.

Fig. 14.16. The cam floor of the South Battery of the South Mill.—Photo courtesy of the Black Hills Mining Museum.

Each 5-stamp battery in the South Mill was driven by one 25-horsepower General Electric motor. The weight of a single-stamp assembly was 1,570 pounds, compared to 900-1,100 pounds in the older mills. Mill capacity for the 120 stamps in the South Mill was 1,900-2,200 tons per day compared to 1,200-1,400 tons per day for the 240 stamps at the Amicus Mill. Each stem at the South Mill was 4 inches in diameter, 14 feet long, and weighed 600 pounds. Because of the increased stamp weight, forged-steel shoes weighing 250 pounds each were selected instead of the conventional white-iron castings. White-iron dies weighing 165 pounds each were cast in the Homestake Foundry. Mortars were also cast as white iron in the Homestake Foundry and weighed 7,000 pounds.[58]

As part of the new design, the rate of drop for a stamp was increased from eighty-eight to one hundred blows per minute, and the length of drop was decreased from 10 to 8 inches. Openings in the mortar screen were initially sized at ½ inch but varied from ⅜ inches to ⅞ inches in later years. Ore was fed to the stamp mill batteries at about minus 1½ inches. Product from the stamp batteries was classified in six 7-foot-diameter-by-7-foot-deep dewatering cones. The sand discharge from the cones was ground in six 5×10-foot Allis Chalmers rod mills, each of which was rotated at 22 revolutions per minute using a 100-horsepower motor (see fig. 14.17). Each rod mill was charged with 24,000 pounds of 3-inch-diameter carbon steel rods 9 feet 11 inches long that pulverized the ore as the mill rotated. For 1924, the cost of stamping averaged 8.6 cents per ton. An additional 7.7 cent cost was attributable to the rod mills.[59]

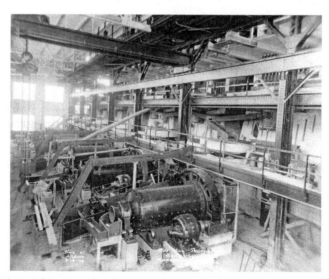

Fig. 14.17. The original rod mills in the South Mill, 1922. The coarse sand product from the twenty-four stamp batteries in the South Mill were pulverized in six 5×10-foot Allis Chalmers rod mills located in the South Mill.—Photo courtesy of the Black Hills Mining Museum.

Quicksilver (mercury) was added to the rod-mill feed at the rate of about one-tenth troy ounce per ounce of gold. Pulp from the rod mill passed to the classifier through a launder containing a copperplate 1 foot wide by 7 feet long. The plate was dressed with 2 ounces of quicksilver per square foot. After every four hours, the amalgam-laden plate was replaced with a fresh plate. Approximately, 80 percent of the gold recovered by amalgamation was recovered on this plate. Overall gold recovered by amalgamation at the South Mill averaged 63 percent over the first four years of operation.[60]

Each of the six rod mills operated in closed circuit with a Dorr duplex classifier that separated sands from slimes. Overflow from the classifiers combined with overflow from the dewatering cones and passed over forty-eight amalgamating plates, each of which was ¼ inch thick, 4½ feet wide, and 12 feet long. These plates were sloped at 1½ inches per foot and arranged in blocks of four, creating twelve amalgamating units 9 feet wide by 24 feet long (see fig. 14.18). This table-and-plate arrangement provided 21.6 square feet of plate per stamp. The copperplates were dressed with quicksilver at the rate of 3 ounces per square foot and were cleaned and dressed every twenty-four hours.

Fig. 14.18. Construction of the plate floor at the South Mill, 1922.—Photo courtesy of the Black Hills Mining Museum.

The pulp stream from each pair of amalgamating plates was directed to four 7×7-foot dewatering cones. Coarse sands associated with the cone underflow were directed to one of two 5×14-foot Allis Chalmers tube mills for regrinding. Each of the tube mills utilized ribbed liners and was charged with 16,000 pounds of 3-inch-diameter pebbles for grinding media. Some 100 pounds of pebbles were consumed in each mill every twenty-four hours. The reground pulp was then passed over two 9×12-foot amalgamating plates. After dewatering in 7-foot cones, the sand fraction underflow was delivered to the Cyanide Plant No. 1 for further classification into sands and slimes. Sands were processed at Cyanide Plant No. 1, and slimes were delivered to the Slime Plant in Deadwood for final treatment. The capacity of the South Mill, when first constructed was 1,800 to 2,000 tons per day, depending on the type and hardness of ore.[61]

Based on the successful startup of the South Mill, the 220-stamp Homestake Mill was shut down on November 18, 1922. The 200-stamp Golden Star Mill was decommissioned next in March 1923. After the two mill buildings were dismantled, the soils were excavated and processed at the South Mill to recover gold values and mercury. In 1925, the 100-stamp Monroe Mill was shut down.[62] With these changes, the active mills were reduced to the Pocahontas, Amicus, and South mills (see fig. 14.19).

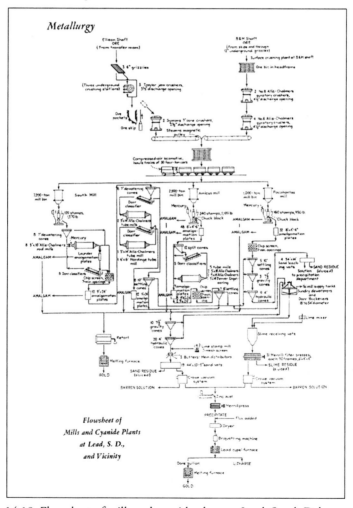

Fig. 14.19. Flow sheet of mills and cyanide plants at Lead, South Dakota, and vicinity, 1931. The Pocahontas and Amicus mills were shut down shortly after the South Mill was expanded in December 1934.—Source: A. W. Allen, "The Homestake Enterprise," *Engineering and Mining Journal* 132, no. 7 (October 12, 1931): 350. Reproduced by permission.

In 1927, two rod mills were added to the South Mill, bringing the total number to eight. In addition, one 5×14-foot Allis Chalmers tube mill was added along with a 6×6-foot Hardinge tube mill. By 1929, gold recovery was reported to be 93 percent of which 62.8 percent was recovered by amalgamation, 20.6 percent by cyanide treatment of sand, and 9.6 percent by cyanide treatment of slime. Total treatment costs were $0.503 per ton, including $0.038 for crushing, $0.276 for milling, and $0.189 for cyaniding. Total ore crushed and processed in 1929 amounted to 1,437,935 tons from which 313,635 troy ounces of gold were recovered.[63]

Precipitation and Refining

By 1929, the active milling and treatment systems included the Pocahontas Mill, Amicus Mill, South Mill, Cyanide Plants No. 1 and No. 2, and Slime Plant. Precipitation of gold and silver at the cyanide plants was accomplished using zinc dust as part of the Merrill-Crowe process. The two cyanide plants were both equipped with two Merrill precipitation presses, each of which held thirty-two triangular plates covered with a double layer of cotton cloth, known as No. 1.30 Sateen.[64]

At the cyanide plants, precipitate was removed from the two Merrill precipitation presses monthly. Prior to opening, the presses were blown with air to displace any remaining solution. A shallow box, mounted on castors, was rolled under the press. Each press was opened, and precipitate was emptied from the filter frames, one at a time. Typically, a press contained up to 2,000 pounds of wet precipitate. After being agitated with acid in a lead-lined tank, the precipitate was placed in an open box and dried over a steam chamber. Once the water content was reduced to about 17 percent, the precipitate was mixed with a flux of borax and litharge and pressed into briquettes about 5 inches in diameter and 4 inches thick. The briquettes were then placed in pans and dried in a drying furnace. Next, the briquettes were placed in a cupel furnace. After fusion and cupellation, the dore' was melted and refined into gold and silver bars.[65]

The Merrill cyanide process continued to be used at the Homestake Mine through mine closure in 2001. Modifications to the process were made in 1973, when Merrill's process for treating slimes was replaced by a carbon-in-pulp process developed by the U.S. Bureau of Mines. In the early 1990s, the Merrill-Crowe process was replaced by carbon adsorption. From 1901 through 2001, Homestake benefited greatly from Charles Merrill's contributions.

Chapter 15
The Conversion to Electricity

Edison Dynamos

The first means of providing electricity at the Homestake Mine was initiated on September 28, 1888, when the Hearst Mercantile Company purchased a 75-lamp dynamo from the Edison United Manufacturing Company of New York (see fig. 15.1). The dynamo, which cost $1,680, had a capacity of seventy-five lamps at sixteen-candle power each when operated at a speed of 1,800 revolutions per minute. The dynamo was first used to power sixty-four lights at the Hearst Mercantile Store on North Mill Street and sixteen lights over the plate floor in the 80-stamp Homestake Mill. The lights in the store were illuminated for the first time on Christmas Eve. The event was said to have made quite an impression on everyone present.[1]

Fig. 15.1. The Edison 75-lamp dynamo that provided the first electrical lighting at the Hearst Mercantile Store and part of the 80-stamp Homestake Mill in 1888.—Photo by author.

Based on the success of the 75-lamp Edison dynamo, Homestake purchased a 150-lamp unit to illuminate the Homestake, Golden Star, and Amicus mills. However, the two dynamos were soon found to be inadequate and were replaced with a single 15-kilowatt Edison dynamo, which was installed in the Amicus Mill. The 75-lamp dynamo was then moved to the Monroe Mill in Terraville, and the 150-lamp unit was moved to Cyanide Plant No. 2 in Gayville. In 1890, an incandescent light machine manufactured by the Brush Arc Light Company was purchased and installed in the Pocahontas Mill in Terraville.

Two 60-kilowatt Edison dynamos were installed in the Old Abe compressor room in 1895 (see fig. 15.2). In addition to providing illumination, these two dynamos powered the first few electric motors at the Homestake Mine. The dynamos also provided lighting for many of the Homestake buildings north of Main Street in Lead, including the Homestake Hotel and Hearst Mercantile Brick Store. The Amicus and Golden Star mills continued to make use of the 15-kilowatt dynamo. Two additional 20-kilowatt dynamos were purchased and installed at the Highland Mining Company's Golden Prospect Shaft and hoist house. By the late 1890s, most all of Homestake's operations were illuminated by electricity.[2]

Fig. 15.2. Two 60-kilowatt Edison dynamos at the Old Abe compressor room, ca. 1895. The dynamos provided electricity to many of the Homestake buildings north of Main Street, including the Homestake Hotel and the Hearst Mercantile Store.—Photo courtesy of the Black Hills Mining Museum.

With the dynamos operating quite successfully, Homestake began investigating how larger amounts of electricity might be cost-effectively generated at one or more hydroelectric plants. As an experiment, a small hydroelectric plant was constructed about one-half mile above the mouth of Spearfish Canyon in the fall of 1904 (see fig. 15.3). In order to procure the necessary water rights to operate the plant, Homestake agreed to supply power to the town of Spearfish. At the same time, Homestake purchased the Spearfish Electric Light and Power Company from George Favorite and some Chicago capitalists, who had been supplying power to Spearfish since 1893.[3]

Fig. 15.3. Homestake's first hydroelectric plant, located approximately one-half mile above the mouth of Spearfish Canyon. The small hydroelectric plant provided electricity for the city of Spearfish and power for the ventilation fans and electric drills used in driving the 23,862-foot-long stream-diversion tunnel between Maurice Dam and the Forebay for Hydroelectric Plant No. 1.—Photo by author.

The new hydroelectric plant housed a single Westinghouse generator that provided 100 kilowatts of power at 2,300 volts. The small plant was deemed a success but wasn't large enough to justify the cost of a power-transmission line to Lead. Instead, Homestake officials decided to construct a larger plant closer to Lead.

Englewood Hydroelectric Plant

A 400-kilowatt plant, called the Englewood Hydroelectric Plant, was constructed on upper Whitewood Creek approximately three-fourths of a mile below the Englewood camp and very near the Black Hills and Fort Pierre Railroad (see fig. 15.4). The plant was commissioned in July 1906. Water was supplied to the plant via the "high-pressure pipeline," which received its water from the Little Rapid Ditch, Peake Ditch, and Spearfish Ditch. An elevation difference of 485 feet in the section of pipeline between the Peake Weir House and the Englewood Hydroelectric Plant powered a 500-horsepower water wheel that drove a 3-phase, 60-cycle, 11,000-volt generator.

Fig. 15.4. The inside of the Englewood Hydroelectric Plant. The plant was commissioned in 1906 and operated until 2002, when it was shut down as part of the mine closure.—Photo courtesy of the Black Hills Mining Museum.

Electricity from the Englewood Hydroelectric Plant was initially used to replace steam power at the Monroe Mill in Terraville and the Mineral Point Mill in Golden Gate. These two mills were selected because of the difficulty in getting coal to them. The 440-volt, 25-horsepower electric motors that were installed powered the belts and pulleys that drove the camshafts of the stamp mill batteries. Each motor powered two 5-stamp batteries. Except for a few minor interruptions due to lightning storms, the electric drives operated reliably for up to sixteen months without interruption. Based on this success, the steam plants and lines were removed from the two mills.[4]

Between 1906 and 2001, the Englewood Hydroelectric Plant produced 177.2 million kilowatt-hours of electricity at 400 kilowatts for the benefit of the Homestake Mine.[5]

When construction of the Slime Plant was nearing completion in January 1907, a decision was made to fit the plant with electric drives. This arrangement proved to be very cost-effective, compared to the alternative of having to install long runs of steam lines throughout the three main buildings. Electricity to operate the plant was initially purchased from Consolidated Power and Light Company, the local power company created in June 1905 as a consolidation of the Belt Light and Power Company of Lead and the Black Hills Electric Company of Deadwood.[6]

In 1910, based on the operating success of the electric drives at the Slime Plant, the stamp batteries in the Pocahontas Mill were converted to electric drives. The timing was right since the boilers were too old to be operated any longer. The regrind plant, pattern shop, and foundry on the north side of Gold Run Gulch were also refitted with electric motors at about the same time, as were the Cyanide No. 1 and Cyanide No. 2 sand treatment plants in Lead and Gayville.[7]

After the Slime Plant was commissioned, Homestake quickly began to recognize the benefits of electric-driven equipment over steam-driven equipment. Except for the Englewood Plant and the small dynamos, the Consolidated Power and Light Company, with its 3,500-kilowatt steam-electric plant located in Pluma, furnished most of the electricity for Homestake's new electric-powered equipment.[8] Homestake's engineers began to devise ways to expand its hydroelectric-generating capacity, particularly since the Englewood Hydroelectric Plant was continuing to generate electricity quite successfully and cost-effectively.

Hydroelectric Plant No. 1

In July 1908, Homestake engineers completed a conceptual plan for a large hydroelectric plant in lower Spearfish Canyon. Judging the plan to be a good one, Superintendent Thomas J. Grier directed Homestake's surveyors to complete a topographic survey of the lower canyon to determine the specific area that might be amenable to the project. Aided by topographic maps drawn by the U.S. Geological Survey, the surveyors and engineers commenced to complete their preliminary survey and engineering work. A site for the stream diversion was proposed at an elevation of 4,375 feet at which point water could be diverted into a conduit and allowed to flow, by gravity, to the mouth of the canyon where a generating plant could be located. The engineers concluded that the recommended endpoints would provide sufficient elevation difference and corresponding water pressure to drive the generators.

Upon approval of the updated plan, the surveyors staked the 4,375-foot contour on both flanks of Spearfish Canyon. The line on the west side of the canyon was determined to be the more suitable side for the conduit. Because of the difficult terrain, it was decided that a tunnel would be preferable to a flume or pipeline for conveying water to the planned location for the hydroelectric plant.[9]

Over the summer and fall of 1908, many townspeople began to speculate about why Homestake was doing all the survey work in Spearfish Canyon. The *Deadwood Daily Pioneer-Times* posed the following question:

> Is the Homestake Mining Company to build a large hydro-electric power plant on Spearfish Creek? This question has been asked of us so often during the past few months that our sleuth was instructed a short time ago to find out and let our readers know something about it. The general office of the Homestake doesn't seem to know where Spearfish Creek is, when asked about it. When the Homestake survey office is asked what a few of its employees have been doing in the canyon all summer and fall, it sagely replies, "Fishing and washing their feet."[10]

By the spring of 1909, all of the activity became clear as the company began placing orders for materials and equipment and hiring people in earnest to work on the hydroelectric project.

Construction work commenced on the Hydroelectric Plant No. 1 in May 1909 under the overall direction of Richard Blackstone, chief engineer and assistant superintendent of Homestake Mining Company. The resident engineer and site superintendent for the project was Alex J. Blackstone, son of Richard.

The overall system included an intake dam (see fig. 15.5), diversion tunnel (see fig. 15.6), forebay (see fig. 15.7), standpipes and penstocks (see fig. 15.8), and the powerhouse (see fig. 15.9), located at the mouth of Spearfish Canyon on Spearfish Creek. Water was diverted from a small dam on Spearfish Creek at Maurice into a concrete-lined tunnel some 23,862 feet long that extended through the west flank of Spearfish Canyon. Water exited the tunnel and flowed across a 40×70-foot concrete forebay. The water then passed through two 48-inch-diameter redwood stave pipes made of California redwood that were supplied by the Redwood Manufacturers Company.

The direct line between the forebay and powerhouse passed over a small hill approximately 110 feet higher than the theoretical gradeline of the pipes. Because of this, the flow pipelines were graded to closely follow the natural ground surface. The pipelines connected to a large cylindrical pipe that formed the base of four standpipes located on top of the hill.

Fig. 15.5. The Maurice Intake Dam for Hydroelectric Plant No. 1. Water from Spearfish Creek entered a concrete-lined tunnel at Maurice. The water flowed through a tunnel 23,862 feet long and exited at Forebay.—Photo courtesy of Gary Lillehaug.

Fig. 15.6. The 23,862-foot-long diversion tunnel for Hydroelectric Plant No. 1. The tunnel was routed along the west wall of Spearfish Canyon from the Maurice Dam to the Forebay, located above the power plant.—Photo courtesy of Gary Lillehaug.

Fig. 15.7. The Forebay, located at the downstream end of the diversion tunnel for Hydroelectric Plant No. 1. The people in the boat have just completed an inspection of the tunnel, which was typically performed every five years. Robert Haag, one of the longtime operators of the hydro plants, is seated in the back of the boat. On February 21, 1930, while attempting to repair a communication line in the tunnel, Sidney Case drowned when the boat in which he and Fred Langhoff were riding capsized. The water current carried Langhoff to Forebay, but Case succumbed to the cold water and drowned.—Photo courtesy of Gary Lillehaug.

Fig. 15.8. The standpipes above Hydroelectric Plant No. 1.—Photo courtesy of Gary Lillehaug.

Fig. 15.9. Homestake's Hydroelectric Plant No. 1. The Hydro Plant No. 1 was commissioned in April 1912. In 2004, Homestake sold the plant to the city of Spearfish for $250,000.—Photo courtesy of Black Hills Mining Museum.

The standpipes were connected to a steel cylinder 74 inches in diameter and 25 feet long. Each of the standpipes was 36 inches in diameter and 54 feet high. The standpipes served to equalize water pressure between the flow pipelines and the pressure pipelines, provide for the escape of air, and dampen any surges in the flow. The tops of the standpipes were slightly higher than the intakes of the flow pipelines at the Forebay. The two redwood flow pipelines connected to the ends of the cylinder through tapered, steel-adapter elbows.

Three pressure pipelines or penstocks had a combined length of 12,164 feet. The penstocks were attached to the side of the cylinder and carried the water to the two 2,000-kilowatt generators in the power plant. Excavation for the penstocks was extremely difficult because of a vast amount of bedrock with little or no soil cover (see fig. 15.10). Outward from the cylinder, the pressure pipelines were 34 inches in diameter. The pipes reduced to 32 inches in diameter about midway to the powerhouse. The lowest sections were 30 inches in diameter and were designed to withstand the 288 pounds per square inch of water pressure associated with 665 feet of elevation difference.

Fig. 15.10. Excavation for the penstocks between the standpipes and the Hydroelectric Plant No. 1, ca. 1910.—Photo courtesy of the Black Hills Mining Museum.

Water entered the powerhouse through the three 30-inch-diameter pressure pipelines. Each of the pipelines branched into two smaller pipes fitted with 6-inch-diameter nozzles at their ends. The nozzles directed the high-pressure water onto Pelton wheels that drove the two generators (see fig. 15.11). The generators were 3-phase, 60-cycle, 2,200-volt units rated at 400 revolutions per minute, each of which was driven by two 1,500-horsepower Pelton waterwheels attached to the ends of each turbine. Electricity was delivered to the Homestake Substation in Lead over an 11-mile-long powerline.[11]

Fig. 15.11. Westinghouse generators at the Hydroelectric Plant No. 1. These two Westinghouse generators have operated continuously at the Hydroelectric Plant No. 1 since 1912 when the plant was commissioned. Between 1912 and 2004, the plant generated some 1.59 billion kilowatt-hours of electricity for the benefit of the Homestake Mine.—Photo courtesy of Gary Lillehaug.

Originally, three generators were installed at the Hydroelectric No. 1 powerhouse. When the Hydroelectric Plant No. 2 was constructed in 1917, one of three generating units at Hydro No. 1 was moved to Hydro No. 2 after it was determined that a third unit was not needed at the No. 1 Plant. After May 5, 2004, when Homestake sold the Hydroelectric Plant No. 1 to the city of Spearfish, hydroelectric power was redirected to the grid of the local utility company. The appropriate credit was issued to the city monthly for the power generated.

Some three hundred workers were originally hired or assigned to construct the Hydroelectric No. 1 system.[12] Three camps were established to accommodate the workers. One camp was located at an old boardinghouse near the Spearfish Fish Hatchery. A second camp was established near the planned intake below Maurice. A third camp was established between the two other camps.[13]

The baseline surveys that were completed showed that natural draws and ravines crossed over the projected centerline of the tunnel at various locations along the west side of Spearfish Canyon. Blackstone took advantage of this topography by dividing the tunnel work into eight separate segments in order to facilitate completion of the tunnel in a timely manner.

Each of the eight tunnel segments varied in length from 1,284 feet to 4,154 feet, depending on the surface topography. Crosscuts were driven laterally into the wall of the canyon to the projected centerline of the tunnel at the proper elevation for the respective tunnel segment. Each "intersection" was called an angle point and was fixed by a permanent survey hub. Extremely accurate surveying and mining controls were required to ensure that the eight tunnels would be driven on the proper line and grade from the crosscuts located high up on the rugged hillside of Spearfish Canyon. Any error in line or grade in one of the tunnel segments would have resulted in a tremendous amount of rework and cost to realign the tunnel.

Richard Blackstone summarized the survey work as follows:

> *The [survey] hubs being established, a traverse was made from one hub to the next succeeding one, repeating angles and steel tape measurements. This work was duplicated by a second party. The closing line was calculated, and its course determined sufficiently accurate to set out the center line of the tunnel for starting.*
>
> *The center lines over tunnels Nos. 1, 2, 3, 4, and 5, were run over the surface, the closing error noted, the ranging out repeated until exact closing direction was determined, and this final angle into the tunnel line set out and permanently plugged in the roof. The exact distance was not of sufficient importance to waste time and effort upon, reliance being placed on the calculated distance given by the traverse.*
>
> *Tunnels Nos. 6, 7, and 8, coming nearer the surface, and over clear and fairly even ground, their center lines were determined by ranging first, a random line, and connecting to true line. The lines being determined and approximate grades set out from preliminary levels, repeated levels were made until agreements were arrived at and deemed sufficiently exact for final. The Burlington railroad tracks, following the valley from the intake to the power house, gave a clear base for the measurements in the traverse, and also for the leveling. Several U.S. Geological Survey bench-marks came within the range of this work, and were used as additional checks.*[14]

Each of the eight tunnel segments was completed by driving opposing headings from each of the eight crosscuts previously driven from the side of Spearfish Canyon. Tunnel work was performed over three eight-hour shifts per day. The 23,862-foot-long tunnel was completed between July 9, 1909, and October 8, 1910.[15]

On a given shift, from any one of the eight access crosscuts, drilling and blasting operations were performed at one face of the 8×9-foot tunnel while hand mucking was concurrently performed at the opposite face. With this arrangement, sixteen headings were worked simultaneously and efficiently. Fourteen electric-powered Temple-Ingersoll drills rated at 5 horsepower each were purchased to drill the twelve to fourteen blastholes required for a drift round 5 feet deep (see fig. 15.12). Broken rock was hand-loaded into 1-ton-capacity end-dump cars borrowed from the mine. The cars were hand-trammed to the dumping points at the end of the crosscuts.

Fig. 15.12. Drilling a drift round in Tunnel No. 3 using electric Templeton drills, September 19, 1910. Note the candles that provided illumination.—Photo courtesy of the Black Hills Mining Museum.

Power for the drills and Sirocco ventilation fans was supplied by the small hydroelectric plant constructed in 1904 near the mouth of Spearfish Canyon. A powerline was constructed up the canyon from the small hydroelectric plant to provide electrical power to each of the access crosscuts. Transformers were placed near the entrances of the crosscuts and supplied power to the drills at 220 volts. Supplemental steam power was supplied by a wood-fired steam plant.

Table 15.1 shows tunnel completion as follows:

Table 15.1 Hydroelectric Plant No. 1 stream diversion tunnel (Maurice Dam to Forebay)

Tunnel Segment	Date Started	Date Finished	Length	Daily Avg. (feet)
No. 1	Jul 9, 1909	Jun 23, 1910	1,965	5.65
No. 2	Jul 9, 1909	Oct. 8, 1910	4,152	9.10
No. 3	Jul 22, 1909	Aug. 13, 1910	4,082	10.60
No. 4	Jul 10, 1909	Jul 29, 1910	4,154	10.82
No. 5	Aug. 22, 1909	Jan. 5, 1910	1,284	9.45
No. 6	Sep. 10, 1909	Aug. 5, 1910	3,762	11.40
No. 7	Sep. 30, 1908	Jun 23, 1910	2,074	7.77
No. 8	Dec. 8, 1909	Sep. 4, 1910	2,389	9.00
Total:			23,862	

Source: Richard Blackstone, "The Hydro-Electric Power Plant of the Homestake Mining Co.," *Mining and Engineering World* XLI, no. 1 (July 4, 1914).

Concrete batch plants using Smith mixers were located at the tunnel intake and Station No. 5. Stream sand and gravel obtained from excavation of the dam were used as aggregates in the concrete mix along with Atlas Portland cement. The wet mix was loaded into 22-inch-gauge cars, each of which had a capacity of ½ cubic yard. The cars were trammed in trains powered by the same wood-fired steam locomotives originally used on the mill tramways in Lead. The steam locomotives worked well for the tunnel-lining work, except for the wood smoke that had to be flushed from the tunnel using the Sirocco ventilation fans.

Concrete forms made in the company's shops were reused as the lining work proceeded throughout the tunnel. The tunnel was lined to a finished width of 6½ feet. Vertical sidewalls extended to the spring line of the tunnel at a height of 5 feet. An arched roof with a radius of 3 feet 3 inches was formed using concrete blocks poured at the batch plants (see fig. 15.13).[16]

Fig. 15.13. The completed tunnel with concrete arch between the Maurice Intake and the Hydroelectric Plant No. 1, 1911. Water from Spearfish Creek is diverted into this tunnel at the Maurice Intake and flows to Forebay, some 23,862 feet downstream.—Photo courtesy of the Black Hills Mining Museum.

The Hydroelectric Plant No. 1 was commissioned in April 1912. Except for minor interruptions, and depending on seasonal stream flow, the plant provided approximately 3,000-4,000 kilovolt-amperes of continuous power to the Homestake Mine from 1912 until 2004. On May 5, 2004, subsequent to closure and decommissioning of the mine, the Hydroelectric No. 1 system was sold to the city of Spearfish for $250,000. Between 1912 and 2004, the Hydroelectric Plant No. 1 produced approximately 1.59 billion kilowatt-hours of electricity for the benefit of the Homestake Mine.[17]

Lead Substation

The substation in Lead was completed in 1911 to receive electricity from the hydroelectric plants and redistribute the same at the appropriate voltages to the mills and plant facilities in Lead and Terraville. Purchased power was also switched at the substation to supplement the hydroelectric power. Electricity was received at 33,000, 11,000, and 2,300 volts and was distributed at 11,000, 2,300, and 440 volts. The substation housed four banks of three transformers, each rated at 667 kilovolt-amperes. The low-voltage side of one bank was connected to an 11,000-volt bus. Two banks were connected in parallel to a 2,300-volt bus, and one bank was connected to a 440-volt bus.

Beginning in May 1912, one month after commissioning the Hydroelectric No. 1 Plant, the Homestake, Golden Star, and Amicus mills were converted to electric drives (see fig. 15.14), as were the 35-horsepower General Electric motors that powered the Gates No. 6 crushers at the Old Brig and Golden Prospect shafts. At about the same time, the Gates No. 6 crushers at the Ellison, B&M (Old Abe), and Golden Star shafts were also converted to electric drives.

Fig. 15.14. The cam floor of the Golden Star Mill. This photo, by Lease, shows the electric motors and belt drives that powered pairs of 5-stamp batteries in Homestake's stamp mills.—Photo courtesy of Evelyn (Schnitzel) Murdy.

Prior to 1913, a Cornish Pump installed at the B&M Shaft was used to dewater the mine (see fig. 15.15). The surface-mounted pump was powered by a steam engine located near the collar of the shaft. The steam engine was connected to the pump rod with a connecting rod and bell crank. The pump rod was housed in a series of 12×12-inch timbers joined end to end from the surface to the 800-foot level. When the mine was deepened to the 1,100-foot level, a condensing steam pump was installed in a brick-lined chamber on that level near the Golden Star Shaft (see fig. 15.16). Steam power for the pump was supplied through a pipe routed in the shaft from the surface to the 1,100-foot level. The pump on the 1,100-foot level and the Cornish Pump at the B&M Shaft fulfilled the mine's dewatering needs until new electric pumps were installed at the B&M Shaft in 1913.[18]

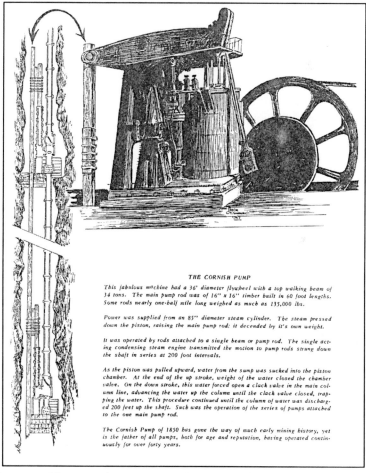

Fig. 15.15. The Cornish Pump. A smaller version of this pump was originally installed at the Old Abe Shaft to dewater the mine.—Illustration courtesy of the Black Hills Mining Museum.

Fig. 15.16. The B&M pump room on the 1,100-foot level.—Photo courtesy of Homestake Mining Company.

In 1913, with a large amount of hydroelectric power still available, Homestake replaced the Cornish Pump at the B&M Shaft with a series of three-stage centrifugal pumps directly connected to 125-horsepower Westinghouse motors. Each pair of pumps was located in underground pump rooms spaced 300 feet vertically. The lowest pump room was on the 1,400-foot level. Pump operation was automatic, based on floats placed in the pump sumps. The two pumps at each station were capable of operating singly or as a pair, depending on the water level in the sump. Capacity of each pump station was 1,000 gallons per minute against a vertical head of 315 feet.

The Pocahontas Mill and the Cyanide No. 1 and Cyanide No. 2 sand treatment plants were converted to hydroelectric power in 1914. These plants had been operating with electrical power purchased from the local power company since 1910. In 1916, the Slime Plant in Deadwood was switched from electricity supplied by the local utility to electricity supplied by Homestake's hydroelectric plants. The power was routed across an 11,000-volt powerline constructed from the Lead Substation to the Slime Plant.[19]

In 1917, the pumps at the Hanna Pump Station were converted to electric drives using hydroelectric power.[20] Previously, the Hanna Pump Station was powered by steam generated in a coal-fired boiler. Coal was brought in by rail across a spur line from the Burlington Railroad at Dumont.

Five new centrifugal pumps directly coupled to 2,200-volt motors were installed at Hanna. The pumps operated against a 400-foot head. Three of the pumps each had a capacity of 2,000 gallons per minute and were connected to 300-horsepower motors. One pump was connected to a 150-horsepower motor and had a capacity of 1,000 gallons per minute. The fifth pump was fitted with a 75-horsepower motor and had a capacity of 500 gallons per minute.

Hydroelectric power was originally supplied to the Hanna Pump Station at 11,000 volts through two circuits from the Englewood Hydroelectric Plant. The

Hanna Pump Station provided much of the water supply for the Homestake Mine and the city of Lead and is still used today to supplement water supplied by a gravity-fed system.[21]

Central Steam Plant

In 1914, all of the individual steam plants in Lead were replaced with a Central Steam Plant constructed just south of the Lead Substation (see fig. 15.17). Work was also started on a 4,000-kilovolt-ampere steam-turbine power station called the Auxiliary Steam-Electric Plant. The Central Steam Plant was equipped with six 600-horsepower boilers. Upon commissioning, the Central Steam Plant supplied the B&M and Ellison complexes with steam for the hoists, compressors, and building heat. The Central Steam Plant later became headquarters for Homestake's Water, Surface Construction, and Transportation groups.

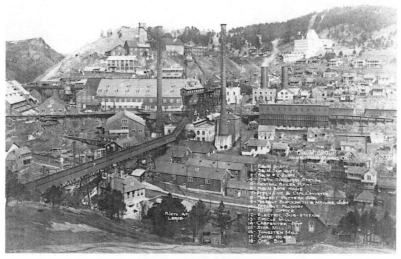

Fig. 15.17. The Central Steam (Boiler) Plant and other surface facilities of the B&M Group. The B&M facilities were later replaced by the new South Mill and Shop Yards buildings and facilities in 1934.—Photo courtesy of Homestake Mining Company.

In addition to the Central Steam Plant, work was completed on installation of a new steam-powered ore hoist at the B&M Shaft in 1915 (see figs. 15.18 and 15.19). The new hoist was designed to handle two 6-ton capacity skips in counterbalance from a depth of 4,000 feet. Ore was dumped from the skips directly into two No. 8 Gates crushers.[22] Steam was supplied to the Fraser and Chalmers cage hoist installed in 1879 and the new Nordberg ore hoist through a 12-inch-diameter steam line laid in a 7×8-foot Steam Tunnel. The 850-foot-long tunnel was driven from the Central Steam Plant to the bottom of a raise that was driven upward some 180 feet to the B&M hoisting works.[23] In 1919, the fifty-drill Ingersoll-Sergeant steam-powered compressor that had been installed in 1895 (see fig. 15.20) was replaced with two new air compressors fitted with electric drives at the B&M compressor room.[24]

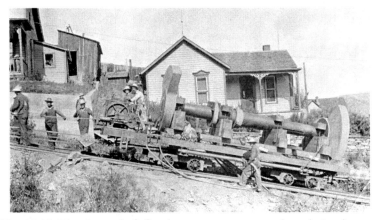

Fig. 15.18. The drum shaft for the new steam-powered ore hoist that was installed at the B&M Shaft in 1915. Note the winches mounted on the front of the railcar that were used to pull the load up Spring Street in Lead.—Photo courtesy of the Black Hills Mining Museum.

Fig. 15.19. The steam-powered ore hoist at the B&M Shaft, 1917. The four-cylinder ore hoist, manufactured by the Nordberg Company, was installed at the B&M No. 1 Shaft in 1914 and 1915. The hoist utilized ⅝-inch-by-7¾-inch flat hoist rope. Most of Homestake's flat hoist ropes were made and repaired at one of two shops; one was located near the Ellison Shaft and the other at the B&M Shaft. The hoist operated with two skips in balance. Each skip held 6 tons of ore.—Photo courtesy of the Phoebe Apperson Hearst Library.

Fig. 15.20. James Meddaugh photo of the Ingersoll-Sergeant five-drill steam compressor installed at the B&M Shaft in 1895.—Photo courtesy of the Black Hills Mining Museum.

On February 26, 1914, about the time the order was being placed for the new B&M ore hoist, Sydney J. Staple received a letter from William Lilly, a former Homestake mechanic who previously worked for John Spargo and was responsible for operation and maintenance of Homestake's hoists. While working for Homestake, Lilly became intrigued with the idea of designing some type of hoist control or safety device that could provide safeguards against overspeed of the hoists or overtravel of the skips and cages at the top or bottom of the shaft. Before he could develop the safety device, Lilly quit Homestake and went to work for the Anaconda Mining Company in Butte, Montana. There, he completed work on the device.

In his letter, Lilly informed Staple that he had developed a hoist regulator that would prevent a hoisting engineer from accidentally raising a cage into the sheave wheel or lowering it into the sump at the bottom of the shaft. In addition, Lilly said that the device could regulate the speed of the skip or cage. In an emergency, the device would automatically interrupt the power supply to the hoist. Lilly suggested installing the safety device on the new B&M ore hoist once the hoist was installed. Staple declined the proposal. By June, Lilly obtained a patent on the hoist regulator. He wrote another letter to Staple, Spargo, and T. J. Grier suggesting that the device be installed on all of Homestake's hoists. Again, the Homestake men turned down the proposal.[25]

In 1915, Henry J. Logan, a coworker with Lilly at Anaconda, improved Lilly's controller. Logan moved to Chicago in 1917 and began marketing the Lilly Hoist Safety Controller through a company called Duro Metal Products.[26] In July of the following year, Sydney Staple replaced John Spargo, who retired from Homestake. Almost immediately, Staple wrote to William Lilly requesting current information about the Lilly hoist controller. With approval from Superintendent Bruce C. Yates, arrangements were made with Henry Logan to test a Lilly controller on the B&M ore hoist, and one was installed a short time later.

On January 2, 1921, Burke Lovejoy was operating the B&M ore hoist when the hoist rope suddenly broke near the hoist room as a loaded skip was being raised. One end of the broken hoist rope flew violently backward, striking and killing Lovejoy instantly. An investigation into the accident revealed that the Lilly controller was not at fault and likely prevented the hoist from incurring serious damage. A Lilly controller was subsequently installed on the new Ellison cage hoist. When the Ross and Yates shafts were constructed in the 1930s, the ore and cage hoists at both shafts were also equipped with Lilly controllers (see fig. 15.21).[27]

Fig. 15.21. The Lilly Hoist Controller, similar to the one developed by William J. Lilly, in about 1914. The first controller at Homestake was installed on the B&M No. 1 ore hoist in 1918. The Lilly controller shown in this photo was installed on the Yates ore hoist.—Photo courtesy of Homestake Mining Company.

On January 1, 1935, Logan formed Logan Engineering Company in Chicago and continued to market several versions of the Lilly Hoist Controller. As business volume increased, Logan expanded the company and constructed a new 14,000-square-foot manufacturing plant, which he was able to pay for within one year.[28]

Auxiliary Steam-Electric Plant

The Auxiliary Steam-Electric Plant was commissioned in June 1916. The purpose of the plant was to generate supplemental electricity for the mine operation during periods when hydroelectric generation was lower because of reduced stream flows in Spearfish Creek. The plant was equipped with two 1,800-kilowatt Westinghouse-Parsons steam turbines (see fig. 15.22). Generating capacity of the plant was 3,000 kilovolt-amperes. The turbines were fed with 100-degree steam supplied by boilers at the Central Steam Plant. Condensate from the turbines was pumped to two cooling towers and returned to a cold well. Power was transmitted to the 2,300-volt bus in the Lead Substation through a tunnel that connected the two buildings.[29]

Fig. 15.22. The Westinghouse-Parsons steam turbines inside the Auxiliary Steam-Electric Plant, 1932. Located at the northeast corner of Washington and Spring streets in Lead, the Auxiliary Steam-Electric Plant operated between June 1916 and December 1935.—Photo courtesy of the Black Hills Mining Museum.

In 1931, an additional steam turbine with a capacity of 5,000 kilowatts was added to the plant.[30] After the Kirk Power Plant was commissioned in 1935, the Auxiliary Steam-Electric Plant was shut down on December 9. Between 1916 and 1935, the Steam-Electric Plant produced approximately 202.8 million kilowatt-hours of electricity.[31]

Hydroelectric Plant No. 2

During the latter part of 1916, work was started on the Hydroelectric Plant No. 2 in Spearfish Canyon. This plant, located at Maurice, was commissioned in January 1918 (see fig. 15.23). Water was collected in dams constructed on Spearfish Creek and Little Spearfish Creek just above Savoy. Water from the Spearfish Creek dam was diverted into a 48-inch-diameter redwood-stave pipe 4,800 feet long.

At Savoy, the 48-inch pipeline was joined by a 30-inch pipeline 1,650 feet long that originated at a dam on Little Spearfish Creek. The two pipes were merged into a single 52-inch-diameter redwood-stave pipeline that carried the water 19,000 feet to a surge tank and penstocks above the Hydroelectric No. 2 Plant. The pipeline was laid on a grade of 2¾ inches per 100 feet to facilitate gravity flow. The pipeline passed through two tunnels that were mined through the limestone bluffs to accommodate the grade of the pipeline. One of the tunnels was 1,300 feet long, and the other was 3,900 feet long. Long trestles were constructed across the mouths of Johnson Gulch, Calamity Gulch, and Long Valley to support the pipeline.

Fig. 15.23. Homestake's Hydroelectric Plant No. 2, 2008. The Hydro No. 2 Plant was commissioned in January 1918 and operated until October 2003.—Photo by author.

Water was delivered from the surge tank to the Hydroelectric Plant No. 2 through two penstocks 2,000 feet long. The vertical drop through the penstocks amounted to 548 feet. The generating equipment in the Hydro No. 2 Plant was nearly identical to the equipment in the Hydro No. 1 Plant. Power generated at the Hydro No. 2 Plant was wyed into the powerline that extended from the Hydro No. 1 Plant to the Lead Substation. The Hydro No. 2 Plant was shut down in October 2003 as part of the decommissioning and closure work for the Homestake Mine. The powerline was removed during the next two years.[32]

The Hydroelectric No. 2 Plant provided approximately 2,500 kilovolt-amperes of continuous power to the Homestake Mine from 1918 until 2003 when the plant was shut down. During this period, the Hydroelectric Plant No. 2 produced some 867 million kilowatt-hours of electricity.[33]

Following completion of the Hydroelectric No. 2 Plant, the Old Brig and Golden Gate hoists were converted from steam to electric power in 1918. In 1919 two new low-pressure compressors driven by 800-horsepower motors were installed at the Ellison to replace the steam-driven compressors. The compressors each had a capacity of 4,500 cubic feet per minute. A third compressor with a 750-horsepower motor was installed in 1919.[34]

A new 525-horsepower, electric-powered, high-pressure compressor (see fig. 15.24) was installed in 1922 to replace the old steam-powered high-pressure compressor. The high-pressure compressor delivered compressed air at 1,000 pounds per square inch to power the underground locomotives. The compressor featured a special regulator that allowed it to deliver low-pressure air whenever the high-pressure demand was satisfied. The low-pressure compressors supplied compressed air at 100 pounds per square inch to power the rock drills and other pneumatic tools and equipment.[35]

Fig. 15.24. Installation of an electric-powered, high-pressure Ingersoll-Rand compressor at the Ellison hoisting works, 1922. The compressor provided air at 1,000 pounds per square inch to the high-pressure pipelines and charging stations underground for use in the air-powered locomotives.—Photo courtesy of the Black Hills Mining Museum.

With abundant hydroelectric power now available, a new electric ore hoist (see fig. 15.25) was installed at the Ellison in 1921 to replace the steam hoist. The new hoist was powered by a 1,400-horsepower motor with a 1,200-horsepower motor generator set.[36] The hoist was capable of raising a skip with a 7-ton payload at a maximum speed of 2,000 feet per minute from the 3,200-foot-level loading pocket. Six gyratory crushers were also installed with 35-horsepower, 2,200-volt motors.[37]

Fig. 15.25. The 1,400-horsepower, two-drum electric ore hoist at the Ellison Hoist Room, 1932. The hoist was installed in 1921 as part of Homestake's conversion from steam power to electrical power.—Photo courtesy of Black Hills Mining Museum.

Concurrent with the conversions to electricity, Homestake began converting most of its boilers to coal to reduce the demand for costly cordwood. The Central Steam Plant and the Auxiliary Steam-Electric Plant alone required large amounts of coal. In order to meet the ever-increasing demand for coal, Homestake purchased the Wyodak coal property near Gillette, Wyoming, in 1921 and formed the Wyodak Coal and Manufacturing Company as a wholly owned subsidiary of Homestake Mining Company (see fig. 15.26).[38] Homestake completed construction of a new coal-fired power plant and tipple at its Wyodak Coal Mine in Gillette in October 1928. A 1,000-kilowatt generator was added in 1929.[39]

Fig. 15.26. The Wyodak Mine near Gillette, Wyoming, 1928.—Photo courtesy of the Black Hills Mining Museum.

In 1921, the 7×14-foot No. 1 Air Raise was completed between the 900-foot level and the surface. The raise was located east of the assay office near the west end of Park Avenue.[40] With electricity and electric-powered equipment now available, the first surface ventilation fan used at the mine was installed over the raise in 1923. The 300-horsepower 2,200-volt motor was capable of moving 225,000 cubic feet of air against $4\frac{1}{3}$ inches of water gauge pressure. With proper positioning of air doors, the fan could be operated to draw exhaust air out of the mine or draw fresh air into the mine.[41] In 1933, No. 1 Air Raise was abandoned because of ground problems, and No. 2 Air Raise was commissioned as a replacement airway. No. 2 Air Raise was located near the west end of Sand Street and was used to ventilate areas above the 1,700-foot level.[42]

By 1925, the three hydroelectric plants and the auxiliary steam-electric plants were supplying all the mine's electrical needs. A total of 44.84 million kilowatt-hours of electricity were generated that year. Of this total, 88.4 percent was generated at the hydroelectric plants, and 11.6 percent was generated at the auxiliary steam plant.[43]

Communications Systems

Homestake installed a 100-drop, magneto-telephone switchboard at the mine operation in 1906 (see fig. 15.27). Within a few years, a second 100-drop section was added to the main switchboard. Most all of the surface buildings, including the mills, shops, hoist rooms, Black Hills and Fort Pierre Railroad stations, superintendent's residence, and foreman's residences were furnished with a magneto phone. The Hanna Pump Station and each of the hydroelectric plants also had small switchboards with phone lines to the operators' residences located nearby. Each underground level of the mine had at least one telephone. By about 1910, an employee could call out to anywhere among the surface and underground facilities at the Homestake Mine. The magneto system was used in conjunction with more modern communications systems until the mid-1980s.[44]

Fig. 15.27. Homestake's first 100-drop magneto switchboard installed in 1906. A second 100-drop section was added a few years later to meet the mine's communications demands. Although each of the phone lines on the property was individually connected to the switchboard, the lines had to be properly "switched" on a twenty-four-hour basis by a crew of switchboard operators.—Photo courtesy of the Black Hills Mining Museum.

Prior to 1916, cage tenders and hoist operators communicated with each other using mechanically operated bells that were activated by hand using steel rods or cables suspended in the shaft. Unfortunately, the mechanical system failed to work properly once the shafts reached a depth of about 1,000 feet. On an experimental basis, an electric signaling system was installed in the skip compartments of the B&M Shaft. This system consisted of a pair of wires that extended from the bells in the hoist room to pull boxes located at each loading pocket in the shaft. Based on the success of this system, a similar system was installed in the cage compartment of the shaft. Similar systems were installed in the Ellison Shaft when the electric hoist was installed in 1921. The Ellison systems included a separate wire to each pull box, which allowed the hoist engineer to determine which level the signal was transmitted from.

In the 1930s, Homestake began investigating the possibilities for using radio transmission for shaft communications between the cage tender and the hoist operator. Initial efforts failed because of the enclosed underground environment that prevented transmission of the radio signals. A short time later, a plan was devised that utilized the hoist rope to transmit a high-frequency electric current from the cage to the hoist room. This system worked well, even from a moving cage.

When Homestake applied for a patent on the carrier-current system using the hoist rope, it discovered that a patent had already been issued to Edwin H. Colpitts on April 26, 1921, for a signaling system substantially identical to the one Homestake had just developed. However, a patent search revealed that the patent had expired, and Homestake was able to continue using its system. By the time the Yates and Ross shafts were commissioned, the shaft signaling system had advanced to two-way voice communication between the cage tender and hoist operator.[45]

Kirk Power Plant

After the South Mill complex and the Ross complex were commissioned in 1934, Homestake once again found it needed additional power. The company was gradually moving away from the B&M complex because of subsidence issues, and the Auxiliary Steam-Electric Plant was in need of an upgrade. As a result, Homestake constructed the Kirk Power Plant in 1934-1935 at a cost of $1,750,000 (see fig. 15.28). As part of the construction, Homestake rerouted and relocated its hydroelectric power, primary substation, and switching facilities to the Kirk Plant. The coal-fired steam turbines in this plant had a capacity of 12,000 kilowatts. The Auxiliary Steam-Electric Plant was subsequently decommissioned.[46]

Fig. 15.28. The Kirk Power Plant, 1935. Homestake completed construction of the Kirk Power Plant in 1935 to meet the electrical demands of the new South Mill and Ross Shaft.—Photo courtesy of the Black Hills Mining Museum.

Coal was supplied to the Kirk Plant at the rate of 60,000 tons per year from the company's Wyodak Mine at Gillette. Under Homestake's ownership, coal production reached a high of 120,502 tons in 1935. Of this amount, 70,072 tons were shipped to the Homestake Mine, 42,030 tons were sold commercially, and the balance was consumed as fuel at the Wyodak Mine.[47]

On October 16, 1954, Homestake transferred ownership and operation of the Kirk Power Plant to Black Hills Power and Light Company. Between 1935 and 1954, the Kirk Power Plant produced 640 million kilowatt-hours of electricity for the benefit of Homestake.[48] After reassessing its business needs, Homestake also sold the Wyodak Mine to Wyodak Development Resources Corporation, a subsidiary of Black Hills Power and Light on November 1, 1956.[49]

In the 1980s and 1990s, Homestake's electrical demand increased significantly for several reasons. The introduction of diesel-powered equipment in the mid-1970s and subsequent additions in the 1980s added significantly to the mine's ventilation requirement. Power costs increased proportionately with the depth of hoisting and pumping. The cost of ventilation and cooling increased dramatically simply because of the breadth and depth of the mine. Measurements showed that the geothermal gradient of the earth's crust was approximately 1.2 degrees Fahrenheit per 100 vertical feet in the Homestake Mine. Extrapolated, the natural rock temperature was observed to be approximately 133 degrees on the 8,000-foot level.

To improve working conditions, a 300-ton-capacity chilled-water system was installed on the 5,000-foot level. This system supplied 42-degree water to work areas in the west ledges below the 5,000-foot level. A new 3,000-horsepower centrifugal fan was installed on the Oro Hondo Shaft in the 1980s to expand the ventilation capacity in the lower part of the mine as more and more diesel-powered equipment was purchased. Additional 30-ton and 60-ton portable refrigeration units were purchased to cool exploration drifts and stopes in the lower reaches of the mine.

Numerous 15-, 30-, and 40-horsepower fans were also purchased to blow fresh air into stopes and drifts. Smaller chilled-water systems were installed on the 7,400- and 8,000-foot levels to provide cooling in the more remote areas. A 2,300-ton refrigeration plant was commissioned on the 6,950-foot level in 1988 to provide bulk air cooling in the west-ledge areas between the 6,950- and 8,000-foot levels. Additional refrigeration required additional service water. Soon, the mine's dewatering system reached its maximum of 2,300 gallons per minute, which added to the strain on the electrical and cooling systems.

At about the same time the ventilation and cooling upgrades were being made, pneumatic-powered diamond drills were replaced with electric-hydraulic drills. Additionally, pneumatic face-jumbo drills were replaced with electric-hydraulic jumbos, pneumatic-powered longhole drills were replaced with electric drills, and electric roof bolters were purchased to improve safety and efficiency of miners performing rock-bolting work. On the surface, a regrind system was added at the mill. A water treatment plant was commissioned in 1984 to meet the more stringent environmental standards. Additional water from mining and milling dictated that more water had to be pumped to and from the Grizzly Gulch Tailing Storage Facility. All of these improvements added significantly to the demand for electricity.

Almost simultaneously with the significant increase in Homestake's electrical demand, Black Hills Power and Light announced its intention to close the costly Kirk Power Plant. Recognizing that Homestake needed more electricity not less, Black Hills Power offered to sell Homestake primary power at a reduced price if Homestake would upgrade and expand its electrical infrastructure to accept electricity at 69 kilovolts instead of 12 kilovolts. As a result, Black Hills Power constructed a 69-kilovolt substation behind the Kirk Power Plant and Homestake upgraded and expanded its Ross and Oro Hondo substations to accept primary power at 69 kilovolts.

As part of the electrical upgrade, Homestake constructed the East Substation in 1994, which significantly increased the mine's electrical capacity. Upgrades were also completed at Homestake's Ross and Oro Hondo substations to facilitate primary power at 69 kilovolts. By 1999, switching facilities were subsequently phased out at the Kirk Plant and relocated elsewhere. Concurrently, Homestake's hydroelectric power was rerouted to the Ross Substation.

After most of the ventilation, cooling, pumping, and mining equipment upgrades had been completed, the mine's electrical demand increased to record highs in the early 1990s. In 1992 alone, some 237.7 million kilowatt-hours of electricity were consumed at the Homestake Mine. Of this amount, 210.1 million kilowatt-hours were purchased at a cost of $10.41 million. The remainder was generated at minimal cost by the three hydroelectric plants and two turbines located on the 2,450- and 5,000-foot levels in the Ross Shaft.[50]

In 1998, a decision was made to scale back ore production at the mine in response to a continuing decline in gold prices and the volume of higher quality ore reserves. The demand for purchased power was substantially reduced as a result of lower production and milling requirements. Fortunately, Homestake's three hydroelectric plants continued to supply much of the power requirement for mining and milling operations. Under Homestake's ownership, the hydroelectric plants generated some 2.63 billion kilowatt-hours of electricity for the Homestake Mine.[51]

Chapter 16
The Lockout of 1909-1910

Lead Miners' Union

The Lead Miners' Union was organized in October 1877 "for the mutual protection and for the purpose of securing for the men engaged in the hazardous occupation of mining, for wages, a just compensation for their labors, and the right to use the fruits of their toil, without let or hindrance, or dictation from their employers, and to otherwise protect their mutual interests."[1] Principal organizers included Thomas Tully, Charles Lyford, and A. J. White. Tully was elected first president, White and Charles Reed, vice presidents, and Lyford, secretary.[2]

The first Lead City Miners' Union Hall, a two-story framed structure, was completed in 1879 at a cost of $6,000. Located on the northwest corner of Main and Bleeker streets, the building was dedicated on March 10, 1880. During the same year, the Lead Miners' Union was incorporated as a benevolent and charitable organization, and a charter was issued by the territory of Dakota. The lower floor of the building was used for business purposes while the upper floor was used as a meeting place and opera house. In December 1894, a new Miners' Union Hall was completed at the southwest corner of Wall and Main streets. The lower two floors of the four-story sandstone structure were devoted to businesses. The third floor housed the Lead Opera House. The fourth floor was divided into lodge rooms, one of which was used for union business.[3]

By the 1890s, the Lead Miners' Union had established a sickness benefit of $6 per week for its members. The sickness benefit was increased to $10 per week by 1909. The union also established a funeral benefit of $75. Union members were charged dues of $1 per month. After it was decided to construct the new Union Hall, an additional assessment of $5 per month was made until each employee had contributed a total of $50. At this point, a $50 bond yielding 8 percent interest was issued to the employee. The bonds were used to pay off the new building.[4]

Around 1891, the Lead Miners' Union tried to convince Superintendent Thomas J. Grier to agree to a closed union shop for the Homestake Mine. Under a closed shop, all employees would be required to belong to the union. Grier refused, claiming that a closed shop would be unfair to those employees who did not wish to belong to the union. Furthermore, it would mean the union would have a voice in the management of the company.

In 1893, the Lead Miners' Union participated in the formation of the Western Federation of Miners. The Butte Miners' Union was listed as WFM No. 1, the Lead Miners' Union was No. 2, and the Central City Miners' Union was No. 3. Members of the Lead Miners' Union, which was also called Lead Miners' Union No. 2 because of its affiliation with the Western Federation of Miners, held positions within the Western Federation at various times. Some of these people included James Kirwan, Richard Bunny, Thomas Ryan, and J. C. McLemore.[5] The town of Terry organized its miners in August 1891. By 1899, approximately, 650 workers were members of the Terry Miners' Union.[6]

The Unions Pressure for an Eight-hour Day and a Closed Shop

Toward the end of 1906, the Lead Miners' Union and the Central City Miner's Union started putting pressure on the mining companies for an eight-hour workday. Miners were working ten-hour shifts, seven days per week. The seven-day week was not much of an issue, but the ten-hour day was. At first, Superintendent Grier refused to consider the request. About the time the union was preparing to send a written request to the Homestake Board of Directors, Grier agreed to the request, and the eight-hour day became effective on January 1, 1907. The union considered this a great victory, particularly since the miners did not have to take a wage cut with the shorter workday.[7]

Eighteen months later, the Terry Miners' Union and mining companies in the Ruby Basin and Bald Mountain Mining districts followed suit with an eight-hour workday. To celebrate the change, the town of Terry held a large celebration at the Miners' Union Hall on June 6, 1908. Some five hundred miners and their families attended. Music was furnished by the Ancient Order of United Workmen's (AOUW) band of Terry. Boxing matches were also held.[8]

In February 1909, William E. Tracy of the Western Federation of Miners made a trip to Lead. He found the Lead Miners' Union No. 2 was growing rapidly and was amazingly free of debt. Tracy returned to Lead in September to conduct an organizing campaign in the other mining camps of the northern Black Hills. He also worked hard to increase membership in the Lead and Central City unions. Yanco Terzich agreed to work to organize more of the Slavonians. Mr. Lowney agreed to work with the Irish, and Mr. Davidson was in charge of organizing the Cornish. The organization effort began to gain momentum.

On October 11, 1909, Emma F. Langdon, representing the *Miners' Magazine*, attended a union meeting and demanded that all former members make up their arrears and become active again. She emphatically told the membership, "Anyone who ignores this will be dealt with as the union shall determine."[9] This resulted in adoption of the following resolutions by the membership:

WHEREAS, owing to the efforts of the Lead City Miners' Union No. 2, W. F. M., acting in conjunction with Central City Miners' Union No. 3, W. F. M., certain improvements and betterment of conditions obtained to all men employed in and around the various mines, mills, and surface work within the jurisdiction of the said unions, among which may be noted the establishment and maintenance of the 8-hour work day at the same scale of wages which formerly prevailed when ten or 12 hours' labor was required from such employee for a shift's work. And,

WHEREAS, the policy of these unions has ever been lenient in the extreme, inviting all honorable miners, millmen, mechanics, and other laborers connected with the mining and milling industry to join and alliliate with them in their respective unions, for the purpose of maintaining the same in a position of strength and efficiency and in order that all might be entitled to full and equal participation in all benefits and benefactions accruing from membership in such union. And,

WHEREAS, it now appears that on account of such liberal policy on our part there are at this time a large number of men employed within our jurisdiction who have neglected to unite with our organization and many who have fallen in arrears to the extent of severing their connection with our union and many others who are vowed enemies of unionism, scabs and spies and strike-breakers, among whom and with whom we cannot as self-respecting union men long continue to labor or associate with. Therefore be it

RESOLVED, That we demand of all ex-members who are now in arrears that they shall at once re-instate and place themselves in good standing in our union, and that we demand and require all eligible men who are employed within our jurisdiction to at once obtain cards certifying their membership in the proper local of the Western Federation of Miners.

RESOLVED, That any and all men within our jurisdiction who shall see fit to neglect or to ignore these just demands and requirements shall be dealt with in the near future accordingly, as we may determine to act in each of their respective cases. And be it further

RESOLVED, That a full and sufficient number of these resolutions be ordered printed and posted in conspicuous public places, and be distributed thoroughly throughout our jurisdiction, and that such public notices be printed in several different languages, viz: Croation, Finnish, Italian and English.[10]

Grier's Lockout Precludes a Strike

By October 24, the Lead Miners' Union No. 2 claimed that 98 percent of the eligible employees of Homestake were now union members and that there were only six nonunion members working at the mine. Absent Grier's involvement or stated objection, the union had successfully increased its membership by about one thousand in just two months. The union's goal was to achieve a closed shop with or without the involvement of company management.[11]

At a union meeting held on Sunday afternoon, October 24, the members passed additional resolutions as follows:

> *WHEREAS, A resolution adopted on Oct. 10, calling upon all workers in the jurisdiction to join the W. F. of M., has been quite generally complied with, therefore, be it*
>
> *RESOLVED, By us, the members of the Lead City Miners' Union No. 2, W. F. of M., and Central City Miners' Union No. 3, W. F. of M., in joint session assembled, that all men neglecting or refusing to become members in good standing of the local in whose jurisdiction they may be working on or before November 25, 1909, will be declared unfair to the W. F. of M. And be it further,*
>
> *RESOLVED, That we, the members of the aforesaid unions, refuse to work with any and all men who become unfair to our organization by or through refusing to comply with the provisions of this resolution.*[12]

The following day, the union asked Grier for a list of Homestake employees, but Grier denied the request. The union proceeded to publish the resolutions in the local newspapers.

On November 10, Homestake filed a complaint against the Lead City Miners' Union in the Eighth Judicial Circuit Court of South Dakota. Homestake claimed damages in the amount of $10,000 as a result of the union's actions that "greatly impaired the efficiency and value of the services and labors of numerous of plaintiff's employees in various departments." The union, in turn, filed a demurrer claiming that Homestake's complaint did not "state or set forth facts sufficient to constitute a cause of action against defendant" and that the court did not have "jurisdiction of the alleged cause of action set forth" in said complaint.[13]

The *Lantern* called Homestake's suit a bluff and attacked Grier as a hypocrite who posed as a friend of labor in order to destroy it. On November 17, 1909, Grier placed a notice in the *Lead Daily Call* notifying employees and the general public that the company would only be hiring nonunion men after January 1, 1910.[14]

> *Notice is hereby given that the Homestake Mining Company will employ only non-union men after January 1st, 1910. The present scale of wages and the eight (8) hour shift will be maintained. All employees who desire to remain in the company's service must register at the general office of the company on or before December 15, 1909.*[15]

The notice was a shock to the union and the community. John Mayo, owner of a local saloon, called a mass meeting that evening at the union hall. A resolution was passed that condemned the actions of the company. The membership decided to telegraph copies of its resolution to Phoebe Hearst, William Randolph Hearst, James Haggin, and the board of directors of Homestake. Grier, who was present at the union meeting, stated, "Men come to see me and ask if they must join at the end of a rope." Grier stated that union men were entering company property without permission and threatening workers. After determining that the boisterous meeting was going nowhere, Grier left as William Tracy was attempting to quiet the miners and restore order to the meeting.

On November 21, the union met with Superintendent Grier, Chambers Kellar, company attorney, and Richard Blackstone, assistant superintendent. The union representatives asked the company officials to withdraw the nonunion order. Grier expressed sorrow over the ordeal but refused to withdraw his notice. Tracy then called a union meeting on November 23 to decide whether to call for a strike. Unfortunately, such an action required a three-fourths majority vote of the membership, and only one-third of the members were able to attend the hastily called meeting. However, the membership was advised that with their permission, the district union officer James Kirwan could call a strike himself. The membership present consented to the arrangement.

The following day, before Kirwan had acted to do anything, Superintendent Grier published a notice in the newspapers that said, "The Homestake Mining Company will cease operating its property this evening." That evening, at 6:00 p.m., the 1,000 stamps at the mills in Lead, Golden Gate, and Terraville were shut down. One resident recounted, "You could even hear the dogs bark." For the two-week period preceding the shutdown, Homestake had been preparing for either a strike or a lockout. Powder stored in the mine was moved to an upper level. Pumps were shut down, water walls were constructed, and general cleanups were made in the mills.[16]

Grier believed he had no other option but to order a lockout. The nonunion hourly and salaried workers only amounted to about five hundred people. Grier later said it would not be profitable to operate at a significantly reduced capacity. Almost immediately, the Lead miners moved to places such as Globe, Arizona, to find mining or other work. The first to move out were the single men. The miners with families followed close behind.

In order to protect its property and nonunion employees, Homestake brought in Pinkerton detectives from Denver. The sheriff of Lawrence County swore most of these people in as deputies. As deputy sheriffs, the detectives had the authority to arrest anyone found breaking the law. A large spotlight was mounted on the Ellison headframe and was used to help patrol the surface works at night. On December 9, when the union demanded that the nonunion watchmen and maintenance workers stop working, Homestake replaced these people with additional detectives.[17] At the peak of the lockout, more than one hundred special detectives were working under the direction of G. A. Northam, Homestake's head of security.[18]

The *Deadwood Daily Pioneer-Times* reported that the International Detective Agency had sixty men guarding company property in addition to the Pinkerton detectives. The *Deadwood Daily Pioneer-Times* estimated that Homestake had probably spent more than $100,000 for the services of the detective agencies over a six-month period.[19]

A large percentage of union members, particularly the Slavonians and Finnish, were staunch union supporters. The Finnish Temperance Society expelled at least one member for working for the company during the lockout. The Slavonians exerted pressure on any of their wavering peers by writing letters to the person's relatives living in the homeland.

In December, other unions began sending cash to help replenish the strike benefit fund of the Lead Miners' Union. The Butte Miners' Union No. 1 sent a check for $25,000. The Western Federation of Miners sent a total of $228,832.25 throughout the lockout until 1912, when the cash contributions were finally discontinued. Barbers, bartenders, musicians, teamsters, carpenters, and electrical workers, who were members of other local unions, also contributed to the strike fund of the Lead Miners' Union. Many of these unions believed that if the Western Federation of Miners was defeated, all unionism in the Black Hills would soon come to an end.

Even the local newspapers took sides. The *Lead Daily Call* and *Deadwood Daily Pioneer-Times* supported the position of the company with their articles. The *Black Hills Daily Register* and the *Lantern* supported the union's position. In late December, Homestake provided area newspapers with copies of new employment cards that applicants were expected to sign if they wanted a job with Homestake. The card read as follows: "I am not a member of any labor union and in consideration of my being employed by the Homestake Mining Company agree that I will not become such while in its service."[20]

In January 1910, thirteen other mining companies in the northern Black Hills joined Homestake and shut down their operations. Like Homestake, these companies also announced they would not open until they had a sufficient number of nonunion workers. The coalition of mining companies began running the following notice in local newspapers:

> In view of the fact that the Mining Industry in the Black Hills District is the source from which all the other business interests in said District derive their main support, and that, said industry intends to establish, permanently, in said District what are commonly called Non-Union labor conditions, it is respectfully suggested to all such other business interests that their actions should be vigorously in support of the aforesaid expressed intention.[21]

The strike had a detrimental effect on local businesses. Numerous businesses started advertising going-out-of-business sales. Some fifty-three businesses, owned mostly by Slavonian and Finnish families, formed a coalition in support of their union-member customers.

Although there were minor skirmishes in the saloons and one incident where a union man named Jack Butler was arrested for throwing an incendiary bomb into a cyanide mill at Whitetail, no major occurrences of violence or destruction of property were reported throughout the lockout. William Tracy, district representative for the Western Federation of Miners, proclaimed, "The Union could have destroyed the Homestake works, but didn't."

One episode involved Homestake's chief counsel, Chambers Kellar. Kellar was furious over some of the newspaper articles written by Freeman Knowles in the *Lantern*. One day, Kellar waited for Knowles to arrive at the Lawrence County Commissioners Room at the county courthouse in Deadwood. When Knowles arrived, Kellar pulled out a small riding whip and proceeded to give Knowles a few lashes. In an act of self-defense, Knowles punched Kellar in the face, giving him a black eye. The state's attorney, who was present, watched the ordeal with great amusement and did nothing to stop the fracas. Knowles made no attempt to press charges against Kellar, but instead took the opportunity to continue writing articles about the "chief pugilist of the Homestake Mining Company."[22]

The Return to Work with Nonunion Employees

Bruce C. Yates, A. J. M. Ross, and Richard Blackstone were sent to Kansas, Colorado, Wisconsin, North Carolina, Tennessee, Georgia, Michigan, and other states to try and find new workers from those states. While Yates was busy hiring "scab labor" in North Carolina and Tennessee, he became aware that a warrant was out for his arrest since state law prohibited what he was doing. With haste, Yates hired a rowboat, crossed the Tennessee River, and arrived safely in Georgia at 1:00 a.m. the next day.

On January 6, 1910, forty-nine men met in the Lead Society Hall and formed the Loyal Legion in support of the nonunion effort of the mining companies. George D. McClellan was elected president, W. J. McMakin, vice president, Will Treweek, secretary, and William Royce, treasurer. Members of the Loyal Legion approached Superintendent Grier and requested permission to go back to work. Grier informed the men he intended to begin reopening the mine immediately with nonunion workers. Such work was initiated on January 9, 1910, using a small crew of nonunion men.[23]

On January 10, Grier officially announced that the mine would reopen. The same day, several machinists were called to work to ready the hoists and other equipment. Other people were sent to bring the horses back into town so they could be lowered into the mine. Motormen were assigned to begin staging supplies using the compressed-air locomotives. Freeman Knowles reported the workers to be "cripples, dead-beats, pimps, preachers, doctors, lawyers, and school children" in the January 13 issue of the *Lantern*.

A work crew was lowered at the Star Shaft on January 13 to start the dewatering pumps. By mid-January, a large number of people had arrived in Lead to begin work alongside the growing number of nonunion workers. Membership in the Loyal Legion increased to five hundred members. On January 19, crews at the Ellison Shaft began hoisting ore that had been stored in underground raises and bins. Crews at the Amicus Mill had its 240 stamps crushing ore on January 21. By February 1, ore was being hoisted at the Golden Star, Old Abe, and Ellison shafts; and all 640 stamps were in operation at the Homestake, Golden Star, and Amicus mills. The 360 stamps at the Mineral Point, Pocahontas, and Monroe mills were restarted by March 3. In a short time, the Homestake Mine was back to full production.[24]

Superintendent Grier estimated that he put approximately one thousand ex-union members back to work. The Golden Reward and Mogul Mining companies also resumed operations with nonunion workers. On January 13, 1913, Charles Moyer cancelled the charter for the Lead Miners' Union No. 2 for nonpayment of dues to the Western Federation of Miners. By 1914, all of the miners' unions in the Black Hills were inactive.[25]

Subsequent Efforts to Organize the Workforce

In the absence of a written contract between the unions and the companies, and without the federal labor laws of today, there was no legal recourse for the unions through negotiation, mediation, or the right of states to allow a closed union shop. The Taft-Hartley Act passed by Congress in 1947 changed all of this.

In June 1947, organizers for District 50 of the United Mine Workers petitioned for an election at the Homestake Mine. The vote was 208 in favor and 938 against having the United Mine Workers as the bargaining representative for the hourly workforce. On March 9, 1951, an election was held to determine whether the workforce wanted to be represented by the United Steelworkers of America, AFL-CIO. The workforce voted against representation with a vote of 901-587. Additional attempts by the Steelworkers to organize Homestake workers failed on June 17, 1953, March 8, 1955, and March 27, 1957.

On June 6, 1966, by a vote of 841 for unionization and 512 against, the United Steelworkers of America became the bargaining representative for the hourly workers at the Homestake Mine. Ray Povandra was elected first president of the Local 7044. Linus Wampler, district representative for the Steelworkers, moved to Lead and established a union hall on Main Street. In November, the Steelworkers and Homestake reached agreement on the first three-year labor agreement.[26]

The Steelworkers continued as bargaining representative for the Homestake hourly workforce until June 2003, when decommissioning, cleanup, and closure activities were completed pursuant to the company's announcement that the mine would close on December 31, 2001. Labor relations were generally amiable between the Steelworkers and Homestake management over this thirty-seven-year period, although the union conducted labor strikes in 1972 and 1982.

Chapter 17
Wages, Benefits, and Community Support

The Early Workforce

Shortly after Congress passed the Homestead Act of 1862, Dakota Territory rapidly became known as a frontier area where one could obtain 160 acres of land for a very nominal fee, build a home, and raise a family. The long Depression that began in 1873 served as further impetus for many Americans to pursue a new way of life in Dakota. Concurrently, wars, famines, poverty, and religious differences created social disruption throughout many of the European countries. As a result, Americans and Europeans alike flocked to Dakota Territory and later to the Black Hills to partake in the American Dream.[1]

By 1879, as the Homestake Mine and all of its affiliated companies began to establish a major presence in the northern Black Hills, the workforce began to naturally transition. Many of Homestake's first employees were the prospectors, opportunists, and thrill seekers who constituted the local labor pool. Gradually, these types of people moved on, hoping to strike it rich elsewhere. In their place came the immigrants, who made their way to the Black Hills in pursuit of good-paying jobs.

Large developing mines such as the Homestake were a perfect match for the skilled and experienced miners from Cornwall, a county in England. The presence of the Cousin Jacks, as they were called, was indicative that the mining camp of Lead City was more than a flash in the pan. Homestake welcomed the Cornish miners, since they brought with them a century's worth of mining knowledge and expertise. The Cousin Jacks and Cousin Jennies also brought with them the Cornish pastie, which became a favorite personal meal of miners and townsfolk alike. The Cornish pastie consisted of meat, potatoes, and onions wrapped in pastry dough, which was baked in an oven.

The other English-speaking people, such as the English, Scottish, Welsh, and Irish, were drawn to the Homestake Mine for the same reasons as the Cornish. Because most all of the immigrants with direct or indirect ties to Great Britain spoke English, they assimilated into the workforce and community with relative ease.[2] By 1880, Lead's population was 1,440. Some 62 percent of the population consisted of English, Irish, Scottish, and Canadian people. Many of these people took up residence in South Lead in the Bleeker, Mill, Sawyer, Hill, High, and Barclay Street areas.[3]

A second wave of immigrants made their way to the northern Black Hills in the 1890s. These people included Swedes, Norwegians, Finns, Italians, and Slavonians. The Italians settled in the Sunnyhill addition to Lead and on Miners and Railroad avenues. The Finns settled in the Hiawatha Park addition of Lead on such streets as Park Avenue and Parkdale. Some of the Finns resided on North Bleeker and Pine streets. The Finns, Italians, and Slavonians were generally strong supporters of the Western Federation of Miners.

The Slavonians were comprised of Austrians, Serbians, Dalmatians, Croatians, Lithuanians, and Montenegrans who lived in Slavonian Alley or the Gwinn Avenue area. Other Slavonians lived nearby on Columbus, Wall, Siever, Stone, Spark, and Addie streets.[4] In 1896, a reporter for the *Lead Evening Call* concluded that "Lead is undoubtedly the most cosmopolitan city of its size in the west or any other part of America. Nearly every nationality on the globe is represented here."[5]

For the most part, the immigrants who spoke little or no English were at a severe disadvantage in being able to perform their jobs or adapt to the community simply because of language barriers. Some immigrants found it difficult to adapt to or accept the local customs and works of local governments. In the mine, some of the employees were fortunate to have a partner of the same nationality. In most cases, however, an employee was assigned work by his shift boss or foreman based on the employee's skill level and specific job classification, which meant he spent much of his shift working with people he could not converse with. After shift, each employee would return home to a neighborhood shared by his fellow countrymen, purchase his groceries at a neighborhood store owned and operated by a person who spoke the same language, and patronize the saloons and restaurants that catered to people of the same nationality.

Prior to 1904, most employees were under the exclusive control of their immediate supervisor. In the shops and mills, the employee reported directly to a foreman. Underground, the employee reported to a shift boss, who in turn reported to a foreman. The surface foremen and underground shift bosses were highly empowered individuals who exerted great authority over the employees who reported to them. The shift boss and surface foreman had the authority to issue an unpaid suspension in the form of days off to a member of his crew for just about any reason. Similarly, they had the authority to fire an employee for little or no reason and could hire replacement workers however they decided. Employees were free to appeal to the foreman or superintendent, but few were brave enough to do so.[6]

Homestake's employment practices changed dramatically in 1904 with the creation of a formal employment department that fell under the management of E. F. Irwin. Shift bosses and surface foremen directed their requests for new hires to the employment office. Shift bosses technically did not have any say in which employees were actually hired, but did have the authority to fire an employee for cause after a probationary period. Applicants had to be sixteen years of age to be employed on the surface. For underground work, employees had to be eighteen years old. Employment preference was given to those applicants who were citizens of the United States, married, owned a house, were affiliated with some religion, and had their family in the area.

Notwithstanding, in June 1914, the total workforce consisted of 2,109 employees of which 1,214 were American-born citizens. The underground complement consisted of 653 American-born citizens, 120 Englishmen, 112 Slavonians, 93 Italians, 92 Finns, and 216 people from other nationalities. As of that date, 114 of the employees had 21 years of service or more with Homestake, which made them eligible to be members of the Homestake Veterans Association.[7]

Wages

For about the first forty years of operation, the daily wage rates for Homestake employees remained fixed. There were several reasons for this: first, the local cost of living was generally not affected by the high inflation that was plaguing much of the rest of the United States. Second, when the union's request for an eight-hour workday was granted and became effective on January 1, 1907, no reduction in the daily wage was made. Prior to this date, employees had been working ten-hour shifts in the mine and shops and twelve-hour shifts in the mills. Third, in 1912 the company started paying a 7 percent annual bonus to every employee on the payroll as of December 31.[8]

The wage scale in place at the Homestake Mine from 1877 to 1917 is shown in table 17.1:

Table 17.1 Wage scale at the Homestake Mine (1877-1917)

Miner	$3.50/shift
Car Man/Shoveler	$3/shift
Timberman	$4/shift
Shift Boss	$4.50/shift
Foreman	to $7/shift
Head Amalgamator	$5/shift
Amalgamator	$3.50-$4/shift
Screen Cleaning Mill Laborer	$1.50/shift
Plate Cleaning Mill Laborer	$2.50/shift
Foundry	$2.50-$5/shift

Machine Shop	$3-$5/shift
Chief Metallurgist	$600/mo.
Assistant Chief Engineer	$350/mo.
Chief Electrician	$240/mo.
Chief Surgeon	$500/mo.
Mine Superintendent	$2,500/mo.
Director of Kindergarten	$100/mo.
First Assistant to Director of Kindergarten	$75/mo.

Source: U.S. Commission on Industrial Relations, *Proceedings of Public Hearing Held at Lead, South Dakota, August 3-4, 1914* (Kansas City, Missouri: Shorthand Reporting Company, 1914).

Stope miners were not paid any type of bonus incentive during the first half century or more at Homestake. Superintendent Thomas J. Grier experimented with a bonus plan for miners in the mid-1890s but quickly abandoned the idea because he believed it caused the miners to work unsafely. However, some twenty miners assigned to drift development were paid on a contract basis and received pay for actual work completed in lieu of any "day's pay" wage. A drift crew consisted of four men, two per shift. Under the drift contract, the crew was responsible for purchasing the necessary amount of powder, caps, fuse, and carbide. The drift miners were compensated for their work on a unit price basis, per lineal foot completed. The compensation averaged $4.59 per man shift in 1914.

In 1914, the total payroll for approximately twenty-two hundred employees was $2.7 million. Some 1,419 of these employees were assigned to underground work. Of the underground employees, there were approximately 535 miners and 743 carmen and shovelers. Shovelers were expected to muck (shovel) anywhere from five to twenty-one cars per shift, depending on the rock pile and the judgment and disposition of the shift boss. A typical mucking expectation for a shoveler was sixteen cars per shift. A tally man was utilized at each ore dump to keep track of the number of cars dumped by the carmen and the motormen.[9]

On April 1, 1917, wages were increased at the Homestake Mine for the first time since the company was organized in November 1877. Employees who received $3 or less per shift were given a 16 percent raise. Employees who received more than $3, but less than $4 per shift, were given a 12 percent raise. The wage for shovelers was increased from $3 per shift to $3.60 per shift. The wage for miners was increased from $3.50 per shift to $4 per shift. Employees who received more than $4, but less than $5 per shift, were given a 10 percent raise. Wage rates were frozen for those employees who received more than $5 per hour. The annual bonus of 7 percent was discontinued for all employees except those earning more than $5 per shift. On March 1, 1918, wages were temporarily increased for some job classifications for the duration of World War I. The wage for a miner was increased to $4.25, and the wage for a shoveler was increased to $3.85 per shift.

With many of its employees absent because of the war effort, Homestake was forced to shut down the Mineral Point and Golden Star mills for several months in 1918 and 1919. This reduced milling capacity from 1,020 stamps to 720 stamps. The labor shortage was impacted even more as a result of the Spanish flu epidemic that arose in 1918. By year's end, the death toll rose to ninety-four with more than two thousand people requiring medical attention. In early 1919, the conditions within the community improved considerably, and the Golden Star Mill was restarted.[10] The Mineral Point Mill wasn't restarted until 1921, using an experimental ball mill.[11]

The labor shortage continued into 1920. Effective March 1, 1920, Superintendent B. C. Yates announced another wage increase. Underground miners would receive $5 per shift; open-cut miners, $4.75; surface miners, $4.50; underground shovelers, $4.50; underground laborers, $4-$4.25; and surface laborers $3.75. Other surface employees received wage increases of twenty-five to fifty cents per shift.[12]

The Fair Labor Standards Act (FLSA) became effective on October 24, 1938. This law required employees to be paid at the rate of "time and one-half" for all hours worked in excess of forty-four hours per week. The forty-eight-hour workweek in effect at the mine was continued, and employees were paid the overtime rate for all hours worked in excess of forty-four hours per week.[13]

On October 24, 1939, provisions of the FLSA necessitated another downward revision of the standard workweek from forty-four hours to forty-two hours.[14] A few years later, the regular forty-eight-hour workweek at the mine was reduced to forty-two hours in an effort to reduce the cost of statutory overtime. On October 24, 1945, the FLSA reduced the regular workweek from forty-two hours to forty hours. The regular workweek at the mine was restored to forty-eight hours in March 1952, based on a six-day workweek, and overtime pay was resumed for the additional eight hours.[15] The six-day forty-eight-hour workweek remained in effect until April 1973, when it was reduced to forty hours.[16]

Homestake Hospitals

In 1877, George Hearst, one of the founders and original directors of Homestake Mining Company, saw the need to provide medical services to employees of the company. A contract was subsequently made with Dr. D. K. Dickinson for providing medical and surgical services for the miners in Lead. A similar contract was made with Dr. J. O. Gunsolly for providing services for the employees of the Deadwood, Golden Terra, Caledonia, and Father DeSmet mines. Each employee was charged a nominal fee that was deposited into a hospital fund.

Homestake constructed a four-room log cabin for use as its first hospital in April 1879. The building was located on the north side of Main Street near Gold Street (see fig. 17.1). This building was used until 1886, when it was replaced with a new two-story building constructed at the northeast corner of Main and North Siever streets (see fig. 17.2). The building was later modified to include a ground floor at street level. Dr. John W. Freeman, who had been a partner with Dr. D. K. Dickinson in Lead City and Terraville since January 1, 1884, was hired as chief surgeon for the hospital in 1904 to replace Dickinson, who moved to California.[17]

Fig. 17.1. The first Homestake Hospital, constructed in April 1879. It was located near the northwest corner of Main and Gold streets.—Photo courtesy of the Black Hills Mining Museum.

Fig. 17.2. Homestake's second hospital, constructed in 1886 at the northeast corner of Siever and Main streets. A ground-level floor with new offices for the doctors was completed in 1896. The horses belonged to the doctors.—Photo courtesy of the Black Hills Mining Museum.

Homestake discontinued charging its employees and dependents for medical services, hospitalization, and prescription drugs in 1910. The company initiated preemployment physicals in September 1911. By 1914, the hospital staff included a surgeon, six physicians, six graduate nurses, six assistants, and various other helpers (see figs. 17.3 and 17.4).[18]

Fig. 17.3. The Homestake Hospital Staff with the company's 1916 ambulance. The photo was taken in the courtyard of the Homestake Opera House and Recreation Building.—Photo courtesy of the Black Hills Mining Museum.

Fig. 17.4. The Homestake ambulance crew, 1931. This photo, by Black Hills Studios Inc. shows various Homestake employees around the 1916 ambulance. A short time later, a new 1931 Studebaker-Belleview ambulance was purchased. A J. M. Ross was mine superintendent at the time.—Photo courtesy of Johnny Sumners and the Black Hills Mining Museum.

During the month of September 1922, work was started on a $125,000 expansion of the hospital. The wood structure was transformed into a twenty-five-bed, three-story brick structure that was commissioned on May 12, 1923 (see fig. 17.5). The new hospital was staffed with six full-time physicians headed by Dr. Clough. In 1928, Dr. Robert B. Fleeger was hired as chief surgeon to replace Dr. Clough, who resigned. An addition was completed on the east end of the hospital in the 1960s.[19]

Fig. 17.5. Homestake's new brick hospital, 1922. Work was completed on the three-story brick hospital in May 1923. The brick structure was constructed around and to the east of the hospital constructed in 1886. The Homestake Hospital was razed in 1985 as part of the first major expansion of surface mining in the Open Cut.—Photo courtesy of the Black Hills Mining Museum.

The newly remodeled hospital supported Homestake and Lead until September 1, 1974, when inpatient care was transferred to St. Joseph's Hospital in Deadwood.[20] In the early 1980s, Homestake funded construction of the Black Hills Medical Center adjacent to St. Joseph's Hospital in Deadwood. Remaining patient care was subsequently transferred to the Medical Center. In 1983, Homestake converted the upper two floors of the old hospital in Lead to offices for its land, exploration, and Open Cut groups.

Hearst Mercantile Company

The first Hearst Mercantile Store was constructed by George Hearst in 1879. The store was located at 100-110 North Mill Street in Lead (see fig. 17.6). At the time the building was constructed, it was the only brick building in town; hence, it became known as the Brick Store. The 38,000-square-foot brick store stocked everything a family could possibly need, including shoes, dry goods, clothing, groceries, hardware, and household furnishings.[21] Upon George Hearst's death in March 1891, Phoebe Hearst incorporated the Hearst Mercantile Company and named Thomas J. Grier president.[22]

Fig. 17.6. The Hearst Mercantile Company's Brick Store on North Mill Street in Lead, 1898. The building also housed the Hearst Free Library, Lead Post Office, and the general offices of Homestake.—Photo courtesy of the Black Hills Mining Museum.

The Hearst stores offered credit accounts to new hires so they could purchase personal items such as overalls, work boots, a dinner pail, carbide lamp, or cap. Employees of Homestake could elect to receive their paychecks at the Homestake Office or sign an order to have their paycheck signed over to the Hearst Mercantile Company. Under the latter arrangement, employees could charge their purchases to an interest-free credit account at the store. On paydays, the employee would instruct the Hearst cashier how much of the employee's paycheck would go toward paying down his account. The balance of the paycheck was remitted to the employee in cash.

The Hearst Mercantile offered the same privileges to employees of other mines outside of Lead. In 1914, some twelve hundred to thirteen hundred employees of Homestake and six hundred to seven hundred employees of other mines around Lead had credit accounts at the Brick Store. Employees with excellent credit history could draw their checks at the Homestake Office without having to assign their checks to the store. Such employees could cash their paychecks at the bank or at the Brick Store. In 1914, cash transactions involving paycheck distributions averaged about $100,000 per month at the Lead store.[23]

Prices at the Brick Store were considered to be fair and reasonable. At times, businesses in Lead found they couldn't compete with the Hearst Mercantile and were forced to close. Over the ten-year period prior to 1914, it was estimated that the cost of living in the United States had increased 59 percent. The cost of living in Lead over the same period exhibited little or no change.[24]

The Hearst Mercantile on North Mill Street also housed the Hearst Free Library and the Homestake Offices. Lead's post office was also located in the Hearst Mercantile building from June 1, 1897 to 1912, when a new post office was completed at 329 West Main Street. In addition to the Hearst store in Lead, Homestake operated branch stores at Terraville, Brownsville, Elk Creek, Piedmont, and Nemo to support its employees and contractors working near these locations.[25]

When the Brick Store on North Mill Street had to be razed because of subsidence caused by underground mine operations, a new store was built on Main Street between the post office and the Homestake Recreation Building (see fig. 17.7). This building, completed in 1922, also housed the general and pay offices of Homestake Mining Company. Disaster struck on August 31, 1942, when a fire started in the basement of the Seely Drug Store adjacent to the Hearst Store. The fire spread very quickly into the basement of the Hearst Store. In a short time, the entire building was engulfed in flames. The store ended up being a total loss and was never rebuilt.[26]

Fig. 17.7. The Hearst Mercantile Store on Lead's Main Street, 1934. This photo, taken on April 18, 1934, shows the Hearst Mercantile building that was constructed on Main Street in 1922 to replace the original building that had to be razed because of the subsidence problem. The new store was located between the post office and the Homestake Recreation building.—Photo courtesy of the Black Hills Mining Museum.

Homestake Veterans Association

On December 14, 1905, many of the senior employees of Homestake Mining Company met at the No. 2 Hose Parlors in Lead and decided to create the Homestake Veterans Association. Membership eligibility was based on a minimum 21 years of employment at the Homestake Mine. Richard Blackstone was elected temporary chairman, and Robert Fraser was elected temporary secretary. A committee consisting of Richard Blackstone, Charles Eckland, J. W. Kissack, Frank Heitler, and A. H. Lundin was elected to draft rules and regulations for the association. One hundred and three employees became charter members of the organization.

On February 6, 1906, the first regular meeting of the Homestake Veterans Association was held. Richard Blackstone was elected president, Robert Fraser, secretary, and W. J. McMakin, vice president. T. J. Grier and J. A. Spargo were elected trustees. One of the first acts of the association involved a contribution to the Salvation Army for the purpose of purchasing Christmas trees and gifts for needy families. During World War I, the association donated thousands of dollars to the American Red Cross and YMCA.[27]

Between 1906 and 2005, a total of 1,705 employees with at least 21 years of employment service at the Homestake Mine had become members of the Homestake Veterans Association. William Lang earned the distinction of having the longest tenure of employment with Homestake, having worked sixty-two years from 1880 through 1942. In 2006, the bylaws for membership were amended to allow any former employee to be deemed eligible for membership in the association regardless of tenure.[28]

Homestake Baseball

As the population of the Lead area increased after about 1900, so did the need for recreation. On June 8, 1908, Homestake began work on a new ballpark for its employees and the residents of Lead. Because there were no available locations in Lead that had flat ground, a site was selected near the west end of Summit Street, and the top of the hill was leveled by Homestake's miners. The work effort involved drilling, blasting, and hand-shoveling activities over a two-month period. The *Deadwood Daily Pioneer-Times* reported, "Four miners from Homestake Mining Company started work yesterday on the new ballpark and have been busily engaged in drilling holes. A large force of shovelers and carmen will now begin work. It is expected to start work with fifty ordinary cars. The top of the hill will be a busy place for the next four to five weeks. The park should be ready for play in about forty days."[29]

More than 13,700 cubic yards of rock were mined and shoveled from the hilltop to create the ballpark.[30] The ballpark featured grandstand seating for one thousand to fifteen hundred people. A dedication for the $15,000 ballpark, later called Mountain Top Field, was held on August 23, 1908 (see fig. 17.8). Two ball games were held after the dedication ceremony. The opening game was between the Upper Mine and Lower Mine teams of Homestake. The second game was between teams representing Lead and Deadwood.[31]

Fig. 17.8. The $15,000 Mountain Top ball park, 1909.—Photo courtesy of the Black Hills Mining Museum.

Through the 1930s and 1940s, Homestake hosted a baseball league for the benefit of its employees. In 1946, four teams comprised the league, including the Surface, Ross Shaft, Yates Shaft, and Spearfish Sawmill teams. Twenty-four games were held between June 11 and August 30.[32] The end-of-season playoffs were typically held during Lead's annual Labor Day celebration. The Surface team captured the league title in 1946[33] while the Spearfish Sawmill claimed the title in 1947.[34]

Povandra Champ Batter

Batting champion Charles Povandra of the Spearfish Sawmill team racked up a .413 average to beat other players in the Homestake league. Surface's Albert "Bud" Mitchell was runnerup with a .411, just two points behind Povandra. Third place winner is Dale Grove, also from Surface, with an even .400. Tied for fourth with .353 apiece are J. Richards of Ross Shaft and Ausmann from Surface. Other averages among the league players are: Perkovich, .349; Richard Enderby, .349; Ledyard, .346; Hardy, .339; Kallenberger, .333; W. Mitchell, .325; Thomas, .319; Laurenti, .316; Goodrich, .310; M. Povandra, .308; Welch, .308.[35]

Over the ensuing years, numerous improvements and upgrades were made to the Mountain Top Field. In 1986, Homestake completed approximately

$243,000 of site improvements at Mountain Top to help offset the loss of the ball field at Mooney Park, which had to be abandoned because of the first expansion of the Open Cut surface mining operation. The improvements consisted of regrading the football field, improving the baseball field, installing a metal-halide lighting system, and constructing an all-weather rubberized running track. Because of the large number of residential structures that were purchased in Terraville and Lead to facilitate the Open Cut Mine operation, Homestake also donated land or assisted in creation of the Hearst housing subdivision, Lead Mall, Golden Hills Resort, Caledonia Apartments, Manuel Brothers Park, and the Mile-High Mobile Home Park.

Homestake Aid Association

On April 25, 1909, employees of Homestake met at the Opera House in the Lead Miners' Union Hall to discuss forming a mutual insurance organization. Chairman George D. McClellan called the meeting to order. A. J. Corum was elected secretary. McClellan informed those gathered that between seventeen hundred and eighteen hundred employees had signed petitions to hold the organizational meeting. Thomas J. Grier, Homestake superintendent, also attended the meeting and endorsed the petitions.

McClellan explained that employee dues for the proposed insurance organization would be $1 per month. Benefits would include an accident disability benefit of $1.50 per day, effective the day of the accident, and a death benefit of $1,000. It was decided that an organizational committee would be formed and would meet with Grier to draw up bylaws and definite plans for the organization. Each department would select one representative to represent that department on the committee. Unfortunately, the union and the company became preoccupied with the lockout of 1909-1910, and creation of the organization was delayed.[36]

Following the lockout that ended in March 1910, Homestake created the Homestake Aid Association on August 1, 1910. The Aid Fund, as it was called, was structured almost identical to what was proposed by the employees at the union meeting held on April 25 of the previous year. A board of directors was elected to manage the fund. Initially, the board consisted of one representative from the hospital, mining, metallurgical, mechanical, and surface departments. At the onset of the plan, each employee contributed $1 per month to the Aid Fund. Homestake contributed $1,000 per month. Benefits were paid to an employee on the sixth day following an absence from work due to illness or from the date of any lost-time accident. A benefit was also paid for a death that resulted from natural causes or an accident. Also in 1910, Homestake discontinued any charge to the employee and his or her dependents for medical services, hospitalization, and prescription drugs.[37]

On June 1, 1989, Homestake and the United Steelworkers of America AFL-CIO-CLC agreed to suspend contributions to the Aid Fund because it was sufficiently funded. The parties agreed that if contributions needed to be resumed in the future, Homestake would pay 75 percent and the employee 25 percent of the monthly contributions required to maintain an adequate funding level. It was agreed that if such contributions needed to be resumed at some point in the future, the employee's contribution would not exceed $5 per month. Subsequent contributions were not needed, and the Aid Fund continued to provide benefits to eligible employees until shortly after the mine's closure on December 31, 2001.[38]

Mine Safety

Homestake established a Mine Rescue team (see fig. 17.9) and first-aid training for employees in 1911. A formal safety department with a full-time safety engineer was established in 1916. By 1931, the Mine Rescue team was equipped with twelve McCaa self-contained breathing apparatus, six self-rescuers, one high-pressure oxygen pump, one carbon monoxide detector, safety lamps, and various first-aid supplies.

Fig. 17.9. Homestake's Mine Rescue team, 1927. Homestake established its first Mine Rescue team in 1911 and a Safety Department in 1916. Over the next ninety years, Homestake's Mine Rescue teams won numerous national mine rescue competitions.—Photo courtesy of the Black Hills Mining Museum.

A safety bonus system was incorporated in 1923. Under the system, an underground employee who completed three hundred shifts in a year with no lost-time injury was awarded a safety bonus of $10 in gold coin. An underground employee who worked five years with no lost-time injury was given a bonus of $20 in gold coin. An underground foreman was eligible to receive a bonus of $5 in gold coin if his crew completed three hundred shifts without a lost-time injury. If the crew had an injury frequency rate below 62.5 injuries per million man-hours worked, a frequency bonus of $20 was paid to the foreman. If the crew's severity rate was less than 0.5 shift lost per 1,000 hours worked, a $20 severity bonus was paid to the foreman of the crew. In 1930, 360 underground employees received the $10 individual bonus and twelve received the $20 bonus.[39]

In 1953, the U.S. Bureau of Mines trained twenty-five employees in mine rescue. Each Mine Rescue team member was furnished with a McCaa two-hour breathing apparatus that afforded protection against carbon monoxide during a mine fire. A stench warning system utilizing ethyl mercaptan was installed that same year. In the event of a mine fire or some emergency that required an evacuation of the mine, ethyl mercaptan was injected into the compressed-air pipelines and the mine ventilation system. Upon smelling the stench, employees working on the various levels underground would immediately stop work and proceed to the nearest shaft station, where they were hoisted to the surface on the cage.[40]

Watch fobs were introduced in 1939 as a safety award to recognize those employees with accident-free work records. Mine Department employees received a bronze fob after five years of accident-free service, a silver fob after ten years, and a gold fob after fifteen years. Rates were double for surface employees. Between 1939 and 1955, 430 gold fobs were awarded to surface and underground employees. The award was later changed to a bolo tie.

By 1953, there were two principal safety committees, including a Central Safety Committee and a Workmen's Safety Committee. The Workmen's Committee met bimonthly and conducted safety inspections on the surface and underground. Recommendations from this committee were presented to the Central Committee, which met on a monthly basis.

On January 1, 1954, safety glasses became mandatory for all employees working underground. The rule was extended to all surface employees on January 1, 1955. Over the six-year period prior to 1954, employees were experiencing an average of nineteen eye injuries per year. After the safety glasses rule was instituted, the number of eye injuries was reduced to three in 1954 and zero through April 1955.[41]

The commitment of Homestake and its employees to improve safety performance at the mine paid off after about 1918. Prior to that time,

Homestake and other companies in the Black Hills were incurring as many as eighteen fatalities per year, with an average of about seven to eight. After 1918, the average was reduced to one or two per year. Throughout the Black Hills, 419 fatalities occurred in the mining industry between 1876 and 1995. Of that number, 270 fatalities occurred at the Homestake Mine and its lumber and timber operations between 1876 and June 2003, when underground decommissioning work was completed and the mine was closed. Approximately, 119 other fatalities occurred at other mines along the Homestake Belt, including those that Homestake either controlled or later acquired. The numbers of fatalities that occurred at mines along the Homestake Belt are shown in table 17.2.

Table 17.2 Mine fatalities along the Homestake Belt (1876-2003)

	No.
Homestake Mining Company	270
Highland Mining Company	47
Deadwood-Terra Mining Company	33
Father DeSmet Consolidated Mining Company	19
Caledonia Mining Company	17
Homestake No. 2	1
Giant and Old Abe Mining Company	1
Hawkeye Gold Mining Company	1
Total:	389

Source: Donald D. Toms, *In the Midst of Life: Mining Company Fatalities, Black Hills, South Dakota, 1876-1995,* I (1996), 1-20.

Of the 419 fatalities in the Black Hills, 41 percent were attributable to a fall of rock, rolling rock, or rockslide. Approximately, 20 percent were attributable to a fall of the person; 14 percent involved explosives; 8 percent involved powered haulage; 5 percent were attributable to some type of falling object (other than rock) such as a loose sinking bucket, timber support, or tree; and 4 percent were attributable to the person being struck by some object in motion such as a skip, cage, or broken cable. The remaining fatalities were attributable to a variety of other direct causes such as electrocution, asphyxiation, drowning, or crushing.[42]

Recreation Building and Opera House

Following the end of the lockout in March 1910, Homestake embarked to establish various other types of employee benefits. In addition to the Employee's Aid Fund, Homestake discontinued charging employees or their dependents for medical services, hospitalization, and prescription drugs.

Homestake initiated construction on the Homestake Opera House and Recreation Building on Lead's Main Street in July 1912. The facilities were opened on August 31, 1914. Homestake's superintendent, Thomas J. Grier, presented use of the facilities as a gift to the residents of Lead. The state-of-the-art facilities consisted of an Opera House with a seating capacity of more than one thousand, a bowling alley, swimming pool, library, social rooms, and billiard rooms. The Opera House hosted many varied events including opera, ballet, vaudeville, concerts, plays, movies, boxing matches, festivals, and high school commencement exercises. All of these amenities, except the Opera House, were free to the residents of Lead.

In 1972, Homestake turned ownership and management of the facilities over to the city of Lead.[43] The Opera House was tragically gutted by a fire on April 2, 1984. The building has since been undergoing major reconstructive work to restore the theater and recreation building to its former grandeur.[44]

U.S. Commission on Industrial Relations

On August 23, 1912, Congress passed legislation to create a nine-member Commission on Industrial Relations. Congress directed the commission to "inquire into the general condition of labor in the principal industries of the United States, including agriculture, and especially in those which are carried on in corporate forms." The commission was also directed to assess "the growth of associations of employers and of wage earners and the effect of such associations upon the relations between employers and employees . . . [and] methods for avoiding or adjusting labor disputes through peaceful and conciliatory mediation and negotiations."

From 1912 to 1915, the commission held 154 days of public hearings in various towns and cities throughout the United States, including Lead, South Dakota. The commissioners interviewed some 740 people from all walks of life who represented employers, employees, and the public at large. Agents of the commission selected witnesses based on input and recommendations from employers, representatives of labor, and associated localities. Witnesses included managers, foremen, attorneys, union officials, workers, public officials, educators, clergy, physicians, and representatives from many other disciplines.[45]

Four members of the Commission on Industrial Relations visited Lead, South Dakota, on August 3-4, 1914. Commission members who participated in this visit included John R. Commons, acting chairman, Austin B. Garretson, John Lennon, and James O'Connell. The commissioners deposed such people as Thomas J. Grier, Homestake superintendent; Thomas D. Murrin, manager of the Hearst Mercantile Company; Chambers Kellar, chief council for Homestake; Father Joseph F. Busch, Catholic bishop; James Kirwan, former executive member of the Western Federation of Miners and former member of the Terry Peak Miners' Union; Thomas J. Ryan, former secretary of Lead Miners' Union; Warren E. Scoggan, district representative of the Western Federation of Miners and former member of the Lead Miners' Union; G. A. Northam, special agent of Homestake; J. L. Neary, credit man for Hearst Mercantile Company; and several other people.

Bishop Busch testified at great length, denouncing the seven-day workweek in use at the Homestake Mine. He stated that the Sabbath Day was meant to be a day of rest and religious observation. Busch was dismayed at the effects of the lockout, claiming that between 75 and 80 percent of the Catholics had left town. Superintendent Grier testified that he had instructed the employees and advised the shift bosses that an employee could take Sunday off by merely advising his shift boss to that effect as late as Saturday. Grier also stated there was ample opportunity for churches to schedule Sunday services between shifts in the morning or afternoon. Others testified that most employees could not afford to take a day off.[46]

At the conclusion of the two-day hearing, Chairman Commons summarized by saying, "You have here the most remarkable business organization that I have come across in the country." Chairman Commons went on to say as follows:

> You have developed welfare features which are beyond anything that I know of, and are given with a liberal hand. You have a high scale of wages and reasonable hours, very fair hours. There has been evidently great progress made in taking care of employees, and in the hospital service. And you have taken care of the cost of living, have kept that down beyond what employees in other communities have been forced to pay.
>
> You have practically been able by your great strength here as a huge corporation, dominating the whole community, and looking out for the welfare of your employees, to bring in a desirable class of citizen. It seems also that you are influential in politics, that you secure a good class of officials, and that you secure the enforcement of law and the reduction of immorality.
>
> It seems also that you make an effort to build up the religious life of the community, and that your policy is broad and liberal in all respects.

> *In all of the reports that are made here, and all of the talk that I hear about town, indicates that you have wielded your power with the greatest fairness and with the confidence of the community and with the idea of building up a good community.*[47]

After Chairman Commons finished with his remarks, Commissioner John Lennon, the American Federation of Labor (AFL) union representative commented, "I would like to have inserted in the record that the expressions given by Professor Commons are by Professor Commons individually as a Commissioner, and are not collective for the Board of Commissioners present." Commissioner O'Connell, another AFL union representative, offered his assessment of Chairman Commons' remarks, saying, "Well, for all the Commissioners . . . there are some things stated I would not agree with under any circumstances." When asked if he had anything to say, Commissioner Garretson, the third AFL union representative, replied, "Nothing, except that the Chairman is speaking for the chair."[48]

The differing opinions and representation of special interests prevailed to the end among the entire nine-person Commission on Industrial Relations. The commission's final report to Congress, also known as the Walsh Report, was submitted in 1916 in eleven volumes containing tens of thousands of pages of testimony. Volume 1 is the report of Basil M. Manly, director of Research and Investigation, who provided support for the commission. Manly's report was signed by Frank P. Walsh, chairman of the commission and a labor lawyer, along with John Lennon, James O'Connell, and Austin Garretson, AFL union representatives.

One of the findings in the Manly report was that "communities which are either wholly or in large part owned or controlled by a single corporation or individual employer, present every aspect of a state of feudalism except the recognition of special duties on the part of the employer. . . . Furthermore, the employer in many cases controls the social and political life of such communities, either by the complete absorption of local political powers or by domination of the local authorities."

Their report went on to say, "We hold that efforts to stay the organization of labor or to restrict the right of employees to organize should not be tolerated, but that the opposite policy should prevail, and the organization of the trade unions and of the employer's organizations should be promoted. . . . This country is no longer a field for slavery, and where men and women are compelled, in order that they may live, to work under conditions in determining which they have no voice, they are not far removed from a condition existing under feudalism or slavery."[49]

Lead Country Club

Homestake's efforts to expand its employee benefits and contributions to the community didn't end after the Commission on Industrial Relations left town. The first retirement plan for employees was established in 1917. In 1922, at the urging of Dr. A. S. Jackson, an organizational meeting was held for the purpose of establishing a golf course near Lead. The Lead Country Club was subsequently formed with Chambers Kellar as the club's first chairman. After evaluating several potential sites for a nine-hole course, the former McNabb Ranch, owned by Homestake Mining Company, was selected. Superintendent Bruce C. Yates, of the Homestake Company, offered free use of the ranch for purposes of building and maintaining a golf course.

John Bland, a golf-course architect who had recently completed a course at Hot Springs, was contracted to lead a large force of men in constructing the course at Lead. By June, the men had the Lead course in playable condition. A clubhouse was completed in the spring of 1923. Homestake furnished the materials for the building while the club members, the First National Bank, and the Miners and Merchants Bank raised the necessary funding for construction. Electric lights and water were added later, courtesy of Homestake.[50] In 2007, Homestake sold its real-property interests in Lead Country Club to a group of private investors.

Homestake Band

The Homestake Band was created in about 1894, when it hosted a grand ball in Lead. A photograph dated July 20, 1918 shows fifteen members of the band posing in the town of Newell. In about 1928, the Lead Kiwanis Club and the Lead Commercial Club decided a community band was needed. Representatives of the two organizations approached Homestake to see if it would be willing to support a community band. Homestake consented, and the Homestake Band was soon reorganized. Henry Elster served as director for several decades and was followed by Paul Hedge in the late 1960s and early 1970s. Three charter members, Henry Phillips, Bob Koontz, and W. J. Mitchell, played in the band well into the late 1960s and early 1970s.

During the summer months, the Homestake Band played weekly concerts in the Band Shell, which was located on the southwest side of Mill and Main streets. The band participated in annual events such as the Belle Fourche Roundup parade and rodeo, Deadwood Days of '76 parade and rodeo, Custer Gold Discovery Days parade and pageant, Rapid City Range Days parade, Lead Labor Day parade and celebrations, Mount Rushmore celebrations, as well as many other celebrations, festivities, and events (see fig. 17.10).[51]

Fig. 17.10. The Homestake Band, 1931. This photo of the Homestake Band was taken in 1931 just before the Days of '76 parade in Deadwood. *Standing, left to right,* Henry Phillips, Rashleigh Ball, Ralph Brothers, Bob Koontz, John Rowland, Ray Rockhold, Avery Woodward, Ed Murray, W. J. Mitchell, Henry P. Elster, director, Jack Dunstan, Sam Tretheway, Clarence Hodges, Dick Murray, Clyde Rowland, Charles Cousins, Frank Cetto, William Morcom, Harold Arthur, Archie Williams, Jack Logeman, Walt Daniels, Herbert Steir, A. L. Coolidge, and Russell Wayland; *kneeling, left to right,* Herbert Rowland, John Castonguay, Ted Fahey, Francis Green, John Morgando Sr., George Kulpaca, Roland Engel, Charles Rossio, Antone Giachetto, Ed Andrews, Nesto Sonza, and Darwin Holway.—Photo courtesy of the Black Hills Mining Museum.

Employee Housing

During the early 1930s, Homestake began increasing the size of its workforce to facilitate various construction and development projects such as the South Mill complex and Ross Shaft complex. Labor was plentiful because of the national Depression. A housing shortage soon developed in Lead. To help remedy the problem, Superintendent Bruce C. Yates decided to spearhead a program to construct houses and sell them at cost to employees. A financing plan was also established with which employees could purchase a company house or a noncompany house and have the installments deducted from their pay.

In 1933 and 1934, Homestake constructed fifty-nine houses at an average cost of $3,198. All of the houses were sold to employees at Homestake's cost. Seven of the houses were purchased in full with cash. The remaining fifty-two houses were financed to the employees at nominal interest rates. Homestake also financed an additional fourteen noncompany homes.[52]

Homestake embarked on another housing project, called the Grier Addition, in May 1963 at the southeast corner of Main and Mill streets. The westerly portion of the site included an area that was primarily being used at the time as a little-league baseball field. As part of the construction work, the ball field was relocated north of Main Street and renamed Mooney Park. The easterly portion of the new housing site was known for many years as Nob Hill.

Between 1895 and 1933, Nob Hill was adorned by a very stately residence that was home to three different Homestake superintendents, Thomas J. Grier, Richard Blackstone, and Bruce C. Yates (see fig. 17.11).[53] After surface subsidence forced abandonment of the mansion on Nob Hill, a new twenty-seven-room, English Tudor mansion, dubbed the Big House, was constructed on Fairview Avenue in Lead at a cost of $200,000 (see fig. 17.12). Don Delicate was the last Homestake general manager to reside in the Big House. Delicate and his family moved to Custer when he retired in 1978.

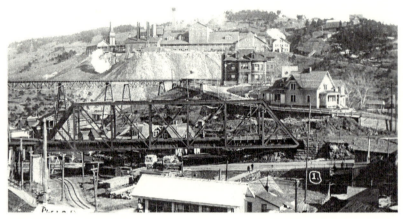

Fig. 17.11. The Homestake superintendent's mansion (left) on Nob Hill.—Photo courtesy of the Black Hills Mining Museum.

Fig. 17.12. The Homestake Mansion on Fairview Avenue in Lead.—Photo courtesy of Johnny Sumners and the Black Hills Mining Museum.

The initial phase of construction included the completion of twenty homes designed and constructed by Knecht's Master Craft Homes Corporation on Homestake's foundations. An open house for the Grier Addition was held on October 21-22, 1963. More than eight hundred people attended the first day of the grand opening, which included tours of a model home. Employees with one or more years of service could purchase a home for $7,800. With a 10 percent down payment, Homestake financed the balance at the rate of 6 percent interest over a ten-year period.[54] Homestake completed ten additional homes within the Grier Addition in 1973 and six more in 1976.[55] As late as 1977, Homestake would also cosign a bank-issued mortgage for an employee, which enabled the employee to obtain financing for the purchase of an existing home situated on company-owned land in Lead pursuant to a right-to-occupancy permit.

The *Sharp Bits* Magazine

Sharp Bits is a publication that was started by Homestake in February 1950 to keep its employees "more fully advised as to the details of the operations of the Company in which they have so active a part."[56] The "bulletin," as it was originally referred to, was issued to employees monthly from February 1950 through January 1967 at which time it was published on a quarterly basis through about 1970. *Sharp Bits* was a very people-oriented publication with each month's edition devoted to one or more aspects of the mine operation. Throughout the life of the publication, virtually, every department, process, method, building, job task, or aspect of the company's operation was succinctly described and illustrated.

The masthead for *Sharp Bits* first appeared on the cover of the March 1950 issue. Jack Bulat, an employee working in the blacksmith shop at the time, designed the masthead. The masthead features a miner, the Yates hoist room and headframe, an ore train on the Ellison Tramway, the Open Cut, a surveyor with his transit, a railroad crane, and the South Mill.[57]

Nicknames

The mostly dark, damp, and hot working conditions underground were tempered somewhat by the creativity and ability of the workforce to invoke a little humor, "horse play," and levity into the workplace. The assignment of nicknames to fellow employees was no exception.

Everyone at the Homestake Mine had a nickname, whether they realized it or not. Some of the nicknames were a play on a person's given name or surname, such as Hill Bill, Lady Di, Six Pack, Special Ed, or Wingnut. One newly hired employee became known as Jack, simply because the boss thought there were "too damn many Bills and Williams on the crew already" and that the person appeared to be a jack-of-all-trades. The person is still known today as Jack although his legal name is William.

Other nicknames were based on where a person worked or where they could usually be found in the mine, such as the Duke of Earl, Lower Lipp, Upper Lipp, and Mileaway. Some employees were assigned a nickname based on their job, such as Plumb Bob, Rock Witch, Wind Witch, or Easy Money.

Most people were assigned a nickname based on their physical attributes, mannerisms, or "reputation." These types of nicknames included Big Time, Bones, Boss Hog, Breezy, Colonel Klink, Talkin' Bull, or the Duck. More than a few employees remarked how they had worked their entire career at Homestake and never knew who certain people were, other than by their nicknames.

Chapter 18
Mine Fires and the Subsidence Issue

The Need for a New Shaft

By about 1895, Homestake recognized that the Golden Prospect, Golden Star, and B&M shafts were at risk because of ground-control problems and the expanding mine operations (see figs. 18.1 and 18.2). All of these shafts were located within ore or very near ore. The Homestake Vertical Shaft had previously been abandoned because of its proximity to surface- and underground-mining operations (see fig. 18.3). Company officials also recognized that a new, deeper shaft was needed on the south side of Gold Run Gulch to access the same ledges that were plunging southward. Therefore, in the fall of 1895, work was started on a new shaft. The shaft, located on the General Ellison mining claim, was appropriately named the Ellison Shaft.

Fig. 18.1. The Golden Prospect (Highland) Shaft of the Highland Mining Company.—Photo courtesy of Bob Otto.

Fig. 18.2. The Homestake Mine, ca. mid-1880s. This photo shows how the Homestake Vertical Shaft (upper left) and the Golden Star Shaft (immediately below the Vertical Shaft) were about to be engulfed by the open cuts and glory holes. The 120-stamp Golden Star Mill is shown on the right.—Photo courtesy of the Black Hills Mining Museum.

Fig. 18.3. The Homestake Vertical Shaft. The Vertical Shaft, located toward the north end of the Homestake claim, was abandoned in the early 1890s to facilitate expansion of the mine.—Photo courtesy of Phoebe Apperson Hearst Library.

Construction of the headframe, hoist house, and crusher room for the Ellison Shaft was started in 1897. These facilities were housed in a steel building 200 feet long, 90 feet wide, and 80 feet high.[1] The Union Iron Works of San Francisco, California, constructed the steam-powered hoist.[2] By June 1900, the bottom of the Ellison Shaft was at the 600-foot level. The Old Abe was to the 1,000-foot level, the Golden Prospect to the 600-foot level, and the Golden Star to the 1,100-foot level. The Ellison hoist, crushing plant, and tramway system were completed on January 1, 1902, at a cost of $1 million.[3]

The Ellison Shaft was originally configured with three compartments (see fig. 18.4). Double-deck cages were installed in two of the compartments. The third compartment housed the service pipelines for compressed air and water. Each cage accommodated two 1-ton-capacity ore cars per deck. The hoist was designed to raise 4 tons of ore per cage load from the shaft's ultimate depth of 3,200 feet. Total hoisting capacity from the skip-loading pocket on the 800-foot level was 5,700 tons per twenty-four hours. Three men were required at the top landing to push the ore cars off the cage and dump the ore into a bin that fed four Gates crushers. Crushed ore passed into a 3,000-ton-capacity bin situated above the tramway.[4] At the tramway level, located 65 feet below the shaft collar, ore was transferred from the crusher bin to the tramway cars. The ore cars were trammed to the Golden Star Mill over a trestle 600 feet long that spanned Gold Run Gulch at a height of 115 feet (see fig. 18.5).

Fig. 18.4. Unloading 1-ton ore cars at the Ellison Shaft. The Ellison Shaft was configured with three main compartments, two of which were equipped with double-deck cages. Note the flattened-strand hoist cable that was used in conjunction with the original steam-powered hoist.—Photo courtesy of the Black Hills Mining Museum.

Fig. 18.5. The Ellison Tramway, July 1, 1920. Prior to commissioning the South Mill in 1922, ore was hoisted at the Ellison Shaft and trammed to the Golden Star Mill, partially shown in the foreground. Because of the surface-subsidence problem, use of the Ellison Tramway was discontinued on March 25, 1934, and all of the haulage was directed to the recently expanded South Mill. The First Finnish Evangelical Lutheran Church, completed on March 13, 1891, is shown in the lower left. —Photo courtesy of Black Hills Mining Museum.

Over the next two decades, several other buildings and facilities were constructed to support the new Ellison hoisting works. These included an engineer's office, fan house, car repair shop, drill shop, compressor house, and Ellison Boiler House (see fig. 18.6).

Fig. 18.6. Some of the Ellison support facilities, 1923. Prior to construction of the Ellison Boiler Plant, the Central Steam Plant supplied steam to the Ellison works through steam pipelines that passed through the trestle located near the Fan House and assay office.—Photo courtesy of the Black Hills Mining Museum.

As part of the Ellison construction project, a 1.25-million-gallon water storage reservoir, called the Ellison Reservoir, was constructed at the top of the hill just south of the Ellison Shaft. The reservoir was 155 feet long, 65 feet wide, and 20 feet deep. Water was fed to the reservoir by gravity through a pipeline called the Hearst Ditch. The Hearst Ditch sourced water from Whitewood Creek below the Englewood Hydroelectric Plant. The water was used for mine and mill operations as well as domestic use for the city of Lead.[5]

Mine Fire of 1907

On March 25, 1907, a fire was discovered in the mine between the 500- and 600-foot levels. The fire is believed to have started after two miners blasted a drift round into the top of No. 5 Stope, which was being prepared for waste backfill. Shrinkage and square-set stoping methods had previously been extensively used to mine the entire area, and considerable caving had occurred above the 500-foot level. For the next three weeks, mine crews wearing special breathing apparatus attempted to put the fire out using a system of water pipes, water hoses, and steam (see fig. 18.7). However, the efforts were largely fruitless, and a decision was subsequently made to flood the mine from its bottom on the 1,400-foot level to the 300-foot level.[6]

Fig. 18.7. One of the fire teams at the Star Shaft, 1907. Note the breathing apparatus and air tank on the cage. The men include (left to right) Bert Fuller, Walt Tilson, Bill Bunney, and Jake Thompson.—Photo courtesy of the Black Hills Mining Museum.

While preparations were being made to flood the mine, a valiant attempt was made to bring all of the horses and mules out from underground. Two horses and two mules could not be reached because of heavy smoke and carbon monoxide. Most all of the mine equipment, including the newer compressed-air locomotives, was left underground. Flooding commenced on April 18. Some 1,280 cubic feet per minute of water from Lead's water mains and the pipelines and ditches that normally provided water for the mills was diverted into the Open Cut in Lead. In addition, a flume was quickly constructed between Whitewood Creek and the mouth of the Savage Tunnel, which was situated approximately 30 feet below the 300-foot level. Some 916 cubic feet per minute of water was diverted into the mine from Whitewood Creek through the Savage Tunnel (see fig. 18.8).

Fig. 18.8. Diverting water from Whitewood Creek into the Savage Tunnel, 1907. After the decision was made to flood the mine to extinguish the 1907 mine fire, a long flume was quickly constructed to divert water from Whitewood Creek into the Savage Tunnel. The Savage Tunnel was located near the present-day access road to the Kirk Fans.—Photo courtesy of the Black Hills Mining Museum.

After about forty days, the water level reached the 300-foot level. An inspection team was sent into the mine on the 300-foot level to assess the air quality in the mine. Unfortunately, the team found that high concentrations of carbon monoxide were still being generated and that the temperature of the water was 120 degrees Fahrenheit. The fire was still active. As a result, a decision was made to flood the mine to the 200-foot level.[7]

By May 29, the water level had risen to approximately 80 feet above the 300-foot level. The carbon monoxide levels, which were being monitored almost continuously from the 200-foot level, were now down to 1 percent. The fire was subsequently deemed extinguished. On May 30, about one-half of the stamp mills were restarted with a limited amount of low-grade ore that was obtained from the open cuts.[8]

Now that the mine fire was extinguished, some 600 million gallons of water had to be removed from the mine before underground operations could be resumed. To accomplish this, the B&M and Golden Prospect shafts were equipped with modified skips designed to bail water at the rate of 1,000 gallons per skip. At the Ellison Shaft, special skips were also used to bail water to the surface. Once the water level was lowered to below the 300-foot level at the Ellison, water from the skips was dumped onto the 300-foot level, whereupon the water drained down to the Savage Tunnel and out to Whitewood Creek. Three lifts using compressed air were also constructed at the Ellison and successfully raised about 4.4 million gallons of water every twenty-four hours. Once the miners were able to return underground in early July, they found that the mining equipment was largely operable with only minor damage to some of it. By July 12, mine production was restored to full capacity.[9]

By 1908, the Old Abe was to the 1,250-foot level, Golden Prospect (Highland) to the 1,000-foot level, Ellison to the 1,700-foot level, and the Golden Star to the 1,100-foot level. In January of that year, a small service hoist was installed on the 1,250-foot level of the Ellison Shaft and a 4½×5-foot cage was installed in the third compartment of the shaft. This arrangement allowed miners and bosses to travel between the 1,250- and 1,700-foot levels without interrupting the large double-deck cages that were needed for hoisting cars of ore.[10] In the Terraville area, the Golden Gate and Old Brig shafts continued to hoist ore from the 800-foot level. Ore from the latter two shafts was milled at the Mineral Point, Pocahontas, and Monroe mills.[11]

Surface Cave-Ins

Expansion of the mine continued to have very unfavorable effects on most all of the surface buildings and facilities situated above the underground workings on the south side of the Open Cut. An article in the *Deadwood Daily Pioneer-Times* illustrates the magnitude of the ground-control problem that had previously forced abandonment of the Homestake Vertical Shaft and was now jeopardizing the Golden Star Shaft:

> *The cave south of the Star hoist, which several months ago let out and made a hole all the way down to the 400-foot level, has not finished its sinking yet, and every once in awhile the earth gives away and another hole has to be filled up. The holes keep a gang of men busy hauling cars of earth and rock to satisfy its cravings for filling material. The old stope to which the cave leads should be filled up in a short time and then the sinking will cease.*[12]

In 1910 and 1911, two cave-ins occurred on South Gold Street. The first cave happened on September 25, 1910, under the Bertolero residence at 110 South Gold Street. Investigation revealed that the No. 3 Independence Stope on the 300-foot level had caved. This particular stope, which was approximately 190 feet long and 45 feet wide, was mined in 1904 and 1905 to a height of 100 feet. A contract for hauling backfill rock from the Golden Star bin to the caved area was awarded to W. H. Dacey. The hauling and filling work commenced on September 29 and concluded on October 25. A total of 4,712 cubic yards of rock were required to fill the void.

The second cave associated with the Independence Stope on the 300-foot level occurred on March 2, 1911, after a water-main burst on South Gold Street between Julius and Main streets. Within a short period, the broken water main caused the ground to drop, exposing a hole that eventually grew to 75 feet in diameter. The ground failure caused the corners of the Campbell House and the Clark Flats to collapse. A conveyor system was installed from a wagon trap on Main Street to the area of the cave (see fig. 18.9). By March 15, some 5,957 cubic yards of rock were conveyed, which filled the hole.[13]

Fig. 18.9. Conveying rock into the cave-in on South Gold Street, March 1911. The large building on the right is the Campbell House, located on the southeast corner of Main and Gold streets. Note the failure of the brick walls of the building on the far right.—Photo courtesy of the Phoebe Apperson Hearst Library.

Mine Fire of 1919

On September 25, 1919, the timber caught fire on the sixth floor of a square-set stope in No. 3 Pillar North on the 800-foot level after a round was blasted. Smoke and carbon monoxide gas quickly contaminated the underground mine environment. Mine personnel were evacuated, and the mine was shut down. By October 5, a decision was made to flood the mine, as was done in 1907 when a fire broke out between the 500- and 600-foot levels.

Two 12-inch-diameter pipelines 600 feet long were laid out on North Mill Street to the Open Cut. As was done in 1907, a flume 3,200 feet long was constructed along Whitewood Creek to the mouth of the Savage Tunnel in Kirk Canyon. The Golden Reward Shaft on Aztec Hill southwest of Lead was dewatered using its skips in an effort to provide additional water in Whitewood Creek. Water was also pumped from Deadwood Creek into the mine through the North End Tunnel that had its portal on the 300-foot level on the south side of Deadwood Gulch. By October 12, some 1,300 cubic feet per minute of water was flowing into the mine. The next day, two men were temporarily overcome by carbon monoxide while removing forms from a concrete bulkhead that had been constructed on the 700-foot level.

On December 1, 1919, the fire was declared extinguished with the mine flooded to the 500-foot level. Immediately, pumping systems were restarted, and the skips at each shaft were used to bail water out of the mine.[14] By December 9, limited ore skipping was resumed from above the 500-foot level at the Old Brig Shaft, and the Pocahontas and Monroe mills were started up with 260 stamps crushing ore. On January 18, 1920, the Amicus Mill was restarted, followed by the Golden Star Mill on February 22 and the Homestake Mill on March 11.[15] Total production for 1919 was severely impacted, with only 1.19 million tons milled compared to 1.63 million tons in 1918.[16]

During the mine fire and shutdown in 1919, the Ellison Shaft was isolated from the rest of the mine with walls and barricades to facilitate an upgrade of the shaft. The shaft was subsequently reconfigured to accommodate skips for hoisting ore instead of caging carloads of ore. The 12×20-foot shaft was reconfigured from two cage compartments and one pipe compartment to one cage compartment, two skip compartments, and one pipe compartment. A new double-drum electric ore hoist manufactured by Nordberg was installed as part of the shaft upgrade.

Concurrently, underground crushing stations were mined and constructed immediately below the 800-, 1,400-, and 2,000-foot levels (see fig. 18.10). Upon completion, ore from each of the levels was dumped from railcars onto a grizzly. Oversize rock passed over the grizzly onto an apron feeder that fed a 36×48-inch Traylor jaw crusher (see fig. 18.11). The crusher reduced the size of run-of-mine ore to minus 6 inches. The minus 6-inch ore passed through the grizzly into a 1,500-ton loading pocket above the skip loader. Crushed ore and fine ore was fed by gravity into the skip loaders, which transferred a measured volume of ore into the new 7-ton-capacity skips. On the surface, oversize ore was fed through four primary gyratory crushers and two secondary gyratory crushers set to two inches. Minus 2-inch ore was transferred to a storage bin from which the ore was loaded into 4-ton cars and hauled to the South Mill on the tramway.[17]

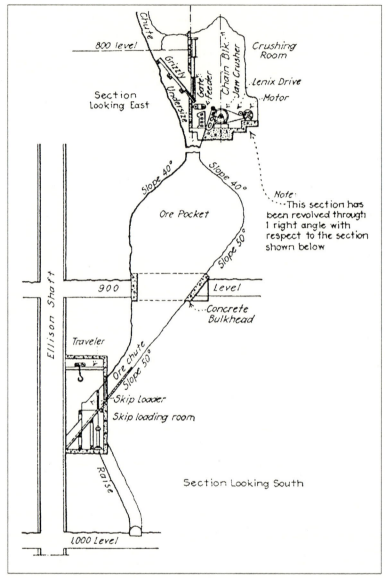

Fig. 18.10. A typical underground crushing and skip-loading station at the Ellison Shaft. Underground crushing stations and skip loaders were constructed below the 800-, 1,400-, and 2,000-foot levels at the Ellison Shaft during the mine fire of 1919. Each station was equipped with a 36×48-inch Traylor jaw crusher that reduced the ore to minus 6 inches.—Source: Felix Edgar Wormser, "Gold-Mining Developments in the Black Hills," *Engineering and Mining Journal* 114, no. 18 (October 28, 1922): 759. Reproduced by permission.

Fig. 18.11. The 36×48-inch Traylor jaw crusher on the 2,000-foot level at the Ellison Shaft, 1923.—Photo courtesy of the Black Hills Mining Museum.

B&M No. 2 as a Replacement Shaft

In 1920, the Golden Prospect (Highland) Shaft was abandoned because of ground problems. Homestake officials were also concerned that the B&M No. 1 (Old Abe) Shaft was in danger because of its proximity to the Open Cut. A decision was made to construct a new shaft, called the B&M No. 2 Shaft (see fig. 18.12), east of the B&M No. 1. The hoist from the Golden Prospect Shaft, built by Fraser and Chalmers in 1892, was moved to the B&M No. 2 Shaft in 1923. Drifts were driven from the B&M No. 1 Shaft to the location for the B&M No. 2 on the 900-foot level and levels above.

The new shaft was mined by driving a series of pilot raises from each level to the surface in 1924. The raises were then stripped to facilitate a three-compartment shaft 18 feet 10 inches long by 7 feet 2 inches wide. Two of the compartments were equipped with interchangeable 4-ton skips and double-deck cages. By 1928, the B&M No. 2 was stripped and timbered to a depth of 60 feet below the 1,550-foot level.[18] Hoisting of ore was subsequently discontinued at the B&M No. 1 Shaft on December 27, 1934. All B&M No. 1 Shaft operations were decommissioned early in 1935.[19]

Fig. 18.12. The B&M No. 2 Shaft, ca. 1925. Directly below the B&M No. 2 Shaft (top) is the Auxiliary Steam-Electric Plant (middle). Also shown are the Amicus Mill and Amicus Plate House (left center), Tungsten Mill (foreground), Central Steam (Boiler) Plant (lower center), and Coal Bins (right center).—Homestake Collection, Adams Museum, Deadwood.

In 1919, Homestake management became increasingly concerned that the Main Ledge structure was pinching out below the 2,000-foot level. The workforce and the townspeople soon became aware of this concern, and many people sold their homes and businesses and left the community. To make matters worse, efforts to relocate Lead's business district were also well underway because of the caving and subsidence problem (see fig. 18.13). Major areas of the mine above the 500-foot level had to be abandoned because of caving. Gradually, changes in mining methods and sequencing of stopes helped to stabilize the ground and prevent additional movement. Permanent ore pillars were left in place below the 500-foot level. Shrinkage and square-set stopes were backfilled with waste rock as quickly as possible.

Fig. 18.13. The caving and subsidence that was plaguing Lead's business district north of Main Street, ca. 1920s. The large building in the lower right is the Cotton and Andrews building, which was located on the southwest corner of Main and Mill streets.—Photo courtesy of Black Hills Mining Museum.

Fear that the mine might close because of insufficient ore reserves contributed to the labor shortage that was already impacting the mine operation because of the effects of World War I and the Spanish flu epidemic of 1918. To address the ore reserve issue, Dr. Donald H. McLaughlin, from Harvard University, was hired to organize a geology department and evaluate the Main Ledge structure.

After several years of site investigation and geologic mapping, McLaughlin issued his report of findings and recommendations in 1927. First, he recommended that the practice of mining 100 percent of the ledge from hanging wall to footwall should be discontinued because the Homestake ledges contained significant amounts of barren or low-grade rock that only served to dilute the grade of the ore delivered to the mills. By not mining the low-grade rock, or mining it separately if necessary, the overall grade of the ore milled would be increased and mining, hoisting, and milling costs would be reduced. The recommendation was implemented in 1928. In October 1929, the level of production from the upper-level drawhole stopes was significantly reduced. Both of these changes resulted in an increase in the grade of ore delivered to the mill from $4.40 (0.213 ounces per ton) in 1926 to $7.52 (0.364 ounces per ton) in 1932.[20]

Milliken (No. 1) Winze

Secondly, McLaughlin recommended sinking a winze from a point 1,450 feet south of the Ellison Shaft on the 2,300-foot level from which exploration work could be conducted. Sinking of the No. 1 Winze was started in 1928 from the 2,300-foot level (see fig. 18.14). By the end of 1933, the Milliken Winze, as it was called, was developed from the 2,000-foot level to 71 feet below the 3,500-foot level. A Nordberg double-drum hoist was installed in the hoist room on the 2,000-foot level and operated two cages in balance in the timbered shaft. Based on the success of drifting and diamond drilling from the Milliken Winze, a decision was made to deepen the Ellison Shaft to the 3,200-foot level.

Fig. 18.14. The original hoist chamber for the Milliken Winze. The Milliken (No. 1) Winze was originally sunk from the 2,300-foot level using this small sinking hoist. Later, a raise was driven from the 2,300-foot level to the 2,000-foot level, where a double-drum hoist with two cages was installed. By 1933, the Milliken was the deepest winze in the mine, with its bottom 71 feet below the 3,500-foot level.—Photo courtesy of Black Hills Mining Museum.

Ellison Fire of 1930

Disaster struck on the evening of July 10, 1930, when a fire broke out in the Ellison headframe. Except for the hoist room, the entire Ellison complex, including the compressor room, headframe, and change room, was ablaze (see fig. 18.15). Two men were trapped on the cage between the 2,000- and 2,150-foot levels. In a short while, the intensity of the fire burned through the hoist cable, causing more than 2,000 feet of hoist cable to plummet down the shaft. The enormous weight and impact of the hoist cable overcame the holding power of the safety dogs on the cage, driving it to the bottom of the shaft. The two men on the cage were killed instantly. The rest of the miners working underground were evacuated through the B&M shafts from the 1,550-foot level. Within hours, the headframe, compressor room, and change house were destroyed (see fig. 18.16).

Fig. 18.15. The Ellison fire, 1930. On the evening of July 10, 1930, a fire started in the Ellison headframe, which resulted in the deaths of two employees who were on the cage at the time.—Photo courtesy of the Black Hills Mining Museum.

Fig. 18.16. The aftermath of the Ellison fire, 1930. The Ellison headframe, compressor room, and change room were completely destroyed in a fire that started on the evening of July 10, 1930. The facilities were quickly rebuilt.—Photo courtesy of the Black Hills Mining Museum.

The following day, cleanup work and plans for reconstruction of the Ellison surface plant were initiated. Fortunately, the cage hoist and the ore hoist were not significantly damaged. By August 16, 1930, the Ellison Shaft was recommissioned on a limited basis. The shaft resumed full-production status in February 1931. That same year, the steam-powered cage hoist was replaced with a new, 1,000-horsepower electric hoist. Drifts were driven from the Milliken Winze on the 3,200-foot level and other levels to the downward projection of the Ellison Shaft. Pilot raises were driven upward to connect to the bottom of the shaft on the 2,300-foot level. The pilot raises were subsequently stripped to full dimension, and a skip pocket was established just below the 2,900-foot level. Hoisting of ore from this pocket was started in August 1934. The Ellison continued to serve as a primary production shaft until shortly after the Yates Shaft was commissioned in 1941.[21]

The persistent subsidence problem culminated on September 2, 1932, when Homestake issued notices to business owners on both sides of Main Street around Mill and Bleeker streets that required them to vacate their buildings because of unsafe conditions. This notice impacted businesses such as the May Grocery Store, Black Hills Studios, Brown Drug Store, Gushurst Grocery, Fink Jewelry Store, Clark Clothing Company, Searle Meat Market, Lead Bakery, Woolworth Store, Red Owl, Lead Drug Store, Robbins Barber Shop, North Star Dairy, Kulpaca Grocery, and many others.[22]

Introduction of Underground Sandfill

In 1932, after nearly three decades of subsidence that caused the areas around North Mill Street and North Bleeker Street to subside by as much as 32 feet, Homestake began experimenting with spent sand tailings from the sand treatment plants for use as backfill material in stopes. The spent sand fraction represented the coarser (plus-200 mesh) product after the rock mass had been drilled, blasted, hoisted, crushed, pulverized to sand-sized grains, and treated with cyanide to dissolve and liberate the gold.

Prior to 1932, all of the spent sand was discharged into Gold Run Creek. The smaller, minus 200-mesh fraction, called the slime fraction, was generally unsuitable for backfill because of its inability to decant water and provide a stable fill. In addition, the slime fraction still contained recoverable gold and was delivered by pipeline to the Slime Plant in Deadwood for further gold recovery.

The first sandfill test was initiated on September 7, 1932. Spent sand was continuously pumped underground from the Cyanide Plant No. 1 to one of the stopes in the Pierce Ledge via pipelines and drill holes until October 7. Upon completion of the test, some 32,875 tons of sand had been pumped underground. Except for a few worn-out pipelines, the experiment was deemed a success.[23] Testing was resumed in the spring of 1933 by sandfilling other Pierce stopes above the 800-foot level. A total of 114,420 tons of sand was directed into several stopes over a seventy-six-day pumping period in 1933.[24]

On October 6, 1934, the sand-pumping system at Cyanide Plant No. 1 was replaced with a new gravity system from Cyanide Plant No. 3, which was commissioned in January. Spent sand from Cyanide Plant No. 3 was sluiced into a sandfill pipe installed in a forty-five-degree raise that had been driven from the 500-foot level to a point on the surface just north of Cyanide Plant No. 3. On the 500-foot level, the sandfill flowed through 1,300 feet of pipe to the Ellison Shaft. Pipelines installed in the shaft allowed sandfill to be directed to the large No. 18 stope on the 1,850-foot level. Concurrently, several other stopes were also prepared for sandfill in an attempt to protect the shaft from the ground-control problem that now had the Ellison Shaft in jeopardy. After 1934, it became standard practice, wherever possible, to sandfill every stope whether or not the stope was backfilled with waste rock.[25]

Sandfilling was introduced to the square-set method of mining on a limited basis in January 1937. Later that year, the North Sand Raise was completed between the 1,100-foot level at a point just east of Cyanide Plant No. 1. A sand dam (see fig. 18.17) was constructed at that location to accept spent sands from both sand treatment plants. Two 6-inch-diameter rubber-lined pipelines, designated as the North Sand Line and the South Sand Line, were installed in the North Sand Raise to the 1,100-foot level. There, one pipeline was routed to the Ellison Shaft and the other to the South Sand Raise.

At these two locations, the sandfill pipelines were extended downward to the various other levels of the mine. After these two systems were commissioned in 1938, the old sandfill system to the 500-foot level was abandoned. Spent sand tailings from Cyanide Plant No. 1 (East Sand Treatment Plant) and Cyanide Plant No. 3 (West Sand Treatment Plant) were gravity fed and stored in the sand dam. Whenever sandfill was needed underground, water cannons were used to slurry the sand from the sand dam into one or both pipelines at 45 to 55 percent density.[26]

Fig. 18.17. The sand dam and cement silo.—Photo courtesy of Homestake Mining Company.

Preparations were also made for establishing a waste-handling system for the upper levels. The intent was to mine rhyolite, an igneous intrusive rock that forms much of the hill north of the B&M No. 2 Shaft, for use as backfill material. The rhyolite would be used, as needed, to backfill stopes in Main Ledge between the 1,100- and 2,600-foot levels. In December 1935, a 7×10-foot timbered cage raise, called the H-Raise, was driven upward 900 feet from the 1,100-foot level to intersect the rhyolite sill area east of the Open Cut. A chamber was mined and a grizzly was installed over the top of the raise.

On the 1,100-foot level, a north-south drift 4,600 feet long was completed between the H-Raise and another system of raises, collectively called the H-2 Raise, which extended to the 2,600-foot level in Main Ledge. The plan was to mine waste rock on or near the surface and transfer it through the H-Raise to the 1,100-foot level. An 8-ton trolley locomotive with Granby-style cars was purchased to transfer the waste rock across the 1,100-foot level to the H-2 Raise. From there, the waste rock would be transferred, as needed, to stopes in Main Ledge below the 1,100-foot level. Crosscuts and chutes were installed at the H-2 waste raise on the 1,550-, 2,000-, 2,150-, 2,300-, 2,450-, and 2,600-foot levels. However, based on the success and relative cost-effectiveness of the hydraulic sandfill system, the elaborate waste-handling system was never fully utilized as originally intended.[27]

After the Ellison Shaft caved in 1975, that part of the North Sand System that was routed through the shaft was replaced with a series of 7⅞-9-inch-diameter holes that were drilled through the rock from the 1,100-foot level to the 3,050-foot level. On key levels, pipelines were installed laterally across the level to the respective stoping areas. In 1978 and 1979, existing pipelines in the South Sand Raise were similarly replaced with holes drilled through the rock. The North and South Sandfill systems enabled sandfill to be delivered by gravity to most anywhere in the mine below the 1,100-foot level.

Through the 1940s, square-set stopes continued to be mined without sandfill even though sand backfill was being used elsewhere throughout the mine. Upon completion of a square-set stope, the ends of the timberline at the bottom of the stope were sealed, and the entire stope was backfilled with sand. Similarly, after all of the ore had been emptied from a shrinkage stope, it was sealed and filled with waste rock generated from development drifts or waste stopes located on levels above the stope. Concurrently, sand was slurried into the stope to provide a tight fill. As the stope was being backfilled, the slurry propagated through the waste fill, causing the sand to settle out in the voids of the waste fill. The water drained out of the stope and was pumped back to the surface where it was reused in the milling processes.

In 1955, more than 685,000 tons of sand were returned to the mine as backfill,[28] compared to 215,600 tons in 1935.[29] By 1969, the number exceeded 989,000 tons.[30] Sand was delivered underground as a slurry at the rate of about

150-190 tons per hour at a slurry density of 45 to 55 percent solids by weight. Because of differences in the bulk densities between the sands and the in situ ore, along with the waste fill that was generated from development drifting underground, only about half as much sand was returned to the stopes as was originally mined.

In 1965, a 200-ton-capacity cement silo was erected near the top of the North Sand Raise at the sand dam. Bulk cement was purchased from the South Dakota State Cement Plant in Rapid City. Cement was added to the sandfill slurry whenever a 6-inch-thick, 10:1 sand-to-cement "cap" was needed to complete a "lift" or mining cycle in an open cut-and-fill stope. The cement cap, which also provided a solid working floor for miners, was placed over the drained sandfill in a stope in lieu of the wood floor that had historically been placed over the sandfill. Both types of floors were incorporated into the cut-and-fill mining cycle to minimize the dilution of ore that would otherwise result when a new lift or ore was mined from the back and removed from the stope.[31]

In the 1980s, a 30:1 sand-cement fill was often utilized to consolidate waste rock fill in vertical crater retreat (VCR) stopes prior to mining ore in adjacent pillar panels. The practice was discontinued in the 1990s because of the high cost and the diminishing number of ore reserves that were amenable to the VCR method.

The sandfill methods and practices introduced in 1932 proved to be highly effective in helping to stabilize underground stopes and stoping areas through cessation of mine operations in December 2001.

Chapter 19
Later Acquisitions Along the Homestake Belt

Starting in about 1912, Homestake Mining Company acquired the properties of various other mining companies that were situated on or near the Homestake Belt. Some of the more noteworthy companies included the Columbus Consolidated Gold Mining Company, which had previously acquired the Hidden Fortune Mining Company on May 1, 1907, the Homestake South Extension Mining Company, and the Oro Hondo Mining Company.

Columbus Gold Mining Company

The Sir Roderick Dhu claim was located on October 23, 1876, when ore was discovered in Sawpit Gulch immediately west of Central City. The claim was situated north of the Belcher and Father DeSmet claims. Chris Ruth and William Lardner, owners of the Sir Roderick Dhu, constructed a 10-stamp mill in Sawpit Gulch and sank two shallow shafts (see fig. 19.1).[1] By early 1880, the company had developed two shafts to a depth of 100 feet. Some 2,500 feet of drifting was completed to develop the ore body that was found to be 90 feet wide. The efforts were not profitable, however, and the mine closed later that year.

Fig. 19.1. The Ruth and Lardner 10-stamp mill in Sawpit Gulch near Central City. The Columbus Shaft, located a short distance up the gulch from the stamp mill, was backfilled by Homestake Mining Company in 1934. In the 1990s, Homestake created the Sawpit Waste Rock Facility as part of its Open Cut mining operations. The eastern toe of the waste rock facility covered the shaft.—Photo courtesy of the Black Hills Mining Museum.

Later in 1880, the property was restaked as the Columbus and Columbus No. 2 claims. A new company called the Columbus Gold Mining Company was organized the same year under the laws of New York with Henry F. Herkner, president. Immediately, the company began issuing capital stock at $10 per share.[2] A new 10-stamp mill was constructed in Sawpit Gulch. A two-compartment 8×10-foot shaft, called the Columbus Shaft, was sunk to a depth of 200 feet. Mining and milling operations were shut down in 1896 because of a shortage of water for the mill.[3]

Columbus Consolidated Gold Mining Company

In 1902, H. J. Mayham, on behalf of a New York syndicate, purchased the Columbus property. Mayham and his investors organized the Columbus Consolidated Gold Mining Company under the laws of the state of South Dakota. Mayham was elected president of the company, and Moses Thompson was named superintendent. The Columbus Shaft was enlarged to three compartments, each of which was 4 feet 9 inches by 5 feet. A new steam hoist, an air compressor, and a pump were also installed.[4] Mayham touted that "the Homestake is the greatest gold mine in the world, and the Columbus is the mine next to it."[5]

In fall 1902, the Columbus Consolidated Company purchased the property of the Baltimore and Deadwood Mining Company, which included a 20-stamp cyanide mill at Blacktail (see fig. 19.2). Some 645 acres of ground between Deadwood and Blacktail gulches were also purchased. Six additional cyanide tanks were added to the mill in 1903, which increased the mill capacity to 100 tons per day. With these new properties, Cambrian ores of the Deadwood Formation were mined at numerous locations around Deadwood Gulch, Blacktail Gulch, and Sheeptail Gulch.

Fig. 19.2. The Columbus Consolidated Mill in Deadwood Gulch, ca. 1902. The Columbus Consolidated Gold Mining Company acquired the 20-stamp cyanide mill of the Baltimore and Deadwood Company in 1902. The mill, which was completed on February 1, 1899, for the Wooley-Peacho Mine, was located on the north side of Deadwood Gulch and west of the mouth of Blacktail Gulch.—Waterland Collection, Devereaux Library Archives, South Dakota School of Mines and Technology.

By 1905, the Columbus Shaft in Sawpit Gulch had been deepened to 500 feet and drifting was underway on the 200- and 500-foot levels. A Precambrian ore body 247 feet wide was discovered on the 200-foot level. Samples indicated the ore averaged $4.96 per ton. Several veins were discovered on the 500-foot level that sampled $3.75 per ton.

On September 10, 1906, the shareholders of the Columbus Consolidated Mining Company met to discuss a possible consolidation with the Hidden Fortune Gold Mining Company. After the matter was approved, assets of the Hidden Fortune Company were transferred to the Columbus Consolidated Mining Company on May 1, 1907. Both mining properties remained idle until 1912 when a group of capitalists from Montreal, Canada, secured an option on all of the Columbus properties. The capitalists dewatered the Columbus and Hidden Fortune shafts and conducted an extensive underground sampling program at both mines. No subsequent mining operations were started at either property. The mills of the Columbus Consolidated Mining Company were later sold to satisfy judgments against the company.[6] Homestake acquired the land of the Columbus Company in 1912 shortly after the option expired.[7]

Hidden Fortune Mining Company

Numerous mines were in operation in the late 1870s in the area between Poorman Gulch and the head of Bobtail Gulch. Some of the operations mined the near-surface conglomerate ores of the Deadwood Formation while others pursued Homestake-like ore in the Precambrian schists. Some of the mines included the Bingham, Cheyenne, Durango, Harrison, and the Hoodlebug, among others. Most of these mines closed, or were being operated very sporadically by the early 1880s.

In about 1889, G. E. Marvine purchased the Cheyenne Consolidated Mill and Mining Company and hired Otto Grantz as the company's resident agent and superintendent. Grantz drove a tunnel from Poorman Gulch across the Hoodlebug, Bingham, Baltic, and Marvine claims. Several ledges of pyritic ore were encountered that assayed from $62 to $163 per ton. Very little ore was ever mined during Marvine's ownership of the company.[8]

Otto Grantz and Dave Wolzmuth located the Hidden Fortune mining claim in 1886. The claim was located on the ridge between the head of Bobtail Gulch and Poorman Gulch. After the Cheyenne Company failed, Grantz acquired the Marvine and Swamp Eagle claims and purchased Wolzmuth's interest in the Hidden Fortune claim. In August 1899, Grantz discovered a fabulously rich pocket of ore on the surface of the Hidden Fortune claim (see fig. 19.3). He purchased a large number of sacks and commenced to fill the sacks with 200 tons of high-grade ore. The sacks were hauled to Deadwood in wagons, reloaded onto a railroad car, and shipped to Denver. After the ore was processed, Grantz netted $60,000 for his work.

Fig. 19.3. The famous Hidden Fortune gold discovery site, 1899. Otto Grantz is standing next to his wife at the site of his original discovery on the Hidden Fortune claim. In 1901, Grantz sold the Hidden Fortune claim and several others to a group of investors from Denver who formed the Hidden Fortune Mining Company. The Hidden Fortune claim is located just north of the Radio Tower in Lead.—Fielder Collection, Devereaux Library Archives, South Dakota School of Mines and Technology.

In 1900, Grantz drove a drift 175 feet long into the hill under the pocket of rich ore. No downward extension of the ore was found. He then mined several additional carloads of rock from around the original discovery site and had the material processed at a smelter in Denver. That same year, he purchased the land holdings of the Cheyenne Consolidated Mill and Mining Company at a sheriff's sale. The following year, Grantz sold all of his property holdings in the area to a group of investors from Denver for $100,000. Immediately, the investors formed the Hidden Fortune Mining Company and began selling stock in the company (see fig. 19.4).[9]

Fig. 19.4. Eastern investors inspect the Hidden Fortune Mine discovery, 1902. In August 1902, 12 tons of rich ore were handpicked from the Hidden Fortune No. 2 Claim and shipped to the National Smelter in Rapid City. The ore yielded $6,598.43 in gold. Some 1,200 tons of additional ore were then shipped, which averaged $23 per ton.—Fielder Collection, Devereaux Library Archives, South Dakota School of Mines and Technology.

Over the next five to six years, the Hidden Fortune Gold Mining Company completed a considerable amount of development work. A drift was driven into the side of the hill from the Bingham Claim and intersected 60 feet of Precambrian ledge. Based on these results, a new 7×12-foot, double-tracked tunnel, called the Baltic Tunnel, was driven 1,888 feet into the side of the hill from Poorman Gulch in 1902 and 1903. Crosscuts were driven at intervals of 150, 500, 650, 1,000, 1,213, and 1,750 feet from the mouth of the Baltic Tunnel. A winze was sunk to a depth of 34 feet from the first crosscut to the west.

To access the new Bingham ore body, grading was completed for a shaft on the Bingham claim not too far from the Baltic Tunnel. A raise 75 feet long was driven from the No. 1 crosscut of the Baltic Tunnel to the surface, which formed the upper part of the new Bingham Shaft. A hoist, a water heater, and two 100-horsepower boilers were purchased for the new shaft. The building that housed the former

Golden Crown Mill was dismantled, moved, and erected over the Bingham Shaft (see fig. 19.5). By February 1903, the shaft was 205 feet deep.[10]

Fig. 19.5. The Bingham Shafthouse of the Hidden Fortune Gold Mining Company. The Bingham Shaft was located in Poorman Gulch, opposite the road to the Cutting Mine. The shafthouse had to be reconstructed in 1903 after a fire destroyed the original building on March 24.—Waterland Collection, Devereaux Library Archives, South Dakota School of Mines and Technology.

In November 1902, separate tunnels were started on the Harrison and St. Patrick claims for the purpose of mining near-surface, siliceous ore from the Harrison, Golden Crown, Durango, Reddy, Golden Summit, and Sula claims. Work was also started on a new 60-stamp cyanide mill that was located on the hillside just above Whitewood Creek, approximately 1 mile below Deadwood. A spur of the Fremont, Elkhorn, and Missouri River Railroad was constructed to the top of the mill bins, which allowed ore to be shipped by rail. Mill construction was delayed several times because of cost overruns and financing issues. The mill, which had a capacity of 175 tons per day, was commissioned on October 1, 1903.[11]

Concurrent with construction of the mill, disaster struck on March 24, 1903, when the Bingham Shafthouse was destroyed by fire. The hoist, compressor, and other equipment in the building were badly damaged. The shaft timbers themselves burned to the 75-foot level. Immediately, a new headframe and a 60×120-foot shafthouse were constructed at a cost of $5,060.19. An additional $2,172.28 was spent on equipment repairs. After the shaft was deepened to 319 feet by September, some 298 feet of crosscut development was completed on the 300-foot level to access the Bingham vein. Two carloads of ore were mined, which yielded an ore value of $11 per ton.

Throughout 1903, development was also conducted on the Bingham, Durango, Harrison, Grantz, Cheyenne, St. Patrick, Golden Crown, Iowa, and Grand Deposit claims. In July, a two-compartment shaft was started on the Golden Crown Claim. By year's end, the shaft was 66 feet deep. The Harrison-Durango Tunnel was driven 695 feet to access the siliceous and cement ores on seven claims west of the Homestake and Highland mines.

Additionally, the St. Patrick Tunnel was extended 791 feet at which point crews encountered ore and started production. Some of the older workings on the Harrison, Iowa, and Hidden Fortune No. 2 were also reopened to access additional cement ore. Several car-loading stations were constructed along the Black Hills and Fort Pierre Railroad (see fig. 19.6). Each station was designed to accommodate eight to twelve railcars. A tramway 1,684 feet long was constructed to transfer 75-100 tons per day of ore from the Sula open cut to the Black Hills and Fort Pierre Railroad.

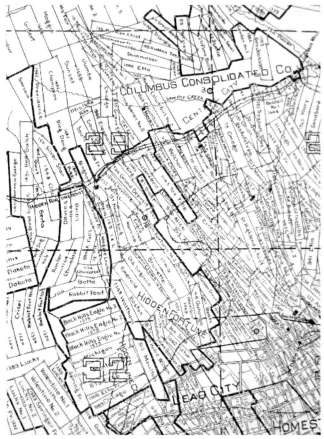

Fig. 19.6. A portion of the 1904 "Map of the Ore District of the Northern Black Hills," by Frank S. Peck, showing the claim holdings of the Hidden Fortune Gold Mining Company and the Columbus Consolidated Gold Mining Company. The Deadwood Central Railroad is shown along Deadwood Creek. The Black Hills and Fort Pierre Railroad is shown crossing the Hidden Fortune claims and extending to points above the Father DeSmet Mill and the stamp mills in Terraville.—Illustration by author.

On February 20, 1904, J. "Henry" Schnitzel was hired to manage the company's mines and new mill after A. H. Danforth resigned because of ill health. Schnitzel was considered by the company to have "more experience than any other man in the Black Hills in managing mining properties . . . is thoroughly acquainted with every foot of our property, has made thousands of assays of the ore of our different mines; is a successful business man, and has made a success of mining in the Black Hills District. In addition to this, he is a chemist of fine technical knowledge and thoroughly conversant with the operation of stamp and cyanide mills."[12]

Unfortunately, despite Schnitzel's qualifications and best efforts, the company was already in deep trouble financially. Ore was being mined and milled faster than it could be replaced with new reserves. In the company's annual report for the period ending February 20, 1904, A. M. Stevenson, president of the company, announced to shareholders that the company would be unable to pay the debt that would mature prior to June 1, 1904. Stevenson informed the shareholders that he had looked after the affairs of the company for the last three years without any compensation whatsoever. In fact, Stevenson said that he, along with Messers Mayham and Shaw had advanced $100,707.52 in order to carry on operations. Stevenson concluded his report to the shareholders by informing them that he was resigning his position as an officer of the company.[13]

H. J. Mayham was subsequently elected president of the company on March 4, 1904. Performance of the mine and mill was steadily declining, and the company was $200,000 in debt. Around January 15, 1905, Mayham learned that some of the directors and shareholders were taking action to elect a new board of directors. Mayham decided to resign as president and director on January 30. Messers, Freeman, Begole, and Bottum also resigned as directors. Judge John P. Allison of Sioux Falls was subsequently elected president of the company, and Nathan E. Franklin, John H. Morcom, Henry Frawley, and Thomas J. Steele were elected directors. Allison, A. M. Stevenson, Otto Grantz, and Herbert S. Shaw continued as directors. Steele was appointed general manager.[14]

Gold production for 1904 and 1905 amounted to $108,000 and $80,724.24, respectively.[15] In 1906, Kirk G. Phillips brought suit against Allison, as president, for $90,000 claiming Allison had failed to properly account for the stock, bonds, and funds of the company. The bondholders subsequently foreclosed on the company. At that time, shareholders of the Hidden Fortune and Columbus Consolidated companies decided to transfer property of the Hidden Fortune Company to the Columbus Company for $230,000 in Columbus bonds and two million shares of Columbus stock. The property was transferred on May 1, 1907.[16]

Homestake Mining Company acquired the claims of the Columbus Consolidated Gold Mining Company in 1912. The properties included the former Hidden Fortune mining claims. In 1916, based on the results of a sampling program conducted around the old Harrison Mine on some of the former Hidden Fortune claims, it was discovered that tungsten was present in economic quantities as wolframite ore. A 5-stamp mill was erected in the old Regrind Plant building in

Lead early in 1916. Ore was carefully mined, hand-sorted, and hauled to the new tungsten mill in wagons. Mill concentrates were dried, sacked, and shipped to an outside facility. Mill tailings were delivered to the Cyanide Plant No. 1 for gold recovery. During the first year of operation, some $229,915 worth of tungsten was recovered. The tungsten operation was shut down in June 1918.[17]

Homestake South Extension Mining Company

The Homestake South Extension Mining Company consisted of 50 acres of patented claims situated south of Homestake's property and east of the property owned by the Oro Hondo Mining Company. The 50 acres straddled Whitewood Creek, with most of the acreage situated on the south side. Andrew H. Oleson of Deadwood was general manager of the company in 1905. Oleson hypothesized that the main Homestake ledges would be too deep to be accessed from his company's property but believed that smaller veins might be found near the surface.

Oleson purchased a small steam-hoisting plant and started sinking a two-compartment shaft near the toe of the hill on the northwest side of Whitewood Creek (see fig. 19.7). By the time the shaft reached a depth of 100 feet in the first part of 1906, it was found that additional boilers were needed to provide sufficient power to the hoist. Two 60-horsepower boilers were installed. A coal shed was constructed alongside the Burlington railroad, which provided easy access to coal fuel. The workforce was increased to twenty-four people. By the end of 1906, the shaft was 165 feet deep. Crosscuts were driven east and west from the 150-foot level. One of the crosscuts encountered a ledge of ore 15 feet wide that averaged $2 per ton in gold.

Fig. 19.7. The Homestake South Extension Company located along the northwest side of Whitewood Creek. The company operated from about 1905 to 1909 during which time a shaft was sunk to a depth of 200 feet. Homestake acquired the mining claims on November 20, 1913. The shaft is now covered by the reclaimed Yates Waste Rock Pile.—Photo courtesy of the Black Hills Mining Museum.

In 1907, the steam hoist was replaced with an electric hoist. A compressor and several air-powered drills were purchased. Thirty acres of additional ground were purchased for $125,000. By year's end, the shaft was 175 feet deep.[18] In 1908, the shaft was deepened and timbered to a depth of 200 feet. Drifting and crosscutting continued on the 150-foot level, bringing the total development on the level to about 2,000 feet. The westerly crosscut on the level extended outward 750 feet and encountered several veins, the largest of which was about 45 feet wide. The ledges were reported to be quite similar to those found within the Homestake Mine.[19]

A short time later, the mine was closed and was acquired by the Black Hills Consolidated Mines. Homestake Mining Company purchased the 80 acres of the Homestake South Extension Mining Company from Black Hills Consolidated Mines on November 20, 1913.[20]

Oro Hondo Mining Company

Early in 1902, George M. Nix of Lead, Otto P. Grantz of Deadwood, and H. J. Mayham of Denver bonded 700 acres of property that adjoined Homestake's property to the south.[21] Nix, Grantz, and Mayham were all very experienced and successful mining men. The men knew that Homestake's Main Ledge was plunging to the south, based on Homestake's planning and work to strategically position the Ellison Shaft in 1895. After carefully calculating the depth of the Homestake ore body at varying distances from its original discovery relative to the Ellison, the men decided to locate a shaft some 1,600 feet south of the Ellison Shaft. Nix, Grantz, and Mayham were confident the shaft would penetrate the Homestake ledges from this location. The men procured options for three possible mill sites, each of which had water and could be accessed by the Burlington Railroad that passed within 200 feet of the shaft site.[22]

With Nix as superintendent, work was started on a three-compartment shaft measuring 7×17½ feet outside the timber. The shaft was located on the south side of Whitewood Creek and upstream a short distance from the 2,500-foot-long Savage Tunnel that Homestake purchased from James Milliken in 1898. Shortly after work was started on the shaft, the Oro Hondo Mining Company was organized with a capitalization of $10 million. Immediately, the property was transferred to the new company. Governor James B. Orman of Colorado was elected president, George M. Nix, vice president and general manager, R. H. Driscoll, treasurer, and George D. Begole, secretary.[23]

Using a windlass, crews sunk the Oro Hondo Shaft to a depth of 80 feet. While waiting for delivery of hoisting equipment, a 46×140-foot shafthouse was constructed (see fig. 19.8). By fall, the hoisting and mining equipment arrived. Crews also constructed a blacksmith shop and carpenter shop. On December 10, 1902, shaft sinking was resumed on a three-shift basis using the new hoisting equipment. A spur from the Burlington Railroad was constructed for delivery of coal and supplies. After the shaft reached a depth of 600 feet, a station was cut and a large steam-driven pump was installed.

Fig. 19.8. The shafthouse of the Oro Hondo Mining Company. By 1918, the Oro Hondo Mining Company had sunk the shaft to a depth of approximately 2,300 feet in an attempt to locate the down-plunge extension of Homestake's Main Ledge ore body. Diamond drill holes drilled from the 2,000-foot level of the Oro Hondo Mine fell short of the Main Ledge structure by about 100 feet. Homestake acquired the property in 1925. Between 1937 and 1941, Homestake extended the shaft to the 4,100-foot level by driving a series of raises from the top of No. 2 Winze on the 3,500-foot level. From that point forward, the shaft was used as an exhaust shaft.—Photo courtesy of the State Archives of the South Dakota State Historical Society.

An annual meeting of the Oro Hondo Mining Company was held on June 28, 1903. At the meeting, company officials announced that the shaft was now 1,000 feet deep and that ore had been discovered 70 feet above that level.[24]

During 1904, the shaft was deepened to 1,050 feet. Approximately, 1,100 feet of crosscutting was completed on the 1,000-foot level in an effort to locate the down-plunge extension of the Homestake ledges. Results were not encouraging. To make matters worse, the company was experiencing financial difficulties. In 1906, the Oro Hondo Mining Company defaulted, and J. W. Sparks and Company purchased it. The Black Hills Consolidated Mines, which also owned the Homestake South Extension Mine just to the east along Whitewood Creek, purchased the Sparks Company's interest in the Oro Hondo Mine for about $75,000 in 1908.

In 1914, John T. Milliken acquired several other undivided interests in the Oro Hondo property and reopened the mine. Because the shaft was flooded to within 30 feet of the collar, Milliken installed skips with a capacity of 600 gallons each and commenced to bail water from the shaft. Two new 100-horsepower boilers were added to the plant to increase steam capacity. A sinking hoist was installed on the 1,000-foot level, and the shaft was deepened to approximately 2,300 feet. Development drifting and diamond drilling were completed on the 1,500- and 2,000-foot levels.

Unbeknownst to Milliken, diamond drilling from the 2,000-foot level stopped short of Homestake's Main Ledge ore body by a mere 100 feet. Faced with growing expenses and little income, the mine was forced to close during World War I. On February 16, 1925, Homestake Mining Company purchased Sparks's interest in the Oro Hondo property. Homestake also purchased the Milliken interest on April 6 the same year.[25]

After acquiring the properties of the Oro Hondo Mining Company, Homestake had no use for the Oro Hondo Shaft as a production shaft since the Main Ledge ore body was already being developed at depth from the Ellison Shaft. The Ellison could supply ore to the South Mill via the existing tramway; the Oro Hondo could not. The biggest consolation was the fact that Homestake now owned the Oro Hondo properties within which significant volumes of down-plunge ore soon would be delineated in Main Ledge.

Chapter 20

What Depression?

While the rest of the nation was reeling from the stock-market crash of 1929 and the Great Depression that followed, Homestake and the northern Black Hills communities were prospering from the increase in gold prices under the Gold Reserve Act of 1934 and Homestake's newly found ore in Main Ledge and 9 Ledge. The price of Homestake stock increased steadily from about $80 per share in October 1929 to $495 per share in December 1935. In 1936, the stock price reached $544 per share. The stock subsequently split eight for one.[1]

Fortunately, for Homestake, the market timing was right. With the great Homestake ledges plunging deeper and deeper to the south, coupled with the loss of the stamp mills and shafts in the Open Cut area, large amounts of capital were needed to expand the mine's surface and underground infrastructure. New surface shafts were needed, extensive underground development and exploration work was required, and mill facilities at the South Mill needed to be expanded to replace the antiquated stamp mills in Lead and Terraville.

The Golden Star Shaft had already been decommissioned because of the subsidence issue. The B&M No. 1 Shaft was now in jeopardy for the same reason. The B&M No. 2 Shaft was in better ground, but it was too far from the south-plunging ledges. Without the B&M shafts, the Ellison Shaft alone would not be able to meet the minimum production requirement for the Homestake Mine. Moreover, the mine staff was concerned about the integrity of the upper portion of the Ellison Shaft because of the expanding subsidence problem.

Ross Shaft

Therefore, in November 1932, Homestake officials decided to construct a new shaft, expand mill facilities, and expand the shops and yards. The new shaft would be located south of the Ellison Shaft and would be named the Ross Shaft, after Superintendent Alec J. M. Ross (see fig. 20.1). In order to accommodate production from the Ross Shaft, ore-treatment facilities would be relocated and expanded by constructing a new sand treatment plant west of the Cyanide Plant No. 1 and adding three rod mills to the South Mill. It was also decided that additional shops and support facilities would be constructed between the Ellison Shaft and the South Mill.

Fig. 20.1. The Ross Shaft Complex. The Ross Shaft was constructed to replace production from the Golden Star and B&M No. 1 shafts, both of which had to be abandoned because of ground problems and expansion of the mine. From left to right are the boiler, hoist room, headframe, crusher room, and dry.—Photo courtesy of the Black Hills Mining Museum.

Conventional shaft-sinking methods were employed to sink the Ross Shaft 137 feet to the tramway level. Below the tramway level, drifts were driven from the Ellison Shaft to the downward projection of the Ross on the 300-, 800-, 1,400-, 1,700-, 2,000-, 2,750-, 2,900-, and 3,050-foot levels. An interconnecting pilot raise was driven upward from each of these levels, with the uppermost raise reaching the tramway level in 1933. To facilitate timely completion of the shaft, the overall series of completed raises was divided into three work sections. Bulkheads were installed at the lower end of each work section to isolate each section. The three sections of pilot raise were subsequently stripped to an overall shaft dimension of 19 feet 3 inches by 14 feet to accommodate a cage compartment, a cage counterweight compartment, two skip compartments, a manway and power cable compartment, and a pipe compartment.

By the end of 1934, the Ellison Shaft was completed to the 3,200-foot level, and the Ross Shaft was completed to a point 108 feet below the 3,050-foot level. Ore raises were driven and connected between the 1,400- and 3,200-foot levels at the Ross during the year. Skip loaders were installed in newly mined loading pockets located just below the 2,000- and 2,600-foot levels. From the surface to a depth of 308 feet, the shaft was lined with concrete to provide ground support. With the efficient raising and stripping method that was employed, the cost to mine and furnish the Ross Shaft from the surface to the 3,200-foot level averaged only $132.31 per linear foot. The steel shaft sets alone accounted for $50.30 of the cost.

The new crushing plant included an 84-inch Allis Chalmers gyratory crusher and two Symons 7-foot cone crushers. Two hoists with bicylindrical-conical drums, manufactured by the Nordberg Manufacturing Company, were installed in a new brick hoist house. The drive motors for the hoists were 1,500 horsepower each, with one on the cage hoist and two on the ore hoist. The first ore was hoisted at the Ross Shaft on November 19, 1934.

In November 1934, an exploration drift was started from the north side of the Ross Shaft on the 3,200-foot level. Subsequent diamond drilling indicated the presence of an enormous ledge of ore to the west of the drift. Based on these results, a similar drift was driven on the 2,900-foot level. A raise was then driven between the two levels at No. 13 stope line. Subsill drifts were driven from the raise. Additional diamond drilling from these subsill drifts confirmed the presence of ore in the 9 Ledge structure.[2]

The Ross was deepened to a point between the 3,650- and 3,800-foot levels in 1935. A skip-loading pocket was completed just below the 3,200-foot level late that year and was commissioned on January 1, 1936.[3] Later that year, the shaft was completed to the 3,800-foot level, and a drift was driven from No. 2 Winze to the Ross on that level. At about the 3,500-foot level, ore-grade mineralization was discovered in the Ross Shaft. One of the first samples taken assayed $9 per ton. Because of poor ground conditions, the shaft had to be lined with concrete between the 3,200- and 3,800-foot levels.

In 1937, a drift was driven from the bottom of No. 2 Winze on the 4,100-foot level, and a pilot raise was driven to the 3,800-foot level for the Ross Shaft. Upon removal of a rock pillar just below the 3,800-foot level, the Ross Shaft was stripped and completed to the 4,100-foot level in March 1938. Ore raises were completed between the 3,350- and 3,950-foot levels, and a skip-loading pocket was mined and completed below the 3,950-foot level. Work was also completed on the Ross Compressor Plant in 1937.[4]

By the end of 1938, exploration work between the 2,900- and 3,650-foot levels had identified 6.446 million tons of ore in 9 Ledge at an average grade of $5.55 (0.269 ounces per ton), based on the historic gold price of $20.67 per ounce. Ore reserves now totaled 18.558 million tons.[5]

With the bottom of No. 2 Winze and the Ross Shaft at the 4,100-foot level, a new mine dewatering system was completed at the Ross in 1938. The system included twin pumps installed at pumping stations located on the 3,650-, 2,450-, and 1,250-foot levels. The new pumping system had a dewatering capacity of approximately 2,300 gallons per minute.[6]

In 1955, a pilot raise was completed for the Ross Shaft from the 5,000-foot level to 312 feet above the 4,850-foot level. The 5,000-foot level had previously been developed from the bottom of No. 3 Winze. In 1956, the pilot raise was completed to the bottom of the shaft just below the 4,100-foot level. The pilot raise was stripped to full dimension, and a skip-loading pocket was mined below the 4,850-foot level. By year's end, shaft steel was installed to the 5,000-foot level, and the skip loader was installed and commissioned.

A pumping station was also installed on the 5,000-foot level and operated in conjunction with the three pumping stations located higher in the shaft (see fig. 20.2).[7] Mine drainage water from the entire mine was directed to the Ross Shaft and pumped to the surface. There, the water was transferred to the South Mill and sand treatment plants and reused as process water.

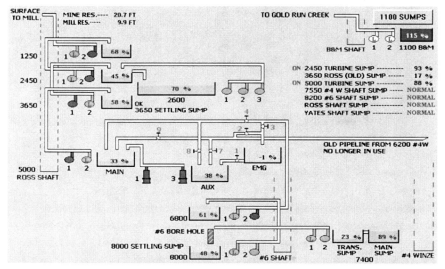

Fig. 20.2. A schematic of the mine pumping system. With the mine monitoring system, Homestake's "load dispatcher" was able to monitor and control the mine pumping system manually or let the system run in auto mode.—Illustration courtesy of Homestake Mining Company.

Cyanide Plant No. 3

As part of the surface relocation and expansion project, a third sand treatment plant, called Cyanide Plant No. 3 (later called the West Sand Plant), was constructed (see fig. 20.3). The plant was commissioned on January 10, 1934. During its first year of operation, an additional 1 percent of gold recovery was achieved as a result of additional fine grinding in the South Mill and additional leaching time in the sand treatment plants. With the additional capacity gained by construction of the South Mill and Cyanide Plant No. 3, the Amicus Mill and the stamp mills at Terraville were no longer needed as planned. The Pocahontas and Amicus mills were decommissioned in December 1934 as was Cyanide Plant No. 2 at Gayville.[8]

Fig. 20.3. Cyanide Plant No. 3, 1933.—Photo courtesy of Homestake Mining Company.

A total of 1.379 million tons of ore was milled at the South Mill in 1935. Revenue from gold production amounted to $19.191 million or $13.91 per ton milled. This included $544,496 from decommissioning and cleanup work at the Amicus and Pocahontas mills. Recovered grade was $8.41 (0.407 ounces per ton), based on the historic gold price of $20.67 per ounce. Gold recovery for the year reached an all-time high at 95.489 percent. Total costs for mining, treatment, general expense, and taxes amounted to $4.716 per ton. Mine management personnel considered the results of operations for 1935 to be "highly satisfactory." In addition, the new refinery and assay office were ready for commissioning.[9]

Production for 1936 and 1937 was equally good. Milled tonnage in 1936 was 1.384 million tons, which resulted in revenue of $19.507 million or $14.095 per ton. This amount included $166,880 derived from cleanup operations at the old refinery. Total costs, including local and state taxes, amounted to $5.006 per ton milled. The average grade of ore produced from underground was $8.70 (0.421 ounces per ton). Overall mill recovery was 95.243 percent. The total treatment cost, including crushing and surface haulage, was $0.768 per ton milled. Final cleanup at the sites of former stamp mills continued in 1937 (see fig. 20.4), resulting in $33,530 from the Pocahontas and Mineral Point mill sites.[10]

Fig. 20.4. Homestake's gold cleanup operation at the Pocahontas Mill site in Terraville, 1937. The gold cleanup was accomplished by a crew of summer students working for Joe Dunmire. The person with the hose is Albert "Bud" Mitchell. Dunmire is seated, panning the concentrates. The other people are not identified.—Homestake Collection, Adams Museum, Deadwood.

South Dakota Ore Tax

As initial work for the Ross Shaft and Cyanide Plant No. 3 was being completed, President Franklin D. Roosevelt signed Executive Order 6102 on April 5, 1933. The order was part of a continuing effort of the U.S. government to provide relief for the national emergency in banking. This order required private citizens to remit all gold coins, gold bullion, and gold certificates exceeding $100 in the aggregate to the Federal Reserve by May 1, 1933. The gold was exchanged for Federal Reserve Note currency at the rate of $20.67 per ounce. The order became known as the Gold Confiscation of 1933.[11]

On January 30, 1934, Congress passed the Gold Reserve Act to protect the currency system of the United States. The act put an end to the private ownership of gold, circulation of gold coins, redemption of currency in gold, and coinage of gold. The act required the Federal Reserve Banks to transfer all gold bullion and coins to the U.S. Department of the Treasury. In return, the banks received gold certificates that served as a reserve against deposits and Federal Reserve Notes.[12]

One day after the Gold Reserve Act was passed, President Franklin D. Roosevelt signed Presidential Proclamation 2072, which refixed the weight of the gold dollar from 25.8 grains of gold nine-tenths fine to 15-5/21 grains. This devaluation of the dollar caused a corresponding increase in the price of gold from $20.67 per troy ounce to $35 per ounce—a value that would not change until 1971 when the government abandoned the gold standard.[13]

Unfortunately, as was the case most everywhere outside of the Black Hills, the state of South Dakota was adversely impacted by the Depression and the drought throughout the early 1930s. Because Homestake was prospering as a result of the increase in the price of gold, state legislators believed Homestake could afford to help fund the state's annual budget for a few years. After legislators proposed a 10 percent "ore tax," Homestake lobbied hard against the tax and appealed support from the general public. Finally, Homestake realized that an ore tax was inevitable and that its only recourse was to lobby for a lower tax rate. With able assistance from Lawrence County's legislators, House Bill No. 4 was passed, which imposed a 4 percent ore tax on gross income effective July 1, 1935. Although other gold mines were operating in the Black Hills at the time, all but Homestake were exempted from the ore tax, based on the production levels specified in the bill.[14]

In 1936, Homestake's state tax bill rose to $1.122 million for property taxes, the ore tax, and unemployment taxes. The ore tax alone amounted to $751,186. However, this amount didn't create too much of a financial hardship for Homestake, since the total operating cost, including state and local taxes, was $5.006 per ton milled, compared to revenues of $14.095 per ton.[15]

Once again, the state legislators saw an opportunity to increase revenue by increasing the ore tax. Despite Homestake's best lobbying efforts, the ore tax was increased to 6 percent of gross income, effective July 1, 1937. Homestake's state taxes increased to $1.302 million that year, including $907,486 for the ore tax. Counting federal taxes, Homestake paid 38 percent of its income for taxes.[16] For 1937 and 1938, Homestake's taxes payable to the state amounted to approximately 49 percent of the state's general fund receipts and 89 percent of disbursements from total appropriations accounts.[17]

Beginning in 1938, Homestake embarked on a widespread information campaign to educate the people of the state about the high costs of mining and exploration. The company referenced the large capital expenditures it was incurring in sinking the Yates and Ross shafts, expanding the South Mill, and searching for new ore reserves to extend the life of the mine. Legislative action to have the ore tax reduced failed.[18]

In 1944, after the mine had been shut down for about two years under the War Production Board's Order L-208 for World War II, Homestake again made an appeal to the general public, stating that the company had preferentially hired five hundred farmers over nonresident miners from 1930 to 1936, at a time when drought conditions were at their worst in South Dakota. Now, Homestake needed support. Homestake said its net income had decreased by an average of 6.5 percent per year from 1935 through 1942. Over the same period, federal, state, and local taxes had increased at an average rate of about 14 percent per year, based on a total of $2.157 million in 1935 and $4.269 million in 1942. The appeal and lobbying efforts were eventually effective. In consideration of the high cost of reopening the mine, the state legislature passed a bill in January 1945 to reduce the ore tax to 4 percent of gross income.[19]

The 4 percent state ore tax remained in effect until 1957, when Homestake's lobbyists and district representatives were successful in having legislation passed that reduced the tax to 2½ percent. In 1964, the tax was reduced to 1 percent because of adverse conditions within the mining industry. Homestake's financial problems continued through 1967, primarily because of continued increases in the cost of operations and a gold price that remained fixed at $35 per ounce. Again, the state legislators were somewhat sympathetic, and a moratorium was placed on the ore tax for one year.[20]

In 1994, state legislators modified the ore tax by establishing a mineral severance tax "for the privilege of severing precious metals in this state." The mineral severance tax was levied at $4 per ounce severed plus 10 percent of the net profits from the sale of the precious metals. If the average gold price averaged $800 per ounce for a calendar quarter, an additional tax of $4 per ounce was imposed.[21]

No. 2 and No. 3 Winzes

Despite the ore tax, the increase in the price of gold to $35 per ounce in 1934 couldn't have been better timed for Homestake. Large amounts of capital were needed to replace the shafts and mills that were being adversely impacted by ground problems in the Open Cut area. Other capital was needed to fund deep-level exploration efforts to search for badly needed replacement ore. Exploration was starting to pay off with confirmation of ore in 9 Ledge and recognition that Main Ledge hadn't bottomed out. Based on this encouragement, Homestake decided to pursue the downward-plunge extensions of Main Ledge and 9 Ledge.

By the end of 1934, the Ellison Shaft was completed to its ultimate depth of 3,200 feet. The Ross Shaft was deepened to the 3,200-foot level, and the Milliken (No. 1) Winze was bottomed 71 feet below the 3,500-foot level.[22] Unfortunately, the downward extension of Main Ledge was found to be too far from the Milliken Winze. Consequently, a drift was driven on the 3,500-foot level from the Milliken to a location for a new winze called No. 2 Winze. The 7×9-foot winze was intentionally located directly below, and in line with, the projection of the Oro Hondo Shaft, which bottomed at about 2,300 feet. No. 2 Winze was sunk from the 3,500-foot level to the 4,100-foot level in 1935 and 1936.[23]

From the bottom of No. 2 Winze, a drift was driven south to a location for No. 3 Winze. After the hoist room, rope raise, and shaft station were completed, the single-drum sinking hoist was moved from No. 2 Winze to No. 3 Winze on the 4,100-foot level. Sinking operations commenced in December 1937. After Winze No. 3 reached the 4,550-foot level in 1939, an exploration drift was driven to the Main Ledge ore body. In 1940, the winze was deepened to the 4,700-foot level, and another drift was driven to Main Ledge. No. 3 Winze reached its ultimate bottom, 25 feet below the 5,000-foot level, in October 1941.[24]

A series of inclined raises was driven from the 4,700-foot level to the 4,100-foot level in Main Ledge. Collectively, the raise system was called the supply raise. A winch was installed on the 4,100-foot level, and mine rail was laid on the floor of the raise. With this system, railcars full of supplies could be lowered to the various levels between the 4,100- and 4,700-foot levels (see fig. 20.5). The supply raise was phased out in the late 1970s after more of the levels were connected to the Ross Shaft.

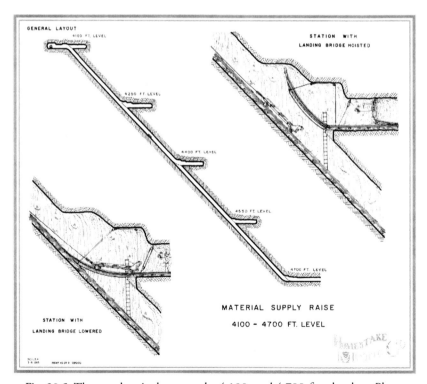

Fig. 20.5. The supply raise between the 4,100- and 4,700-foot levels.—Photo courtesy of Homestake Mining Company.

The B&M No. 2, Ellison, and Ross shafts provided primary and secondary access from the 1,550-foot level to the surface. As mine development was completed within the Main Ledge area, a "safety" stairway, called the Golden Stairway, was completed between the 1,550- and 2,600-foot levels in 1936 (see fig. 20.6). The stairway was completed between the 4,100- and 2,600-foot levels in Main Ledge in 1937.[25] Wooden stairs and pipe handrails were installed throughout the Golden Stairway, thereby providing a travel way and secondary escape way for people working on levels below the 1,550-foot level.[26]

Fig. 20.6. The Golden Stairway that extended between the 1,550- and 4,100-foot levels.—Photo courtesy of Homestake Mining Company.

Many employees became quite proficient at straddling the two handrails and sliding down the steeply inclined stairway from level to level with their lunch pail, bit rack, or other tools fastened to their mine belts or the shoulder straps on their bib overalls. The popular means of downward travel was frowned upon by company management and was mostly brought to an end by applying cement to the handrails.

Shortly after the Ross Shaft was completed to the 3,500-foot level in March 1936, the Milliken (No. 1) Winze was abandoned.[27] Similarly, after the Ross Shaft was completed to the 4,100-foot level in March 1938, No. 2 Winze was abandoned.[28] A series of 8×10-foot raises were driven and connected between the top of No. 2 Winze on the 3,500-foot level and the bottom of the Oro Hondo Shaft near the 2,300-foot level. A 700-horsepower Jeffrey Aerodyne fan with an operating capacity of about 276,000 cubic feet per minute was installed on the surface over the Oro Hondo Shaft in 1941 (see fig. 20.7), which greatly improved ventilation flow through mine areas to the 4,100-foot level.[29]

Fig. 20.7. The new 700-horsepower Jeffrey Aerodyne fan installed over the Oro Hondo Shaft, 1941.—Photo courtesy of the Black Hills Mining Museum.

Yates Shaft

On May 3, 1938, the Homestake board of directors approved another new shaft. The decision was based on site management's belief that existing ground-control problems would force abandonment and closure of the Ellison Shaft in a few years. The Ross Shaft alone did not have the hoisting capacity to meet the production level required to facilitate economic mine operations. The new shaft was named the Yates in memory of Bruce C. Yates, who was general manager of the mine from 1918 to the time of his death on August 10, 1936. In view of topographical limitations and the location of the new South Mill, a decision was made to position the Yates Shaft on the ridgetop called Flag Rocks. Work was immediately started to mine a level area on the ridgetop for the hoist building, headframe, and new mine office.

By the end of 1939, construction work was mostly completed on the hoist building, headframe, and crusher room. The remaining buildings and facilities were completed by December 1942. Total cost of the Yates complex was $1.707 million. Of this total, the cost of the Shafthouse and Crusher Building was $136,174. The Hoist House cost $95,054. The installed cost of the hoists, ropes, bridge crane, and crushing equipment amounted to $736,223, including $618,360 for the two hoists. The cost of the change house was $94,572. The sawmill cost $17,105 while the Mine and Engineering Office cost $94,110. The Yates Compressor Plant cost $235,319. Miscellaneous utility systems were also completed.[30] Over the same time frame, a significant amount of other construction work was completed at additional cost in relocating shops and facilities to the south side of Gold Run Gulch (see fig. 20.8).[31]

Fig. 20.8. The Yates, Ellison, and South Mill complexes, ca. 1942.—Photo courtesy of the Black Hills Mining Museum.

The Yates Shaft was designed as a timbered shaft with six main compartments configured to an overall outside dimension of 27 feet 8 inches by 15 feet. This configuration provided for two cage compartments, two skip compartments, one runabout cage compartment, and a utility compartment. A cage hoist and an ore hoist were purchased from the Nordberg Manufacturing Company at a cost of $309,180 each (see fig. 20.9). The hoists were nearly identical to the two hoists installed at the Ross Shaft. The ore hoist was capable of hoisting 10 tons of rock from a depth of 5,000 feet, whereas the cage hoist was designed to raise a 6-ton load on the cage from the same depth. The ore hoist was driven by two Westinghouse 1,500-horsepower direct-current motors with a 2,000-horsepower motor generator set. The cage hoist was powered by two Westinghouse 1,500-horsepower direct-current motors with a 1,000-horsepower motor generator set.[32]

Fig. 20.9. The Yates ore hoist (top) and Yates cage hoist (bottom).—Photo courtesy of the Black Hills Mining Museum.

Access drifts were driven from the Ross Shaft to the location for the Yates Shaft on the tramway 800-, 1,100-, 1,700-, 2,300-, 3,050-, 3,650-, and 4,100-foot levels. As was done with the Ross Shaft, the Yates was developed by driving a series of 7×9-foot pilot raises upward from the various underground levels and then stripping the pilot raises to full dimension. Shaft sets were installed from the top down. The shaft was concrete lined for a distance of 305 feet below the shaft collar. Skip-loading pockets were mined and installed below the 2,600-, 3,350-, and 4,100-foot levels. Ore raises were mined above each of the loading pockets, with the upper-most ore dump located on the 1,700-foot level. The first ore was hoisted at the shaft on October 1, 1941.[33]

An 8×8-foot drift was started on the 4,850-foot level from No. 3 Winze toward the Yates in June 1952. The drift reached the downward projection of the Yates Shaft in September 1953. The first rock bolts used in the mine were installed as needed throughout this drift. Round by round, all of the rock mined from this drift had to be loaded into 1-ton-capacity end-dump cars and individually caged at No. 3 Winze to the 4,100-foot level. The cars were hauled in trains to the Yates Shaft, where they were hoisted to the surface and dumped at the Yates Waste Rock Facility.[34]

A 7×9-foot pilot raise was driven for the Yates Shaft from the 4,850-foot level to the temporary bottom of the shaft just below the 4,100-foot level skip loader. The shaft was stripped to full dimension (see fig. 20.10), and shaft sets were completed to the 4,850-foot level in 1954.[35] In 1955, sinking operations were completed at the Yates Shaft to a point 211 feet below the 4,850-foot level, whereupon a skip-loading pocket was mined just above the shaft bottom.[36] The skip loader was constructed and commissioned during the first half of 1956.[37]

Fig. 20.10. Deepening the Yates Shaft, 1954. A 7×9-foot pilot raise driven from the 4,850-foot level to the 4,250-foot elevation (where the men are standing) was stripped to the full shaft dimension of 17½ feet by 30 feet by drilling and blasting individual bench rounds into the pilot raise. A bench round consisted of sixty holes, seven feet deep. From left to right are Joel Deeble, William "Blackie" Enderby, Arley Iverson, and Perry "Jack" Deatherage.—Photo courtesy of Homestake Mining Company.

South Mill Expansion

In 1952, a Symons 7-foot shorthead cone crusher was installed at both the Yates and Ross crushing plants. This addition permitted ore to be supplied to the rod mills at minus ½ inch. The smaller feed size allowed three of the old Allis Chalmers rod mills and one Marcy rod mill to be removed from operation. The remaining rod mills were converted to open circuit. Sixty of the 180 stamps at the South Mill were also decommissioned and removed.[38]

A year later, three new ball mills were added in series with the three rod mills at the South Mill. The remaining stamp batteries were removed from service toward the end of 1953. Power consumption was reduced from 10.5 kilowatt-hours per ton to 7.5 kilowatt-hours using the new crushing and grinding equipment. By year's end, mill capacity increased to about 4,000 tons per day, compared to 3,315 tons per day in 1952. With the finer grind, overall gold recovery increased to 96.66 percent in 1953.[39] For 1955, mill production averaged 4,247 tons per day, and overall recovery was 97.08 percent.[40]

On February 15, 1956, a fourth ball mill was commissioned in the South Mill grinding circuit. Four cyclones were added, and a rake classifier was removed from service. With these additions, an additional 77,603 tons of ore were mined and milled compared to 1955.[41]

With a new state-of-the-art mill, two new shafts, modern shop facilities, and 15.2 million tons of ore reserves at the end of 1956,[42] the Homestake Mine was well positioned for continued success. Attention could now be devoted toward "deep-level" exploration below the 4,850-foot level.

Chapter 21
The Deep-Level Projects

No. 4 Winze

In 1953, drifting was started southward on the 4,850-foot level from No. 3 Winze toward the new No. 4 Winze. The purpose of this winze was to provide access to the downward extensions of Main, 9, and 11 ledges that were known to exist between the 4,100- and 4,850-foot levels.[1] Mining for the first hoist room and installation of the cage hoist was completed in 1955. By year's end, shaft sinking was completed to the 5,150-foot level. The winze was configured much like the Ross, with a cage compartment, counterweight compartment, two skip compartments, utility compartment, and manway.[2]

A typical shaft-sinking cycle consisted of drilling and blasting a round of holes 7-9 feet deep, using a shaft-sinking jumbo designed and fabricated by Homestake (see fig. 21.1). Broken rock from the sinking operation was loaded into 20-cubic-foot sinking buckets using a hydraulically operated clamshell called a hydro-mucker. Three miners were needed to operate the hydro-mucker (see fig. 21.2).

Fig. 21.1. No. 4 Winze shaft-sinking jumbo. This jumbo was equipped with multiple drills and feeds for drilling a shaft-sinking round 9 feet deep. After the entire round was drilled, the holes were loaded and blasted using gelatin dynamite.—Photo courtesy of Homestake Mining Company.

Fig. 21.2. Mucking the bottom of No. 4 Winze using sinking buckets and a hydro-mucker clamshell (not shown), 1962. Using one set of cables, William Burk is operating a single-drum winch to raise and lower the empty bucket, which is being positioned by Kenneth Stevens and Orville Fritz. The winch cable was then reattached to the hydro-mucker so it could be raised and lowered to muck the shaft bottom. Two additional sets of cables are hanging to Burk's right. One set controlled the hydraulics for the hydro-mucker, and the other set controlled the two-drum slusher hoist for positioning the hydro-mucker across the shaft bottom. Operation of the hydro-mucker required all three men, each operating a set of control cables.—Photo courtesy of Homestake Mining Company.

Prior to completion of the skip-loading pocket below the 4,850-foot level at the Yates Shaft in mid-1956, development rock from the sinking operations at No. 4 Winze was loaded into 1-ton-capacity end-dump cars and transferred to No. 3 Winze. Here, the cars were hoisted on the cage one car at a time to the 4,100-foot level. The cars were transferred to the Yates Shaft and hoisted to the surface where the rock was conveyed to the Yates Waste Rock Facility.

After the broken rock was mucked from the sinking round, a steel shaft set was hung from the previously installed set using angle-iron studdles (see fig. 21.3). At each shaft station, 12-inch steel bearing beams were laid horizontally across the shaft and cemented into hitches mined into the walls or station sill. The bearing beams provided primary support for the steel shaft sets between levels. Each of the shaft sets between levels was aligned using plumb lines and braced to the ribs of the shaft. The perimeter of the shaft steel was enclosed with galvanized, corrugated metal sheeting, called lacing, to prevent wall rock from entering the shaft compartments. The full perimeter of the skip compartments was enclosed with lacing to prevent rock from falling into the cage and pipe compartments during skipping operations.

Fig. 21.3. Hanging a set of shaft steel at the 6,350-foot level in No. 4 Winze, 1962. The men include, left to right, Freddie Olson, William "Woody" Worsham, and Harold "Bud" Bennett.—Photo courtesy of Homestake Mining Company.

Toward the end of 1956, the winze was completed to the 5,700-foot elevation. Shaft-sinking operations were suspended at that point to allow drifting and diamond-drilling work to take place on the 5,300- and 5,600-foot levels to assess the down-plunge ore potential in Main Ledge and 9 Ledge.[3] Upon completion of the level-development work in May 1958, sinking operations were resumed. The winze was completed to a point 53 feet below the 6,200-foot level in February 1959. Sinking operations were again suspended to allow drift development and diamond drilling to take place on the 5,900- and 6,200-foot levels.

In January 1960, diamond drilling encountered "high-grade mineralization" in 9 Ledge on the 6,200-foot level. Based on these encouraging results, a decision was made to add a new mill unit and increase production to 5,800 tons per day by mid-1960. Homestake officials also decided to sink No. 4 Winze to the 6,800-foot level and develop the 6,500- and 6,800-foot levels to the Main Ledge and 9 Ledge ore areas.[4]

Installation of the ore hoist (see fig. 21.4) and sheave wheel (see fig. 21.5) was completed on February 12, 1962. After installing a temporary sinking hoist on the 6,200-foot level, shaft sinking was resumed in July 1962. Below the 6,200-foot level, each bucket was hoisted to the 6,200-foot level using the cage hoist. Here, the rock was dumped into rocker cars. The cars were hoisted on the cage to the 5,900-foot level, where the rock was dumped into a trench and slushed to a skip loader. The rock was then hoisted to the 4,850-foot level, dumped into a bin, and transferred into 8-ton-capacity bottom-dump railcars. A trolley locomotive hauled the railcars some 7,000 feet to the Yates Shaft, where the rock was dumped into the newly commissioned skip pocket just below the 4,850-foot level. As a final step, the waste rock was hoisted to the surface and conveyed to the Yates Waste Rock Facility.

Fig. 21.4. The newly installed ore hoist at No. 4 Winze, 1962. Hoist operators William Parsons and Paul Talley are familiarizing themselves with the hoist, which was powered by dual 300-horsepower motors.—Photo courtesy of the Black Hills Mining Museum.

Fig. 21.5. Final inspection of the newly installed sheave wheel and rope at No. 4 Winze, 1962.—Photo courtesy of Homestake Mining Company.

The winze was completed to a point just below the 6,800-foot level in July 1964 at which time a slusher trench and skip-loading pocket were mined and installed on the 6,800-foot level (see fig. 21.6). Conventional ore and waste raises were driven between the 6,800- and 6,050-foot levels. Pumping stations were installed on the 6,200- and 6,800-foot levels. Mine drainage water was pumped up No. 4 Winze to the 5,000-foot level, where the water flowed across the level to the Ross pumping system.[5]

Fig. 21.6. Slushing ore in the skip-loader trench on the 6,800-foot level at No. 4 Winze, 1966.—Photo courtesy of Homestake Mining Company.

In late 1986, work commenced to sink No. 4 Winze from the 6,800-foot level to the 7,590-foot elevation. Drifts were driven from No. 6 Winze on the 6,950- and 7,400-foot levels to the downward projection of No. 4 Winze to connect the mine infrastructure and facilitate disposal of waste rock from development work. Sinking operations were completed during the first quarter of 1988. In 1989, a skip loader was completed at the 7,500-foot level. Ore and waste raises were bored between the skip pocket and the 6,800-foot level. With lower Main, 9, 11, and 13 ledges mostly depleted by 1990, the majority of subsequent development and production from No. 4 Winze was focused in 15 Ledge.[6]

No. 5 Shaft

During 1955, as the Yates and Ross shafts were being extended below the 4,100-foot level and sinking was underway at No. 4 Winze, management personnel decided that additional mine ventilation was needed to facilitate exploration, development, and production work below the 4,850-foot level. Therefore, Homestake officials gave their approval to sink a new shaft from the surface to the 5,000-foot level. The new shaft was named No. 5 Shaft. A location in Grizzly Gulch was selected for No. 5 Shaft so that it could supply fresh air almost directly to No. 4 Winze on the 5,000-foot level.

During 1955, drifts were started toward the downward projection of No. 5 Shaft on the 2,000-, 2,900-, and 3,800-foot levels. Highball drift crews were selected to drive the three 1-mile-long access drifts. An experienced three-man highball crew could complete a drift cycle of mucking, tramming, installing track, drilling, and blasting in one eight-hour shift compared with two shifts for a cycle using two-man crews in other drifts. From No. 3 Winze, the 4,700-, 4,850-, and 5,000-foot levels had already been developed to No. 4 Winze, which was approximately 185 feet from the planned projection for No. 5 Shaft. With the upper access drifts well underway, site preparation work for the new shaft was started in Grizzly Gulch on May 11, 1956.

Upon completion of the access drifts, pilot raises for the shaft were driven upward from the 5,000-, 4,700-, 3,800-, 2,900-, and 2,000-foot levels. The latter raise was driven to a point approximately 1,100 feet below the shaft collar on the surface. In July 1957, the hoist house and headframe were completed (see fig. 21.7) and sinking operations were started. A shaft jumbo drill was used to drill the shaft bottom rounds. The jumbo was lowered down the shaft by a special attachment to the bottom of the men-and-materials cage. Once in position at the bottom of the shaft, anchor jacks were extended outward to the walls of the shaft to support the jumbo during drilling operations. The jumbo incorporated six air drills mounted on 10-foot-long feed shells. Blast holes were drilled 9 feet deep. Some 110 holes were required to complete one 9-foot-deep sinking round for the 19-foot-diameter shaft. The holes were detonated using electric blasting caps and 45 percent gelatin dynamite.

Fig. 21.7. No. 5 Shaft and hoist house, 2006. This photo, taken from the face of the Grizzly Gulch Tailing Storage Facility, shows the hoist house and headframe for No. 5 Shaft. The Seepage Collection Pumphouse for the tailing dam is shown to the right of No. 5 Shaft. The Ross Shaft (upper left) and Yates Shaft (upper right) are also shown.—Photo courtesy of Homestake Mining Company.

Blasted rock was mucked into 72-cubic-foot-capacity sinking buckets using a track-mounted power shovel. The crawler was lowered down the shaft using a special crosshead mounted to the bottom of the sinking bucket. Between sixty-two and seventy-four buckets of rock were required to muck out a 9-foot-deep sinking round. After mucking was completed, a steel shaft set was suspended from the completed set above. The shaft steel consisted of four 6-inch H-beams attached to a circular 4-piece H-beam wall plate. Shaft sets were installed on 10-foot centers with bearing beams installed every 300 feet. Some twenty-nine sheets of galvanized, corrugated steel were used to lace each 10-foot shaft set.

With accurate survey work and careful mining, all of the individual pilot raises were connected as planned. The uppermost pilot raise intersected that portion of the shaft that was sunk from the surface. Round by round, the sinking crew stripped the pilot raises to the full diameter of the shaft and installed steel shaft sets and corrugated metal lacing as they progressed downward. Some 917 tons of structural steel were installed in the shaft at a cost of $251,000 along with 953 tons of corrugated metal lacing at a cost of $142,000.[7]

In 1964, No. 5 Shaft was deepened to the 5,600-foot level. This was accomplished by first driving a drift from No. 4 Winze to the downward projection of No. 5 Shaft on the 5,600-foot level. At this location, an Alimak Raise Climber (see fig. 21.8) was used to drive a 7×7-foot pilot raise between the 5,600- and 5,000-foot levels. The 600-foot-long raise was stripped to 11 feet in diameter.[8] No. 5 Shaft was deepened to the 6,200-foot level in 1970. The shaft reached its ultimate bottom at the 6,800-foot level in 1983 when a 13-foot-diameter raise was bored between the 6,200- and 6,800-foot levels.

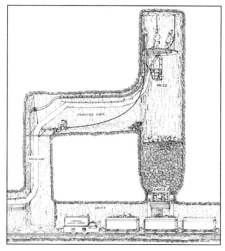

Fig. 21.8. Alimak full-face method for driving the No. 5 Air Shaft extension. An Alimak Raise Climber was used to drive the pilot raise for No. 5 Shaft between the 5,600- and 5,000-foot levels. The raise was subsequently stripped to 11 feet in diameter.—Courtesy of Homestake Mining Company.

Oro Hondo Shaft

With the Yates, Ross, and No. 5 shafts as primary air-intake shafts, additional exhaust-air capacity was needed to match the potential of the mine's intake-air capacity. The Ellison and Oro Hondo shafts were already being used as exhaust shafts, but their capacity was limited because of size. Without additional exhaust-air capacity, the full potential of the air-intake shafts could not be achieved. To remedy this problem, Homestake officials decided to enlarge the cross section of the Oro Hondo Shaft from the surface to the 3,950-foot level.

Stripping of the Oro Hondo Shaft was completed in four phases, beginning with the lowest section in 1960. This phase of work involved enlarging the existing 8×10-foot shaft to 14×14-foot between the 3,200- and 3,950-foot levels. Crews completed the enlargement by extending the 3,200-, 3,350-, 3,500-, 3,650-, 3,800-, and 3,950-foot levels to the Oro Hondo Shaft. At each shaft station, a small bench was mined around the perimeter of the shaft to facilitate stripping operations using a diamond drill. From the first drill station on the 3,800-foot level, approximately fifteen vertical holes were drilled around the perimeter of the shaft to the 3,950-foot level. The 150-foot-long holes were loaded and blasted using detonating cord and dynamite. Broken rock cascaded to the 4,100-foot level where the rock was loaded into railcars and dumped as waste backfill elsewhere. This process was repeated in an upward fashion from the 3,650-, 3,500-, 3,350-, and 3,200-foot levels.

To facilitate completion of the final three phases of shaft stripping, a hoist and headframe were installed over the shaft collar. From the surface to the shaft's original depth of about 2,300 feet, crews completed the second phase of work by refurbishing the 7×17½-foot timbered shaft that the Oro Hondo Mining Company had installed in 1915. After the 2,300-foot level was reached, a temporary safety bulkhead was installed across the shaft to protect miners during the next phase of work.

Between the 2,300- and 3,200-foot levels, the 8×10-foot pilot raise driven in 1939 and 1940 was enlarged to 14×14 feet. Miners completed this third phase of work by drilling and blasting successive vertical rounds 8-10 feet deep from a 10×10-foot work deck. The work deck was suspended from the main cage with 1,000 feet of 1⅛-inch-diameter wire rope. This arrangement kept the main cage above the temporary bulkhead in the shaft but allowed the work deck to advance to the 3,200-foot level as the shaft-stripping work advanced downward. The blasted rock was allowed to fall to the 4,100-foot level where it was removed with the aid of rail equipment.

To facilitate the final phase of work, the timber bulkhead at the 2,300-foot level was removed. Starting at the 2,300-foot level and working upward, the miners systematically removed all of the shaft timber, leaving a nominal 8×18½-foot bare shaft. Upon completion of the work, a second 700-horsepower Jeffrey Aerodyne fan with an operating capacity of about 250,000 cubic feet per minute was installed over the top of the Oro Hondo Shaft in September 1964.

The completed work allowed the Oro Hondo Shaft to be used as a primary exhaust shaft throughout the remaining life of the mine.[9]

Upon completion of the shaft-enlargement work, the exhaust capacity of the Oro Hondo Shaft increased to 432,000 cubic feet per minute. The exhaust fan over the Ellison Shaft removed 219,000 cubic feet per minute of exhaust air from the upper levels. No. 5 Shaft supplied 410,000 cubic feet per minute of fresh air to the lower mine areas around No. 4 Winze. The Yates and Ross shafts supplied the balance of the fresh air for the mine. With the Oro Hondo Shaft and No. 5 Shaft, the mine now had adequate ventilating capacity to facilitate exploration and mining work to the 6,800-foot level.[10]

Bored Raise Drilling

Raise-boring technology was introduced at the Homestake Mine in May 1967 when a 6-foot-diameter ventilation borehole was completed at 27 stope line in 19 Ledge between the 5,900- and 6,800-foot levels (see fig. 21.9). Many problems were encountered in attempting to drill the exceptionally long hole, which wasn't completed until February 1, 1968. Excessive deviation of the first pilot hole necessitated drilling a second pilot hole. With a little "Kentucky windage," the second hole was determined to be acceptable although it was 70 feet off target. Despite problems with the hydraulic motor on the drill, broken drill rods, and excessive cutter wear, the borehole proved to be a good learning experience. The average cost of the first raise was $126 per foot compared to $100 per foot for a 7×9½-foot cage raise.[11]

Fig. 21.9. The first borehole raise drill at the Homestake Mine, May 1967. This photo shows a Model 7200 Security-Dresser raise drill as it is being set up and aligned to drill the first borehole between the 5,900- and 6,800-foot levels at 27 stope line in 19 Ledge. One of the "stabilizer" rods is shown in the "rod racker" and is ready to be lifted into position and added to the drill string. From left to right are Olin Hart, chief geologist; Albert "Bud" Mitchell, chief surveyor; and Deane Pock, transit man.—Photo courtesy of Homestake Mining Company.

From that point forward, the state-of-the-art method for creating raises using borehole raise drills instead of the more conventional methods played a major role in completion of the deep-level project for the west ledges, as well as regular mine operations. Based on the favorable health and safety aspects that a bored raise offered compared to conventional raising, the latter method was largely phased out in the early 1970s. A borehole raise drill was subsequently purchased for boring 54-, 60-, and 72-inch-diameter raises. By 1971, the cost for a 5-foot-diameter bored raise was reduced to $58.13 per foot.

Bored raises 5 feet in diameter were routinely used to control and direct mine ventilation air in the mining areas. The 5-foot borehole also provided manway access into cut-and-fill stopes. The borehole manway was furnished with a ladder, safety-landing screens every 30 feet, a wooden slide for moving slushers, stope jumbos, other supplies and materials, a power cable, and pipelines for compressed air, water, and sandfill. Some 8,586 feet of bored raise was completed throughout the mine by July 1971.[12]

Slightly larger borehole raise drills were procured beginning in March 1971 as part of the west-ledge, deep-level project. These drills were used to bore two parallel series of 7-foot-diameter boreholes between the 5,600- and 8,000-foot levels. One series of boreholes was used to transfer waste rock to the 8,000-foot level; the other series was used to transfer ore. This system of ore and waste raises, called the No. 52 Raise System, was completed in 1977. On the 8,000-foot level, the rock was transferred into railcars, hauled to No. 6 Winze, hoisted to the 4,850-foot level, and transferred to the Yates or Ross Shaft.

In 1982, Homestake procured a Dresser 900 borehole raise drill that had a capacity of 900,000 pounds of thrust. This drill was first used to drill a 13-foot-diameter ventilation raise for the extension of No. 5 Shaft between the 6,200- and 6,800-foot levels. In 1983, the No. 31 Exhaust Raise System was extended by boring a series of 13-foot-diameter raises between the 6,800- and 7,700-foot levels.[13]

Preparatory work for a bored raise included mining a chamber 25 feet long, 9 feet wide, and 18 feet high at the upper location for the proposed raise. The borehole raise drill was assembled and bolted to a concrete pad inside the chamber. Depending on the specific model of drill and desired borehole diameter, a 9⅞-, 12¼-, or 13¾-inch-diameter pilot hole was drilled downward from one level to another up to 900 feet.

Upon completion of the pilot hole to the target area at the lower level, the pilot bit was located and removed from the string of drill rods. Mining was usually required to expose the pilot hole and enlarge the area to facilitate replacement of the pilot bit with the reaming head. The reaming head contained an array of tungsten-carbide "cutters" (see fig. 21.10). By applying slow rotation and upward thrust to the drill rods, the operator of the raise drill caused the reaming head to slowly ream the pilot hole from the lower level to the upper level. The thrust of the tungsten-carbide cutters against the solid face of rock in the raise caused the rock to break in tension. The broken chips of rock fell to the bottom of the raise where they were mucked using rail or rubber-tired loading equipment.

Fig. 21.10. Installing the borehole reaming head and track for mucking the drill "cuttings." This photo, taken in 1967, shows the first 6-foot-diameter reaming head that was used to bore raises at the Homestake Mine. The crew was installing rail so an air-powered rail mucker or "power shovel" could be used to muck the cuttings generated during back-reaming operations.—Photo courtesy of Homestake Mining Company.

Diamond Drilling

Air-powered diamond drills were introduced into the mine on a sustained basis in about 1916 to drill small-diameter core holes from existing raises and drifts.[14] Geologists used the core of rock obtained to determine geologic structure, develop ore estimates for stopes, and characterize the overall mineral inventory, including the ore reserve. Mining engineers subsequently used this information to plan and develop the mine's infrastructure and ore areas.

Diamond drill stations were typically located on 50-foot intervals on the various mine levels. Sometimes, a tighter spacing was used in raises. An array of holes was drilled in a vertical plane from each drill station to define geologic structure (see fig. 21.11). Each time a hole was advanced either 5 or 10 feet, depending on the length of the core barrel, the rods were pulled, and the cylindrical core of rock was removed from the core barrel.

Upon completion of a hole, or at other desired times, the drillers used a relatively simple and fairly accurate method to determine the inclination of the hole at intervals of 100 feet. This was done by inserting a glass vial containing hydrofluoric acid into a container called a clinometer. One or more clinometers were positioned within the string of drill rods at 100-foot intervals. The drill string was pushed to the end of the hole and allowed to remain at rest for approximately twenty-five minutes. During this time, the acid caused a "level" mark to be etched on the inside of the glass vial. After the vials were retrieved, the angle of the etch line on each vial was measured to determine the vertical inclination at each measurement point along the completed hole.

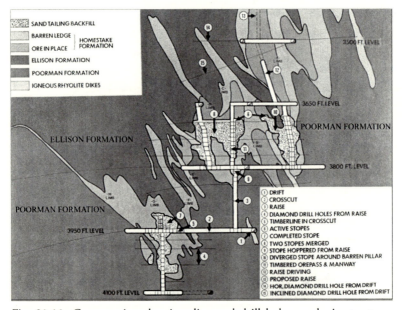

Fig. 21.11. Cross section showing diamond drill holes, geologic structure, mine development, and open cut-and-fill stoping. Diamond drilling aided the geologists in defining geologic structure and obtaining core samples of rock for logging, assaying, estimating ore reserves, and mapping. Information supplied by the geologists enabled mining engineers to prepare plans, schedules, and budgets for the mine infrastructure, stope development, ore production, and backfilling.—Illustration courtesy of Homestake Mining Company.

Core from the Homestake Formation was logged by a geologist, divided into 5-foot-long samples, and delivered to the assay office. Here, each of the samples was crushed, pulverized, and fire assayed to determine average gold content. Using the assays and information obtained from visual core logging, geologic plans and sections were created from which discrete ore reserve estimates and mining estimates were calculated. Collectively, all of this information was used to define a mineral inventory for the mine from which the ore reserve was determined.

In the 1980s, modern electric-hydraulic core drills were purchased to replace the aging pneumatic drills. The electric-hydraulic drills (see fig. 21.12) were capable of drilling core holes thousands of feet long, depending on the type of drill and desired hole diameter. The drills utilized wireline methods, which were introduced at Homestake in 1973.[15] This technology allowed the operator to extract the core without having to pull the rods each time the core barrel and rods were advanced the length of the core barrel.

Using the wireline method, an overshot device attached to the wireline cable was pumped through the inside of the rods to the core barrel. After the overshot latched onto the end of the inner tube, which contained the rock core, the overshot and inner-tube assembly was winched to the collar of the hole at the drill station. Some 6.945 million feet of diamond core drilling was completed at the Homestake Mine from 1916 through 2001.[16]

Fig. 21.12. A Longyear electric-hydraulic drill with a wireline hoist. Steve Albrecht is the drill operator.—Photo courtesy of Homestake Mining Company.

No. 6 Winze

In the 1960s, No. 26 Service Raise was completed between the 4,850- and 5,900-foot levels to provide access to those geologic structures that were referred to as the west ledges or, collectively, 19 Ledge. A short time later, No. 28 Service Raise was completed between the 5,900- and 6,800-foot levels for the same purpose. Subsequent diamond drilling and geologic mapping in these areas defined ore-grade mineralization within three separate fold structures. These structures were named 17, 19, and 21 Ledge. Based on favorable exploration results, approval was given to develop options for a new deep-level development program to define, develop, and exploit the west ledges to the 7,400-foot level.

Six options were investigated for deepening the mine to the 7,400-foot level. The preferred option involved mining a new winze from the 4,550-foot level to the 7,550-foot level. The proposed winze, named No. 6 Winze, would be located approximately 175 feet from the Ross Shaft. At this location, ore from the west ledges could be hoisted at No. 6 Winze and transferred by gravity into the Ross skip-loading pocket below the 4,850-foot level (see fig. 21.13). Alternatively, the ore hoisted at No. 6 Winze could be fed into a transfer raise, loaded into 8-ton bottom-dump railcars, and hauled by trolley locomotive to the Yates or Ross shafts on the 4,850-foot level. Based on the good operating flexibility that this plan offered compared to the other five options, approval was given in 1969 to proceed with the $8.032-million deep-level project, which would include further enhancements to the mine ventilation system.[17]

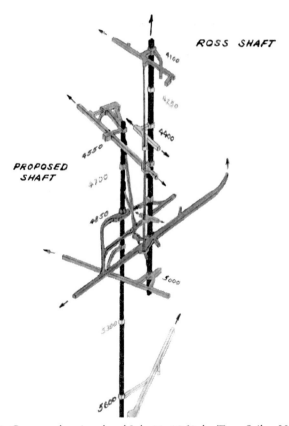

Fig. 21.13. Concept drawing dated July 11, 1969, by Tony Seiler, Homestake mining engineer, showing the proposed location and configuration for No. 6 Winze relative to the Ross Shaft. The winze was constructed as shown, except the hoist room was modified to accept Koepe friction hoists manufactured by ASEA.—Illustration courtesy of Charles Tesch Sr.

No. 6 Winze was developed by driving access drifts on the 5,900- and 6,800-foot levels from No. 26 and No. 28 service raises, respectively, to the downward projection for No. 6 Winze. More than 4,000 feet of drifting was required on each level. From the northeast end of the 6,800-foot level, a 7-foot 10-inch by 9-foot 6-inch pilot raise was driven to the 5,900-foot level. Concurrently, a pilot raise was driven from the 5,900-foot level to the 5,000-foot level. Shorter pilot raises were concurrently driven from the 5,000-foot level to the 4,850-foot level and from the latter level to the hoist room on the 4,550-foot level. The pilot raises were subsequently stripped to the full shaft dimension in a "top-down" manner. By the end of 1971, shaft stripping had advanced to 153 feet below the shaft station for the 5,600-foot level.[18]

Upon completion of the hoist room on the 4,550-foot level, two Koepe friction hoists were installed on that level in 1972. The ore hoist included two 10-ton-capacity skips in balance. The cage hoist operated in balance using a cage and counterweight. A substantial bulkhead was installed just above the 6,800-foot level to facilitate shaft sinking operations below the bulkhead concurrent with shaft-stripping operations above the bulkhead. Conventional shaft-sinking methods were used to sink the winze from the 6,800-foot level to a point 112 feet below the 7,100-foot level in 1972.

Stripping of No. 6 Winze was completed to the 6,800-foot level in January 1973. Shaft steel was configured to an overall dimension of 14 feet by 17 feet 1 inch, which accommodated the cage, counterweight, skips, manway, and utilities compartments. A skip loader was installed just below the 6,800-foot level, and a rock-handling raise was completed between the pocket and the main level.[19]

The sinking hoist was moved to the 7,100-foot level early in 1974. The sheaves and bucket dump were also installed. In April, a contract was awarded to Canadian Mine Services Inc. and sinking operations were initiated from the temporary shaft bottom at the 7,212-foot elevation.[20] Because the new Koepe friction hoists were designed with extra capacity, it was decided to extend the winze beyond the planned bottom of 7,550 feet and establish a skip-loading pocket below the 8,000-foot level. In September 1975, the winze reached its ultimate bottom 251 feet below the 8,000-foot level.[21]

Upon completion of shaft sinking, the skip-loading pocket was mined at the 8,100-foot location. A series of ore and waste raises 7 feet in diameter were bored between the 5,900- and 8,000-foot levels. Installation of the 8,100 skip loader was completed in October 1977. Following installation of longer hoist ropes in October, skipping was initiated from the 8,100 skip pocket in November. Pumping stations were installed on the 6,800- and 8,000-foot levels. Mine drainage water from No. 6 Winze was pumped to the 5,000-foot level and transferred to the Ross pumping system.[22]

No. 7 Winze

As part of the deep-level development project, work commenced on No. 7 Winze in 1974 (see fig. 21.14). The purpose of this winze was to provide secondary access between the 6,800- and 8,000-foot levels. Work completed that year included driving "40 crosscut" on the 6,800-foot level to the planned location for the winze and mining the hoist room.[23] Conventional methods were used to sink the winze to the 6,950-foot level. Because of the extremely adverse ground conditions that were encountered over the first 150 feet of shaft sinking, a decision was made to convert the rectangular shaft configuration to a concrete-lined shaft 15 feet in diameter. The winze was advanced to the 6,992-foot elevation by the end of 1975.[24]

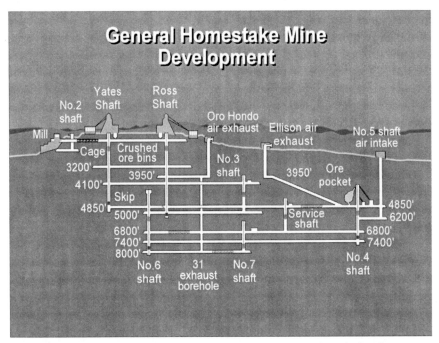

Fig. 21.14. Idealized section of the Homestake Mine. This idealized cross section shows the general mine infrastructure after No. 4 Winze was deepened to the 7,550-foot level and No. 7 Winze was completed between the 6,800- and 8,000-foot levels. The mine included eleven historic levels between the surface and the 1,100-foot level and forty-six levels between the 1,100- and 8,000-foot levels. Ore was hoisted at No. 4 Winze, No. 6 Winze, Yates Shaft, and Ross Shaft.—Illustration courtesy of Homestake Mining Company.

As a result of operating losses at the mine during the first quarter of 1976, sinking operations at No. 7 Winze were suspended at a point 52 feet below the

7,100-foot level on April 30.[25] In 1977, it was decided to complete the winze. A 7-foot-diameter pilot raise was drilled between the bottom of the winze and the 7,400-foot level. A second borehole was completed between the 7,400- and 8,000-foot levels.[26] In October 1980, a contract was awarded to American Mine Services (AMS) to strip and concrete line the winze from the 7,152-foot elevation to the 8,000-foot level. AMS completed the winze in November 1981. AMS also completed drift development on the 7,100-, 7,550-, 7,700-, and 7,850-foot levels.

No. 7 Winze was used as a service shaft to facilitate delivery of supplies and materials to work areas on the 6,950-, 7,250-, 7,550-, and 7,850-foot levels south of No. 6 Winze. No. 7 Winze also served as a secondary escape route for people working below the 6,800-foot level. Upon completion of the rubber-tired ramp system from the 6,800- to 8,000-foot levels in the early 1990s, No. 7 Winze was decommissioned.

Deep-Level Ventilation Project

In conjunction with the deep-level development project to the 8,000-foot level, additional airflow and cooling were also needed to facilitate development and mining to such depths. Below the 800-foot level, the geothermal gradient in the mine resulted in a temperature rise of approximately 1.2 degrees Fahrenheit per 100 vertical feet. Accordingly, the natural rock temperature was observed to be approximately 119 degrees at the 6,800-foot level and 133 degrees at the 8,000-foot level.[27]

Initially, mine planners thought the entire mine could best be ventilated by converting No. 5 Shaft to an exhaust shaft. All other shafts would be converted to air-intake shafts, providing 800,000 cubic feet per minute of airflow through the mine.[28] After evaluating other options, it was decided to continue bringing fresh air into the mine through the Yates, Ross, and No. 5 shafts. In addition to the No. 2 ventilation circuit, which ventilated mine areas above the 2,300-foot level, two other independent exhaust circuits would be created, which would direct exhaust air to the Ellison and Oro Hondo shafts. The first phase of the major ventilation project was initiated in 1971.

The Ellison circuit was designed to ventilate the east ledges, which at the time consisted of Main, 9, and 11 ledges. The Ellison exhaust circuit was comprised of the Ellison Shaft, Milliken Shaft between the 2,750- and 3,500-foot levels, and new 13×18-foot airway drifts on the 3,500-, 3,800-, 3,950-, 4,850-, and 5,300-foot levels. Boreholes 7 feet in diameter were drilled among these levels and stripped to 16 feet in diameter in 1972.[29] Shaft timber was removed from the Milliken Shaft between the 2,750- and 3,500-foot levels to provide a better airway. A brattice wall was installed on the 3,950-foot level to isolate the Ellison circuit from the Oro Hondo circuit.

Two 350-horsepower booster fans were installed and commissioned on the 3,500-foot level in 1973, which provided 150,000 cubic feet per minute of exhaust airflow. One 300-horsepower booster fan was installed on the 3,800-foot level, which moved 100,000 cubic feet per minute of exhaust air to the Golden Stairway.[30] As part of the fan upgrades completed in 1973, the 350-horsepower motor on the Ellison exhaust fan was replaced with a 700-horsepower motor.[31]

The Oro Hondo circuit was designed to ventilate the west ledges. The Oro Hondo exhaust circuit included the Oro Hondo Shaft and new 13×18-foot airway drifts on the 3,950-, 4,100-, 4,850-, 5,300-, and 5,900-foot levels in 19 Ledge. Boreholes 7 feet in diameter were drilled among these levels and stripped to 16 feet in diameter in 1972.[32] The 700-horsepower motor on one of the two Jeffrey exhaust fans at the Oro Hondo Shaft was replaced with a 1,250-horsepower motor in 1973. The other 700-horsepower fan was decommissioned.[33]

On February 1, 1975, the Ellison Shaft caved. The force and impact of the cave caused the structural timber in the exhaust shaft to collapse. Failure of the shaft timber caused all of the utilities in the shaft to fail, including the sandfill pipes, electrical cable, compressed-air lines, and waterlines. All of this material fell to the bottom of the shaft, creating a pile of debris that filled the shaft from its bottom on the 3,200-foot level to the 2,450-foot level. When the shaft collapsed, the venturi effect of the falling debris caused an in-rush of air that bent the main shaft on the exhaust fan at the top of the shaft. The cave caused a significant portion of the mine's ventilation and other services to be immediately severed, resulting in a significant impact to operations.

Despite the concentrated work effort that was directed to remedy the situation, it took until September 17 to restore the systems that had been destroyed. Work steps to restore the Ellison ventilation circuit were initiated by drawing debris from the shaft on the 2,600-foot level. No attempt was made to remove debris between the 2,600- and 3,200-foot levels. A bulkhead comprised of 24-inch I-beams was constructed across the shaft at the 300-foot level. Some 511 cubic yards of concrete were poured over the I-beams to create a substantial concrete plug. The shaft was then filled with waste rock from the 300-foot level to the surface.

Crews created a new exhaust portal by driving a 14×14-foot exhaust drift from the Kirk Timber Yard to intersect the Ellison Shaft just below the new bulkhead. Two 700-horsepower Trane fans 89 inches in diameter were installed at the portal of the new exhaust drift (see fig. 21.15). Upon startup, exhaust flow through the two fans was measured at 500,000 cubic feet per minute with 7 inches of water gauge pressure.[34]

Fig. 21.15. The two 700-horsepower Kirk fans and isolation doors. The fans were installed in 1975 at the portal of the 300-foot level in the Kirk Timber Yard after the Ellison Shaft caved and damaged the surface fan over the shaft.—Photo courtesy of Homestake Mining Company.

Crews replaced the North Sand Line system that was destroyed in the Ellison cave-in by drilling a series of $7\frac{7}{8}$- to 9-inch diameter holes from level to level between the 1,100- and 3,500-foot levels. The series of sand holes was connected by installing rubber-lined steel pipe and pipe adapters on each of the mine levels.

Other remedial work included relocation of the blasting cap magazine from the Kirk Timber Yard to the 800-foot level and replacement of drainage systems, compressed-air lines, waterlines, and electrical feeders at new locations.

To complicate matters, the Milliken Winze caved on April 29, 1975. The cave was large enough to completely block the airway at the bottom of the winze on the 3,500-foot level. After a large amount of rock was mucked from the bottom of the winze, ventilation was restored through the Milliken on May 6.

The troubles continued. On May 9, a mine fire started in an old timberline in Main Ledge on the 3,050-foot level. The mine was immediately evacuated. Mine Rescue crews equipped with McCaa self-contained breathing apparatus were dispatched to isolate the fire and install air and water foggers. Their efforts were successful. On May 22, inspectors from the Mine Enforcement and Safety Administration lifted their closure order, and miners were allowed to return to work on May 23.[35]

In 1977, work was started on another upgrade to the deep-level ventilation system. The first phase of work consisted of installing a chilled-water system on the 5,000-foot level near the bottom of the Ross Shaft. A 700,000-gallon-capacity chilled-water sump and a 300,000-gallon-capacity condenser-water sump were mined on the 5,000-foot level to facilitate the system.[36] The mining work was completed in 1978. Two 290-ton-capacity water chillers, pumps, and a power recovery turbine were installed and commissioned in 1979. The system cooled city water to approximately 42 degrees Fahrenheit. The "chilled water" was directed to the various mining areas in the west ledges below the 5,000-foot level, where the water was used in portable refrigeration units (see fig. 21.16), stope-cooling coils, drills, and most other applications where service water was needed.[37]

Fig. 21.16. A portable refrigeration unit called a spot cooler. Some twenty-five to thirty portable spot coolers of 30-ton and 60-ton capacity were utilized to provide cool air to isolated areas such as exploration drifts proximate to the main ventilation circuit.—Photo courtesy of Homestake Mining Company.

In 1986, a $6.3-million expansion of the deep-level ventilation project was implemented to increase airflow through the west ledges between the 4,850- and 8,000-foot levels. The first phase of work included procurement of a new, 3,000-horsepower, variable-speed centrifugal fan from the American Davidson Company. After this fan was installed over the Oro Hondo Shaft (see fig. 21.17), the existing 1,250-horsepower Jeffrey Aerodyne fan was configured as a ready spare in case the centrifugal fan had to be shut down for maintenance.[38]

The Oro Hondo exhaust system was extended by boring a series of 13- and 14-foot-diameter ventilation raises near 31 stope line between the 5,900- and 8,000-foot levels. Upon completion of the system, exhaust airflow through the Oro Hondo circuit from the west ledges increased to more than 500,000 cubic feet per minute. The Ellison and No. 2 ventilation circuits served to ventilate the east ledges of the mine. Collectively, the exhaust flow of the mine increased to approximately 950,000-1,050,000 cubic feet of air per minute.[39]

Fig. 21.17. The 3,000-horsepower American Davidson centrifugal fan, *left*, and the 1,250-horsepower Jeffrey Aerodyne backup fan, *right*, over the Oro Hondo Shaft. The centrifugal fan was used to remove exhaust air from the mining areas in the west ledges between the 4,850- and 8,000-foot levels. The fan was capable of removing up to 520,000 cubic feet per minute of exhaust air at 25 inches of water gauge pressure.—Photo courtesy of Homestake Mining Company.

The next phase of the deep-level ventilation project involved mining for and installing a refrigeration plant and controlled recirculation system on the 6,950- and 7,100-foot levels. Mining for the plant and ancillary facilities started in 1986 and was completed in 1987. The refrigeration plant and controlled recirculation system were commissioned during the fourth quarter of 1988.[40]

The controlled recirculation system was engineered by John Marks, Homestake's ventilation engineer. The system was designed to clean and cool up to 300,000 cubic feet per minute of mine air and return it to work areas in the west ledges below the 6,800-foot level. Four York helical rotary-screw shell-and-tube heat exchangers with 800-horsepower motors provided a total of 2,300 tons of refrigeration in the form of chilled water (see fig. 21.18). Exhaust air from the area was scrubbed and cooled to approximately 64-67 degrees Fahrenheit using chilled-water sprays positioned in bulk air drifts on the 6,950- and 7,100-foot levels. The air was then returned to active mining areas below the 6,800-foot level.

Heat from the plant-condenser water was transferred to approximately 300,000 cubic feet per minute of hot reject air that passed through four parallel cooling towers 10 feet in diameter. From here, the reject air was directed to the series of No. 31 exhaust raises, where the air was routed to the surface via the Oro Hondo Shaft on the 3,950-foot level.[41]

Fig. 21.18. The 2,300-ton-capacity refrigeration plant on the 6,950-foot level. The plant utilized four 575-ton-capacity shell-and-tube heat exchangers with 800-horsepower motors to clean, cool, and recirculate approximately 300,000 cubic feet per minute of air at 64-67 degrees Fahrenheit to work areas between the 6,800- and 8,000-foot levels.—Photo courtesy of Homestake Mining Company.

A mine-monitoring system was also installed as part of the ventilation upgrade (see fig. 21.19). This system provided remote monitoring and control of surface fans, dewatering pumps, and the underground ventilation systems. Carbon monoxide (CO) sensors were strategically placed in key intake and exhaust airways to detect and measure the concentration of the gas and help determine the general source location. Carbon monoxide produced from blasting operations was readily identified on the mine-monitoring system, based on a characteristic trend line that spiked rather quickly and attenuated slowly, concurrent with scheduled blasting times.

Conversely, carbon monoxide associated with a mine fire caused by spontaneous combustion in a historic timber stope yielded a trend line that increased in intensity very slowly at first, then spiked upward, and remained high for an extended period. The mine-monitoring system proved invaluable in the early detection of underground mine fires, the orderly and safe evacuation of people, and timely location of each fire.

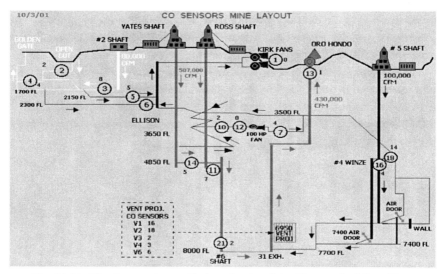

Fig. 21.19. The three main ventilation circuits of the Homestake Mine. Note the array of carbon monoxide sensors, denoted by circles, which were integrated into the mine-monitoring system for early detection of mine fires.—Illustration courtesy of Homestake Mining Company.

North Homestake Exploration Project

Based on the results of an extensive surface exploration program, Homestake initiated the North Homestake Exploration Project in 1990. The purpose of the project was to locate a new ore ledge north of the existing mine. A new ledge was critically needed to establish a new centroid of mining and help extend the life of the mine. A five-year, $23-million program was approved, which included driving an 11,300-foot-long drift to the north of the existing mine from No. 6 Winze on the 6,800-foot level.[42]

The 12×12-foot exploration drift was driven on a continuous, seven-day work schedule using conventional methods. A 290-ton refrigeration unit manufactured by the Dunham-Bush Company was purchased and installed, which provided cooling for the drift. Notwithstanding, advancement of the drift was hampered because of highly stressed rock conditions and the extreme heat load that emanated from hot groundwater inflows into the drift. Mining crews advanced the drift to the planned target area late in 1993.

The site conditions presented additional challenges for diamond-drill crews in terms of heat and zones of groundwater that exerted up to 2,700 pounds per square inch of water pressure on the drill rods. By early 1994, a decision was made to terminate the project based on unfavorable drilling results. Although the underground drilling substantiated geologic projections and the presence of the Homestake Formation, assay results failed to indicate sufficient ore-grade mineralization to develop any new ore reserves or warrant additional work. The project was concluded after a substantial concrete plug was constructed in the drift a short distance from No. 6 Winze.[43]

Upon completion of the North Homestake Project, the 290-ton refrigeration unit was moved to the 7,400-foot level to ventilate the mining areas in 15 Ledge. A 350-ton Dunham-Bush refrigeration unit was purchased and installed on the 8,000-foot level in 1996. This unit provided cooling to mining areas around the south end of the level.[44]

Chapter 22
The End of Amalgamation

Grizzly Gulch Tailing Storage Facility

Prior to 1977, tailings from the Homestake mills and raw sewage from the city of Lead were disposed of by discharging the wastes into Gold Run Creek. Over the same one-hundred-year period, raw sewage from Central City and Deadwood were discharged into Deadwood Creek and Whitewood Creek, as were the tailings from all of the historic stamp mills situated along the two creeks. In 1959, the South Dakota State Department of Health completed two studies of the pollution problems in the Lead-Deadwood communities. The cities of Lead and Deadwood conducted independent investigations for solutions to the raw-waste problem in 1963.

During the next three years, Homestake Mining Company contracted with two different engineering firms to investigate possible solutions to the solid-waste issues in the community. In 1968, citizens of Lead and Deadwood voted to form the Lead Deadwood Sanitary District to pursue remedies for the raw sewage problem. The Lead Deadwood Sanitary District became functional in April 1969. Homestake and the sanitary district subsequently agreed that the best way to resolve the pollution problems was by mutual effort.[1]

In 1969, the Lead Deadwood Sanitary District contracted with Brady Consultants to draft a report to resolve the raw sewage and tailings issues. Pursuant to the National Environmental Policy Act of 1969, the report and plan were submitted to the newly created Environmental Protection Agency (EPA) in 1970. At about the same time, the Lead Deadwood Sanitary District purchased 204 acres in Centennial Valley for the purpose of constructing a holding pond and treatment plant for the solid wastes. An environmental impact hearing was held in Deadwood in December 1971. Shortly after an environmental impact statement was approved for the project in March 1972, the Lead Deadwood Sanitary District received $4.7 million in grants toward construction of the $6.4-million facility.

Residents of Central City, Pluma, and Grizzly Gulch joined the Lead Deadwood Sanitary District in 1972. For the next two years, the Lead Deadwood Sanitary District and Homestake contracted with several different engineering firms to conduct soils and other engineering studies relative to the proposed construction site in Centennial Valley. By August 1973, the estimated cost of the pollution control facility had increased to $13.4 million. The Save Centennial Valley Association was formed, which opposed any such construction in Centennial Valley. In November, the Lead Deadwood Sanitary District received notice from the EPA to reevaluate the Centennial Valley Project and examine other alternative sites. By December, the estimated cost to construct a tailing and treatment facility in Centennial Valley was revised to $18 million.

The Lead Deadwood Sanitary District contracted with the engineering firms of Wright-McLaughlin Engineers and Bauer, Sheaffer, and Lear to investigate alternative plans. The Colorado School of Mines Research Institute was also contracted to investigate the waste products of Gold Run Creek. Homestake contracted independently with the engineering firm of Dames and Moore to assess the Centennial Valley proposal and investigate possible alternatives. In the interim, the communities of Terraville, Fantail, Whitetail, Terry Gulch, Nevada Gulch, Aztec Hill, Terry Peak, and other outlying areas were also annexed into the Lead Deadwood Sanitary District.

In March 1974, the proposal prepared by Bauer, Sheaffer, and Lear for constructing a storage facility in Grizzly Gulch was presented to the communities and submitted to the EPA and State Department of Environmental Protection. As a result, the Lead Deadwood Sanitary District abandoned the Centennial Valley plan and contracted with Dames and Moore to develop a detailed design plan for a combined solid-waste storage facility in Grizzly Gulch. At the request of the United Steelworkers Union, the Mining Enforcement and Safety Administration of the Department of Interior reviewed the proposed plan for a tailing facility at Grizzly Gulch, since the tailing storage facility was planned to be located upstream from No. 5 Shaft.

On July 5, 1974, the Lead Deadwood Sanitary District received a written response from the EPA relative to the Bauer, Sheaffer, and Lear proposal for a facility in Grizzly Gulch. The EPA concluded that the cost of constructing the pipelines, pumping systems, and tailing storage facility should be borne by Homestake and that the Lead Deadwood Sanitary District should investigate a separate, conventional treatment facility for treating raw sewage. Following a meeting with the EPA and others on July 30, the sanitary district decided to contract with the engineering firm of CH_2M Hill to investigate other treatment and storage options.[2]

CH_2M Hill presented its proposal for additional waste treatment options in 1975. At a subsequent review meeting, state and federal officials indicated their preference toward a split system with industrial wastes being handled internally within the Homestake plant and a separate treatment lagoon for treatment of raw sewage wastes. As a result of that meeting, Homestake contracted with the engineering firms of Dames and Moore, Wright McLaughlin Engineers, and Woodward Clyde to design a tailings-handling and storage system at Grizzly Gulch. In a similar fashion, the Lead Deadwood Sanitary District directed

CH$_2$M Hill to design a conventional wastewater treatment plant, which would tentatively be located in Deadwood.

Toward the end of 1975, Homestake submitted a $10.06-million plan to its corporate officers and board of directors for construction of a tailing storage facility that would be used exclusively by the company at Grizzly Gulch. The Homestake Board approved the plan on January 9, 1976. Construction commenced on March 29, 1976. A joint venture of Summit Inc. and Delzer Construction started work on the 2-million-cubic-yard, 235-foot-high tailings dam in April 1976. By year's end, the dam was within 80 vertical feet of being completed.

Other related work completed in 1976 included enlargement of the sand dam in Lead and completion of a 2.35-million-gallon water reservoir above the South Mill. Work was also started on construction of two slurry pipelines and a decant pipeline from the East Sand Plant and Mill Reservoir to the Grizzly Gulch Tailing Storage Facility. The completion date specified by EPA for the project was July 1, 1977. By the end of 1976, the project cost was reestimated to be $12.223 million.[3]

During the winter of 1976-77, Lead received a record 258.2 inches of snowfall, which adversely impacted the completion date, July 1, 1977, mandated by EPA. After pleading its case for an extension of time, Homestake and the EPA signed a consent decree and stipulation on October 11, 1977, which extended the completion date to December 4, 1977. The consent decree stipulated that failure to complete the project by that date would result in a $1,000 per day fine through January 4, 1978, and $10,000 per day thereafter.

Brablec Construction completed the Slurry Pumphouse in Lead on August 31, 1977. Grading for the 15,700-foot-long pipeline route and installation of the three pipelines were completed on October 29. The tailings dam was completed on August 26 to an elevation of 5,350 feet, which was 229 feet above the valley floor of Grizzly Gulch. The crest of the starter dam was 1,350 feet long and 29 feet wide.

A shotcrete-lined interceptor canal was completed on November 7. The interceptor canal was designed to intercept storm water runoff and route the water around the dam to the creek in Grizzly Gulch. The project was declared completed on December 2, 1977, after the slurry pumps were started and tailings were pumped to the impoundment facility in Grizzly Gulch. Final cost for the system and starter dam was $14.226 million.[4]

The Grizzly Gulch Tailing Storage Facility (see fig. 22.1) was utilized for permanent storage of spent mill tailings not needed as backfill underground. The tailings were pumped as a sand-slurry some 2.9 miles to the impoundment. The impoundment allowed the tailings to settle out, creating a decant pool of water over the tailings. A floating barge supported the decant pumps, which directed the water to a booster pump. The booster pump returned the water through a 12-inch-diameter pipeline to the Mill Reservoir, where water was removed and reused throughout the treatment facilities as process makeup water.

Fig. 22.1. The Grizzly Gulch Tailing Storage Facility. The tailing facility was constructed in four phases, beginning with a 229-foot-high starter dam completed in 1977. Three additional raises were completed in 1983, 1989, and 1999, which raised the dam an additional 150 feet. More than 40 million tons of mill tailings were pumped to the facility from 1977 through the first half of 2002. Evaporation sprays were being used when this photo was taken. The barge pumps are also shown.—Photo courtesy of Homestake Mining Company.

Carbon-in-Pulp Process

As one of the first steps taken to reduce the pollution of Gold Run and Whitewood creeks, amalgamation and the use of mercury were discontinued at the Homestake Mine in 1970 pursuant to the National Environmental Policy Act. This change had an immediate adverse effect on gold recovery and associated revenue from gold sales. Overall recovery decreased from 95.02 percent in 1970 to 93.03 percent in 1971, which had a very unfavorable effect on income. As a result, various metallurgical tests were undertaken to find an economic means for improving recovery.[5]

The Carbon-in-Pulp (CIP) method was subsequently selected, which meant that the Slime Plant could be shut down following construction of a CIP Plant. The Homestake board of directors gave their approval to construct the new plant on February 4, 1972.[6] Construction was started immediately. The CIP Plant was commissioned on March 21, 1973. After the new plant was successfully operating at the required capacity, the Deadwood Slime Plant was shut down on August 10, 1973.[7] The new CIP Plant, which was the first large-scale plant of its kind constructed in the world, worked well. By the end of 1974, overall gold recovery increased to 95.12 percent, which was the nominal rate of recovery being achieved before amalgamation was discontinued.[8]

Using the new CIP process, much of the milling and treatment process remained the same (see fig. 22.2) despite abandonment of the Slime Plant that Charles Merrill designed and helped commission in 1906. As before, run-of-mine ore was hoisted at the Yates and Ross shafts and reduced to one-half inch through three-stage crushing systems at the Yates and Ross crushing plants. Each crushing plant consisted of a 30-inch Allis Chalmers gyratory crusher, a Symons 7-foot standard cone crusher, a Symons 7-foot shorthead cone crusher, and Tyler screens. Minus ½-inch crushed ore was transferred from the Yates and Ross ore bins to the South Mill bins on the tramway.

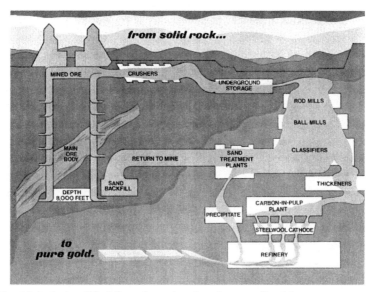

Fig. 22.2. A simplified flow sheet showing the overall gold-recovery process used at the Homestake Mine after the Slime Plant was decommissioned in 1973.—Source: Homestake Mining Company, *1876-1976 Homestake Centennial* (1976).

Ore was drawn from the mill bins and fed through one of four rod-and-ball mill circuits. Three of the grinding circuits consisted of a 6×12-foot Marcy Rod Mill in series with a 9×11-foot Allis Chalmers Ball Mill. A fourth grinding circuit consisted of a 7½×15-foot Marcy Rod Mill followed by a 9×11-foot Allis Chalmers Ball Mill. Using this conventional milling arrangement, the ore was reduced to an optimal size in two stages. Size reduction was critical in order to achieve a high degree of gold recovery using gravity-separation techniques in the milling circuits and subsequent chemical and electrolytic treatment in the CIP and sand plants.

Krebs cyclones were used to classify the ore pulp in the milling circuits. The slime fraction, consisting of approximately 65 percent, minus 200-mesh product, was treated in the CIP Plant to optimize gold recovery. The sand fraction was directed to the East and West sand plants for gold recovery.[9]

The sand fraction was delivered to the sand plants at approximately 70 to 75 percent, plus-200 mesh. The East Sand Plant (formerly called Cyanide Plant No. 1) housed twenty-three redwood leaching vats, each of which held 750 tons of sand. The West Sand Plant (formerly called Cyanide Plant No. 3) housed twelve vats, each with a capacity of 759 tons. The vats were filled through Butters-Mein sand distributors. Fill time for a vat averaged 9 hours at the East Plant and 12-13 hours at the West Plant. After an 18-hour drain period, the sands were aerated for 24-32 hours using compressed air at 9 pounds per square inch.

Gold was leached from the sand fraction using a strong sodium cyanide solution over a 24-hour period. The leach period was followed by a second drain cycle and aeration period totaling 32 hours. A second leach cycle using a strong cyanide solution was completed over a 50-hour period, which included draining and aeration. The sands were then leached one final time using a weak cyanide solution for fourteen hours. Following a 25-hour rinse period, the gold-bearing solution was pumped to the CIP Plant. The barren sands were sluiced out of the vat and directed to the sand dams for use as backfill underground. Excess sand was pumped to the Grizzly Gulch Tailing Storage Facility. The sand vat was then refilled with clarified water, completing the 216-hour sand treatment cycle.

Prior to 1990, recovery of gold from the pregnant cyanide solution obtained from the sand plants was accomplished using the Merrill-Crowe zinc precipitation process. Using this process, zinc dust was added to the gold-bearing cyanide solution to replace the gold in the aurocyanide complex. This caused the gold to precipitate from solution. The "precipitate" was filtered from the solution in Merrill's plate-and-frame filters and treated in the refinery to recover the gold.[10]

With the new carbon-in-pulp system, the slime fraction from the grinding circuit, classified to approximately 97 percent minus-200 mesh, was delivered to two Dorr slime thickeners 125 feet in diameter (see fig. 22.3). The slimes were thickened to 42 to 45 percent solids and pumped to Denver-Dillon trash screens. Following conditioning with lime in a tank 18 feet in diameter and 20 feet deep, the pulp flowed to a feed sump where sodium cyanide was added.

From here, the pulp was pumped to the first of seven dissolution tanks, 30 feet in diameter and 20 feet deep (see fig. 22.4). The pulp flowed by gravity from tank to tank, where the pulp was agitated and aerated. The gold was dissolved by the cyanide, forming a soluble aurocyanide complex. From the last dissolution tank, the slimes were pumped to an adsorption circuit, where the gold was adsorbed onto activated carbon in a series of five adsorption tanks. Carbon was advanced through the tanks in the opposite direction of the flow of pulp. "Loaded" carbon was screened from the pulp and sent to the desorption vessels, where the gold was stripped from the carbon in Zadra electrolytic cells.[11]

Fig. 22.3. The 125-foot-diameter Dorr slime thickeners.—Photo courtesy of Homestake Mining Company.

Fig. 22.4. The dissolution agitator tanks at the Carbon-in-Pulp Plant.—Photo courtesy of Homestake Mining Company.

The Wastewater Treatment Plant

In 1981, Homestake initiated pilot plant testing to investigate the feasibility of using bacteria to treat excess process water and ensure that the water could be discharged in compliance with the latest federal and state stream standards. Based on successful results in the pilot plant, construction was started in February 1983 on a full-scale, biological water treatment plant. The biological treatment process incorporated a strain of bacteria dubbed "Mudlock" that was discovered by and named after Terry Mudder and Jim Whitlock, research chemists at Homestake. The full-scale water treatment plant was commissioned in August 1984 (see fig. 22.5).[12]

The treatment plant included forty-eight Rotating Biological Contactors (RBCs) on which the biomass resided. Influent to the plant consisted of a blend of decant water from the tailings dam and water pumped from the mine, which optimized the nutrient feed for the biomass. As the water passed through the first stages of RBCs (see fig. 22.6), cyanide was converted to ammonia. Subsequent stages of RBCs converted the ammonia to nitrate. Toxic metals and biomass were precipitated using ferric chloride and a polymer. A sludge containing metals and biomass was settled in the clarifier. The effluent (treated water) was polished in sand filters and discharged into Gold Run Creek under a National Pollutant Discharge Elimination System (NPDES) permit.[13]

Fig. 22.5. Aerial view of Homestake's Wastewater Treatment Plant. The plant is located on the divide between Gold Run Gulch and Whitewood Creek near the site of the former Lead City Dump.—Photo courtesy of Homestake Mining Company.

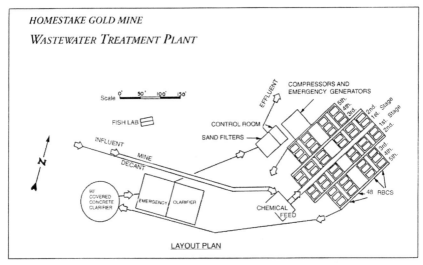

Fig. 22.6. Plan illustration of Homestake's Wastewater Treatment Plant. Bacteria were utilized at the plant to remove metal contaminants from mine water and decant water pumped from the Grizzly Gulch Tailing Storage Facility.—Illustration courtesy of Homestake Mining Company.

Expansions of the Grizzly Gulch Tailing Storage Facility

Homestake contracted with Banner Associates and Woodward Clyde in June 1981 to design the first 50-foot-high downstream raise to the tailings dam.[14] A construction contract was awarded to Summit Inc. on February 17, 1982, to begin work on the first 50-foot raise to the dam to increase storage capacity. Between December 2, 1977, and December 31, 1982, some 5.642 million tons of excess sands and slime tailings were pumped to the Grizzly Gulch Tailing Storage Facility.[15] Work was completed on the first raise to the 5,400-foot elevation in November 1983.

In July 1987, work was started on the second 50-foot-high downstream raise to the dam. This work, which raised the crest of the dam to the 5,450-foot elevation, was completed in 1989. A new shotcrete-lined interceptor canal, 4.8 miles long, was also completed to intercept and route storm water runoff around the tailing storage facility at a higher elevation.

Work was started on the third and final raise of the dam in 1998. A "centerline raise," completed during the fall of 1999, raised the clay core of the dam to the 5,469-foot elevation and the crest of the dam to the 5,500-foot elevation. Between December 1977 and March 2002, when final milling operations were completed, more than 40 million tons of tailings were deposited in the tailing storage facility.

The storage facility was designed to stringent regulatory requirements. One of the permit requirements specified that the impoundment had to be managed to maintain sufficient storage capacity above the tailing level and free-water pond to impound a "probable maximum flood event," plus a one-hundred-year storm event. Such back-to-back storm events would result in direct-fall precipitation and storm water runoff into the impoundment of approximately 42 inches of water, equivalent to about 1 million gallons of water.[16]

Mill Optimization Projects

Between 1987 and 1990, Homestake embarked on a series of optimization projects designed to improve gold recovery and reduce treatment costs. These projects included adding a gravity gold-recovery system and a sulfide regrind system at the South Mill, a carbon adsorption system at the Grizzly Gulch Tailing Storage Facility, an oxygen-vacuum system at the sand plants, and a carbon adsorption system to replace the Merrill-Crowe zinc precipitation process.

Construction work was started at the South Mill on the gravity gold-recovery system in January 1987. The system was completed in May 1987 at a cost of $1.8 million. The overall system consisted of a three-stage gravity circuit that was added to each of the four milling circuits. The newly installed equipment consisted of four 24×36-inch Hazen-Quinn duplex jigs, four strake-concentrating belts, and four Deister tables. Concentrates from the four strakes flowed to two rougher tables, a cleaner table, and a finishing table. Final concentrates, consisting of 50 percent gold, were sent to the refinery for direct smelting. Coarse gold recovered in the gravity circuits had a very favorable effect in reducing gold values in the tailings from the sand plants.[17]

Unfortunately, a gold theft occurred during commissioning of the gravity gold-recovery system. A subsequent investigation resulted in the arrest of nine employees of the contractor that was installing the equipment. Officials believed that more than $130,000 in gold was stolen over the course of the project. Fortunately, most of the gold was eventually recovered. Upon completion of the project, gravity recovery in the South Mill increased to 44.49 percent in 1987 compared to 28.92 percent in 1986. Total recovery for 1987 increased slightly to 94.75 percent.[18]

In 1990, as part of the mill optimization project, a sulfide regrind system was installed in the South Mill. This system was designed to optimize the recovery of coarse gold from the gravity circuits and increase mill capacity by 500 tons per day to take advantage of increased ore production from surface mining operations in the Open Cut. To facilitate the additional tonnage, two leaching tanks and one carbon-in-leach tank were added to the Carbon-in-Pulp circuit. An 8×12-foot Allis Chalmers ball mill was also added to the No. 4 milling circuit to reduce the size of the gold-bearing sulfide concentrates from that circuit.

Product from the regrind mill was delivered to Carpco-Humphrey primary spirals (see fig. 22.7), which replaced the jig and strake equipment. Gold and sulfide concentrates from the primary and secondary spirals in each of the four milling circuits were directed to two Deister rougher tables. Gold concentrates from the rougher tables were further classified on cleaner and finishing tables.

Fig. 22.7. Carpco-Humphrey spirals.—Photo courtesy of Homestake Mining Company.

Concentrates from the cleaning table were delivered to the refinery for direct smelting. Sulfide "middlings" from the rougher tables were returned to the cyclones where underflow material was directed to the regrind circuit. Overflow from the primary cyclones was delivered to secondary cyclones. Overflow from the secondary cyclones was sent to the two 125-foot-diameter thickeners, where slimes were thickened and directed to the Carbon-in-Pulp Plant. Underflow from the secondary cyclones was sent to the sand plant for leaching in vats. Final gold concentrates from the finishing table were delivered to the refinery for direct smelting. The sulfide regrind system increased gold recovery from 50 to 55 percent in the No. 4 milling circuit.[19]

On March 15, 1989, a new carbon adsorption system was commissioned at the Grizzly Gulch Tailing Storage Facility to recover gold from pooled decant water in the tailing facility. During the first forty-one weeks of operation some 6,800 ounces of gold were recovered from the system.[20] In 1992, the carbon adsorption system at Grizzly Gulch recovered 12,346 ounces of gold at a cost of $24.25 per ounce. An average of 10-12,000 ounces of gold was recovered annually for the next several years.[21]

In 1990, as part of the mill optimization project, the Merrill-Crowe zinc precipitation process was abandoned, and the gold-bearing cyanide solution from the sand plants was treated in a new five-tank carbon adsorption circuit. Using this process, some 1,300 pounds of activated carbon was advanced through the tanks countercurrent to the flow of the gold-bearing cyanide solution. Gold was adsorbed from solution onto the activated carbon at the rate of 175-250 ounces per ton. Loaded carbon was pumped to desorption vessels where the gold and silver were stripped from the carbon using a hot, caustic cyanide solution. The gold and silver were plated on steel wool cathodes in Zadra electrolytic cells. Cathodes were delivered to the refinery where the gold and silver were parted and refined to 99.99 percent fineness. The stripped carbon was thermally reactivated and recycled.[22]

Following completion of the mill optimization project, the sand treatment cycle was reduced from about 216 hours to 148 hours. This reduction was achieved as a result of the new gravity-recovery equipment in the South Mill, which increased the amount of coarse gold removed prior to sand treatment. In addition, an oxygen-vacuum system was installed to oxygenate the cyanide solutions in the sand plants. These two changes allowed one of the leach cycles to be eliminated from the sand treatment cycle.[23] More importantly, overall gold recovery increased to 95.15 percent in 1992. Of this amount, 50 percent was recovered by gravity in the South Mill, 32 percent by cyanidation in the sand plants, 9 percent in the Carbon-in-Pulp Plant, 3 percent by carbon adsorption at the Grizzly Gulch Tailing Facility, and 1 percent from byproducts.[24]

The National Environmental Policy Act forced a time frame for Homestake and the surrounding communities to resolve the solid-waste issues that were being investigated prior to January 1, 1970, when the act was signed into law. The company responded to the challenges in earnest by constructing the first large-scale Carbon-in-Pulp Plant in August 1973. Over the next two decades, tens of millions of dollars were spent in constructing and enlarging a tailing storage facility, developing and constructing a wastewater treatment plant, and adapting the latest metallurgical processes and equipment. All of this work enabled Homestake to remain in compliance with the new environmental laws and restore the relatively high rate of gold recovery previously attained using amalgamation.

Chapter 23
Mechanization of Mining

Square-Set Cut-and-Fill Stoping

Between 1937 and 1951, square-set stopes were filled with sandfill, but only after the stope had been mined upward to the next level. Starting in 1951, Homestake began experimenting with two types of cut-and-fill mining methods. One method was called square-set cut-and-fill mining, and the other was called open cut-and-fill mining. Both methods incorporated sandfill as an active part of the mining cycle as each floor or lift was completed.

Essentially, the square-set cut-and-fill method was the same as open timber stoping, except that one or more timbered chimneys were carried within the stope and each floor was sandfilled as it was completed. One-half of the chimney served as a manway between the active stope floor and the sill floor. The other half of the chimney, called the "binline," was used to transfer broken ore from the active floor to a chute on the haulage level.

Wood walls, called gob fences, were constructed lift by lift along the predetermined mining limits of each primary shrinkage or open cut-and-fill stope. After the primary stopes were completed, the intermediate, highly stressed ore pillars were mined using square-set cut-and-fill mining methods. The gob fences prevented the sandfill in the primary stopes from sloughing into the intermediate square-set stopes and diluting the grade of the pillar ore as it was mined and removed (see fig. 23.1). To improve production efficiencies, air-powered two-drum slushers were introduced into the square-set stopes to minimize the amount of hand shoveling. After a horizontal lift of rock was completely mined on a set-by-set basis, the entire "floor" was filled with sand.

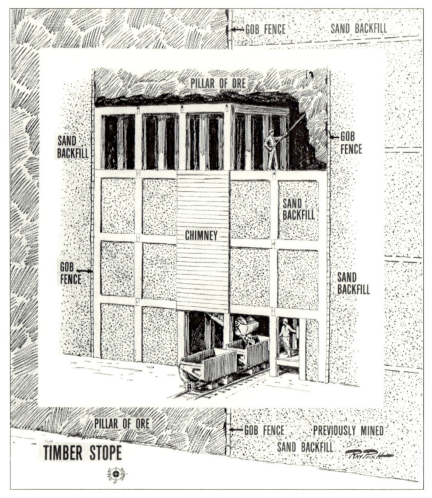

Fig. 23.1. Illustration showing the square-set cut-and-fill method. By 1956, open-panel, square-set stoping was phased out and replaced with square-set cut-and-fill stoping. The latter method, first introduced in 1951, involved sandfilling each floor after it was completed, which resulted in fewer ground-support problems.—Illustration courtesy of Homestake Mining Company.

The cut-and-fill method proved to be significant in terms of providing additional support in active mining areas. Because of this, Homestake was able to replace the 12×12-inch square-set timbers with 10×10-inch timbers. By 1955, square-set cut-and-fill stoping accounted for 24 percent of the total mine production compared to only 1 percent for open square-set stoping.[1]

In 1977, Homestake adopted a ground-support technique called cablebolting to reduce or eliminate the costly square-set method of mining. A square-set stope called 33-34 DE, 19 Ledge, on the 5,750-foot level was selected as a test. From the center of each square set on the active floor of the stope, a 2½-inch-diameter hole was drilled upward through the ore to a height of 60 feet. A ⅝-inch-diameter high-strength cable was then inserted into each hole and grouted in place.

The cablebolts were effective in holding the rock mass together, and the stope was successfully converted to an open cut-and-fill stope. A skid-mounted longhole drilling rig was subsequently designed and fabricated by Homestake personnel for drilling the 60-foot-long cablebolt holes.[2] By 1980, square-set cut-and-fill contributed only 1 percent of the mine's total ore production. The method was completely phased out in 1983.[3]

Open Cut-and-Fill Stoping

The second method of cut-and-fill mining, called open cut-and-fill stoping, was first used on an experimental basis at the Homestake Mine in 1951. This method proved to be a more efficient and cost-effective method compared to square-set stoping and shrinkage stoping.

Ground-control problems were significantly reduced with the open cut-and-fill method because of the tight sandfill that was integral to the method. As the mine advanced downward, the lithostatic pressure, or overburden pressure, increased proportionately with depth. The increase in rock stress caused the barren or subgrade wallrock to slough or collapse in a shrinkage stope, which diluted the quality of the broken ore pile as it was drawn out of the stope.

As a result, management decided to give preference to the open cut-and-fill method. The contribution of ore production from shrinkage stoping was reduced from 63 percent of total ore production in 1951[4] to 22 percent in 1955. Open cut-and-fill accounted for 13 percent of ore production in 1952[5] and 51 percent in 1955.[6] The shrinkage method was completely phased out in 1984.[7]

An open cut-and-fill stope was started by mining a sill cut and the first lift to a height of approximately 17-18 feet above the sill. A raise may already have been driven or bored in ore between the sill floor and the level above. The raise provided ventilation, utilities, and a travel way into the stope from the level above. After the broken ore was removed from these two cuts, a timberline was constructed across the sill floor. The open cut-and-fill stope was then mined upward in successive "lifts" 9-11 feet high until the level above was reached (see fig. 23.2).

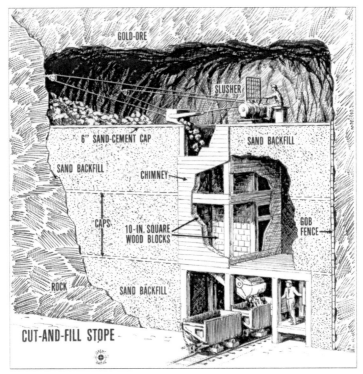

Fig. 23.2. Illustration showing the overhand open cut-and-fill method. The open cut-and-fill method became the predominant mining method from the early 1950s through the 1970s at the Homestake Mine. The end grain of the 10-inch-square binline blocks provided a wear-resistant surface that helped prevent the posts, caps, and ties of the chimney from being damaged by the broken ore.—Illustration courtesy of Homestake Mining Company.

The timberline incorporated a track line for railcars, utility lines, a binline for transferring rock from the stope, a chute for transferring rock into ore cars, and a manway. The manway provided access into the stope from the lower sill level. Wood slabs 2 inches thick were laid across the sill floor to form a "carpet." The carpet provided a sand-containment barrier for the benefit of the stope below.

After the timberline was constructed, its ends were sealed to the ribs of the crosscut at the lateral limits of the stope. The stoping void around and above the timberline was then backfilled with sand tailings that were slurried from the mill. Approximately 7-9 feet of headroom was left open between the newly created sand floor and the back (ceiling) of the stope. A wood carpet was laid across the top of the sandfill. The carpet acted as a barrier to minimize the dilution of broken ore with barren sand as the next lift was mined.

Before mining could proceed, the specific ore area to be mined had to be carefully defined. This was accomplished by sampling the ribs and back of the stope by collecting pick samples and drilling test holes. In later years, a pattern of test holes up to 15 feet long was drilled into the back and ribs. Wet sludge samples were collected from each test hole on 3- or 5-foot intervals for ore-control purposes. The samples were delivered to the assay office, where they were prepared and fire assayed.

Using the assay information and visual determinations from geologic mapping, the shift boss or geologist marked out the specific ore areas to be mined. Any ore in the ribs of the stope was generally mined first. A 9- to 11-foot cut was then mined from the back of the stope. Some of this rock was removed by slushing to create sufficient room to drill for and install rock bolts to support the ribs and back. The remaining broken ore in the stope was removed by a miner who used a small air-powered slusher to slush the ore across the carpet and into the binline. The ore was transferred into ore cars by a second miner, who operated the chute gate at the sill level.

In the early 1950s, several upgrades were made to the mining tools and equipment to improve the efficiency of the new open cut-and-fill method. The quarter-octagon drill steel with a forged integral bit was replaced with sectionalized drill steel that accepted a one-pass blade bit or a tungsten-carbide bit. The latter could be resharpened. Post-and-bar mounted drills were replaced with pneumatic jackleg drills that incorporated telescopic legs (see figs. 23.3 and 23.4).

A short time later, Homestake designed and fabricated its own "stope jumbo" for drilling out or "45-ing" the stope backs (see fig. 23.5). Using the stope jumbo, the backs of open cut-and-fill stopes were drilled using 15-foot-long "flex" drill steel with the jumbo set at 55 degrees from horizontal. Triple-drum electric slushers were also introduced to increase flexibility and production in removing broken ore from the floor of the stope (see fig. 23.6). The tool and equipment upgrades were completed in 1953.[8]

Fig. 23.3. Drilling out a "slab round" in an open cut-and-fill stope using some of the first "jackleg" drills with telescopic legs.—Photo courtesy of Homestake Mining Company.

Fig. 23.4. "45-ing" the back of an open cut-and-fill stope using jackleg drills.—Photo courtesy of Homestake Mining Company.

Fig. 23.5. "45-ing" the back of an open cut-and-fill stope using stope jumbos designed by Homestake. Note the cement cap on the floor of the stope.—Photo courtesy of Homestake Mining Company.

Fig. 23.6. Slushing ore to the binline in an open cut-and-fill stope. The miner is operating a triple-drum electric slusher. Ore is being "slushed" into the binline in front of the slusher. The manway is temporarily covered with wood slabs.—Photo courtesy of Homestake Mining Company.

In 1965, the wood carpets were phased out and replaced with a 6-inch-thick cement cap. After the newly mined lift was sandfilled to the desired height, a cement cap was placed over the sandfill by slurrying a 10:1 sand-to-cement mixture from the sand dam. As with the wood carpet, the cement cap prevented newly mined ore from being diluted with barren mill tailings during slushing operations.

Upon completion of the "lift," an extension was added to the chimney (see fig. 23.7). The chimney was laced with tongue-and-groove lacing slabs that were hand gunnited to form a water seal. The slusher was subsequently hoisted to the top of the chimney using the slusher's own capabilities as a hoist. The stope was backfilled with sand slurry. As the stope was filled, water was allowed to decant from the sand. The decanted water was drained from the stope through drain holes in the lacing slabs of the chimney.

As the level of the sand increased in the stope, drain hole boards were nailed across the drain holes in the chimney to contain the sand. Once the level of the sand was within 6 inches of the top of the chimney, cement was added to the sand-slurry mixture, and the fill was capped with the 10:1 sand-cement fill. After one or two shifts, the sandfill was sufficiently drained, and the cement cap was sufficiently cured to support the miners. The cut-and-fill cycle was then repeated.

Fig. 23.7. An open cut-and-fill stope ready for sandfill. Broken ore has been slushed out of the stope, exposing the cement cap and sandfill from the previous lift. The two chimneys have been extended. The slusher has been hoisted to the top of the new chimney on the left and covered with an old vent bag in preparation for sandfilling. The 6×8-inch piece of lagging suspended from each chimney by wire cables served as a "deadman" or anchor to facilitate slushing operations on the subsequent "lift."—Photo courtesy of Homestake Mining Company.

Open cut-and-fill stopes were typically worked on a two-shift-per-day basis with two miners per shift. Miners were paid on a contract-incentive basis according to the surveyed volume of ore removed from the stope over one complete cycle or "lift." Stope miners also received a regular day's pay wage. A portion of the day's pay wage was deducted from the bonus paid on the volume of ore removed from the stope as an incentive to minimize the time required to complete the lift.

If a dispute or some other problem arose over matters affecting contract-incentive pay, the miners would first try to resolve the problem by talking to the shift boss in his underground office, called a "doghouse," at the end of the shift (see fig. 23.8). If the shift boss was unable to resolve the issue, the miners could present the matter to the Contract Advisory Committee for final resolution.

Fig. 23.8. A typical "doghouse" used by an underground shift boss on the 2,750-foot level, ca. 1941.—Photo courtesy of Homestake Mining Company.

Likewise, if the miners assigned to a particular contract believed that a fellow miner was not fulfilling his share of the work, or was claimed to be incompatible with other members of the crew, the miners assigned to the contract could "sign" a member out of the contract by a simple majority vote. The "sign-out" required approval of the shift boss and foreman and was usually granted.

A miner signed out of a contract could seek to get invited into another contract if an opening was available, bid on posted day's pay jobs, or perform work as an "unassigned" miner. Any unassigned miners a shift boss was fortunate enough to have on his crew were assigned to whatever service tasks needed to be completed in the bosses' geographic area of responsibility or "beat." Unassigned miners were also used to fill in for miners who were absent from their contract or day's pay jobs. Considering the amount of work that was done under the various bonus incentive plans, contract-related "sign-outs" were few and far between probably because of the high level of camaraderie among employees.

As with most jobs, shift work in a stope or drift required good communications based on a work plan everyone agreed upon. The following poem aptly describes the "opposite shift:"

> *I want to tell you boss, I've mined*
> *For twenty years or more;*
> *In all this time I've never seen*
> *A mess like this before.*
> *Just take a look at that air line,*
> *It's broken plumb in half.*
> *You see, the other shift walks out*

And then we stand the gaff.
They didn't cover up the hose,
We can't find all the tools.
No one can get away with this!
I guess they think we're fools.
How anyone in eight short hours
Could get so little done!
They really tore the place apart
And never got a ton.
Unless they teach those farmers how
To timber and to drill
I'll turn my pick and shovel in
And go on down the hill.
About the other shift that worked—
What's that? You're sure you're right?
The hell you say, how could that be?
There was no shift last night?⁹

Blasthole Sublevel Stoping

In 1972, Homestake began investigating the feasibility of using bulk mining methods to exploit some of the lower-grade geologic estimates in the mineral inventory that could not otherwise be mined at a profit using the more selective cut-and-fill methods. The blasthole sublevel open-stoping method was selected.

Development work was started on three different stopes, including 45-51GH, 9 Ledge, 2,300-foot level; 42-46D, 9 Ledge, 4,700-foot level; and 65-71D, 9 Ledge, 6,050-foot level. The general plan for these stopes included drilling out an entire 150-foot-high stope from the top sill, bottom sill, and one or more intermediate subsills. Longhole drills were utilized to drill the blastholes. The drills were equipped with 4-foot-long extension steel and 2¼-inch-diameter button bits. Blasting was performed using ammonium nitrate with fuel oil (ANFO), nonelectric blasting caps, and detonating cord.

After the stope on the 4,700-foot level was put into production, caving problems developed, and the stope had to be converted to a shrinkage stope. Additional diamond drilling showed that the stope on the 6,050-foot level was not amenable to the blasthole method because of poor ore continuity. The stope on the 2,300-foot level was successfully put into production in November 1975.

The cutoff grade for blasthole stoping was set at $2.40 per ton based on the historic gold price of $20.67 per ounce. By comparison, the cutoff grade for square-set cut-and-fill stoping was set at $3.20 per ton, $3 for open cut-and-fill stoping, and $2.40 for shrinkage stoping. Ore was extracted from the blasthole stopes using Wagner ST-2B LHD (load-haul-dump) diesel loaders with 2-cubic-yard buckets. The loader mucked the ore from drawpoint drifts located in the footwall of the stope. Ore was trammed to a grizzly at the transfer raise. Pedestal-mounted Kent pneumatic rock breakers and Allied hydraulic rock breakers were used to reduce oversize rock on the grizzly.[10]

Vertical Crater Retreat Mining

Homestake began its investigation of the vertical crater retreat (VCR) method in 1977 by conducting a series of small-scale cratering tests using class A explosives placed in 4-inch-diameter holes. Initial results were unfavorable simply because a 4-inch-diameter hole was not large enough to allow an explosive charge of sufficient mass and energy to be configured as a "spherical" charge. Based on cratering theories developed by the U.S. Army, a spherical charge mass, as opposed to a columnar charge, was required to produce a crater.

Further testing indicated that a 6-inch-diameter hole was large enough to concentrate enough charge mass and energy to produce the desired cratering effect at the bottom of each hole. Optimal cratering results were obtained by placing the center of a 60-pound charge 6½ feet from the bottom of each hole. Detonation of the charge then created a crater approximately 8-10 feet deep and 16-18 feet in diameter. An initial drilling pattern of 7×9 feet was selected for the first full-scale vertical crater retreat stopes.

Based on the results of the cratering tests, four track-mounted down-the-hole drills were ordered in early 1978 (see fig. 23.9). Each drill was supplied with compressed air at 250 pounds per square inch from two booster compressors driven by 40-horsepower motors. Surveyors marked out the drilling pattern in the top sill by installing survey plugs along the ribs of the stope. Stringlines were installed between the survey plugs by the drillers to ensure proper alignment of the drill.

Fig. 23.9. The first vertical crater retreat drill at Homestake, 1978. Darrell Taggart is operating the drill, which is a Gardner Denver ATD 3100 carrier with a TRW Mission mast and rotating-head assembly.—Homestake Collection, Adams Museum, Deadwood.

An ore area to be mined using the vertical crater retreat method first had to be defined by drilling a pattern of tightly spaced diamond drill holes. Using the information gained from assays and geologic mapping, an analysis was undertaken to affirm that the ore area was sufficiently uniform and amenable to vertical crater retreat mining. The exact limits of the stope to be mined were determined by engineering and geology personnel. The stope was developed by mining 12-foot-high top and undercut sills, using conventional mining equipment. An 8×8-foot pattern of 6½-inch-diameter blastholes was drilled from the top sill to the undercut sill (see fig. 23.10).

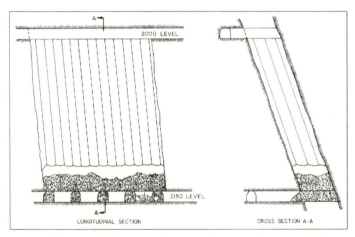

Fig. 23.10. Illustration showing the vertical crater retreat (VCR) mining method. Using this mining method, the bottom of each hole was loaded with a 60-pound explosive charge from the top sill. Broken ore was mucked from the undercut sill using diesel or electric load-haul-dump (LHD) front-end loaders. The miners performed all of their work from the top sill and drawpoint drifts at or below the undercut.—Illustration courtesy of Homestake Mining Company.

Upon completion of drilling, a blasting crew replaced the drilling crew. The blasting crew prepared each hole by lowering a hole-plugging device to the bottom of each premeasured hole and sealing the plug with sand or drill cuttings. A 60-pound explosive charge was lowered down each hole to the exact charge placement depth immediately above the back of the stope. The top of the charge was stemmed with several feet of sand, which served to confine the gases and energy of detonation. At the designated blasting time, a horizontal slice of rock 8-10 feet thick was blasted into the undercut.

Diesel-powered front-end loaders, called load-haul-dumps (LHDs), were used to muck the ore from drawpoint drifts located at the undercut level. The blasting cycle was repeated until a crown pillar 20-30 feet thick remained below the top sill. The entire crown pillar was loaded with double- or triple-decked charges and blasted out in one shot. After all of the ore was removed from the stope using remote-controlled load-haul-dumps, the undercut area was isolated with sandfill containment walls, and the stope was backfilled with development waste rock and sandfill.[11]

In 1982, vertical crater retreat stoping accounted for 47 percent of total mine production; open cut-and-fill stoping, 40 percent; and blasthole stoping, 7 percent. The remaining production was derived from shrinkage and square-set cut-and-fill mining.[12] By the end of 1984, the shrinkage and square-set cut-and-fill methods were phased out because of relatively high operating costs and low efficiencies. Blasthole stoping was also phased out in favor of vertical crater retreat stoping. By the late 1980s, application of the vertical crater retreat method was also significantly reduced because remaining ore reserves were too irregular to be mined using the bulk mining method.

Uphole and Bench Mining

As an alternate to the vertical crater retreat and blasthole methods, numerous "uphole" and "bench" stopes were mined using small-diameter longhole drills. Electric-hydraulic longhole drills (see fig. 23.11) were used to drill out a 25- to 60-foot-thick ore area from a top sill or an undercut using 2½-inch-diameter blastholes. If extended ground control was needed within or above a stope, ⅝-inch-diameter cablebolts up to 60 feet long were inserted and grouted in a separate pattern of longholes. The broken ore was mucked from the undercut level using remote-controlled LHDs.

Fig. 23.11. A Tamrock-Secoma electric-hydraulic longhole drill with diesel-tramming capability. From about 1996 through 2001, Tamrock-Secoma longhole drills were used to drill 2½-inch-diameter longholes up to 60 feet in length for use as blastholes in "bench" and "uphole" stopes. Cablebolts can be seen in this photo in the "back" of the stope above the drill carrier.—Photo courtesy of Homestake Mining Company.

Mechanized Cut-and-Fill Stoping

In 1982, Homestake employees commenced work to mechanize the open cut-and-fill method in an effort to improve safety performance, operating costs, and productivity. Handheld jackleg drills were replaced with 2-boom, rubber-tired, Jarvis Clark drift jumbos for production drilling work (see fig. 23.12). The slusher was replaced with 2-cubic-yard and 3.5-cubic-yard Wagner LHDs. Ramp systems were driven between levels and among stopes, which enabled efficient utilization of the equipment once the ramp system was established.

In many areas, a mechanized mining crew consisted of six miners. Two two-man crews performed the bolting, drilling, and blasting activities on the day and night shifts, while single LHD operators worked on the afternoon and graveyard shifts performing the barring and mucking work. Electric blasting caps were replaced with nonelectric caps and detonating cord.

Fig. 23.12. A 2-boom, Jarvis Clark MJM-20B diesel-powered drift jumbo equipped with pneumatic drills. Prior to the introduction of electric-hydraulic drill jumbos, this jumbo was used to drill drift rounds and breast rounds in mechanized cut-and-fill stopes. The drift jumbo was also used to "drive" development drifts and access ramps.—Photo courtesy of Homestake Mining Company.

In 1995, the efficiency of the mechanized cut-and-fill (MCF) stoping was further enhanced by replacing twenty-four pneumatic drift jumbos with fourteen single-boom, Tamrock-Secoma drift jumbos.[13] The new jumbo featured a single electric-hydraulic drill that was as productive as both of the pneumatic drills on the older jumbo. At the same time, Tamrock roof bolters were purchased, which improved the safety and efficiency of rock-bolting work. The aging LHD fleet was replaced with thirteen 3.5-cubic-yard Elphinstone R1300 LHDs that featured air-conditioned cabs (see fig. 23.13).

Fig. 23.13. An Elphinstone R1300, 3.5-cubic-yard load-haul-dump (LHD).—Photo courtesy of Homestake Mining Company.

With all of these equipment upgrades, coupled with a gradual decline in the number of ore reserve estimates amenable to bulk mining, the mechanized cut-and-fill method became the predominant mining method until the mine closed on December 31, 2001.

Chapter 24
Modern Mining in the Open Cut

Exploration Work

In 1980, at the recommendation of Joel K. Waterland, assistant general manager of production for the Homestake Mine, a study was undertaken to evaluate the potential for remnant ore-grade rock around and below the historic Open Cut. The studies concluded that based on current metals prices and mining technology, significant potential existed for remnant ore above the 800-foot level. A review of the historic mine maps and sections indicated that the caved and waste-gobbed areas likely contained ore-grade rock and that numerous support pillars likely contained ore, such as the large pillar around the Golden Star Shaft.

In May 1981, an underground exploration program was initiated in the Open Cut area to verify the conclusions and recommendations of the study. A heavily timbered crosscut, 320 feet long, was spile driven through a caved and broken portion of the Star Shaft pillar at No. 1 North Pillar on the 800-foot level below the Open Cut.[1] A second crosscut, 230 feet long, was completed in October 1982 at No. 4 North Pillar on the 800-foot level. Both crosscuts provided valuable information relative to ground conditions and confirmed the existence of ore-grade mineralization. Concurrent with the underground work, a surface-drilling program was initiated in August 1981 within the Open Cut. Some twenty-four core- and reverse-circulation exploration holes were completed in 1981. An additional 340 holes were completed in 1982.

Based on initial results of the exploration program, it was decided that a significant amount of ore-grade mineralization did exist in the Open Cut and that the ore could best be mined using modern surface mining methods. Unfortunately, all of the residents of Terraville would have to be moved to facilitate the mining operation. After approvals were given for the project, land title and appraisal work was immediately started, and Homestake began to acquire the properties in Terraville and relocate its residents. By the end of 1982, all but two of the properties in Terraville had been purchased. At the same time, an ore reserve of 6.921 million tons with a contained grade of 0.144 ounces of gold per ore ton was identified by the engineers and geologists.[2]

Terraville Test Pit

In September 1983, a feasibility study was completed for a proposed surface mining operation in the Open Cut area using a mining contractor. The study was based on a mineable ore reserve of 7.531 million tons at a contained grade of 0.116 ounces per ton. Using a design cutoff grade of 0.04 ounces per ton and an internal cutoff grade of 0.025 ounces per ton, it was estimated that about 815,000 ounces of gold could be recovered at the South Mill complex. Based on the proposed design for the open pit, some 59.105 million tons of waste rock would have to be mined to exploit the 7.531 million tons of ore. Cash-operating costs were projected to be $266 per recovered ounce.

To verify the feasibility study, it was decided that a test pit should be mined. The Homestake Board of Directors approved the test mining operation, known as the Terraville Test Pit, on September 23, 1983. Contracts were awarded to Johnson Brothers Company of Spearfish and Summit Inc. of Rapid City. Johnson Brothers would complete the preproduction development work while Summit of Rapid City would complete the mining work. Homestake would provide engineering, geology, and survey control for the mining work. Mining work commenced in the Terraville Test Pit on October 24, 1983,[3] and was completed on June 22, 1985.[4]

Overall results from the test were quite favorable. Compared to the ore reserve within the Terraville Test Pit, 94.2 percent of the planned ore tons were mined at 105.3 percent of the planned head grade. Some 315,689 tons of ore were milled at a head grade of 0.070 ounces per ton from the 5,200 bench through the 5,040 bench compared to the ore reserve of 335,019 tons at a head grade of 0.0665 ounces per ton. Recovered gold totaled 19,731 ounces at a total cost of $309 per ounce.[5]

Pit Expansions

Based on the favorable results of the Terraville Test Pit, the Homestake Board of Directors approved a request for expenditure in May 1985 to expand surface mining operations in the Open Cut using a mining contractor. The Version 17 Plan included 53 million tons of waste rock and 5.048 million tons of ore at a contained grade of 0.118 ounces of gold per ton. On October 22, 1985, Homestake signed a five-year contract with S/J Constructors, which was a joint venture of Summit and Johnson Brothers. The mining plan called for establishing "catch benches" 25-42 feet wide every 80 vertical feet. By year's end, the mining contractor had mobilized mining equipment consisting of track-mounted blasthole drills, 85-ton off-road trucks, and a 13-cubic-yard front-end loader.[6]

A maintenance shop was moved from Homestake's Pitch Mine in Colorado to the Open Cut in 1985. A three-stage crushing plant, designed and constructed by General Steel and Supply, was constructed at a cost of approximately $3.5 million (see fig. 24.1). The plant was commissioned in January 1987. A 50×60-inch Universal jaw crusher, a Symons 7-foot standard cone crusher, and a Symons 7-foot shorthead cone crusher were used to crush the run-of-mine ore to minus ½ inch. A belt magnet proved useful in removing tramp iron, which included such things as horseshoes, hand steel, and other metals from the historic mining areas.

A Japanese Pipe Conveyor, 6,300 feet long, was designed and constructed by Robins Engineers and Constructors at a cost of about $4.4 million. The conveyor, which was the longest of its type in the United States at the time it was constructed, transported ore at the rate of 350 tons per hour from the crushing plant to the South Mill (see fig. 24.2). The pipe conveyor was also commissioned in January 1987.[7]

Fig. 24.1. The Open Cut crushing plant and crushed ore stockpile, 1998.—Photo courtesy of Mark Zwaschka.

Fig. 24.2. The 6,300-foot-long Japanese Pipe Conveyor. The conveyor transported minus ½-inch crushed ore from the Open Cut crushing plant to the stockpile and ore bins at the South Mill.—Photo courtesy of Homestake Mining Company.

A large number of residences and businesses were purchased in Lead to facilitate the first expansion of the Open Cut in 1985. The ball field at Mooney Park had to be abandoned and was replaced by improving existing facilities at Mountain Top Park at a cost of $243,000.[8] In 1992, based on the success of mining in the Open Cut to date, the Homestake Board of Directors approved a second mine expansion. Almost immediately, mining in the bottom of the pit was suspended, and work was concentrated on a pushback of the east, west, and south walls.

Highway 85 was relocated to the south to facilitate the expansion (see fig. 24.3). A portion of this work included completion of an archaeological survey, and resource mitigation for twenty-three structures designated as historic within the original town site for Lead. In an effort to enable displaced residents to relocate in Lead, Homestake partnered with the city of Lead and various developers to create the Mile-High Mobile Home Park, Caledonia Apartments, and the Hearst housing subdivision. Concurrently, Homestake completed the Manuel Brothers Park along the south edge of the Open Cut to replace the Sinking Gardens, which was formerly located along the north side of Main Street. The cost of the new park was approximately $1.1 million.[9]

Fig. 24.3. An aerial view of the Open Cut area, ca. 2001. Modern surface mining operations were conducted in the Open Cut area between October 1983 and September 1998. The Terraville Test Pit was located in the very upper left-hand corner of the photo. The Open Cut was mined to a depth of 40 feet below the 800-foot level of the underground mine.—Photo courtesy of Historical Footprints Inc.

As part of the 1992 expansion, the Homestake Visitor Center was moved to the south edge of the pit, and a viewing platform was established. The historic First Finnish Lutheran Church, which had become the Sweatman Art

Memorial in the 1960s, was moved from the Sinking Gardens to the west side of the Homestake Visitor Center and opened for business as JL's Gift Shop.[10] The newer building for First Lutheran Church, which was constructed in 1962, was moved to Spearfish to facilitate the Open Cut expansion. First Lutheran Church and Bethel Lutheran Church subsequently merged as Shepherd of the Hills Lutheran Church in December 1992.

Over the next six years, mining continued outward and downward through the maze of caved or filled square-set stopes, shrinkage stopes, drawhole stopes, fire areas, drifts, shafts, and pillars that existed adjacent to and below the historic Open Cut. The practical mining limit of the surface mining operation was reached in September 1998 at the 4,380 elevation, which was about 40 feet below the 800-foot level of the underground mine. The last of the ore in the Open Cut was mined on September 8, 1998. Between October 1983 and September 1998, some 130 million tons of waste rock and 13.433 million tons of ore were mined, which resulted in the recovery of 1.267 million ounces of gold (see table 24.1).[11]

Table 24.1 Open Cut surface mining production

Year	(tons) Mined Waste	(tons) Mined Ore/ Low Grade	(tons) Milled Ore	(troy ounces) Gold Recovered	(ounces per ton) Recovered Grade
1983	368,395	4,480	0	0	N/A
1984	3,084,645	310,544	202,150	10,958	0.0542
1985	3,691,264	383,683	420,927	25,458	0.0605
1986	12,818,212	375,956	431,866	31,663	0.0733
1987	13,676,193	482,512	472,786	40,723	0.0861
1988	11,984,866	847,019	516,116	46,175	0.0895
1989	10,331,846	997,050	700,747	68,539	0.0978
1990	9,785,185	642,590	712,498	82,102	0.1152
1991	9,544,769	968,272	976,097	93,566	0.0959
1992	9,683,485	1,133,972	1,086,666	127,674	0.1175
1993	10,516,898	1,048,272	1,223,484	106,265	0.0869
1994	13,200,281	1,191,789	1,243,806	100,665	0.0809
1995	8,616,178	1,532,507	1,008,764	91,846	0.0910
1996	7,706,419	2,134,894	1,198,093	123,198	0.1028
1997	4,398,193	1,893,590	1,219,202	133,091	0.1092
1998	882,775	690,536	1,579,675	159,956	0.1013
1999	0	0	439,892	25,412	0.0578
Total:	130,289,604	14,637,666	13,432,769	1,267,291	0.0943

Source: Homestake Mining Company, "Open Cut Inception-to-Date Production."

Reclamation of Waste Rock Facilities

Waste rock facilities were established at the Highland, Upper Bobtail, Bobtail, East Waste Rock, Sawpit, and Terraville Test Pit areas to receive the 130 million tons of waste rock mined and removed from the Open Cut (see fig. 24.4). The East Waste Rock Facility, including Bobtail, was the largest, holding 90-92 million tons of rock. Some 25-28 million tons of rock were placed in the Sawpit Waste Rock Facility. As surface mining operations were nearing completion in 1997 and 1998, the remaining waste rock was placed in the Terraville Test Pit.

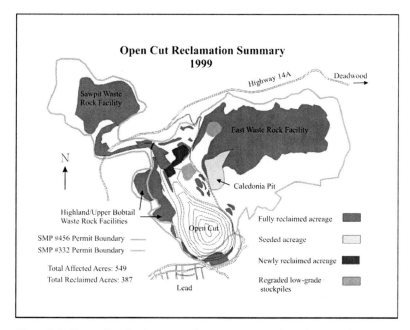

Fig. 24.4. Open Cut Reclamation Summary, 1999. Two large-scale mining permits were issued by the South Dakota Board of Minerals and Environment under the South Dakota Mined Land Reclamation Act that allowed surface mining operations in the Open Cut. Some 130 million tons of waste rock were placed in waste rock facilities that had to be reclaimed to standards specified in the permits.—Illustration courtesy of Homestake Mining Company.

The South Dakota Board of Minerals and Environment issued two large-scale mining permits (Nos. 332 and 456) to Homestake for its surface mining operations in the Open Cut. A total of 658 acres were included within the boundaries specified for the two permits. Of this total, 549 acres were considered affected with 520 of the acres actually disturbed. As of December 31, 1999, the first year following closure of the Open Cut operations, 397 acres were considered fully reclaimed under SDCL-45-6B-97.[12]

On April 20, 2006, state officials released a total of 499.74 acres (see figs. 24.5 and 24.6). Homestake's Reclamation Bond was subsequently reduced to $1.246 million. Following closure of the Homestake Mine in 2001, various other continuing postclosure reclamation projects were completed around the waste rock facilities.[13]

Fig. 24.5. The top of the reclaimed East Waste Rock Facility.—Photo courtesy of Homestake Mining Company.

Fig. 24.6. A portion of the reclaimed Sawpit Waste Rock Facility featuring a wildlife habitat area.—Photo courtesy of Homestake Mining Company.

Chapter 25
How Other Work Was Performed

Mine Surveying

Surveying is one of the many important tasks that required a staff of people to support operations at the mine from its inception through closure. Such work involved staking mining claims; marking boundaries; establishing baselines; determining lines and grades for ditches, flumes, pipelines, and railroads; determining areas and volumes; and transferring the survey control from the surface to the underground levels.

Surveys played a major role in the development of the mine's infrastructure, which included shafts, winzes, drifts, raises, ramps, and drainage systems. In later years, alignment surveys were performed to orient diamond drills, borehole raise drills, longhole drills, and down-the-hole drills as well as to establish sampling patterns in open cut-and-fill stopes. Surveys were also performed to determine the volume of rock removed from cut-and-fill and shrinkage stopes. Considering the amount of surveying that was completed, errors were few and far between.

Richard Blackstone, the chief engineer for Homestake, established baseline, coordinate, and triangulation systems for the mine on November 20, 1879. These systems became the basis for all subsequent survey control at the Homestake Mine for the life of the mine. The baseline was laid out on an orientation of North 35° 40′ West, which closely approximated the strike of the Main Ledge ore body that had been discovered. The triangulation system helped ensure that the various control points were accurate in terms of the coordinates and elevations assigned to them.

Survey control was transferred from the surface triangulation system to the underground levels of the mine using the following methods: (1) through an adit or tunnel, (2) by hanging two wires down a vertical shaft or winze, and (3) by hanging one wire in each of two vertical shafts and surveying between the wires on the surface and underground. The latter method was the one that was successfully employed to establish the starting points for the five pilot raises and shaft-sinking operation that created No. 5 Shaft.

The 19-foot-diameter No. 5 Shaft was completed from the surface to the 5,000-foot level between May 11, 1956, and November 16, 1959, by incorporating concurrent sinking and raising methods. As one crew commenced to sink the shaft from the surface, other crews drove pilot raises concurrently from the 2,000-, 2,900-, 3,800-, 4,700-, and 5,000-foot levels. Accurate survey control was required to establish the precise location at which a pilot raise was to be driven from each of the five underground levels.

The two-wire method for "transferring the meridian" was the most common method for transferring survey control underground. Using this method, the survey control was transferred from the triangulation system to the shaft collar. The control was then transferred down the shaft to one or more mine levels, and it was later transferred from one level to another as the mine areas were developed.

Two platforms were constructed across the shaft or winze at the upper location and one at the lower location. From the uppermost platform, which supported the reels of wire, two high-tensile steel wires, 0.022 inches in diameter, were lowered down the shaft to the lowest platform. The bottom end of each wire was attached to a nominal 60- to 65-pound weight. Each weight was suspended in a bucket of oil and water to dampen the movement of the weight. Upper and lower survey crews recorded the measurements between the two wires. The measurements were communicated between crews to ensure that the plumb lines had dampened out and that each crew had obtained the exact same measurement between the parallel wires.

From the upper and lower shaft stations, each surveyor set up his transit and "wiggled in" on the two wires to align the transit on the exact plane created by the two wires (see fig. 25.1). This process is referred to as coplaning. After each transit was perfectly aligned with the plane of the wires, measurements were taken between the instrument and each wire. The lower crew transferred the survey control by establishing two colinear control points, using wooden plugs and spads, in the back of the drift (see fig. 25.2).[1] The precise angle formed at the transit between the "backsight line" to the wires and the foresight line to the newly established set of survey plugs was determined using the Weisbach method.[2]

Fig. 25.1. Transferring the survey control at No. 3 Winze. Virgil Palmer is shown here "wiggling in" on the wires on the 4,250-foot level at No. 3 Winze. The survey control is being transferred from the 4,100-foot level to the 4,250-foot level. A cap lamp has been placed on one of the shaft timbers to provide illumination behind the wires.—Homestake Collection, Adams Museum, Deadwood.

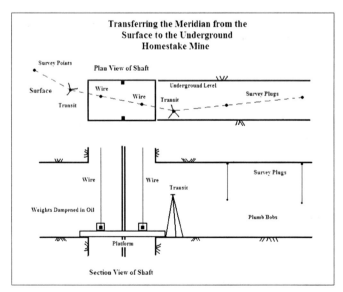

Fig. 25.2. Illustration showing how the survey control was transferred from the surface to an underground level or from one underground level to another.—Illustration by author.

Using this method, the transitman backsighted the wires and turned and accumulated the "angle right" six times—three with the telescope direct and three with the telescope inverted. If the sum of the angles read by taking the repetitions wasn't within the error of the instrument, the process was repeated. After the mine levels in a given area were more fully developed, the survey crews checked their surveys by transferring the control through other shafts, winzes, raises, or ramps to adjust and close the surveys.

The two-wire method was successfully used to develop the Ellison, Yates, and Ross shafts, as well as No. 3 Winze, No. 4 Winze, No. 6 Winze, and the associated mine levels. No. 6 Winze, which was developed starting in 1969, required transferring survey control on the 4,550-, 4,850-, 5,000-, 5,900-, and 6,800-foot levels to facilitate the various mining crews who were performing concurrent sinking and raising operations to create the winze. Survey control was transferred from the 4,850-foot level to the 5,900-foot level through No. 26 service raise. The control was then transferred across the 5,900-foot level for approximately 4,000 feet to the downward projection of No. 6 Winze.

The same survey control was also transferred from the 5,900-foot level to the 6,800-foot level through No. 28 service raise. From here, the control was transferred some 4,000 feet to the same downward projection of No. 6 Winze. Pilot raises, 900 feet long, were concurrently driven from the two levels. The same process was used to establish a pilot raise for the winze from the 5,000-foot level to the hoist room on the 4,550-foot level. The different sections of the winze were successfully connected to within ⅜ inch in one corner and 1½ inches in another.[3] All of these surveys were completed with transits and steel tapes prior to the advent of theodolites and total-station instruments that utilize electronic distance-measuring technology, digital readouts of angles, and electronic data collectors.

A system of stope and pillar lines was originally developed in the 1880s to define 60-foot mining limits for both primary stopes and square-set stopes. The stope and pillar lines were oriented transverse to the North 35° 40′ West baseline that was established for the mine. A few years later, the width of the pillars was reduced from 60 feet to 42 feet and then to 40 feet.

The Main Ledge reference line system had its origin through the Golden Star Shaft. Depending on the mine area, stope lines and intermediate pillar lines in Main Ledge were either 102 feet or 100 feet apart, respectively. The 9 Ledge reference line system had its origin through the Ross Shaft. Stope lines and intermediate pillar lines in 9 Ledge were predominately 95 feet apart, respectively. In the west ledges below the 5,000-foot level, stope lines and pillar lines were established every 100 feet, respectively, based on the 19 Ledge reference line system.

As the centroids of the ledges were gradually depleted, the conventional order and size of primary stopes and pillars was no longer practical. Therefore, geologists began defining individual ore reserve estimates according to what area could efficiently be exploited using the open cut-and-fill method, with consideration to local geologic and ground-control conditions. At about the same time, because of high costs, applications of the square-set method were restricted to those areas exhibiting unstable ground conditions.

With these changes, the various reference line systems evolved into a simple geographical reference relative to the extent and location of stopes, winzes, raises, and facilities. No. 6 Winze, for example, was located along the 52 pillar line. No. 7 Winze was located some 1,200 feet to the south at 40 pillar line. Ore estimates and associated stopes were similarly designated, such as 38-43B, 21L, 7400. This indicated the ore estimate and corresponding stope was primarily located between the 38 and 43 stope lines in the B limb of 21 Ledge on or above the 7,400-foot level.

Sampling and Assaying

Fire assaying is a laboratory procedure commonly used in the mining industry to determine the gold and silver content of rock samples or ore being processed. Because the metal content of most precious metal ore deposits is often highly variable over a few inches or feet, geologists and mining engineers rely heavily on various sampling techniques and statistical methods to determine the average metal concentration or "grade" of a given mass of rock.

Rock samples can be obtained from chip sampling, grab sampling, channel sampling (see fig. 25.3), or various types of drill sampling. Samples are routinely collected throughout the various milling and treatment processes. Assays are routinely used throughout the mining cycle for exploration planning, ore reserve estimation, mining control, and processing control.

Fig. 25.3. Channel sampling at the Homestake Mine, 1923. Channel sampling was a very meticulous and labor-intensive method of sampling. The method was later replaced with dry-drill and test hole sampling methods that provided deeper sample penetration.—Photo courtesy of Homestake Mining Company.

Gold occurrences within the Homestake Formation were mostly microscopic in size. On rare occasions, particularly in Main Ledge, abundant free gold was seen with the naked eye. If certain associated ore minerals such as pyrrhotite and arsenopyrite were absent, even the most experienced geologists oftentimes had difficulty differentiating subeconomic rock from ore. The term *ore* refers to mineral-bearing rock that is valued for its economic metal content. A given volume of rock is properly termed ore only if the rock can be mined and milled at a reasonable profit, based on the recovery and sale of the contained metals and after deductions for exploration, preproduction development, and reclamation work.

Very selective mining methods, such as cut-and-fill, were developed and extensively used at the Homestake Mine to minimize the amount of waste rock or subeconomic rock that might otherwise be mined and milled. If too much of the subeconomic rock was allowed to be mined or if the mineralization within a particular stope was inherently highly variable, the profitability of the entire mine was impacted.

The cost of mining, transporting, and milling a ton of waste rock or "low-grade" was essentially the same as that for a ton of ore. There simply had to be enough gold recovered from each ton of rock processed at the mill to meet operational expenses and pay back any capital investments. Ore "dilution," as it was called, was often very problematic for mine personnel and required not only the selection of the optimal mining method, but strict adherence to established grade-control policies and procedures.

The bulk mining methods, such as vertical crater retreat and longhole open stoping, offered advantages in terms of worker safety since the miners performed their work external to the stope itself. The bulk methods were generally more cost-effective compared to the more labor-intensive methods such as cut-and-fill if they were applied to regular, well-defined ore bodies where internal and external sources of dilution could be minimized.

Application of a bulk mining method to a narrow limb of ore exhibiting major folding or zones of weak mineralization usually resulted in significant ore dilution. Therefore, ore estimates planned to be mined using a bulk method were extensively defined by diamond drilling and drill sampling to characterize the ore area and affirm the choice for the mining method. This process enabled the geologist, mining engineer, and blasting technician to predetermine the limits of the bulk stope with confidence that the stope would not be an economic failure.

With the cut-and-fill method, mining progressed round by round and lift by lift. Samples were collected every round laterally and every lift and were sent to the assay office daily. The assay results, which were generated within twenty-four hours, were used to determine the exact mining limits of the stope on a round-by-round basis. Consequently, the higher mined grade helped offset the higher cost of the method compared to bulk mining.

Rock samples or samples from drill cuttings were collected, bagged, tagged, and delivered to the assay office. There, each sample was crushed and pulverized to minus-100 mesh. The pulverized sample was passed through a splitter, which mixed and divided the sample into smaller but equally representative portions. A very small sample was removed from one of the splits and weighed to establish the desired sample weight.

In 1866, Charles F. Chandler at the Columbia School of Mines developed the assay-ton system of weights for assaying. Chandler's system is based on 7,000 troy grains per pound (avoirdupois). One short ton is equal to 14,000,000 troy grains. With one troy ounce equal to 480 troy grains, one short ton is also equal to 29,166.667 troy ounces. By taking one milligram as the unit, one assay ton is equal to 29,166.667 milligrams or 29.1667 grams. Therefore, 1 milligram bears the same relation to one assay ton as 1 troy ounce bears to 1 short ton. If one assay ton of ore (29.1667 grams) was assayed and the resulting button weighed 1 milligram, the assay value of the ore would be 1 troy ounce per ton of ore.[4]

At the Homestake Mine, the assay-ton system, based on a 29.1667-gram sample, was used until the late 1920s for assaying mine samples. Because Homestake sold most all of its bullion to the U.S. Mint at the then-fixed gold price of $20.67 per troy ounce, management decided to substitute a "dollar-ton" system for the assay-ton system. Based on a ratio of 20.6718 to 20, a dollar-ton factor of 30.1464 grams was subsequently used for assaying mine samples. Using the dollar-ton system, a milligram of gold represented an ore value of $20 per ton, based on a gold price of $20.6718 per ton. The assay-ton system, based on a 29.1667-gram sample, continued to be used for metallurgical process control.[5]

Because rock from the Homestake Formation was high in silica, each 30.1464-gram sample was mixed with soda ash, borax, flour, silica, and a lead oxide reagent called litharge. Silver chloride was added to the flux mix if the samples were suspected to be low in gold content and the assayer or sampler was not interested in the silver content.

During the fusion phase, the entire sample-flux mixture was put into a firebrick crucible (cup) and placed in a furnace where the contents of the crucible were roasted at approximately 1,800 degrees Fahrenheit until the entire contents were fused. The soda ash and borax combined with the silica in the rock to form a glass slag that migrated to the top of the molten lead, gold, and silver as the sample was melted. As the litharge oxidized, the lead combined with the gold and silver at the bottom of the crucible. After the crucible was removed from the furnace, the molten sample was poured into a scorification mold and allowed to dry. The glass slag was broken away from the cone-shaped sample, exposing a button of lead, gold, and silver.

During the cupellation phase, the lead button was flattened and squared with a hammer and placed on a cupel made of bone ash. The button and cupel were reheated in a furnace to about 1,800 degrees Fahrenheit until all of the lead was oxidized and absorbed into the cupel. The gold and silver button that remained was removed from the cupel and weighed.

The next step of the process was called parting. The silver, if any, was "parted" out using hot, diluted nitric acid. The remaining bead of gold was reweighed. Every milligram of gold in the button was equivalent to 1 troy ounce of gold per ton of ore. The assays were used in the determination of mineral inventories, ore reserves, or for quality control during day-to-day mining operations.

Bullion bars were sampled by drilling holes in diagonally opposite corners of the bar using an air drill. An upper and lower sample was collected from each hole. Two 500 milligram splits were measured from each sample and fire assayed to verify bar quality. Prior to construction of the refinery and implementation of refining methods, the desired silver-to-gold ratio for bullion bars sold to the U.S. Mint was 2.5:1.[6]

Smelting and Refining

In 1933, Homestake decided to construct its own refinery on the east side of Haggin Street and begin refining bullion in Lead. The refinery was equipped with five natural gas-fired amalgam-retort furnaces and four coke-fired melting furnaces. A short time later, one of the retort furnaces and the four melting furnaces were removed and replaced with six natural gas-fired furnaces. Eight parting furnaces were also installed.

Gold-zinc "precipitates" from the filter presses in the East Sand Plant and Slime Plant, gold-plated steel wool from the Carbon-in-Pulp Plant, and gravity-gold concentrates from the South Mill were smelted and refined at the refinery. The precipitates contained approximately 4 ounces of gold per pound. After the precipitates were dried to about 10 percent moisture content, a flux containing soda ash, silica, sodium borate, and sodium nitrate was blended in.

The precipitates were then pressed into 10-pound briquettes and placed in a 26-inch Monarch-Rockwell furnace. Here, the briquettes were heated to a temperature of 2,000-2,200 degrees Fahrenheit until they were completely melted. Base metals and other impurities were oxidized and became part of the slag formed by the flux. After the slag was poured off, doré bullion consisting of 800-fine gold and 170-fine silver was obtained. The bullion was weighed, sampled, and stored for further refining.

At the Carbon-in-Pulp Plant, gold was adsorbed onto activated carbon. The gold was stripped from the carbon by passing the carbon through desorption cells containing sodium hydroxide and sodium cyanide. Next, the gold-bearing solution was passed through three stages of Zadra electrolytic cells where the gold was plated on steel wool cathodes.

When the cathodes in the first cell were loaded with approximately 600-700 troy ounces of doré bullion, the cathodes were delivered to the refinery and fluxed in much the same manner as precipitates, excluding the soda ash. The fluxed cathodes were placed in a Monarch tilting furnace, melted, and poured into a cone mold. After a cooling period, the bullion was removed from the mold, and the slag was chipped from the bullion. This bullion assayed approximately 815-fine gold and 170-fine silver. The bullion was weighed, sampled, and stored for further refining.

Concentrates from the South Mill gravity circuit typically contained about 10 ounces of doré' per pound. The concentrates were fluxed in the same manner as the steel wool cathodes and melted in a Monarch tilting furnace. After pouring and quenching, the crude bullion was weighed, sampled, and stored for further refining. This crude bullion consisted of approximately 780-fine gold and 160-fine silver.[7]

The Miller chlorine process was used to separate or part silver from the crude bullion obtained from smelting the precipitates, steel wool cathodes, and gravity concentrates. Homestake adopted the chlorine process in 1933, concurrent with the company's decision to construct a refinery in Lead. Using this process, gaseous chlorine was passed through a bath of molten bullion. The chlorine gas, supplied from bottles stored in a small brick building attached to the refinery, was delivered to the parting furnaces (see fig. 25.4) through copper tubing and seamless steel pipes. In the furnaces, silver and metals other than gold were converted to chlorides and skimmed off the top of the molten materials, leaving only the gold.

Fig. 25.4. One of the chlorine-parting furnaces that was utilized to part silver from the crude bullion.—Fred Mosley collection.

The parting process was initiated by slowly heating a clay crucible in a parting furnace. The crude bullion was concurrently melted in one or more of the large melting furnaces. A small amount of borax was added to the bullion to flux the impurities. After the materials were completely fused, the resulting slag of impurities was skimmed off the top of the molten bullion. The parting crucible was removed from the parting furnace, filled with approximately 650 troy ounces of molten bullion, and returned to the furnace. After a few ounces of borax were added to the crucible, the clay tube used to introduce the chlorine gas was inserted into the molten bullion.

As the chlorine gas was dispersed through the molten bullion, the silver reacted with the chlorine atoms, forming silver chloride. The silver chloride was skimmed off as it rose to the top of the crucible. Approximately, two hours were required to part all of the silver chloride from the molten bullion. After all the silver chloride was skimmed off, the remaining molten product exhibited a distinctively high luster with a greenish hue, indicating very pure gold. After this state was reached, the gold was poured into bars averaging "994" fineness.

After the parting process was completed, the gold was remelted, refined again, and poured into bars of "998" fineness (see figs. 25.5 and 25.6). The bars were weighed, stamped, and sold to the U.S. Mint in Denver. If additional refining was desired, the gold was cast into anodes and refined to "999.9" fineness in Zadra electrolytic cells, each with a capacity of about 800 troy ounces.

To achieve "999.9" fineness, the gold anodes were placed in polypropylene tanks containing 90-100 grams per liter of gold chloride and 15 percent hydrochloric acid. The acid reacted with the small amount of silver in the anode, forming silver chloride. The silver chloride precipitated to the bottom of the cell. Upon being subjected to direct current from a rectifier at 0.5 volts and 100 amperes, the gold was plated on the cathode in the electrolytic cell. The gold cathodes were washed, dried, remelted, and poured into bars of 999.9 fineness.

Fig. 25.5. A gold pour at the Homestake Refinery. Fred Mosley is the person on the right. The other person is not identified.—Fred Mosley collection.

Fig. 25.6. Gold bars at the Homestake Refinery, 1966. This photo, taken on December 28, 1966, shows twenty-four gold bars of "998" fineness that contained a total of 13,628 ounces of gold. Based on a fixed gold price of $35 per ounce at the time, the bars were sold to the U. S. Mint in Denver for about $476,980. At $1,000 per ounce, identical bars would be worth $13.6 million.—Fred Mosley collection.

The silver chloride obtained from the parting furnaces was also remelted. After adding sodium chloride to the molten chloride, the charge was poured into slabs ½ inch thick by 6 inches square. The silver chloride slabs were placed in acid-proof trays containing muriatic acid and iron slugs. Here, the silver was reduced to metallic silver. The metallic silver was remelted in a tilting furnace and poured into 400-500 troy ounce bars of "920-960" fineness (see fig. 25.7). As a final step, the silver bars were refined in electrolytic cells, remelted, and poured into bars of "999.0+" fineness.

Fig. 25.7. Gerald Powers making a silver pour using an electric tilting furnace.—Photo courtesy of Homestake Mining Company.

Slag obtained from the various smelting operations typically averaged about 20 troy ounces of gold per ton. The slag was combined with floor cleanup and melted in a scavenger furnace using coke as a heat source. The products recovered from the furnace included a high-grade gold matte, lead bullion, and nearly barren slag. The gold matte was shipped to a smelter annually. The lead was separated from the gold and silver by cupellation. Slag was recycled in the milling circuit.[8]

Figure 25.8 summarizes the various processes used at the Lead Refinery to refine crude bullion.

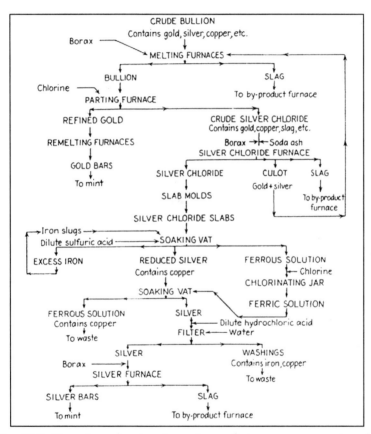

Fig. 25.8. Flow sheet showing the various refining processes used at the refinery.—Source: Nathaniel Herz, "Bullion Parting at the Homestake Mine," technical report, Homestake Mining Company, 1935.

Sand Casting

From inception of the company, Homestake found it had to be self-sufficient to meet the ever-increasing demands of production, maintenance, and profitability. Prior to about 1900, all of the equipment and supplies were delivered to the mine by bull teams under the supervision of a wagon master. As Homestake consolidated other mines in the Open Cut area, it established the B&M complex on the east side of the Open Cut. This complex included a carpenter shop, blacksmith shop, machine shop, rope house, pattern shop, foundry, and other miscellaneous shops. These shops allowed Homestake to be almost self-sufficient. Most any part or piece of equipment could be fabricated or repaired.

In December 1908, Homestake completed work on a new foundry near the B&M Shaft to replace the original one. The new foundry housed a 58×40-foot floor space on the molding floor and a 40×90-foot wing. The building included a 10-ton Niles electric crane that could easily handle the larger castings that weighed 3-4 tons. A 20-ton Whiting cupola was also constructed. For night work, three arc lights and numerous incandescent lights illuminated the foundry. The building featured a unique "ventilation system" where any of the windows in the roof or walls of the building could be opened or closed from a central station on the main floor.[9]

In 1927, another new foundry was constructed between the Ellison Shaft and the South Mill. Tools and replacement parts for most any piece of equipment in the mine or mill were cast in this foundry (see figs. 25.9 and 25.10). A wooden pattern of the desired part to be cast was designed and crafted in a wood shop called the Pattern Shop. Highly skilled wood craftsmen, called patternmakers, were employed by Homestake to create hundreds of different pattern designs and actual wooden patterns. The pattern had to be an exact replica of the desired cast metal part, whether it was a gear, shaft, mine car wheel, or pipe fitting. Dry, straight-grained pine lumber was used to construct the pattern. The patternmaker used shrink scales for their measurements, which included allowances for the different amounts of metal shrinkage that could be expected after the casting was allowed to cool.

Fig. 25.9. Inside view of the foundry. Homestake's original foundry was located between the Amicus Mill and the Old Abe Shaft. In 1927, a new foundry, shown above, was completed between the Ellison Shaft and the South Mill.—Photo courtesy of the Black Hills Mining Museum.

Fig. 25.10. Jim Brosnahan and Lee Wright making a pour from a cupola, 1954.—Photo courtesy of Homestake Mining Company.

The casting process incorporated an appropriately sized molding box, called a flask, to contain the sand mold. The flask consisted of two halves, each with an open top and open bottom. The upper half of the flask was called the cope, and the bottom half was called the drag. The cope and drag contained two or more holes for alignment pins to ensure that the two halves would always assume the same precise fit each time they were reassembled.

The sand mold was prepared by placing the drag upside down on a flat mold board. The pattern to be casted was placed inside the drag and on the mold board. The pattern was encased with a layer of blended sand and pitch that were tightly compacted around the pattern using a rubber tamping tool. Finally, one or more layers of molding sand were placed over and around the packed pitch-sand and tightly tamped until the entire drag was full of sand. The sand was struck level with the top of the drag. A vent hole was created in the sand by inserting a wire through the sand to the pattern. The vent hole allowed gases to release during the casting process. A bottom board was placed over the drag, and the entire assembly was turned over.

The mold board was then removed from the drag. Loose sand was brushed off the upper surface of the pattern. A light coating of parting powder was spread over the compacted sand in the drag. The cope was placed on the drag and pinned to ensure proper alignment between the cope and the drag. A layer of pitch-sand was placed over the area of the pattern and tamped. A runner, or fill tube, was placed against the pattern to create a channel for the molten metal. After the cope was filled and tamped with molding sand, the sand was struck level with the top edge of the cope and the fill tube was removed.

The cope was carefully removed from the drag, turned over, and set aside. The pattern was carefully removed from the drag, which left an exact impression of the pattern in the compacted sand. The cope and the drag were realigned and clamped into place. The mold was cast by pouring molten metal, such as iron, bronze, copper, or aluminum into the hole created by removal of the runner. After a cooling period, the cope and drag were separated. The metal casting was removed and delivered to the machine shop for any necessary grinding, machining, or drilling. (see fig. 25.11).[10]

Fig. 25.11. Removing a metal casting from the sand mold. Bob Campbell and Fritz Teupel have just finished lifting a heavy metal casting from the sand mold.—Homestake Collection, Adams Museum, Deadwood.

In 1934, for example, Homestake crafted 1,051 patterns and cast 34,879 parts and pieces for such things as locomotives, ore cars, pumps, and crushers.[11] In 1937, Homestake began casting its own 2-inch balls for use in the new ball mills in the South Mills.[12]

On October 8, 1942, the War Production Board issued Order L-208, which mandated the closure of all gold mines in the United States for the duration of World War II. Shortly after the order was issued, Homestake filed an appeal with the board, asking for a variance to allow hoisting and milling of the broken ore inventory until June 8, 1943. The appeal was granted. Unfortunately, the order was a death knell for many mines and companies when the owners determined it was not economical to reopen after the order was lifted on July 1, 1945. Fortunately, Homestake's rich Main Ledge ore and proper care and maintenance of the surface and underground mine facilities during the shutdown allowed Homestake Mining Company to successfully resume operations after the war ended.

The Homestake Sawmill in Spearfish was allowed to continue operating at full capacity throughout the war because the lumber was needed for construction work at military bases and other places. Likewise, Homestake's machine shop and foundry were converted to help support the war production effort. During the first year, Homestake gathered and shipped more than 500,000 pounds of scrap metal, 247,000 pounds of manganese scrap, 46,370 pounds of copper and brass scrap, and 30,950 pounds of scrap rubber. Some Homestake personnel and their families were sent to Montpelier, Idaho, to operate a mine that supplied vanadium as a strategic mineral for the war production effort. Other employees were sent to California to develop a manganese property as part of the war effort.[13]

The machine shop and foundry produced a variety of parts for the war effort, including trailer axles, motor end shields, large nets, eye pins, tools and wrenches for tanks, and hand grenades. The hand grenades were cast in the foundry using sand molds made from intricate wooden patterns. After being cast, the hand grenades were delivered to the machine shop. Here, the grenades were placed in a jig and their bases were ground flat (see fig. 25.12).

At the next station, a hole was drilled through the top of each grenade, and its top was machined flat. The hole was tapped to receive the lever and pin at a later stage. After undergoing pressure testing to check for leaks, groups of grenades were screwed onto a plate to facilitate painting. The plate of grenades was turned over and positioned over a tray of green paint. The tray was raised upward until the grenades were fully immersed in paint. Next, the grenades were placed in a drying oven. After the paint on the grenades was dry, a yellow stripe was painted around the tops of some of the grenades, which designated them as high-explosive grenades. Finally, each grenade cast at Homestake was stamped HMC on its side.[14]

Fig. 25.12. Grinding the bases of hand grenades cast at Homestake Foundry during World War II.—Homestake Collection, Adams Museum, Deadwood.

Brookhaven Solar Neutrino Experiment

Part of the underground decommissioning and cleanup work completed in 2003 involved the Brookhaven solar neutrino experiment, also known as the Davis chlorine experiment, which was located on the 4,850-foot level near the Yates Shaft. In December 1964, Brookhaven National Laboratory obtained approval from Homestake Mining Company to construct a solar neutrino detector on the 4,850-foot level of the mine. A team of Brookhaven scientists, headed by Dr. Ray Davis Jr., designed the experiment, which was initially called the Brookhaven-Homestake Solar Neutrino Observatory.

The objective of the experiment was to detect solar neutrinos and learn more about the nuclear fusion and decay processes that were believed to occur within the core of the sun. A deep mine such as the Homestake Mine was considered ideal for such an experiment since nearly 1 mile of bedrock schist rocks above the detector would filter out the naturally occurring cosmic rays, which would otherwise clutter the detector.

The project was divided into three construction phases. Phase 1, called site preparation, involved excavation of the drifts and chambers and construction of the concrete floors by Homestake crews. Phase 2 involved assembly of the storage vessel and installation of pumps, piping and electrical systems, and other control room equipment using crews from Chicago Bridge and Iron Company (CBI) and Homestake. Phase 3 involved filling the tank, testing the equipment, and commissioning the experiment. The latter phase was largely undertaken by the Wiscombe Company, a subcontractor of CBI, in conjunction with personnel from Brookhaven.

With some of its best miners, Homestake commenced mining the drifts and chambers for the facility in January 1965. Mining work consisted of driving a main access drift and mining a control room, tank chamber, decline drift, and pump room (see fig. 25.13). The chamber for the storage vessel was mined 32 feet high, 30 feet wide, and 60 feet long. The floor of the chamber was located 24 feet below the 4,850-foot level. Two crews concurrently mined the tank chamber. One crew mined a top sill on the 4,850-foot level while the other crew mined an undercut from the decline and pump room area. The pillar between the two cuts was drilled and blasted by "45-ing" the back of the undercut. More than 7,000 tons of rock were mined from the drifts and rooms that comprised the experiment. Phase 1 work was completed in May 1965.[15]

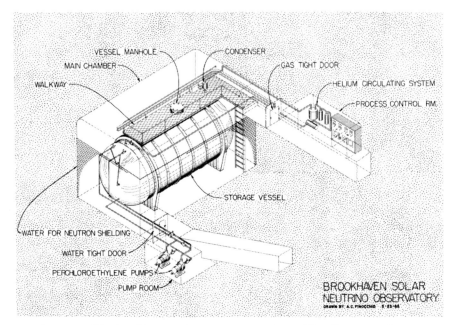

Fig. 25.13. The Brookhaven solar neutrino experiment. This illustration depicts the general layout for the Brookhaven-Homestake Solar Neutrino Observatory (Davis chlorine experiment) that was constructed on the 4,850-foot level of the Homestake Mine beginning in January 1965.—Illustration courtesy of Brookhaven National Laboratory.

Tank sections for the phase 2 work were fabricated in Salt Lake City and sent to the Homestake Mine in June 1965. Because of cage size and weight limitations in the Yates Shaft, the tank had to be designed and fabricated in sections, the largest of which was 5×12×6 feet and weighed 6 tons. Crews from the Chicago Bridge and Iron Company assembled the storage vessel in the tank room (see fig. 25.14). Once constructed, the tank was 20 feet in diameter and 48 feet long.[16]

Upon completion of testing, the storage vessel was filled with approximately 100,000 gallons of perchloroethylene as part of phase 3. The chemical was transferred from ten 10,000-gallon-capacity tank cars on the surface to the storage vessel on the 4,850-foot level using three 700-gallon-capacity tank cars fabricated by Homestake crews. Filling of the vessel started on June 15 and was completed on July 1, 1966.[17] Some 144 tank trips were required to fill the vessel.

Fig. 25.14. Dr. Ray Davis, senior scientist in charge of the Brookhaven-Homestake solar neutrino experiment, inspects progress on assembly of the chlorine tank in the tank chamber, 1965.—Photo courtesy of the Brookhaven National Laboratory.

After the tank was filled with perchloroethylene, some 300,000 gallons of water were fed into the rock chamber to envelope the tank (see fig. 25.15). The water served to help shield the tank from background radiation that could interfere with the argon-37 counting process. Upon completion of testing in late 1967, the Brookhaven solar neutrino experiment was commissioned in early 1968.

Fig. 25.15. Dr. Ray Davis takes a dip in the water surrounding the tank of perchloroethylene in the tank chamber, 1971.—Photo courtesy of Brookhaven National Laboratory.

Perchloroethylene was selected as the primary detection medium because of its high-chlorine content. Whenever a neutrino passed through the perchloroethylene, it could interact with an atom of chlorine-37, causing the chlorine atom to transform into a radioactive isotope of argon-37. Any argon atoms present were periodically mobilized to the control room by injecting helium gas into the storage vessel and allowing the gas to bubble through the perchloroethylene. In the control room, the helium gas and argon-37 were directed into a charcoal filter that was cooled to minus 320 degrees Fahrenheit using liquid nitrogen. The argon-37 atoms were collected in the filter. The number of neutrino interactions was determined by a process of "counting" the number of argon-37 atoms (see fig. 25.16).[18]

Fig. 25.16. Dr. Ray Davis assembles the argon gas extraction and purification system in the control room.—Homestake Collection, Adams Museum, Deadwood.

Unfortunately, over the life of the experiment, Davis was only able to detect about one-third of the neutrinos that John Bahcall, another scientist, calculated should be detected. This led astrophysicists to investigate the "solar neutrino problem." Over the next three decades, scientists performing neutrino experiments at other sites throughout the world determined that three main types of neutrinos (i.e., tau, muon, and electron) existed, and that they did have mass. The Davis detector could only detect the electron variety, which accounted for the difference between the predicted amounts and the actual number detected.

In 1985, after retiring from the Brookhaven National Laboratory, Davis joined the University of Pennsylvania. Other university personnel such as Dr. Kenneth Lande also became more involved in the experiment at the Homestake Mine. Dr. Ray Davis and two other astrophysicists were awarded the Nobel Prize in Physics in 2002. Davis was recognized "for pioneering contributions to astrophysics, in particular for the detection of cosmic neutrinos." The experiment demonstrated that the sun is powered by the nuclear fusion of hydrogen into helium and that the electron neutrinos created during this process oscillate into other types of neutrinos by the time they reach the Earth.[19]

Chapter 26
Production Statistics

Production and Revenue History

From January 1, 1878, through December 31, 2002, surface and underground production through the mills owned by Homestake Mining Company amounted to 167.63 million tons of ore (see table 26.1). From this ore, 39.85 million troy ounces of gold and 9.03 million troy ounces of silver were recovered. Recovered grades averaged 0.238 ounces per ton milled for gold and 0.054 ounces per ton for silver. Over the life of the mine, 1 troy ounce of gold was recovered for every 4.2 tons of ore milled.

Table 26.1 Production statistics for the Homestake Mine (1878-2002)[1]

Years	(tons) Ore Milled	(gold) Ounces Sold	($/troy ounce) Avg. Price	(silver) Ounces Sold	($/troy ounce) Avg. Price	(Estimated) Bullion Revenue
1878-1900[2]	6,686,371	1,492,264	$20.67	354,562	$0.89	$31,161,000
1901-1910[3]	13,028,308	2,265,479	$20.67	623,508	$0.64	$47,226,000
1911-1921	17,582,843	3,280,924	$20.67	966,661	$0.71	$68,503,000
1922-1933	17,820,248	4,279,721	$20.67	1,144,274	$0.53	$89,068,000
1934-1941	11,309,610	4,344,215	$34.87[4]	946,347	$0.70	$152,145,000
1942-1951	7,719,595	3,144,829	$35.00	640,883	$0.85	$110,614,000
1952-1967	27,660,922	9,011,994	$35.00	1,824,956	$1.06	$317,354,000
1968-1981	23,575,346	5,372,515	$139.67	1,113,754	$4.34	$755,213,000
1982-1991	21,763,461	3,294,584	$386.30	603,841	$6.05	$1,276,351,000
1992-2002	20,482,775	3,361,683	$343.68	813,046	$4.75	$1,159,205,000
Total:[1]	167,629,479	39,848,208	$100.11	9,031,832	$1.94	$4,006,880,000

(1) Includes all production from the Open Cut, but excludes production from the Father DeSmet, Highland, Deadwood-Terra, and Caledonia companies prior to June 1, 1901.

(2) Through May 31, 1900.

(3) From June 1, 1900.

(4) Reflects 40,497 ounces of gold sold at $20.67 per ounce in January 1934.

Source: Homestake Mining Company, "Bullion Production Data, January 1, 1878, through December 31, 2002."

The figures in table 26.1 exclude production and bullion revenue prior to June 1, 1901, from the other mining companies situated along the Homestake Belt. After this date, the mining companies that were owned or controlled by the Homestake capitalists were merged into Homestake Mining Company, and production was reported as part of Homestake's production.

The figures shown in table 26.2 reflect the actual bullion revenue that was generated by each of the other major belt mines through May 31, 1901. Production numbers for gold and silver are unavailable. Notwithstanding, with the gold price fixed at $20.67 per ounce and the silver price averaging $1 or less per ounce prior to 1901, the other belt mines likely produced about 1.3 million ounces of gold prior through May 31, 1901, based on reported bullion revenue.[1]

Table 26.2 Product revenue and dividends of the other Homestake Belt Mines through May 31, 1901[1]

Company	Bullion Revenue	Dividends
Highland Mining Company	$12,359,398.95	$3,834,717.69
Caledonia Mining Company	$2,341,130.92	$192,000.00
Deadwood Mining Company	$834,192.28	$275,000.00
Deadwood-Terra Mining Company	$9,001,002.61	$1,350,000.00
Father DeSmet Mining Company	$3,421,199.01	$1,125,000.00
Gopher Mining Company	$71,140.41	$0.00
Clara Consolidated Mining Company	$13,590.56	$0.00
Total:	$28,041,654.74	$6,776,717.69

(1) Mining companies whose production was reported separately and excluded from Homestake's production until June 1, 1901.

Source: Bruce C. Yates, "The Homestake Mine," *The Black Hills Engineer* XIV, no. 3 (May 1926): 143.

Placer production from Deadwood Gulch and neighboring gulches added another 200,000 ounces, based on reported revenue of about $4 million through

1880.[2] With more than 39.8 million ounces recovered at the Homestake Mine, 1.3 million ounces from the other belt mines, and 200,000 ounces from the placers around Deadwood Gulch, some 41.3 million ounces of gold were recovered through 2002. Except for a relatively small amount of production from the "cement ore" in the Deadwood Formation, the gold was primarily derived from the Homestake Formation and its associated placer deposits.

Based on average annual metal prices, the estimated total revenue associated with sales of gold and silver throughout the life of the Homestake Mine amounted to approximately $4.007 billion.[3] Prior to May 31, 1901, product revenue from the other mines along the Homestake Belt amounted to an additional $28.042 million.[4] With $4 million of additional placer production from Deadwood Gulch and neighboring gulches, an estimated $4.039 billion in bullion sales was realized from lode- and placer-mining operations along the Homestake Belt.

Table 26.3 shows the annual bullion revenues that were recorded by T. J. Grier for the Homestake, Highland, and Deadwood-Terra Mining companies from 1878 through 1894. Grier was an accountant and bookkeeper for Homestake at the time. The table reflects the production capacity of the Highland and Deadwood-Terra mines relative to the Homestake Mine over the seventeen-year period.

Table 26.3 Product revenue from the Homestake, Highland, and Deadwood-Terra Mining companies (1878-1894)[1]

Year	Homestake Mine[2]	Highland Mine[3]	Deadwood-Terra Mine[4]
1878	$230,331.82	$13,162.48	$—
1879	$744,557.41	$88,032.51	Unavailable
1880	$1,341,063.40	$340,466.42	Unavailable
1881	$1,102,683.45	$478,995.57	Unavailable
1882	$1,163,796.03	$359,163.41	$639,652.95
1883	$1,156,013.70	$480,812.74	$475,408.70
1884	$1,241,526.82	$497,398.70	$479,980.26
1885	$1,276,268.39	$403,190.51	$436,119.09
1886	$1,349,906.25	$470,977.78	$593,097.53
1887	$928,483.84	$391,311.69	$644,509.56
1888	$913,415.75	$527,889.77	$645,480.40
1889	$973,908.78	$709,413.55	$596,173.39
1890	$1,052,138.02	$758,993.26	$672,675.30

1891	$1,197,581.55	$897,057.65	$673,571.85
1892	$1,142,950.51	$808,805.55	$574,359.22
1893	$1,272,531.01	$811,654.53	$266,574.24
1894	$1,415,274.17	$799,786.31	$164,949.29
Total:	$18,502,430.90	$8,837,112.43	$6,862,551.78

(1) Product revenue was derived from the sale of dore' (gold and silver bullion).
(2) Product revenue was sourced from production through the 80-stamp Homestake Mill and the 120-stamp Golden Star Mill.
(3) Product revenue was sourced from production through the 30-stamp White Mill through May 1880, thereafter, from the 120-stamp Highland Mill.
(4) Product revenue was sourced from production through the Deadwood and Golden Terra Mills. The 60-stamp Deadwood Mill was commissioned in February 1879 and the 60-stamp Golden Terra Mill in June 1879. Product revenue is unavailable for 1879 through 1881.

Source: T. J. Grier Personal Notes, Black Hills Mining Museum.

The Decision to Close the Mine

Throughout much of the life of the Homestake Mine, product revenues from the sale of gold and silver were based on prices that were largely fixed by the U.S government. With passage of the Coinage Act in 1834, new standards were established for minting gold coins. As a result of this act, the price of gold increased from $19.39 per troy ounce to $20.67 per ounce.[5] Gold remained valued at $20.67 per ounce until January 1934, when Congress passed the Gold Reserve Act and President Franklin D. Roosevelt signed Presidential Proclamation 2072.

These actions devalued the dollar from $20.67 per troy ounce to $35 per troy ounce. The gold price remained fixed at $35 per ounce until August 15, 1971, when the federal government abandoned the gold standard, and the price of gold was allowed to fluctuate on the world markets. By the end of 1974, the price of gold was trading at nearly $184 per ounce.[6]

From 1998 through 2001, the price of gold averaged about $282 per ounce.[7] At these market prices, the Homestake Mine was barely able to operate on a break-even basis despite reductions in the workforce, restructuring, implementation of more stringent mining and grade-control policies, and curtailment of exploration work. Essentially, the operation was forced to mine out its developed ore reserves.

Historically, the centroid areas of the Main Ledge ore body were large and rich in gold, compared to most of the other ledges. Whenever additional gold production was needed throughout the 1950s and 1960s or the mill head grade diminished, mine personnel simply started a few more stopes in Main Ledge. From 1962 through 1971, annual gold production averaged 591,229 ounces, most of which were sourced from Main Ledge. The recovered grade during this ten-year period averaged 0.306 ounces per ton. The highest annual gold production achieved during the life of the mine was attained in 1966 when 633,915 ounces of gold were recovered from 2,002,239 tons of ore. The recovered grade for that year averaged 0.317 ounces per ton.[8]

After the Main Ledge ore areas were largely depleted in the early 1970s, the honeypot of above-average grade ore was depleted. Unfortunately, because of their inherent narrowness and unreliable continuity of ore-grade mineralization, most of the other ledges in the mine yielded lower mined head grades compared to Main Ledge. From 1982 through 2001, annual gold production averaged 328,507 ounces at a recovered grade of 0.156 ounces per ton. Although the average grade over this twenty-year period was reduced because of the lower-grade production from the Open Cut, which yielded 0.0943 ounces per ton over its life, the mine may not have otherwise been profitable without low-cost production from the Open Cut.[9]

Table 26.4 lists the total ore production by level and ledge over the life of the Homestake Mine:

Table 26.4 Historic ore production in tons by level and ledge (1878-2001)

	17, 19, and 21 Ledges	11, 13, and 15 Ledges	9 Ledge	5 and 7 Ledges	Main Ledge	Caledonia Ledge	Total
Open Cut	—	—	—	—	13,477,769	—	13,477,769
Historical Surface	—	—	—	—	16,112,189	—	16,112,189
100-800	—	—	—	—	22,428,983	4,602,041	27,031,024
900-1100	—	21,785	—	—	8,614,355	838,232	9,474,372
1250-3350	228	2,385	18,560,242	6,420,712	20,509,315	124,652	45,617,534
3500-4850	1,383,102	4,437,631	16,429,884	585	6,216,099	—	28,467,301
5000-5900	3,850,587	2,787,866	1,917,450	178	1,179,873	—	9,735,954
6050-6800	6,047,333	2,226,059	534,203	—	1,206,277	—	10,013,872
6950-8150	7,193,678	434,334	=	=	71,452	=	7,699,464
Total:[1]	18,474,928	9,910,060	37,441,779	6,421,475	89,816,312	5,564,925	167,629,479
	11.0%	5.9%	22.4%	3.8%	53.6%	3.3%	100.0%

(1) Excludes production from the Father DeSmet, Highland, Deadwood-Terra, and Caledonia Mining companies prior to June 1, 1901.

Source: Homestake Mining Company, "Tonnage Mined as of the end of 2001 by Level/Ledge."

As shown, Main Ledge accounted for the largest amount of ore production over the life of the mine. Including surface mining production from the Open Cut, 89.8 million tons or 53.6 percent of the mine's total ore production were sourced from Main Ledge. The second largest amount of ore production was sourced from 9 Ledge at 37.4 million tons or 22.4 percent of the total. Collectively, all of the other ledges contributed 40.4 million tons or 24 percent of the mine total.

Ironically, on September 11, 2000, when Homestake Mining Company announced its decision to close the Homestake Mine,[10] the underground mineral inventory still contained 4.9 million ounces of gold. Unfortunately, the resource inventory was comprised of hundreds of small irregular resource estimates scattered from the surface to a depth of 8,600 feet. Few centroid areas existed to facilitate a logical core mine operation.[11] Millions of dollars had been infused into the mine and mill facilities each year to modernize equipment and keep the shafts in a safe operating condition. The board of directors was aware that large amounts of capital were going to be required over the next decade to upgrade the shafts and mill facilities, in addition to sustaining capital for ongoing operations. With better opportunities for the capital elsewhere, the board decided in 2000 that the mine would close on December 31, 2001.[12]

Chapter 27
Beyond Closure

Mine Decommissioning and Postclosure Reclamation Work

Following the announcement in September 2000 that it would cease mining operations on December 31, 2001, Homestake's site management personnel held town meetings to inform and update the general public about Homestake's plan to close and reclaim the Homestake Mine at an estimated cost of $66 million dollars. Homestake stated that the work would include decommissioning, closing, and reclaiming the underground areas, Grizzly Gulch Tailing Storage Facility, Open Cut waste rock and ore-handling facilities, mill buildings and facilities, hydroelectric plants, and various other historic or support facilities. Homestake indicated that a water collection and water treatment plant would also be constructed to collect and treat seepage from the Open Cut waste rock facilities.

One important part of the plan included work to preserve and catalog the many thousands of maps, plats, deeds, drawings, photographs, and other historical records that weren't already on loan to the Black Hills Mining Museum in Lead. After evaluating its options, Homestake donated the remaining portion of its recorded history to the Adams Museum & House, which then created the Homestake Adams Research and Cultural Center (HARCC) in Deadwood.

As the newest entity affiliated with the Adams Museum & House, the Homestake Adams Research and Cultural Center was created to protect the Homestake collection and make Homestake's history and legacy accessible to the general public. This charge is now being accomplished through exhibits, research opportunities, and educational programming at the new Homestake Adams Research and Cultural Center in Deadwood. The research center serves as a destination that appeals to engineers, geologists, historians, authors, genealogists, archaeologists, scientists, scholars, tourists, and other people interested in the Homestake legacy.[1]

Much of the physical work to close the Homestake Mine site was required under Homestake's various operating permits previously issued by the South Dakota Department of Environment and Natural Resources (SD-DENR). Notwithstanding, a significant amount of other reclamation work was voluntarily undertaken and completed by Homestake as part of its commitment in the 1970s to improve the environmental, health, and safety aspects of the overall mine operation.

As part of the closure plan, the accessible underground areas were inspected concurrent with, and subsequent to, final operations work. Any supplies, fuel, chemicals, mobile equipment, stationary equipment, or other materials that could potentially impact groundwater were methodically removed from underground areas. Upon satisfactory completion of this level-by-level work, each underground work area was reinspected and signed off as being "clean for closure." Following a series of final inspections by the SD-DENR (see fig. 27.1), the underground portion of the mine was closed on June 30, 2003.

Fig. 27.1. Representatives of Homestake and the South Dakota Department of Environment and Natural Resources who made an inspection of Homestake's underground decommissioning and closure work on June 13, 2003. From left to right are Karl Burke, Mark Nelson, Don Rosowitz, Robert "Bob" Townsend, Jon Epp (kneeling), Mike Cepak, Kevin Christensen, Mark Keenihan, and Steven Mitchell. Tom Regan, Homestake's safety director, also participated in the tour and took the photo.—Photo courtesy of Homestake Mining Company.

On the surface, but away from the immediate mine site, more than two hundred abandoned mined land sites were closed and reclaimed on properties Homestake Mining Company acquired during its 124-year history. In and around Lead, the reclamation work included site cleanup and reclamation of mill facilities, creation of an outdoor interpretive park called Gold Run Park, decommissioning and cleanup of surface shops and yards, reclamation of the former City Dump (see figs. 27.2 and 27.3), and reclamation of the Yates Waste Rock Facility (see figs. 27.4 and 27.5).

Other projects included environmental remediation and reclamation work at the Open Cut waste rock facilities, Open Cut crushing and conveying facilities, Slime Plant and pipelines, Grizzly Gulch Tailing Storage Facility, Englewood Hydroelectric Plant, and Hydroelectric Plant No. 2. Water collection and treatment facilities were also constructed to treat seepage from the Open Cut waste rock facilities. All of the above reclamation work was largely completed by fall 2009.[2]

Fig. 27.2. The City Dump, ca. 1999. The City Dump was located above Gold Run Creek and near Homestake's Water Treatment Plant that was commissioned in 1984. The City Dump, located on Homestake property, was closed in about 1973.—Photo courtesy of Homestake Mining Company.

Fig. 27.3. The City Dump site on July 9, 2002, a year after it was reclaimed by Homestake at a cost of $1.5 million.—Photo courtesy of Homestake Mining Company.

Fig. 27.4. The Yates Waste Rock Facility, November 12, 2004.—Photo courtesy of Homestake Mining Company.

Fig. 27.5. The Yates Waste Rock Facility in July 2009, after it was reclaimed by Homestake at a cost of $3.5 million.—Photo courtesy of Homestake Mining Company.

On April 20, 2006, a milestone was reached when Homestake Mining Company became the first mine operator in the state of South Dakota to have its liability released for 499.74 acres of affected land that Homestake had reclaimed pursuant to the requirements of its two large-scale mining permits. The South Dakota Board of Minerals and Environment issued the first permit in the early 1980s when Homestake requested authorization to conduct large-scale surface mining operations in the Open Cut area. The second permit was issued after Homestake requested approval to expand its Open Cut operations. In 2006, the Board of Minerals and Environment reduced Homestake's reclamation bonding requirement to $1.246 million for reclamation work not fully completed on 108 acres.[3]

In fall 2006, Homestake completed work on an $18.5-million project designed to collect and treat seepage from the East Waste Rock and Sawpit Waste Rock facilities. The two reclaimed facilities contain waste rock that was mined from the Open Cut surface mining operation from late 1983 through 1998. The seepage that emanates from the toe of the waste rock facilities contains elevated concentrations of dissolved solids and selenium.

Seepage is collected from the waste rock facilities and piped to a biological water treatment plant that was constructed at the site of Homestake's former Cyanide Plant No. 2 near Blacktail. There, the water is treated to reduce the concentrations of dissolved solids and selenium in order to meet the surface water quality standards specified in the company's surface water discharge permit. Treated water is discharged to Deadwood Creek, the Lead-Deadwood Sanitary District's treatment plant, or some combination thereof.

In 2006, the South Dakota Board of Minerals and Environment endorsed Homestake's postclosure plan for long-term monitoring and treatment of seepage from the Open Cut waste rock facilities. A postclosure bonding requirement of $39.2 million was established by the board as surety for Homestake's monitoring and water treatment obligations for a one-hundred-year period.[4]

Sanford Underground Science and Engineering Laboratory

Shortly after Homestake Mining Company announced on September 11, 2000, that it would cease production at the Homestake Mine, the Davis chlorine experiment on the 4,850-foot level was decommissioned. All of the chemicals and laboratory equipment, except for the perchloroethylene tank, were brought to the surface as part of the decommissioning and cleanup work for the entire mine.

Recognizing the benefits and potential of the deep underground mine relative to science, Dr. Kenneth Lande, a scientist and professor of physics and astronomy at the University of Pennsylvania, suggested to Homestake and members of the scientific community that the mine should be converted to an underground science laboratory. Homestake was receptive to the idea, but indicated it was concerned about potential longer-term liability issues and what legal entity would actually own and operate those properties that Homestake might be willing to sell or donate. Preliminary discussions were subsequently held with Governor William Janklow of the state of South Dakota relative to the possibility of the state taking over ownership of the mine.

Months later, Homestake Mining Company and Barrick Gold Corporation issued a joint press release on June 25, 2001, that said the board of directors of each company had unanimously approved a plan to merge the companies. The merger plan was subsequently approved by the shareholders of both companies on December 14, 2001. As part of the transaction, Barrick issued approximately 140 million shares of its stock to acquire Homestake at a total value of $2.3 billion.

The same day, Homestake Mining Company was delisted from the New York Stock Exchange, after having been continuously listed since January 16, 1879. After the merger was completed, Barrick personnel assumed Homestake's role in the process of determining whether it was feasible and prudent for Homestake to transfer ownership of the Homestake Mine to some entity for the purpose of creating an underground science laboratory.[5]

During June 2003, Homestake Mining Company, now a wholly owned subsidiary of Barrick Gold Corporation, completed all of the underground decommissioning and cleanup work at the Homestake Mine. Because the business and legal framework for transferring ownership of the mine was far from being completed to Barrick's satisfaction, final steps were taken to decommission the pumping and electrical systems, and the mine was closed on June 30.

Governor Mike Rounds created the Homestake Laboratory Conversion Project (HLCP) on July 1. The management of the Homestake Laboratory Conversion Project was directed to develop a definitive business plan that would outline to the National Science Foundation (NSF) and other stakeholders what would be required to acquire the mine and convert it to a Deep Underground Science and Engineering Laboratory (DUSEL), subject to Barrick's stated concerns.

After the state's plan was developed and further negotiations and legal work were completed, Barrick agreed to an Agreement in Principle on January 12, 2004, wherein it would donate the underground portion of the Homestake Mine and certain other real and personal property to a new legal entity to be created by the state of South Dakota. The agreement excluded transfer of the Open Cut, various waste rock facilities, Grizzly Gulch Tailing Storage Facility, and other properties Homestake wished to retain.

The agreement was subject to official endorsement of the National Science Foundation for creation of a Deep Underground Science and Engineering Laboratory, selection of the Homestake Mine by the NSF as the preferred site for a DUSEL, and satisfactory liability protection for Barrick. The Agreement in Principle also specified that upon creation of the legal entity, called the South Dakota Science and Technology Authority, Barrick and Homestake would turn over the real and personal property to the authority within thirty days after it had obtained the necessary funding, permits, and approvals to construct and operate the underground laboratory.

The South Dakota state legislature created the South Dakota Science and Technology Authority on February 11, 2004. The purpose of the authority was to accept the donation of the Homestake Mine property in Lead and convert it into a science, engineering, and education center. The South Dakota legislature subsequently allocated $14.3 million toward the effort. Additionally, a $10 million Federal Housing and Urban Development grant was secured to provide initial funding to reopen the shafts and reestablish a pumping system to remove water from the mine.

Concurrently, scientists from the Lawrence Berkeley National Laboratory and University of California at Berkeley organized the Homestake Collaboration to solicit letters of interest from scientists for possible experiments at Homestake. The collaboration was also tasked with coordinating its planning efforts with those of the South Dakota Science and Technology Authority and submitting a competitive proposal to the National Science Foundation for construction of a Deep Underground Science and Engineering Laboratory at Homestake. In a similar manner, scientists from other universities and national laboratories submitted competitive proposals to the National Science Foundation for underground labs to be constructed at various other sites around the country.[6]

In July 2005, the National Science Foundation narrowed its list of eight possible sites for a Deep Underground Science and Engineering Laboratory to two sites, which included the Homestake Mine and the Henderson Mine in Colorado. Each collaboration team was awarded a $500,000 grant from the National Science Foundation to produce a conceptual design for a possible Deep Underground Science and Engineering Laboratory at the respective locations.[7]

After the announcement was made, an appeal was filed with the National Science Foundation to open up the selection process again to collaboration teams in Washington, New Mexico, Minnesota, and other locations. The appeal was successful, and several other collaboration teams were given $500,000 grants to prepare and submit conceptual design proposals.[8]

Meanwhile, during a special legislative session called for by Governor Mike Rounds in October 2005, the South Dakota legislature approved $19.9 million to develop an interim laboratory at the former Homestake Mine. On April 14, 2006, Barrick Gold Corporation signed a property donation agreement with the state for the donation and transfer of the underground portion of the Homestake Mine, 7,000 acres of mineral rights, and 186 acres of surface lands to the South Dakota Science and Technology Authority (see fig. 27.6). In May, the South Dakota Science and Technology Authority moved its office from Rapid City to the administration building at Homestake in Lead and commenced to finalize the conceptual design report and complete other planning to establish an interim science laboratory on the 4,850-foot level of the mine.

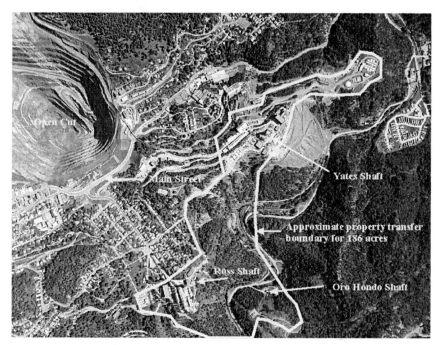

Fig. 27.6. Aerial view showing the 186 acres of land that Barrick Gold Corporation donated to the South Dakota Science and Technology Authority in 2006.—Illustration by author.

On June 26, 2006, Governor Rounds announced the receipt of a $70 million private gift from Sioux Falls philanthropist T. Denny Sanford to help develop the underground science laboratory and establish a science education center. The gift was given in segments, with $35 million used to prepare the mine for science to the 4,850-foot level, $20 million to develop a science education center on the surface of the former Homestake Gold Mine, and another $15 million to reopen the mine for science to the 8,000-foot level. The laboratory was subsequently named the Sanford Underground Science and Engineering Laboratory.

A year later on July 10, 2007, Governor Rounds announced that the National Science Foundation had selected Homestake as the agency's preferred site for the proposed national Deep Underground Science and Engineering Laboratory. One of the main factors in the federal agency's final decision was the state of South Dakota's ability to build an interim laboratory at the 4,850-foot level and establish early science in the mine.

Almost simultaneously with the National Science Foundation's announcement, the South Dakota Science and Technology Authority reopened the mine with a development contractor and began work to reestablish the infrastructure to facilitate creation of an interim science laboratory on the 4,850-foot level using funds already acquired.[9]

By the time a pumping system was reestablished in the Ross Shaft, groundwater inflows had already flooded the lower reaches of the mine to approximately 20 feet above the 4,550-foot level.[10] Dewatering to the 4,850-foot level was completed by May 13, 2009.[11] Concurrent with dewatering, other work was completed to rehabilitate the Yates and Ross shafts, complete inspections of accessible portions of underground mine levels, and begin work to establish an interim laboratory.

Several experiments were established at the Sanford Lab as part of an Early Science Implementation Program. One such experiment is designed to detect very minute changes in ground movement caused by seismic events, dewatering of the lab, rock strain attributable to excavations, or the moon's gravitational effect on the earth. Dr. Larry Stetler, a geological engineer from the South Dakota School of Mines and Technology, and Dr. James Volk, a high-energy physicist from the Fermi National Accelerator Laboratory at Batavia, Illinois, are the primary investigators in charge of the experiment, dubbed the Homestake Hydrostatic Water Level System.[12]

The first two arrays for the experiment were installed on the 2,000-foot level in January 2009. The arrays operate on the principle that water seeks its own level. Each array consists of long water-filled tubes that extend more than 1,000 feet across the mine level. The ends of the tubes are connected to pools of water. Electronic sensors continuously measure minute changes in the water level of each pool and stream the measurement data to researchers over fiber-optic lines.

In an interview held in March 2009, Stetler told a *Black Hills Pioneer* reporter, "We're right now seeing the high tide and low tide effect on the coast with the moon's gravity every twelve hours. Then every fourteen days we see the lunar cycles. So we're seeing the pull of gravity on the earth as it's raising the ground and lowering the ground." Additional arrays are planned for other areas of the Sanford Lab.[13]

Planning for another experiment, called the Large Underground Xenon (LUX) experiment, began in October 2006. This experiment will occupy the same space that Nobel Prize-winning physicist Dr. Ray Davis used for his chlorine experiment on the 4,850-foot level. The LUX experiment is designed to detect traces of fundamental dark matter particles called weakly interacting massive particles (WIMPs). Dark matter is known to exist because its gravitational field affects matter that scientists can observe.

The LUX experiment will be commissioned in 2010. A detector vessel will be filled with 300 kilograms of liquid xenon cooled to minus 165 degrees Fahrenheit. As the WIMPs pass through the detector, instrumentation will record the energy that is released. The 4,850-foot level is considered ideal for the experiment since the overlying bedrock will shield the detector from interfering cosmic radiation that naturally reaches the earth's surface.[14]

The Majorana experiment is a third early-science experiment at the Sanford Lab. This experiment is designed to perform a measurement of neutrinoless double-beta decay in order to determine the mass of neutrinos and whether the elusive neutrinos are their own antiparticle. Other neutrino experiments in the world have determined that the particles have mass and that "the mass of all the neutrinos in the universe is comparable to the mass of all the stars in the universe." Physicists believe that experiments based on neutrinoless double-beta decay are the only practical means of determining whether neutrinos are actually Majorana or Dirac particles. The research involves nuclear physicists, high-energy physicists, and chemists.[15]

Deep Underground Science and Engineering Laboratory

The DUSEL Experiment and Development Coordination Committee (DEDC) was formed to identify the first science experiments that would be installed in the DUSEL at Homestake. The DEDC is comprised of three representatives from the particle physics and astrophysics community and one each from the biology, geoscience, and engineering communities. The DEDC is charged with developing the framework for the early-science experiments, coordinating the efforts of the scientific community in completing the necessary design and evaluation of the experiments, and acting as a liaison among members of the scientific community, the NSF, and the South Dakota Science and Technology Authority.[16]

Through the DUSEL Experiment and Development Coordination Committee, members of the particle physics, astrophysics, and nuclear physics communities have identified the following questions that help support the need for a DUSEL:

1. What are the fundamental symmetries and absolute mass of the neutrino? Can these provide a window into the origin and pattern of particle masses that make up our universe?
2. What are the fundamental properties and interactions of the three families of neutrinos, e.g., CP violation, mass hierarchy, CKM matrix and mixing

angles? What can these and other neutrino studies tell us about the matter/antimatter asymmetry in the universe? Can it reveal new insights into the unification of the fundamental forces that govern physical laws as we now understand them?

3. What is the proton lifetime, and does it decay? Is ordinary matter inherently (un)stable?
4. What is dark matter?
5. What is the spectrum of neutrinos from supernovae and the Big Bang, and what can this tell us about the history and evolution of our universe?[17]

The ultimate objective of the DUSEL Experiment and Development Coordination Committee effort is to ensure timely completion of a funding application to the National Science Foundation for endorsement of and funding for the initial suite of experiments and for the Deep Underground Science and Engineering Laboratory itself.[18]

To facilitate completion of the preliminary design for the DUSEL, the National Science Foundation awarded the University of California-Berkeley, $15 million in September 2007 for completion of the design work over a three-year period. A supplemental funding request of $3 million was subsequently approved.

In 2009, the National Science Foundation announced an additional award of $29 million toward completion of the design for DUSEL. The award raised the NSF's financial commitment for the design work to $47 million. An additional $21 million was awarded by the NSF to various scientists who are tasked with completing designs for some of the first experiments that will be conducted in the DUSEL.[19]

The National Science Foundation indicated that the DUSEL Experiment and Development Coordination Committee's application, if approved, will be funded as a Major Research Equipment and Facilities Construction (MREFC) project. Assuming that the necessary approvals are attained at the federal level, funding will support construction and operation of the Deep Underground Science and Engineering Laboratory and the proposed suite of experiments.[20]

On August 15, 2009, officials of the Sanford Underground Science and Engineering Laboratory announced that the National Science Foundation had signed a cooperative agreement with the U.S. Department of Energy toward joint participation in the Deep Underground Science and Engineering Laboratory. Ultimately, the fully developed DUSEL could receive as much as $400 million from the National Science Foundation for science experiments and improvements to the laboratory infrastructure. The U.S. Department of Energy could spend twice that amount on the project.[21]

Dr. Kevin Lesko, principal investigator for the DUSEL Experiment and Development Coordination Committee, presented a preliminary construction timeline for the DUSEL at a planning workshop in Lead, South Dakota, in October 2009. Lesko informed scientists attending the workshop that the National Science Foundation will review the design proposal for the DUSEL in 2011. Pending the necessary federal approvals, construction of the Deep Underground Science and Engineering Laboratory will start in 2013. The first major experiments are scheduled to be commissioned by 2016. The DUSEL, as envisioned, includes primary campuses on the surface, 4,850-, and 7,400-foot levels.[22]

Under the plan developed by the DUSEL Experiment and Development Coordination Committee, a midlevel campus for science, research, education, and outreach will be established on the 4,850-foot level (see fig. 27.7). Research activities planned for this campus will include physics experiments relating to matter, antimatter, dark matter, astrophysics, double-beta decay, and solar neutrinos. Various laboratory modules will be created in the general area between the Yates and Ross shafts. Large detector chambers, 100-300 kilotons in size, will be mined and furnished using a phased approach.[23]

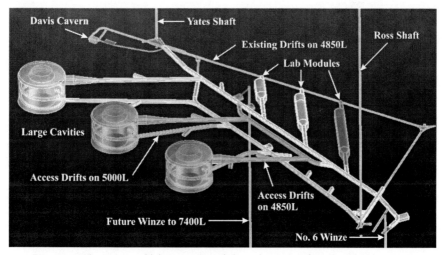

Fig. 27.7. The proposed laboratory module at the 4,850-foot level for the Deep Underground Science and Engineering Laboratory at Homestake.—Illustration courtesy of the Lawrence Berkeley National Laboratory.

One such detector—proposed for construction by 2017—is called the Long Baseline Neutrino Experiment. One 300-kiloton or two 150-kiloton detectors will be located on the 4,850-foot level campus. Each detector chamber will be filled with purified water and equipped with high-speed electronics capable of recording the reactions that are expected to occur when neutrinos pass through the chamber.

A beam of neutrinos 800 miles long will be directed toward the DUSEL detector from an accelerator located at the Fermi National Accelerator Laboratory in Batavia, Illinois. The high-speed electronics will record the neutrino reactions, as they occur, within five hundred billionths of a second. According to Bob Svoboda, principal investigator for the experiment: "We will tell you how many particles there were, what their energies were, what direction they were going, and even what type they were in most cases." The experiment will attempt to explain why there is so much matter in the universe and very little antimatter.[24]

As part of the long-term plan, a deep-level science and research campus will be constructed on the 7,400-foot level in 2017 and 2018 (see fig. 27.8). This campus will be accessed through No. 6 Winze or a new winze that will extend from the 4,850-foot level to the 7,400-foot level. Three 20×20×75-meter lab modules will be created using a phased approach. Research conducted within this campus will include those dark matter, double-beta decay, and solar neutrino experiments that require greater depth for shielding of cosmic rays.[25]

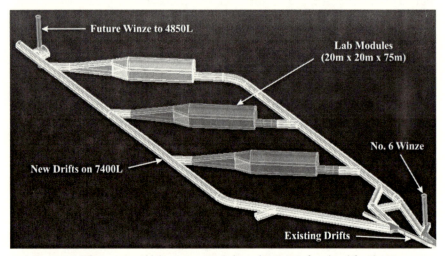

Fig. 27.8. The proposed laboratory module at the 7,400-foot level for the Deep Underground Science and Engineering Laboratory at Homestake.—Illustration courtesy of the Lawrence Berkeley National Laboratory.

Chapter 28
Some Early Homestake VIPs

Fred and Moses Manuel

Ferdinand Frederick Hebert dit Manuel (see fig. 28.1) and Moyse "Mose" Hebert dit Manuel (see fig. 28.2) were the sons of Pierre Hebert dit Manuel and his first wife, Marguerite Pepin. Their Hebert ancestors immigrated to Acadia from La Haye, France, in the early 1600s and later settled in Gentilly, a community within the city of Becancour, Quebec. The Manuel brothers had four siblings, including Marie Ida "Adelaide" Hebert, Marie Phelonise Manuel, Marie Emilie Amelia Hebert, and Anonyme Hebert. Fred was born in 1845. Moyce or "Mose" was born on January 16, 1848, in Plessisville, Quebec.[1]

Fig. 28.1. Fred Manuel.—Homestake Collection, Adams Museum, Deadwood.

Fig. 28.2. Moses Manuel.—Photo courtesy of Homestake Mining Company.

The Manuel family moved to Minnesota in 1855 when Fred and Mose were small children. As teenagers, Fred and Mose spent much of their time hunting and trapping in northwestern Minnesota and Dakota Territory. Most of their outdoor activities were done on "neutral ground" between the battlegrounds of the Chippewa and Sioux, who were the principal inhabitants of the area. In 1862, a company of Civil War soldiers passed by the family farm. Fred decided to join the soldiers and ended up serving in the tenth Minnesota regiment until the war ended. Mose wanted to enlist also, but the soldiers refused to take him, saying he was too young.

In the spring of 1867, Mose, also known as Moses, decided to give up hunting and trapping and join his brother, Fred, who had gone to Helena, Montana Territory, the previous year. Moses joined a party under Judge Ellis, who was leading a train in that direction via the northwestern route to Helena. At Fort Wadsworth, the party was detained for a week until more troops could be sent in to escort the party to Fort Ransom. Upon reaching Fort Ransom, the party was again delayed, this time for two weeks until two companies of infantry could be sent in to escort the wagon train to Fort Stevenson. After arriving there, Moses learned that a party of Sioux warriors had attacked the fort the previous day, killing one or two men.

From Fort Stevenson, the wagon train departed for Fort Buford. The train was comprised of about sixty men and women, most of whom represented twelve families. At Fort Buford, it was learned that a military escort would not be available to accompany the train to Fort Benton. Members of the train decided to continue on without a military escort.

At the Milk River, a large band of Yankton Sioux surrounded the train, which created fear and consternation among the emigrants. The situation was resolved when a band of Red River Indians saw what was happening, rushed to the rescue, and were able to make peace with the Yankton Sioux. Moses and the other members of his party expressed their appreciation to the Red River Indians by giving them provisions, ammunition, and most anything else they wanted. The wagon train was able to move on and reached Helena in September.

At Helena, Moses found Fred and joined him in a placer-mining operation. Later that fall, Fred and Moses decided to set out on a prospecting trip around Montana Territory. They located their first quartz mine about 28 miles west of Helena. After operating their mine quite successfully for a few years, Fred decided to go to Utah to prospect. The following year Moses also went to Utah. There, he met a man named Jack Reynolds. Moses and Jack went to Idaho in search of a lost placer mine but were unable to find it. They subsequently returned to Utah, whereupon Moses decided to go to Nevada and Arizona.

Near a place called Meadow Valley Wash, Nevada, Moses had a harrowing encounter with a band of Indians. In his memoir entitled "Forty-Eight Years in the West," he described his encounter with the Indians as follows:

> When I got close to where this war party had been seen, I camped, with the intention of going through under cover of night. I slept a little too long, however, and was late in starting. When I got close to the camp I expected to find, I tore a blanket into strips and tied up the horse's feet in it to prevent his making any noise, thinking to make a sneak through, but as luck would have it the horse neighed and the Indians were aroused. Two of them jumped and caught hold of the bridle. The rest were on the bank, and looked to me at the time like about a thousand. I knew there was no use shooting at so many. I hit one of those holding the bridle and they let go. The horse started to run and the Indians started to shoot, and they were at very close range. The pack horse, running alongside of my saddle horse, was hit and fell. I kept on until I reached what is known as the Big Muddy [River], where I came across some prospectors camped and secured some provisions from them.[2]

From Meadow Valley Wash, Moses made his way across the Colorado River at Patterson Ferry. After prospecting that area for a while, Moses moved on to California and worked in some of the mines in the Sierra Nevadas. In 1874, he went to Fort Wrangel, Alaska, on a steamer. From there, he and three other prospectors went to a place called Deer Creek. There, he met an Indian woman who spoke fairly good French. She told Moses and his partners a story about a river toward Great Slave Lake where, as a young girl, she had found yellow stones that she used as sinkers. She then drew a map of the area for Moses and his partners, whereupon they went to the place and found rich gold placers. After working the entire season at this location, Moses and his partners decided to go to Portland, sell their placer gold, and take a trip to Africa.

Upon reaching Portland, Moses and his partners sold their gold for about $25,000. While waiting for a ship to take them to Africa, Moses saw a newspaper that told of General Custer's gold discovery in the Black Hills. Moses had always been intrigued with the country called the Black Hills, so he cancelled his trip to Africa and set out for Helena. He soon learned that Fred had also returned to Montana Territory and was mining in the Marysville District.

Fred was dubious about going to the Black Hills. He talked Moses into staying in Montana Territory for a while. In the spring of 1875, Moses and Fred took a contract on the Wipperwill claim in the district, built an arrastra, and milled 400 tons of quartz ore. After finishing that contract in the fall of 1875, Moses and Fred decided to set out for the Black Hills—just in time for winter.

At Green River, Moses and Fred were joined by six other men. The group made their way to Camp Brown, where they waited for a week or two until they could travel with two companies of cavalry to Fort Laramie. At Whiskey Gap, the men encountered a heavy snowstorm with extremely strong and cold winds. With no wood to build a fire, the men dug holes in the snow and waited for the blizzard to subside. Many of the soldiers ended up with frozen hands and feet and had to be taken to Fort Stevens by train. The rest of the group continued on to Fort Laramie. Moses, Fred, and the other six men decided to wait there until the weather warmed up. The prospectors eventually moved on, reaching Cheyenne in December 1875.

While in Cheyenne, the eight men found six others, who also wanted to go to the Black Hills. Fred and Moses purchased a wagon, four horses, and a stock of supplies. The party of fourteen men started for the Hills. Upon reaching Cooney's ranch on the North Platte River, the men found sixty or seventy other people who were waiting for the safety of a larger group to continue on to the Black Hills. After failing to convince the other men that they had a large enough party, Moses and Fred's party of fourteen continued on. Near the mouth of Red Canyon, the men spotted a band of about sixteen Indians but avoided being detected.

At Custer, Moses and Fred met Annie Tallent's husband, D. G. Tallent. Fred and Moses did a little prospecting around Custer but quickly became dissatisfied with the results. They loaded their four horses with provisions and tools and moved to an area near present-day Hill City, where they helped another party sink a shaft. Finding nothing there, they moved to Palmer Gulch near Harney Peak. Here, they staked a few claims.

The next day, a large party of miners from Colorado moved into the gulch and staked much of the remaining ground. After sinking a shaft to bedrock and finding nothing worth pursuing, Moses and Fred moved on to Boxelder Creek. The men prospected there for a few days and found fine color. After a few days, some other men came along and reported that there were rich diggings farther north. Moses and Fred then partnered with Henry "Hank" Harney and Alexander "Alf" Engh and headed north. In Bobtail Gulch and the north fork of Gold Run Gulch, the four men staked several claims, including the Golden Terry, Mammoth, Homestake, and Old Abe. Soon, the claims would make the men quite rich.[3]

By the fall of 1877, after Moses and Fred had sold all of their mining claims in the Black Hills, they decided to return to Minnesota where they had grown up. Right away, Moses married Josephine Beliveau. Six children were born to the couple: Lydia, Enos, Edna, Clarence, Edgar, and Victor Manuel. Fred married Minta "Mittie" Ferris on March 5, 1878, in Henderson, Sibley County, Minnesota. The couple had three children whose names were Charles, Ida, and Robert.[4]

After spending about two years in Minnesota, Moses and his wife moved to California but only resided there for a few months. Moses had gold fever again. Off they went to Montana, where he spent the rest of his life. In 1882, Moses and Fred went to Basin Gulch and discovered the Mittie and Josephine mines, both of which evolved into good mines. Moses sold his interest in the Mittie Mine to Fred and sold the Josephine for $30,000 later that year. Moses purchased Fred's interest in the Minnesota Mine near Helena.

In 1894, Moses located a claim called the Porphyry Dyke, constructed a 10-stamp mill, and milled 25,000 tons of ore before being shut down as a result of an injunction filed by a Helena water company. He subsequently sold the mine to James Breen of British Columbia.[5] Moses was killed in the Minnesota Mine on July 5, 1905, as a result of an explosion of gasoline fumes in the shaft.[6] Fred Manuel died October 29, 1897.[7]

Henry Clay Harney and Alexander Engh

Henry Clay "Hank" Harney (originally spelled Horney) was one of the original locators of the Homestake claim on April 9, 1876 (see fig. 28.3). Henry was born on January 1, 1844, in Hancock County, Illinois, the fourth of ten children born to William Horney and Louisa Ann Haggard. When the Civil War broke out, Henry wanted to join the Illinois Infantry, but the units were full. Instead, he traveled to Missouri and enlisted as a private in Company A of the Tenth Missouri Infantry on March 27, 1862. His regiment was lead by Col. Leonidas Horney, a second cousin. On November 25, 1863, Hank was wounded at Chattanooga and was hospitalized there. He was transferred to Plymouth, Illinois, where he remained hospitalized until June 30, 1864.[8]

Fig. 28.3. Henry Clay "Hank" Harney. Harney and Alexander Engh met the Manuel brothers while prospecting on Boxelder Creek in the Black Hills in January 1876.—Photo courtesy of James W. Harney Sr.

After the war, Hank assumed the spelling of Harney even though his brothers retained the Horney name. Hank decided to become a miner. He partnered with Alexander "Alf" Engh (see fig. 28.4), who was a cousin to Matilda Caroline Foss, Hank's future wife. Harney and Engh went to the Black Hills in January 1876. It was on Boxelder Creek that they met Fred and Moses Manuel.

Fig. 28.4. Alexander "Alf" Engh.—Photo courtesy of James W. Harney Sr.

After the Manuels, Harney, and Engh sold their mining interests in the Black Hills, Hank Harney married Matilda Caroline Foss in Chicago, Illinois, on October 31, 1880. Matilda was born September 8, 1859, in Norway. Within a few months, the couple moved to Eureka, Kansas, where they purchased a thoroughbred horse ranch. Hank and Matilda had seven children, all of whom were born in Eureka, Kansas. He then moved back to Chicago, where he died November 20, 1907. Harney is buried in the Mt. Olive Cemetery in Chicago.[9]

Alexander Engh was born January 24, 1850, in Nittedal, Akershus, Norway. The name is shown as Aksel Elini Engh in early Norwegian records. Alexander was the fourth of six children, born to Martin Aleksander Engh and Eline Jensen Dahl, both of whom were born and died in Norway. Martin's sister Cathrine Kaia Dorthea Engh and her husband, Peter Andreas Foss, had six children, one of whom was Matilda Caroline Foss, who married Henry Clay Horney (Harney).[10]

George Hearst

George Hearst (see fig. 28.5) was born September 3, 1820, on the family farm near Sullivan, Franklin County, Missouri. His parents were William George Hearst and Elizabeth Collins, who were married in 1817 in Franklin County, Missouri. George had a sister Martha and a brother named Jacob. Other siblings may have included Phillip, William, and Elizabeth. George's father died in November 1844, when George was twenty-four years old.[11]

Fig. 28.5. George Hearst, a partner in Hearst, Haggin, Tevis, and Company.—Homestake Collection, Adams Museum, Deadwood.

George's father owned three farms that encompassed between 600 and 800 acres. When George was fifteen, his father placed him in charge of one of the farms. During his teenage years, George and his father delivered and sold hogs to the French people, who owned some of the mines and smelters in the area. George was impressed with the fact that the Frenchmen always seemed to have money and had nice furnishings in their homes. This induced George to become interested in mining, since farming seemed such a slow way to make money.[12]

George graduated from the Franklin County Mining School in 1838.[13] When he was twenty-two, he went to work at one of the large lead mines near his home. He soon became involved in a few copper and iron mines. Between 1842 and 1849, he was able to save enough money from his mining-related work to pay off some $10,000 in debts that encumbered his father's estate.

Upon hearing of the California gold rush, George became stricken with gold fever but was talked out of a move by his friends in Missouri. Around May 1850, George decided he was going to California and joined a party of fifteen who had similar aspirations. The lengthy trip took him through Fort Kearney in Nebraska, Fort Laramie in Wyoming, and Hangtown, now called Placerville, in California.

For the next five or six years, George worked the California gold placers, but never made much money at it. He subsequently opened a general merchandise store in Nevada City, California. After his efforts to open a branch store in Sacramento proved to be a business failure, he became interested in hardrock mining. He and his early associates started the Lecompton Mine near Nevada City in 1857, which turned out to be a good business venture. George sold his interest in this mine in 1859.

After hearing of a new gold find in the Washoe Valley in Nevada, Hearst went to have a look. There, he picked up a tip from a man who speculated that the gold ore being mined might also contain silver. George became excited about this possibility. To satisfy his curiosity, he obtained some samples of the rock and had them assayed at Nevada City. Much to his amazement, Hearst learned the ore was valued at $3,000 per ton, mostly because of its rich silver content. Having already met attorneys J. B. Haggin and Lloyd Tevis, the three men formed an investment company called Hearst, Haggin, Tevis, and Company and quickly moved to acquire the Ophir mining claim where the samples had been collected. Hearst's investment resulted in a one-sixth ownership interest in the Ophir.

To substantiate the assays, Haggin and Tevis agreed that Hearst would arrange to have a bulk sample of rock mined from the claim. The rock would be hauled to San Francisco in wagons and smelted. This he did, which resulted in an $80,000 profit for the company from 45 tons of rock. By March 1860, news of the discovery resulted in a rush that led to development of the famous Comstock Lode and creation of Virginia City, Nevada.[14] Hearst wrote later of the rich silver discovery as follows:

> We... finally struck a German who said he could smelt it. He said he would build a furnace and smelt it for $450 a ton. We said all right. He then built a furnace and smelted the stuff, and we hauled it around to the Mint. People would not believe it was silver, it looked so much like lead in fact it had not been refined, and there was a great deal of lead in it. Everybody said it was lead and of no account, and that these fellows had got the stuff up from Mexico and were trying to work some kind of swindle.... After paying 25 cents a pound to get it over and $450 a ton for smelting we cleared out of that $76,000, and besides that we sold slags for several thousand more, so we may say we cleared $80,000 for that lot. Then the excitement commenced. This was along in March, 1860. The people began to flock there in crowds.[15]

Upon hearing that his mother was seriously ill, George moved back to Missouri in the summer of 1860 to be with her. She died in 1861. During his stay, he met Phoebe Elizabeth Apperson, who was teaching school nearby. Phoebe and George were married in Steelville, Missouri, on June 15, 1862. A short time later, the two moved to California. William Randolph was their only child, born on April 29, 1863.

In 1865, George was elected to the California State Assembly as an assemblyman for the Eight District, which included San Francisco County. He held this office through 1866. While living in San Francisco, he purchased the 48,000-acre Piedra Blanca ranch near San Simeon and expanded the size of the ranch during the next several years. William Randolph Hearst constructed the Hearst Castle on the ranch in 1919.[16]

In 1867, George sold his interest in the Ophir Mine and ventured into the real estate business. At about the same time, he invested $30,000 in a mine in Kern County, California. Over the next two years, his real estate investments garnered an additional $160,000 for him. He invested in the Eureka Mine with another man, whereupon the men made $80,000 each after selling that mine. At Pioche, Nevada, Hearst invested in the Hermes Mining Company and netted $250,000.

Acting on a tip from his friend, Marcus Daly, Hearst invested $33,000 in the Ontario Mine in 1872. The mine was located about 35 miles southeast of Salt Lake City, Utah. Over the next fifteen years, the Ontario Mine netted Hearst, Haggin, Tevis, and Company and other investors about $14 million. The men also invested in the Daly Mine, which produced $30,000 in revenue per month.

Hearst, Haggin, Tevis, and Company purchased the Homestake Mine in 1877. At the same time, Hearst convinced his investors to acquire various water rights and ditches in the northern Black Hills. Their $170,000 investment in water returned an average of $8,000 per month for many years thereafter. The water rights all but guaranteed a supply of water for the mines and mills around the Homestake Belt. Additional revenue and goodwill was gained by selling excess water to the surrounding communities. Hearst's and Haggin's investment in and creation of the Black Hills & Fort Pierre Railroad Company returned an average of $15,000 per month as income. The railroad supplied the mines and mills with wood and provided a much-needed transportation service for the people of the northern Black Hills.

Hearst's next investment with his partners was the Anaconda Mine at Butte, Montana Territory. Marcus Daly was one of the first men to acquire the Anaconda claims in 1881. After sinking the main shaft and discovering rich copper ore at a depth of 200 feet, Daly invited Hearst to form a partnership. Hearst immediately recognized that the mine would be a moneymaker and took Daly up on his offer through Hearst, Haggin, Tevis, and Company. Hearst ended up with thirty shares in the mine; J. B. Haggin, twenty-seven shares; Daly, twenty-five shares; and Lloyd Tevis, seventeen shares.

Within a short time, the shaft was deepened from 200 feet to 1,000 feet. Rich copper ore averaging 30-45 feet wide was found across this entire interval. Based on their initial success with the Black Hills and Fort Pierre Railroad in the Black Hills, the investors started a railroad at Butte, which allowed freight to be delivered to the mine very economically.[17]

In his memoir, Hearst summarized his investment philosophy as follows:

> *The general character of my mining has been to get the ore out, reduce it to bullion, and sell it. We would open a mine, prospect it, build a mill, and work it, and manufacture the ore in gold or silver, lead or copper, in other words, we were engaged in what is called legitimate mining. I never sold any stock in any mine that was not in operation. We would always sell a little then, when we got started, in order to get a little money back, but very often, we would not even do that. I do not think that there is any stock sold in the Anaconda Mine.*
>
> *On the whole, I think that mining is about the best business of all. Of course, there may be chances in some railroad scheme that is better; but for a regular business, I think I would prefer mining. I know that if I were starting to make some money, I would go to mining in preference to anything else, and if I were going to start out to make ten million dollars, I would most assuredly go to mining. As to farming, that is hardly a business at all. The farmer, of course, is sure of making a living, and that is about all.*
>
> *I think that if a man is careful, he can make mining as safe as anything, and I do not see why he cannot pick out mines which will very nearly prove a sure thing. Of course a man could invest his money in corner lots, or buy State bonds, all of which is safe and would pay something, but*

it would be a very limited amount, and a man must have a great deal of money in the first place to make a start in this direction. At all events, as far as I'm concerned, I would prefer mining to anything else.[18]

George Hearst acquired the *San Francisco Examiner* in 1880. Although it wasn't one of his most favorite investments, ownership of the newspaper proved to be useful in furthering Hearst's political ambitions. Seven years later on March 4, 1887, George turned the *San Francisco Examiner* over to his son, William Randolph Hearst, who formed the Hearst Corporation.[19]

Reflecting on his investment in the newspaper business, George Hearst wrote as follows:

As to my connection with the Examiner newspaper, I may say, that I always took a great interest in politics, and when some of my friends came to me and stated that the party needed a paper and offered to subscribe something toward starting one, I said, "There is no use in doing that, why not buy up the old Examiner?" It was at that time an evening paper, and there never was any disgrace attached to it, however, when it came to paying for it, I found that I had to put in all the money myself.

I knew no more about a newspaper than the man in the moon, and when I looked over the property, I said, "This looks very like a quartz mill to me, and it will take a great deal of money to manage it." The boys said they would attend to that, and turn it into a morning paper and it would not cost more than ten or twelve thousand dollars. It ran on for some two or three years, and I kept sinking money into it all the time, until I found that it was worth sixty-five thousand dollars less than nothing. I believe I lost some three or four thousand dollars a month by it. I tried all sorts of people to make it a success, because, of course, I had no time to attend to it, and could not pretend to run it.

After I had lost about a quarter of a million by the paper, my boy Will came out of school, and said he wanted to try his hand at the paper. I thought it was about the very worst thing he could do, and told him so, adding that I could not make it pay. He said, however, that the reason that the paper did not pay was, because it was not the best paper in the country. He said that if he had it he would make it the best paper, and that then it would pay. I then put him in it, and made him a deed of the whole property from top to bottom, and agreed to stand by him for two years. Now, I don't think there is a better paper in the country, and if I wanted to get all the news I would rather have the Examiner than any of them. It has got just as good eastern news as any of the other papers and a great deal better western news. My son, Will, is one of the hardest workers I ever saw, and he certainly has made the paper a decided success.[20]

To help fulfill his political aspirations, Hearst ran for governor of California in 1882 but was defeated by George Stoneman. Governor Stoneman appointed Hearst to the U.S. Senate to fill the vacancy caused by the death of Democrat John F. Miller in 1886. Hearst held that position from March 23, 1886, to August 4, 1886, when a successor was elected. In 1887, he was elected to the U.S. Senate as a Democrat and served in that capacity until his death on February 28, 1891, in Washington, D. C.[21] George Hearst was buried in the Cypress Lawn Cemetery in Colma, San Mateo County, California.[22]

Phoebe Apperson Hearst

Phoebe Elizabeth Apperson (see fig. 28.6) was born December 3, 1842, in a log house on the family farm, which was located along the Meramec River in Franklin County, Missouri.[23] She was the daughter of Randolph Walker Apperson of Virginia and Drucilla Whitmire of South Carolina.[24] Phoebe had the equivalent of an eighth-grade education. She was a very prolific reader, however, and became self-taught in her main areas of interest, which included art, history, and education.

Fig. 28.6. Phoebe Apperson Hearst. After her husband George's death in February 1891, Phoebe provided funds to create the Hearst Free Library in Lead in 1894 and the Hearst Free Kindergarten in 1900. Her estate continued to provide financial support to both facilities for several years after her death in 1919.—Photo courtesy of the Black Hills Mining Museum.

When George Hearst's mother died in 1861, George remained in Missouri to settle her estate. It was during this time that Phoebe met George. Much to her parents' dismay, Phoebe and George eloped to Steelville, Missouri, where they were married on June 15, 1862. George was forty-one years old, and Phoebe was nineteen. Over the next several months, Phoebe traveled with George to the various mining camps in which George and his investors held business interests. Life in the frontier mining camps had to be an eye-opening and sometimes frightening experience for Phoebe, since most of the mining camps at that time were governed by miners and miners' courts. Notwithstanding, Phoebe's exposure to life in the mining camps may very well have been the inspiration for the philanthropic life she chose to live in later years.[25]

William Randolph Hearst was born to George and Phoebe in San Francisco, California, on April 29, 1863. In 1873, Phoebe took William on a yearlong tour to Europe where they visited castles, museums, and cultural centers. When George was elected to the U.S. Senate in 1887, George and Phoebe moved to Washington, D. C. After George died in 1891, Phoebe moved back to California and discovered that William had started construction of a mansion on the family ranch near Pleasanton.[26]

Phoebe assumed oversight of the project from William and hired Julia Morgan as architect. The 48,000-acre ranch was comprised of the former Bernal Rancho and the Alisal Rancheria. The latter ranch had as many as 125 Ohlone Indians living on it when the Hearsts acquired the property. Phoebe named the fifty-three-room mansion Hacienda del Pozo de Verona, meaning "the house of the wellhead of Verona." The mansion courtyard featured a fountain, which Phoebe and William purchased from the city of Verona, Italy. The entire mansion and grounds were designed around Mediterranean and California Mission architecture. The Western Pacific Railroad located the Verona Station not too far from the Hacienda, which allowed Phoebe to host groups of forty or fifty artists, composers, movie stars, and European friends for a weekend.

After Phoebe died in 1919, William maintained the mansion and ranch until 1924, when he sold the property to a group of investors. Much of the ranch was later converted to the Castlewood Country Club. A hillside golf course was created in 1926 and a valley course in 1948. The mansion served as the clubhouse from 1925 until 1969, when it was destroyed by fire. A new clubhouse was built at the same location and of the same architectural design as the original Hacienda.[27]

For almost three decades after George's death, Phoebe managed the multimillion-dollar empire that she and George had created. Phoebe's philosophy was to help others help themselves. As early as 1887, when she and George were living in Washington, D. C., Phoebe established three free kindergartens there and arranged for classes to train kindergarten teachers. Over the next decade, approximately 90 percent of the teachers in the District of Columbia were trained in educational institutions established by Phoebe. She also created three free kindergartens in San Francisco and one in Lead, South Dakota, in 1900.[28]

Phoebe provided much other philanthropic support to the city of Lead. In 1894, Phoebe funded a free public library for Lead. During its first two years, the Hearst Free Library was located in the Miners' Union Opera House. Mary Jane Palethorpe was the first librarian. In July 1896, the library was moved to an upstairs room in the Hearst Mercantile store on North Mill Street. The library was furnished with cushioned chairs, a piano, tables, and more than four thousand books.[29]

On August 5, 1914, upon completion of the Homestake Opera House and Recreation Building, the Hearst Free Library was moved to the top floor of the Recreation Building. For several years after her death, Phoebe Hearst's estate continued to fund the public library. In 1925, Homestake assumed responsibility for the library and operated it until 1972, when it was turned over to the city of Lead.[30]

At the request of a woman named Alice Birney, Phoebe helped create the National Congress of Mothers in 1887. This organization later evolved into the National Parent-Teacher Association. The same year, Phoebe became the first female regent of the University of California and served in that capacity until her death on April 13, 1919. In 1892, Phoebe made a monetary gift to the University of California-Berkeley, which permitted the creation of twenty scholarships for women students. She later donated an additional $60,000 to create an endowment that would fund the scholarships annually.[31]

The Berkeley campus has several buildings that bear the Hearst name, thanks to Phoebe's monetary gifts to the university. In 1899, Phoebe contracted with Architect Bernard Maybeck to build a large reception hall. The following year, the building was moved to the campus and was converted to a womens' gymnasium. After the building was destroyed by fire in 1922, William Randolph Hearst contracted with architects Maybeck and Julia Morgan. A new gymnasium was designed and constructed and became known as the Hearst Memorial Gymnasium for Women. The Hearst Memorial Mining Building was constructed in 1902, to commemorate her late husband, at a cost of $800,000. The building housed one of the country's finest colleges of mining. Phoebe also paid for the Hearst Greek Theater on campus. The theater was completed in 1903.[32]

Phoebe's philanthropic efforts extended to the San Francisco YMCA and YWCA, where no man or woman in need of a place to sleep was turned away. To meet expenses, the two organizations simply sent a bill to Phoebe each month, and a check was written. In 1896, contractors hired by Phoebe Hearst completed work on the National Cathedral School of Girls in Washington, D. C., whereupon the school was donated to the District Episcopal Archdiocese.[33] In the winter of 1919, during a trip to New York, Phoebe contracted the Spanish influenza. She returned to her Hacienda in California but weakened from the influenza and died on April 13, 1919.[34]

James Ben Ali Haggin

James Ben Ali Haggin (see fig. 28.7) was born December 9, 1822 in Harrodsburg, Mercer County, Kentucky. He was the son of Terah Temple Haggin and Adeline Ben Ali. James's paternal grandfather, John Haggin, came to Kentucky in 1775. Ibrahim Ben Ali, James's maternal grandfather, was born in Turkey and immigrated to America as an adult.

Fig. 28.7. James Ben Ali Haggin. Haggin was a mastermind at planning and executing the many brilliant business strategies that were conceptualized by his business partners and himself.—Waterland Collection, Devereaux Library Archives, South Dakota School of Mines and Technology.

Haggin graduated from the Centre College at Danville, Kentucky. After college, he opened a law office in Shelbyville, Kentucky. A short time later, he moved his practice to Natchez, Mississippi. On December 28, 1846, Haggin married Elizabeth Jane Sanders. Sanders's sister, Susan Gano Sanders, was married to Haggin's partner, Lloyd Tevis. The Haggins' children included Louis Terah, James Ben Ali Jr., Margaret Sanders, Adeline Ben Ali, and Edith Hunter. In 1897, Haggin remarried, this time to Margaret S. Voorhies.[35]

In 1849, Haggin decided to join the California gold rush. He went to New Orleans and boarded a steamer bound for San Francisco. While traveling through Panama, Haggin was stricken with malaria, which delayed his arrival in California until 1850. In October that year, Haggin went into partnership with Lloyd Tevis, and the two men opened a law office in Sacramento.[36] The two lawyers and George Hearst formed Hearst, Haggin, Tevis, and Company, which they devoted to mining investments in places such as California, Nevada, Utah, South Dakota, Montana, Mexico, and Peru.

In 1859, Haggin and Tevis acquired the 44,000-acre Rancho Del Paso from Samuel Norris, who had defaulted on a $65,000 mortgage. The Rancho Del Paso, which means the Ranch of the Pass, was located near Sacramento. An early-day trail leading to Emigrant Gap Pass in the Sierra Nevada Mountains crossed diagonally through the ranch. Tevis lived on the ranch for several years.

In 1873, Haggin hired John Mackay as ranch superintendent. Over the next few years, Haggin began breeding horses at the Rancho Del Paso. Eventually, the ranch became one of the largest horse-breeding operations in the world. It had twenty-four stables, some of which housed sixty-four stalls. About one hundred employees cared for as many as six hundred broodmares and forty stallions at the ranch.

Haggin and Tevis arranged to have the Southern Pacific Railroad cross through the ranch. The Arcade was a railroad terminal that was constructed on the ranch near the present-day intersection of Marconi Avenue and Auburn Boulevard in Sacramento. The terminal served as a shipping center from which horses were shipped to and from the ranch on specially designed railroad cars. Some of Haggin's champion horses included Ben Ali, winner of the 1886 Kentucky Derby, Midlothian, Daribin, Tyrant, Salvatore, Ms. Woodford, Firenzi, Star Ruby, Water Boy, Africander, Yellow Tail, Hamburg Bell, and many others.

The ranch was incorporated in 1891 as the Rancho Del Paso Land Company. Ten years after Tevis's death, Haggin sold the property to the Sacramento Valley Colonization Company on December 22, 1910, for $1.5 million. In 1916, some 148 acres of the ranch were sold to Del Paso Country Club for $22,253.80.[37]

Haggin relocated to New York in 1890, where he continued to manage his varied business interests. In 1897, he moved to his native state of Kentucky and purchased the Elmendorf Farm. There, he resumed his passion for breeding and racing thoroughbred horses. Haggin died at the age of ninety-one on September 13, 1914, at his villa in Newport, Rhode Island.[38]

Through a donation made in 1964, Louis Haggin helped create the Haggin Cup in honor of his grandfather, James Ben Ali Haggin. The Haggin Cup is awarded to the horse "in the most superior physical condition" among the top ten finishers in the Western States Trail Ride, also known as the Tevis Cup. The Tevis Cup was created through a donation made by Will Tevis in honor of his grandfather, Lloyd Tevis. The Tevis Cup is awarded to the rider who completes the 100-mile endurance run between Lake Tahoe and Auburn in the shortest amount of time and whose horse is "fit to continue."[39]

Lloyd Tevis

Lloyd Tevis (see fig. 28.8) was born in Shelbyville, Shelby County, Kentucky, on March 20, 1824. He married Susan Gano Sanders on April 20, 1854. The couple had five children: Margaret, Harry Lloyd, Louise, Hugh, and William Sanders. Later, in 1854, Tevis and Haggin established their law office in San Francisco. Tevis purchased a home in South Park while Haggin built a large home on Rincon Hill. In 1872, both men moved their families to Taylor Street on Nob Hill.

Fig. 28.8. Lloyd Tevis. Tevis was a shrewd negotiator and finance magnate for the Hearst, Haggin, and Tevis Company.—Waterland Collection, Devereaux Library Archives, South Dakota School of Mines and Technology.

Tevis's house was located at 1316 Taylor Street, at the corner of Jackson. Millions of dollars were spent on the interior of the home and its furnishings. The home contained eleven bedrooms and a library that was rumored to be the largest in California. Haggin's house, located a block away on the southeast corner of Taylor and Washington streets, boasted sixty-one rooms and eleven baths. The mansion also featured a mansard roof with a tower that extended to a height of 86 feet. The stables could accommodate eighteen carriages and forty horses. On the morning of April 18, 1906, the San Francisco earthquake destroyed Haggin's mansion.

In addition to the Hearst, Haggin, and Tevis Company, which was oriented toward mining interests, Tevis and Haggin invested in numerous other nonmining properties and companies in California. The two men held interests in the Cliff House and the Spring Valley Water Company. Tevis cofounded the California Street Cable Car Railway with Leland Stanford and Darius Ogden Mills, a banker.[40] Tevis held major ownership interests in the California Steam Navigation Company, California State Telegraph Company, Pacific Ice

Company, Risdon Iron Works, Oriental and Occidental Steamship Company, and the Sutro Tunnel Company in Virginia City, Nevada.[41]

In the 1870s, Tevis and Haggin acquired several hundred thousand acres of land in the San Joaquin Valley on which they established the second largest sheep and cattle ranch in California.[42] Much of this land was acquired under the Desert Land Act, which was passed by Congress on March 3, 1877. The purpose of the act was to promote the economic development of arid and semiarid public lands in the western United States.[43] Pursuant to the act, a person could acquire up to 640 acres of such land for twenty-five cents per acre. Upon proof of reclamation of the land by irrigation within three years, the claimant paid an additional $1 per acre.

Much of the "desert land" that Tevis and Haggin had their eyes on was located close to the Kern River. Haggin and Tevis exerted their influence on Senator Sargent of California to have federal legislation passed to open up federal lands for development in the area, which he accomplished with passage of the Desert Land Act. According to the *San Francisco Chronicle*, the president's signature on the act was not even dry when William Boss Carr, acting as agent for Tevis and Haggin, received word on March 31 that the act had been signed. Immediately, Carr started recording land applications and transfer deeds at the Land Office on behalf of dozens of vagabonds who filed claims and immediately sold them to Tevis and Haggin at a tidy profit. All of this was done before most anyone else in the country had even heard of the Desert Land Act. The *San Francisco Chronicle* called the whole affair an "atrocious villainy" and unsuccessfully called for a return of the lands.[44]

In 1881, Tevis and Haggin engaged in a legal battle with two other major landowners—Henry Miller and Charles Lux—in the San Joaquin Valley. The dispute involved an issue of irrigation versus riparian rights to the Kern River. After a trial and an appeal to the state supreme court in 1886, Miller and Lux prevailed in favor of their riparian rights. However, the following year, the state legislature passed the Wright Irrigation Act, which gave landowners the right to form cooperative water districts and acquire riparian rights through condemnation.[45] Tevis and Haggin subsequently formed the Kern County Land Company and transferred their desert lands to the corporation in 1890.[46]

One of Tevis's more famous business strategies involved the Pacific Express Company, Central Pacific Railroad, and the Wells Fargo & Company. As construction of the nation's first transcontinental railroad was underway by the Central Pacific and Union Pacific railroads, Tevis envisioned that the dominance of Wells Fargo's stagecoach empire was destined to wane.

The Big Four group of California investors, which included Charles Crocker, Mark Hopkins, Collis P. Huntington, and Leland Stanford, owned the Central Pacific Railroad.[47] Shortly after the golden spike was driven to complete the transcontinental railway on May 10, 1869, Tevis and Darius O. Mills, who were the largest shareholders in the Southern Pacific Railroad, organized the Pacific Express Company with H. D. Bacon and members of the Big Four in July 1869.

The Pacific Express Company was a separate and distinct company from the Pacific Union Express Company that was one of Wells Fargo's rivals. Assets of the Pacific Express Company consisted of little more than some stationery and cash. Notwithstanding, the company procured a ten-year exclusive contract from the Central Pacific to transport mail, gold, and other express across the railroad. Charles Crocker was instrumental in negotiating a similar contract from the Union Pacific Railroad.

In no time, it became public knowledge that the Pacific Express Company had secured exclusive express contracts across the new transcontinental railway. Almost overnight, the price of Wells Fargo stock plummeted from around $100 to $13 per share. Recognizing another opportunity, Tevis, Crocker, Haggin, and Mills began buying large amounts of Wells Fargo & Company stock. By fall 1869, the men owned a controlling interest in the company.[48]

On October 4, 1869, William G. Fargo, Charles Fargo, and Ashbel H. Barney, president of Wells Fargo & Company, agreed to meet with Tevis, Bacon, and Mills in Omaha, Nebraska. Tevis, the shrewd lawyer and negotiator, once said that he could think five times faster than any other man in San Francisco. His skills were put to the test. Tevis and his partners—now with a controlling interest in Wells Fargo—convinced the Fargos and Barney that Wells Fargo needed the Pacific Express Company in order for Wells Fargo to survive.

An agreement, known as the Omaha Treaty, was reached whereby Wells Fargo would purchase the Pacific Express Company from Tevis and his partners for $5 million in newly issued capital stock—an amount equal to one-half of the existing $10 million of capital stock of the entire Wells Fargo & Company. Additionally, all assets previously owned by Wells Fargo & Company in excess of 3½ percent of its existing $10 million of capital stock were to be distributed to the holders of these shares. The Pacific Express Company, in return, agreed to pay to Wells Fargo 3⅓ percent of $5 million cash or approximately $166,667. The shareholders subsequently approved the agreement on November 25, 1869.[49]

Tevis became vice president and a director of Wells Fargo in 1870. In 1872, he was elected president, replacing William Fargo. That same year J. B. Haggin and Mark Hopkins were elected directors.[50] In 1875, with Tevis as president, Haggin was elected vice president.[51] Under the leadership of Tevis and Haggin, the Wells Fargo & Company flourished. A transcontinental express business was established, and the number of banking and express offices expanded from 436 in 1871 to 2,830 in 1892. Tevis retired from Wells Fargo as president in 1892 and as director in 1893.[52]

Lloyd Tevis also recognized or learned that the lands of California contained oil. Together with Charles Felton, director of the U.S. Mint in San Francisco and assistant treasurer of the United States, and George Loomis, Felton's brother-in-law, the Pacific Coast Oil Company was formed on September 10, 1879. The company purchased the interests of some other oil men who had developed a very good well, but lacked the capital to refine and market the oil. The Pacific Coast Oil Company constructed a six-hundred-barrel-per-day refinery in Alameda. In 1900, Pacific Coast Oil Company was sold to Standard Oil Company, which constructed a much larger refinery in Richmond.[53]

Lloyd Tevis died on July 24, 1899, at the age of seventy-five.[54]

Charles Washington Merrill

Charles Washington Merrill (see fig. 28.9) was born on December 21, 1869, in Concord, New Hampshire, to Sylvester and Clara Lydia (French) Merrill. Clara's grandfather, Robert Morse, was related to Samuel F. B. Morse, inventor of the single-wire telegraph. Charles was named after his two grandfathers, Charles H. French and Washington Merrill. The Merrill family immigrated to America from England in 1633. Merrill's maternal ancestors were also English immigrants who arrived in America two years later. The Merrills of England were descendants of a French Huguenot family named De Merle, who fled France following the St. Bartholomew Day Massacre August 23-24, 1572.

Fig. 28.9. Charles Washington Merrill. This photo of Charles W. Merrill appeared in *Black Hills Illustrated* in 1904. At the time, Merrill was perfecting the Merrill cyanide process at Homestake's Cyanide Plant No. 1 in Lead.—Waterland Collection, Devereaux Library Archives, South Dakota School of Mines and Technology.

After graduating from Phillips Academy, Sylvester Merrill moved to Concord, New Hampshire, and went to work as a clerk and bookkeeper at a furniture manufacturing facility. There, he became a coworker and friend with Nathaniel P. Cole. In 1864 or 1865, Sylvester moved to San Francisco. The trip was a long one by ship around Cape Horn, South America. In San Francisco, Sylvester went to work as a bookkeeper for Cole, who had previously made the same trip and started a furniture business. Sylvester returned to Concord in 1868 and married Clara French. The newlyweds returned to San Francisco by ship and moved into their first home at No. 2 Martha Place, which was located south of Geary Street between Mason and Taylor streets.

In November 1869, Clara returned to Concord by herself to bear the couple's first child. This time, the trip was made in just ten days across the newly constructed transcontinental railroad. On December 21, Charles Washington Merrill was born in the home of Clara's parents. Clara and Charles returned to San Francisco across the Central Pacific Railroad in early March 1870. Sylvester, who remained in San Francisco, was at the train terminal in Oakland to greet them and take them to their new home at 936 Howard Street.[55]

Around the late 1870s, Sylvester and Clara moved across the bay to Alameda, where Charles was enrolled in the Alameda public school system. He graduated from Alameda High School in 1887. That fall, he enrolled in the College of Mines at the University of California. In his early childhood, Charles decided he wanted to pursue a career in mining and metallurgy. After graduating with a bachelor of science degree in metallurgy and mining in May 1891, Merrill took a job for the summer with the U.S. Geological Survey doing work in the Sierra Nevada Mountains.

In October, he returned to San Francisco, where he was introduced to Alexis Janin. Janin was a metallurgical engineer who owned his own consulting business. Janin's brother, Louis, was a mining engineer, who also had an office in San Francisco and had done considerable work for George Hearst, J. B. Haggin, and Lloyd Tevis. Merrill went to work for Alexis Janin in his laboratory in November 1891. Janin allowed Merrill free use of the laboratory during Merrill's off-shift hours. Merrill took advantage of this and worked long days and weeks to further the research of his graduation thesis, which was directed toward the recovery of gold using cyanide. Merrill never claimed to have been the first one to have experimented or successfully used cyanide on gold-bearing ore, but he was one of the first to develop and perfect an economic process outside of the laboratory.

Around February 1893, Merrill informed Alexis Janin that he was resigning to open his own metallurgical office. Janin wrote back to Merrill on February 14, 1893, telling him, "I have no doubt but that you will soon be in a fair way to earn a good professional living and gain reputation. Business brings business . . . and others will be after you. Don't be in a hurry to take hold of a new thing until you have thoroughly mastered it. Then you will know the condition necessary to make a success, and, as a new broom, you can demand the proper assistance, and get it . . . . Wishing you much luck."[56]

Merrill opened his consulting office on the upper floor of a building at 117 Geary Street, which was the furniture business where his father worked. Almost immediately, Merrill met with Superintendent Thomas H. Leggett at the Standard Consolidated Mine at Bodie, California, and proposed that Leggett construct an experimental plant to test Merrill's cyanide process. Merrill was so certain that his process would work that he told Leggett he would work for nothing until the process proved itself. Leggett agreed to the terms, and the plant was constructed. From March 1893 through August 1894, Merrill successfully operated the first commercial cyanide plant in the United States.

Based on his success at Bodie, Merrill already had his next job lined up at the Harqua Hala gold mine near Yuma, Arizona Territory. After looking at the operation with Superintendent Robert M. Raymond and performing some tests in the laboratory, Charlie Merrill was confident that his process would work on Harqua Hala ore. He proposed to Raymond that an experimental cyanide plant be constructed. Again, Merrill proposed that he be paid nothing until the process proved successful.

In a very short time, there was no question that the Harqua Hala ore could be commercially treated using Merrill cyanide process. Construction of a large cyanide plant was started, and Merrill was named chief metallurgist. Toward the middle of 1895, about the time Merrill was satisfied that the new cyanide plant was working as planned, he was informed by two of his friends that a mine operation in Montana was interested in the Merrill cyanide process.

Merrill went to Montana Mining Company at Marysville, Montana, and met with General Manager A. T. Bayliss at the Drumlummon Mine. After performing some lab tests on tailings from the mine, Merrill made his standard proposal for construction of an experimental plant. The plant proved to be successful. Merrill was named resident metallurgical engineer and put in charge of constructing and commissioning a larger cyanide plant. By November 1897, Merrill successfully completed his work for Montana Mining Company. After agreeing to a consulting agreement with the company over the next three years, Merrill returned to San Francisco.[57]

At the age of twenty-eight, Charlie Merrill decided it was time to get married. He married Clara Scott Robinson in Alameda, California, on February 9, 1898.[58] That fall, B. B. Thayer, a consulting engineer for Homestake in San Francisco, California, called upon Charles Merrill to discuss the Merrill cyanide process for possible application at the Homestake Mine in South Dakota.

In January 1898, Merrill arrived in Lead, South Dakota, to discuss the Merrill cyanide process with Superintendent Thomas J. Grier. After testing samples of the mill tailings, Merrill made his standard pitch for no compensation until the process proved successful in the pilot plant and on a commercial basis. While at Homestake, Merrill developed and perfected the Merrill precipitation process, which utilized zinc dust instead of zinc shavings, the Merrill clarifying filter, and the Merrill slime process. Based on Merrill's ten years of work at Homestake, it has been estimated that the value of the ore reserve at that time was increased by approximately $20 million.

In later years, when asked about his practice of not accepting any compensation until the Merrill cyanide process worked, Merrill replied, "Well, I always believed in that old adage, 'nothing ventured, nothing gained.' However, I never ventured until I was as certain as it was humanly possible to be, that I knew what I was doing. Mind you, I'm not talking about 'cock-sureness'; that is something entirely different. What I mean is that I was sure because I had tested the thing out, not once or twice, but dozens of times in every possible way. Of course, even then there was the risk of accruing from the intangibles, but such risk I was always willing to run."

In 1908, with his work and contract finished, Cyanide Charlie sold his home in Lead and returned to San Francisco with the small fortune that he had earned for his countless long days at work. Accompanying him were his wife, Clara, and the three children born to the couple during their ten years in Lead. The couple's children born in Lead included Beatrice Maud, John Lisgar, and Gregor Charles. A fourth son, Bruce Robinson, was born in California. Years later, Merrill remarked, "During the first few years, my wife had to put up with a man whose work kept him away from home so much that he was almost a stranger there. But fortunately that period did not last too long, and when my Homestake work was finished and we settled down in Berkeley, the 'pioneer days' were definitely over."

After leaving the Homestake Mine, Charlie Merrill formed the Merrill Company in San Francisco. Three of his principal Homestake assistants, Charles C. Broadwater, Louis D. Mills, and Frank H. Ricker, joined him as business associates. Herbert Shuey was hired later for his abilities in finance and accounting. Merrill often referred to his associates as the Big Four. Thomas B. Crowe, who helped Merrill refine the Merrill-Crowe precipitation process, was also hired.

On September 13, 1910, Merrill incorporated his company. Merrill's business plan was to make his patented metallurgical processes available to gold and silver mines throughout the world. For the first five years, the company's office was located at 143 Second Street, between Mission and Howard. As the business grew, it was moved into the Rapp building next door.

In 1916, Merrill formed a subsidiary company called the Alloys Company to produce the various grades of zinc dust that were required in the Merrill precipitation process. Milton H. Newell was hired and brought with him a unique process for manufacturing the dust. Eventually, five zinc dust plants were constructed worldwide. The Alloys Company was sold to the Metals Disintegrating Company of Elizabeth, New Jersey, in 1945.

At the age of eighty-three, Charlie Merrill decided to slow down a little. His son, John, who had been with the Merrill Company for twenty years, took over as president of the company. Charlie himself decided to work at the office three days per week, instead of five or six. He also started spending the winter months in Honolulu, where every morning, he enjoyed a swim in the surf. Charles died in Honolulu on February 6, 1956, at the age of eighty-six.[59]

Famous VIPs as Visitors

One of the more noteworthy people to visit the Homestake Mine included President William Howard Taft, who visited Deadwood, Lead, and Sturgis on October 21, 1911. The presidential party arrived in Deadwood at 12:25 p.m. on the Burlington train.[60] After speaking from a platform on the lawn of the Lawrence County Courthouse, President Taft was presented with a 1-pound paperweight made of gold from the Golden Reward Consolidated Gold Mining and Milling Company, the Mogul Mining Company, Lundberg, Dorr, and Wilson Mining Company, the Trojan Mining Company, and the Wasp No. 2 Mining Company.[61]

A luncheon was held for the presidential party and local dignitaries at the Franklin Hotel. During the luncheon, Richard Bunney and W. E. Scroggin, two prominent members of the Western Federation of Miners, presented a petition to Congressman E. W. Martin. The two men asked Congressman Martin to forward the petition to President Taft. The petition asked President Taft to exercise his influence in having Homestake Mining Company discontinue using the "card system" that it had been using since the lockout of 1909-1910. Under the card system, applicants for employment at Homestake were required to sign a card, which signified that the applicant was not a member of any labor union and that they would not join any such organization while employed by the company.[62]

At 3:00 p.m. that afternoon, President Taft traveled to Lead. His first stop was at the Homestake Assay Office, where he observed a gold pour and the process for making gold bricks.[63] His next stop was at Christ Episcopal Church, where he delivered a speech to local-area citizens who had gathered in the courtyard of the Hearst Free Kindergarten.

President Taft and the presidential party were driven to the Ellison Shaft, where Superintendent Thomas J. Grier and other Homestake personnel gave them a tour of the 1,100-foot level. The visitors toured several underground work areas, one of which was a large stope that was illuminated so President Taft could see the air drills at work. The tour group also visited the B&M pump room, which was of special interest to President Taft.

Here, President Taft was presented with a gold-plated miner's candlestick. Several flashlight photos were also taken of the presidential party at the pump room and at the shaft station of the Golden Star Shaft (see fig. 28.10). The tour group returned to the surface via the Star Shaft. After returning to Deadwood, the presidential party boarded a Northwestern train at 5:25 p.m. and traveled to Sturgis, where they visited Fort Meade. From there, the tour group returned to Rapid City.[64]

Fig. 28.10. President William Howard Taft's visit to the Homestake Mine on October 21, 1911. The photo was taken at the Golden Star Shaft on the 1,100-foot level. President Howard Taft is the man in the center of the front row. Superintendent Thomas J. Grier is standing at President Taft's left side. The two men to President Taft's immediate right are Congressman W. E. Martin (with the hat) and Father Joseph F. Busch, Catholic Bishop.—Homestake Collection, Adams Museum, Deadwood.

Gutzon Borglum and Mary Garden were given a tour of the Homestake Mine on October 20, 1931. Borglum was in the process of sculpting the faces of U.S. presidents George Washington, Thomas Jefferson, Theodore Roosevelt, and Abraham Lincoln on Mount Rushmore, a project which started in 1927 and was completed in 1941. Funding for the project was derived in many different ways, one of which included a benefit concert held in Rapid City. Mary Garden, the famous operatic soprano, agreed to sing at the concert. After the concert, Homestake superintendent Bruce C. Yates invited Borglum and Garden and their guests to visit the Homestake Mine. The invitation was accepted.

The mine tour for Borglum and Garden consisted of a cage ride at the newly rebuilt Ellison Shaft as well as a luncheon party that was held in the sill cut of a new stope being developed on the 2,000-foot level (see fig. 28.11). To prepare for the luncheon, the stope was temporarily converted to a reception hall by installing electric lights, tables, benches, and a wooden floor. The main course consisted of Cornish pasties, which were served by the Homestake engineers, geologists, and surveyors. Superintendent Yates provided the special guests with a description of the mine operation. Following the luncheon, the group returned to the surface, where they were treated to a short concert by the Homestake Band.[65]

Fig. 28.11. Gutzon Borglum's and Mary Garden's visit to the Homestake Mine, October 20, 1931. Mount Rushmore sculptor Gutzon Borglum and operatic soprano Mary Garden are shown at far right. Homestake superintendent Bruce C. Yates is seated on the end of the table directly across from Borglum and Garden.—Photo courtesy of the Black Hills Mining Museum.

Some of the other famous visitors to the Homestake Mine included Wendell L. Willkie, Republican candidate for president (1940), U.S. senator Robert A. Taft on May 26, 1952, and producer Cecil B. DeMille in 1953.[66] Taft, a U.S. senator from Ohio and son of President William Howard Taft, visited the mine forty-one years after his father's visit. Senator Taft's trip to the Black Hills was timed with his bid for the Republican presidential nomination, which was narrowly won by Dwight D. Eisenhower.[67]

Epilogue

Nothing ever came easy for the owners and operators of the Homestake Mine—at least not for very long. Relative to other gold mines, the ore at the Homestake Mine was generally low-grade, which meant that continual improvements in mining and milling methods were required to allow the mine to remain viable.

As the ledges were found to plunge deeper, large capital investments were continually required to sink new shafts, develop new levels, expand the ventilation systems, delineate new ore reserves, and improve the milling and support systems. Concurrently, the employees were rightfully given higher pay and better benefits. The only constants at the mine were change and higher costs. Fortunately, Homestake was always blessed with competent, dedicated, and loyal people who, with vision and talent, could effectively respond to just about any change or condition. The exception was sustained low gold prices, for which they had little or no control.

As another chapter of the Homestake story is closed, a new chapter is opened. The former Homestake Mine, now the Sanford Underground Science and Engineering Laboratory at Homestake, is again open, but this time for science. Early-science experiments run concurrent with work to dewater, rehabilitate, and convert the mine to a science laboratory. Former Homestake people now work with other talented people as employees of the South Dakota Science and Technology Authority. World-renowned scientists and planners continue to collaborate to decide on the most appropriate science experiments and infrastructure for the longer-term Deep Underground Science and Engineering Laboratory or DUSEL. Concurrently, the National Science Foundation helps prescribe the process for financial support of the DUSEL.

From nuggets to neutrinos, the Homestake story and legacy continue to unfold.

Appendix A
Water rights acquired by Homestake Mining Company

Water Right/Location of Intake	Date Located	Original Locators	Miner's Inches	Acquired by HMC Agent
Bear Butte Creek				
900 ft. above Deadwood Road	Jun 29, 1879	James Leavy	All	Nov. 5, 1880
Silver Spring w. of Dwd. Rd.	Jun 9, 1880	J. Leavy/ J. VanDyke	All	Nov. 5, 1880
NE Fork above Harrington Camp	Jul 11, 1882	Samuel McMaster	50	Jul 11, 1882
Boxelder Creek				
500 yd. NW from HMC Sawmill	May 10, 1886	Homestake Mining Company	20	May 10, 1886
City Creek				
Deadwood City Water Right	Apr. 25, 1876	Weed/Miller/ McKinley	All	Unknown
Deadwood Creek				
Foster Ditch (from Claim No. 27)	Apr. 24, 1876	Captain W. M. Foster	250	Sept. 14, 1878
Pocket Gulch	March 1877	J. W. Clark/Ames	All	Feb. 17, 1905
DeSmet, Bowie & Palmer Ditch	Nov. 19, 1878	P. Smith/A. Bowie	100	Apr. 23, 1881
¼ mi. above Oro City Sawmill	Oct. 18, 1878	J. A. Pierce/ J. Sapsley	250	Apr. 23, 1881

Poorman Gulch (Placer #1 & #2)	May 13, 1879	John Hose/Elias Roupe	All	Sept. 24, 1881
Sawpit Gulch Spring	Sept. 20, 1879	Isaac B. Hewitt	10	Feb. 17, 1905
Elk Creek				
Elk Creek Springs	Nov. 23, 1877	Buchanan/Preble/ Sutherland	1,000	Apr. 23, 1881
Spring on Johnson Ranch	Unknown	Martin G. Johnson	All	Sept. 7, 1889
Elk Creek	Aug. 4, 1880	L. L. Alexander	All	Apr. 23, 1881
False Bottom Creek				
Columbia Ditch	Dec. 21, 1877	John Moore/Lardner	100	Jan. 6, 1891
Rapid Creek				
Tilson Springs	May 18, 1879	William Tilson	All	Oct. 8, 1885
Little Rapid Springs #1	Aug. 18, 1885	J.B. Haggin/ George Hearst	50	Aug. 18, 1885
Little Rapid Springs #2	Aug. 27, 1885	J.B. Haggin/ George Hearst	40	Aug. 18, 1885
Little Rapid Springs #3	Aug. 22, 1885	J.B. Haggin/ George Hearst	40	Aug. 27, 1885
Redwater River				
Concordia Ditch	Jun 24, 1879	Concordia Ditch Company	1,000	Dec. 23, 1899
Concordia Ditch	Dec. 21, 1889	Concordia Ditch Company	2,000	Dec. 23, 1899
Spearfish Creek				
Evans-Tonn Ditch	May 1876	Robert Evans et al.	173.1	Feb. 17, 1905
Ramsdell Ditch	Jun 1, 1876	Joseph Ramsdell	229	Feb. 17, 1905
Walton Ditch	Mar. 1, 1877	Cooper/Walton/ Maurer	191	Jul 6, 1936
Main Right Fork of East Fork	Jun 4, 1877	Kuykendall/ Thompson/Berry	100	Apr. 23, 1881

Main Left Fork of East Fork	Jun 4, 1877	Kuykendall/Thompson/Berry	800	Apr. 23, 1881
East Fork	Jun 4, 1877	Whitehead/Thompson/Benham	500	Apr. 23, 1881
Little Spearfish	June 1878	Unknown	540	Feb. 17, 1905
Peake #1	Oct. 3, 1889	G. Hearst/J. B. Haggin	50	Oct. 3, 1889
Peake #2	Oct. 5, 1889	G. Hearst/J. B. Haggin	50	Oct. 5, 1889
Raddick (Juniper) Gulch	Nov. 15, 1889	William Raddick	300	Feb. 17, 1905
Greenback (below Crossing)	Feb. 2, 1892	George E. Brettell	3,000	Aug. 17, 1900
Lead Company (above Crossing)	Feb. 2, 1892	Brettell/May/Abt/ et al.	3,000	Aug. 17, 1900
Elk Mountain (1/2 mi. above falls)	Aug. 8, 1892	Ed Reilly/John F. Barry	2,500	Feb. 17, 1905
SD Mining Co. (East Fork)	Nov. 27, 1893	South Dakota Mining Company	500	Aug. 17, 1900
SD Mining Co. (Ice Box)	Feb. 24, 1894	South Dakota Mining Company	250	Aug. 17, 1900
Raddick Hay Ranch	Apr. 29, 1897	Tourney/Clement/Evans	All	Feb. 17, 1905
Gregg or Owens-Gay Ditch	Mar. 22, 1900	Harrison Gregg	22	Feb. 17, 1905
Hydroelectric No. 1	January 1909	Homestake Mining Company	4,800	January 1909
Hydroelectric No. 2	Jun 9, 1917	Homestake Mining Company	3,000	Jun 9, 1917
Whitewood Creek				
Pioneer Ditch (Nevada Gulch)	Mar. 4, 1876	L. Webb/R. McLennan	300	Dec. 26, 1877
French Boys (Old Abe)	May 10, 1876	Harney/Engh/Manuel/Manuel	400	Oct. 12, 1879
Spring Creek (Deadwood)	May 20, 1876	L. Parkhurst/J. Leavy	All	Jun 7, 1897

Boulder (Whitewood) Ditch	Jun 1, 1876	George Atchison	1,000	Apr. 23, 1881
1 mi. above Ten-Mile Station	Aug. 18, 1876	Myers, Belding, Atchison	500	Jan. 5, 1878
Montana Ditch—Reno Gulch	Jun 16, 1877	Aaron Bell/John Roberts	600	Jul 15, 1878
1 mi. above Ten-Mile Station	Jan. 1, 1878	Myers, Belding, Atchison	100	Jan. 5, 1878
Montana Ditch—Whitetail	Jun 16, 1878	Abt/Distelhorst/Budts	800	Jul 15, 1878
Savage Water Right below Kirk	Dec. 2, 1878	P. W. Savage	1,000	Feb. 17, 1905
Strawberry Spring	Sept. 26, 1881	Martin Sands	40	Oct. 11, 1883
Whitewood Springs	Sept. 27, 1881	Martin Sands	20	Oct. 11, 1883
Mineral Springs	Sept. 30, 1881	Martin Sands	30	Oct. 11, 1883
Flat Iron Spring	Sept. 30, 1881	Martin Sands	20	Oct. 11, 1883
Hearst Ditch	Unknown	Young/Corson	All	Jan. 30, 1880

Source: Homestake Mining Company, Water Rights files.

Appendix B
Superintendents and general managers of the Homestake Mine

	Tenure
William A. Farish	1877-1878
Samuel McMaster	1878-1884
Thomas J. Grier	1884-1914
Richard Blackstone	1914-1918
Bruce C. Yates	1918-1936
Guy N. Bjorge	1936-1953
Abbott H. Shoemaker	1953-1957
James O. Harder	1957-1971
Donald T. Delicate	1971-1978
Allen S. Winters	1978-1987
Gary A. Loving	1987-1992
Allen S. Winters	1992-1995
Stephen A. Orr	1995-1997
Vernon C. Baker	1997-1999
Bruce Bried	1999-2001

Sources: Black Hills Mining Museum, Lead, South Dakota; Mildred Fielder, *The Treasure of Homestake Gold* (Aberdeen: North Plains Press, 1970), 43, 51.

Appendix C
Mining Law of 1872

In the midst of the growing dissension over the Fort Laramie Treaty of 1868, President Ulysses S. Grant signed into law the Mining Law of 1872 on May 16, 1872. The Mining Law of 1872 was one of a number of public land laws passed by Congress in the late 1800s to promote the development or settlement of public domain lands in the western United States. The Mining Law of 1872 provided (and still provides) citizens of the United States with the right to explore for, discover, and exploit certain valuable mineral deposits on those federal lands open for these purposes. The mining law specifies the means for (1) discovery of a valuable mineral deposit, (2) location of mining claims and sites, (3) recordation of mining claims and sites, (4) maintenance of mining claims and sites, and (5) mineral patents.[1]

Location of mining claims is available to citizens of the United States and those who have indicated their intention to become citizens. The process of locating an unpatented lode claim on public lands varies slightly according to state and local rules and regulations but generally requires the locator to post a location notice, monument the boundaries of the claim, and file notices at the county courthouse and with the Bureau of Land Management. A claim may be located as a lode claim for mineral deposits found as veins or rocks in place or as a placer claim for placer deposits. The mining law also provides for acquiring a 5-acre claim of nonmineral land for a mill site.

The holder of an unpatented mining claim has possessory rights to the claim, which gives him or her the right to mine and remove minerals from the claim. Legal title to the claim remains with the United States. The original mining law required $100 worth of assessment work to be performed on each claim annually. In 1993, Congress passed legislation that requires payment of a $100 fee per year, per claim, in lieu of the assessment work. A "small miner" exemption is allowed for miners who hold ten or less claims.[2]

The Mining Law of 1872 specifies a procedure for the holder to obtain a patent to the claim, whereby upon meeting certain requirements and paying a filing fee, the government issues fee-simple title to the claim. The requirements for a patent require the applicant to (1) demonstrate the discovery of a valuable (commercial) mineral deposit, (2) possess clear title to the claim, (3) have recorded affidavits of annual assessment work and to have expended $500 of improvements to the claim, and (4) pay the necessary filing fees. Once a patent is applied for, the claim is surveyed and assigned a unique mineral survey number. A patent enables the holder, as owner, to use the claim for mining or nonmining purposes. In 1994, Congress placed a moratorium on patenting a claim. Holders of unpatented mining claims and owners of patented mining claims may transfer title to their claim through a deed.

Mining claims may be located in nineteen states, including South Dakota, where federal lands exist and are open to mineral entry. Once discovery of a mineral has been made, a lode-mining claim may be located. A mining claim may not exceed 1,500 feet in length along the vein or lode and not more than 300 feet on each side of the axis of the vein or lode. The end lines of the claim must be parallel to each other. Generally, the holder's mineral right is limited to those portions of the vein or lode that lie within the vertical planes of the sidelines of the claim.

A placer claim is used to claim unconsolidated mineral deposits such as gold placers. The legal description for a placer claim includes the section, township, range, and principal meridian in which the claim is located. The size of a placer claim is limited to twenty acres per locator. An association may be formed of no more than eight locators, each of whom may claim up to 20 acres, subject to a maximum of 160 acres for the association.[3]

Notes

Chapter One

1. John Paul Gries, *Roadside Geology of South Dakota* (Missoula: Mountain Press Publishing Company, 1996), 2-3, 215.
2. Ibid.
3. Ibid., 210-15.
4. Ibid., 3, 99.
5. Chester R. Longwell, Richard Foster Flint, and John E. Sanders, *Physical Geology* (New York: John Wiley and Sons, Inc., 1969), 123.
6. D. G. Driscoll, J. M. Carter, J. E. Williamson, and L. D. Putnam, *Hydrology of the Black Hills Area, South Dakota*, U.S. Geological Survey Water-Resources Investigations Report 02-4094, (2002), 10-19.
7. Gries, *Roadside Geology*, 292-93.
8. Thomas J. Campbell, "Synopsis of Homestake Mine Geology," Homestake—background material for DUSEL, Deep Underground Science and Engineering Laboratory, http://homestake.sdsmt.edu/Geology/geology.htm (accessed March 20, 2009).
9. Gries, *Roadside Geology*, 2-3, 214-15.
10. Driscoll, Carter, Williamson, and Putnam, *Hydrology of the Black Hills Area*, 20-21.
11. Campbell, "Synopsis of Homestake Mine Geology."
12. Kathy Hart, "Homestake Mine Area Geology," Homestake Mining Company.
13. Ibid.
14. Campbell, "Synopsis of Homestake Mine Geology."
15. U.S. Department of the Interior, *Geology of the Region around Lead*, South Dakota, by Sidney Paige, Bulletin 765, U.S. Geological Survey, 1924.
16. A. B. Yates, "Structure of the Homestake Ore Body" (PhD diss., Harvard Engineering School, March 1931), 52.
17. Ibid., 149-56.
18. Ibid., 141-49.
19. Ibid., 128-39.
20. Joseph P. Connolly, *The Tertiary Mineralization of the Black Hills* (Rapid City: South Dakota School of Mines, September 1927), 55-57.

21. Christopher Hills, *Gold Pans and Broken Picks* (Spearfish: Clark Printing, 1998), 37-38.
22. Connolly, *The Tertiary Mineralization of the Black Hills*, 55-57.

Chapter Two

1. Peter Rosen, *Pa-Ha-Sa-Pah or the Black Hills of South Dakota* (St. Louis: Nixon-Jones Printing Company, 1895), 3.
2. Gail and Michael Evans-Hatch, *Place of Passages: Jewel Cave National Monument Historic Resource Study* (Omaha: Midwestern Region National Park Service, 2006), 31.
3. Doane Robinson, *A Brief History of South Dakota* (New York: American Book Company, 1905), 17-21.
4. George W. Stokes and Howard R. Driggs, *Deadwood Gold: A Story of the Black Hills* (New York: World Book Company, 1927), 58.
5. Badlands Natural History Association, "Early Indians and Explorers," Badlands, History of Badlands National Monument, http://www.nps.gov/history/history/online_books/badl/sec1.htm (accessed January 17, 2009).
6. Sven G. Froiland, *Natural History of the Black Hills* (Iowa: Graphic Publishing Company, 1978), 62-63.
7. Gary McLain, *Indian America* (Santa Fe: John Muir Publications, 1994), 96.
8. Froiland, *Natural History of the Black Hills*, 64-67.
9. Rosen, *Pa-Ha-Sa-Pah: Or the Black Hills*, 17-18.
10. Robinson, *A Brief History of South Dakota*, 17-21.
11. Wikipedia, The Free Encyclopedia, "Lakota People, History," http://en.wikipedia.org/wiki/Lakota_people (accessed January 17, 2009); Froiland, *Natural History of the Black Hills*, 65-68.
12. Evans-Hatch, *Place of Passages*, 32.
13. James Clyman, "Narrative," American Mountain Men, http://www.xmission.com/~drudy/mtman/html/clyman.html (accessed March 20, 2009).
14. Fort Berthold Library, "Treaty with the Arikara Tribe, 1825," The Library Corporation, http://lib.fbcc.bia.edu/FortBerthold/TA25.asp (accessed March 25, 2009).
15. Louis Houck, *The Spanish Regime in Missouri* (Chicago: The Lakeside Press, 1909), 355-61.
16. Weston Arthur Goodspeed, "The Province and the States" (1904, USGenWeb, 2009), VI, chap. III, 271-80, http://files.usgwarchives.net/sd/history/province/b-hills.txt (accessed March 23, 2009).
17. Frank Thomson, *The Thoen Stone: A Saga of the Black Hills* (Detroit: The Harlo Press, 1966), 19-23.

18. Ibid., 25-26.
19. Ibid., 27-28.
20. Ibid., 76-80.
21. Historic St. Mary's Mission, "Fr. De Smet & Historic St. Mary's Mission," Historic St. Mary's Mission, Inc., http://www.saintmarysmission.org/FatherDeSmet.html (accessed March 23, 2009).
22. Rosen, *Pa-Ha-Sa-Pah: Or the Black Hills*, 238-42.
23. Goodspeed, *The Province and the States*, 271-80.
24. Rosen, *Pa-Ha-Sa-Pah: Or the Black Hills*, 245-47.
25. Ibid., 31.
26. John G. Neihardt, *Black Elk Speaks*, 7th ed., (Lincoln: University of Nebraska Press, reprinted in 1993), 79.
27. Thomson, *The Thoen Stone*, 134-35.
28. Watson Parker, *Gold in the Black Hills*, 1st ed. (Lincoln: University of Nebraska Press, 1966), 11-12.
29. Jesse Brown and A. M. Willard, *The Black Hills Trails* (Rapid City: Rapid City Journal Company, 1924), 30.
30. Brown and Willard, *The Black Hills Trails*, 30.
31. Rosen, *Pa-Ha-Sa-Pah: Or the Black Hills*, 253-54.
32. Thomson, *The Thoen Stone*, 133; Parker, Gold in the Black Hills, 13.
33. Parker, *Gold in the Black Hills*, 12.
34. Ibid., 15.
35. Colin Rickards, *Bowler Hats and Stetsons: Stories of Englishmen in the Wild West* (New York: Bonanza Books, 1965), 77-82.
36. Parker, *Gold in the Black Hills*, 14.
37. Mildred Fielder, *Lost Gold* (Aberdeen: North Plains Press, 1978), 98-105.
38. Thomson, *The Thoen Stone*, 55-56.
39. Ibid., 56-57.
40. Ibid., 57-58, 63.
41. Ibid., 112-13.
42. Michael T. Lubragge, "Manifest Destiny," From Revolution to Reconstruction, http://www.let.rug.nl/usa/E/manifest/manif1.htm#int (accessed March 23, 2009).
43. Mary Trotter Kion, "Forts of the West, An Introduction," Suite101.com, http://www.suite101.com/article.cfm/great_american_plains/112946 (accessed March 23, 2009).
44. Robert L. Munkres, "Tales of Old Fort Laramie," Muskingum College, http://www.muskingum.edu/~rmunkres/military/Laramie/Tales.html (accessed March 23, 2009).
45. Answers.com, "Indian Agents," U.S. History Encyclopedia, http://www.answers.com/topic/indian-agents (accessed March 23, 2009)

46. Martin Luschei, *The Black Hills & the Indians* (San Luis Obispo: Niobrara Press, 2007), 21-22.
47. E. Laveille, *The Life of Father De Smet* (New York: P.J. Kenedy & Sons, 1915), 234.
48. Luschei, *The Black Hills & the Indians*, 23-24.
49. Ibid., 25.
50. Canku Luta, "Treaty of Fort Laramie, September 17, 1851," Red Road, Inc., http://www.canku-luta.org/PineRidge/laramie_treaty.html (accessed March 23, 2009).
51. Luschei, *The Black Hills & the Indians*, 25-26.
52. Ibid., 26.
53. Laveille, *The Life of Father De Smet*, 237.
54. Lloyd E. McCann, "The Grattan Massacre," Reprinted from Nebraska History XXXVII, no. 1 (March 1956): 6-20.
55. Edward Lazarus, *Black Hills/White Justice: The Sioux Nation Versus the United States, 1775 to the Present* (New York: HarperCollins Publishers, 1991), 21.
56. McCann, *The Grattan Massacre*, 6-20.
57. Lazarus, *Black Hills/White Justice*, 23-24.

Chapter Three

1. U.S. Department of the Interior, National Park Service, "The Great Reconnaissance," National Park Service, http://www.nps.gov/archive/jeff/LewisClark2/Circa1804/WestwardExpansion/EarlyExplorers/OtherExplorers.htm (accessed March 23, 2009).
2. Bruce Brown, "Winter Count of Crazy Horse's Life," B.F. Communications, Inc., http://www.astonisher.com/archives/museum/crazy_horse_wintercount.html (accessed March 23, 2009); George W. Kingsbury, *History of Dakota Territory I* (Chicago: S. J. Clarke Publishing Company, 1915), 862.
3. Edward Lazarus, *Black Hills/White Justice*, 73.
4. Lieutenant Gouverneur Kemble Warren, *Preliminary Report of Explorations in Nebraska and Dakota in the Years 1855-'56-'57*, Engineer Department, U.S. Army, (Washington, DC: GPO, 1875), 17-22.
5. Ibid., 30.
6. Ibid., 46.
7. A. A. Humphries to Hon. Wm., W. Belknap, Secretary of War, *Preliminary Report of Explorations in Nebraska and Dakota in the Years 1855-'56-'57*, Engineer Department, United States Army, (Washington, DC: GPO, August 20, 1875).
8. Rootsweb, "Report of Captain W. F. Raynolds' Expedition to Explore the Headwaters of the Missouri & Yellowstone Rivers," American Serial Set, 40th Congress, 2nd Session, Senate Executive Document 77, Ancestry.

9. com, http://freepages.history.rootsweb.ancestry.com/~familyinformation/fpk/raynolds_rpt.html (accessed March 23, 2009).
9. Ibid.
10. Fielder, *Lost Gold*, 73.
11. Watson Parker, *Gold in the Black Hills*, 19-20.
12. *Wikipedia Encyclopedia*, Dakota Territory, http://en.wikipedia.org/wiki/Dakota_Territory (accessed March 20, 2009).
13. Wyoming Commentary, "Wyoming Atlas of Historical County Boundaries," The Newberry Library, http://historical-county.newberry.org/website/Wyoming/documents/WY_Commentary.htm (accessed March 23, 2009).
14. U.S. Congress, *Homestead Act of 1862*, 37th Cong., 2nd sess., Public Acts of the Thirty-seventh Congress, ch. 75 (May 20, 1862): 392-93, The Library of Congress, American Memory, http://memory.loc.gov/cgi-bin/ampage?collId=llsl&fileName=012/llsl012.db&recNum=423 (accessed March 24, 2009).
15. U.S. Congress, *Pacific Railway Act of 1862*, 37th Cong., 2nd sess., Public Acts of the Thirty-seventh Congress, ch. 75 (July 1, 1862): 489-99, The Library of Congress, American Memory, http://memory.loc.gov/cgi-bin/ampage (accessed March 24, 2009).
16. Gail and Michael Evans-Hatch, *Place of Passages*, 40.
17. Parker, *Gold in the Black Hills*, 20-21; Kingsbury, *History of Dakota Territory I*, 864-66.
18. Lazarus, *Black Hills/White Justice*, 38-39.
19. Ibid., 43-44.
20. Ibid., 45.
21. Wyoming Commentary, "Wyoming Atlas of Historical County Boundaries."
22. Lazarus, *Black Hills/White Justice*, 45-46.
23. Ibid., 46-47.
24. The Avalon Project at Yale Law School, "Fort Laramie Treaty, 1868," Lillian Goldman Law Library, http://avalon.law.yale.edu/19th_century/nt001.asp (accessed March 23, 2009); Kingsbury, *History of Dakota Territory I*, 862.
25. Ibid.
26. Lazarus, *Black Hills/White Justice*, 49-50.
27. The Avalon Project at Yale Law School, "Fort Laramie Treaty, 1868."
28. Lazarus, *Black Hills/White Justice*, 50-52.
29. Ibid., 50-56.
30. Martin Luschei, *The Black Hills & the Indians*, 40-41.
31. *New York Times*, June 8, 1870.
32. Luschei, *The Black Hills & the Indians*, 42.
33. Ibid., 42-43.
34. Ibid., 43-44.

35. Ibid.
36. Lazarus, *Black Hills/White Justice*, 64.
37. Parker, *Gold in the Black Hills*, 22-23.
38. Luschei, *The Black Hills & the Indians*, 46.
39. Lazarus, *Black Hills/White Justice*, 67-68.
40. Evans-Hatch, *Place of Passages*, 48-49.
41. Luschei, *The Black Hills & the Indians*, 46.
42. Evans-Hatch, *Place of Passages*, 49-50.
43. Field Division of Education, "The History of Scottsbluff, Nebraska: Last Stand of the Indians," National Park Service, http://www.nps.gov/history/history/online_books/berkeley/brand1/brand1n.htm (accessed March 23, 2009).
44. John G. Neihardt, *Black Elk Speaks*, 79-80.
45. U.S. Department of the Interior, Bureau of Land Management, "Introduction: 1872 Mining Law," Energy and Minerals Technical Assistance Program, http://www.blm.gov/iemtap/mineintro.html (accessed March 25, 2009).
46. Parker, *Gold in the Black Hills*, 23.
47. Luschei, *The Black Hills & the Indians*, 50-51.
48. Making of America Books, "Report of a Reconnaissance of the Black Hills of Dakota, made in the Summer of 1874," by William Ludlow, (U.S. Government Printing Office, 1875): 5-8, Andrew W. Mellon Foundation, http://quod.lib.umich.edu/cgi/t/text/text-idx?c=moa;cc=moa;rgn=main;view=text;idno=AFK4434.0001.001 (accessed March 20, 2009).
49. Ibid.
50. Brian D. Dippie, "It's Equal I have Never Seen, Custer Explores the Black Hills in 1874," *Columbia Magazine* 19, no. 2 (Summer 2005): 3.
51. Luschei, *The Black Hills & the Indians*, 51.
52. Donald R. Progulske and Frank J. Shideler, *Following Custer* (Brookings: South Dakota State University, Agricultural Experiment Station, Bulletin 674), 7-8.
53. Cleophas C. O'Harra, "Early Placer Gold Mining in the Black Hills," The Black Hills Engineer XIX, no. 4 (November 1931): 343-344.
54. Evans-Hatch, *Place of Passages*, 51-52.
55. Making of America Books, William Ludlow, "Report of a Reconnaissance of the Black Hills of Dakota, made in the Summer of 1874," 5-8.
56. Luschei, *The Black Hills & the Indians*, 56-57.
57. Ibid.
58. O'Harra, "Early Placer Gold Mining in the Black Hills," 343-44.
59. Dippie, "It's Equal I have Never Seen," 10.
60. Ernest Grafe and Paul Horsted, *Exploring with Custer: The 1874 Black Hills Expedition* (Custer: Golden Valley Press, 2005), 96-157.
61. Dippie, "It's Equal I have Never Seen," 11-12.

62. John D. McDermott, *Gold Rush: The Black Hills Story* (Pierre: South Dakota State Historical Society Press, 2001), 12.
63. Luschei, *The Black Hills & the Indians*, 65-66.
64. *Yankton Press and Dakotan*, "Yankton Got a Brief Dose of Gold Fever—From a Distance," September 15, 1999, http://yankton.net/articles/1999/09/15/business/19990915-archive10.txt (accessed March 23, 2009).

Chapter Four

1. Watson Parker, *Gold in the Black Hills*, 22-23.
2. Ibid., 28-29.
3. Arthur Weston Goodspeed, *The Province and the States*, 271-80.
4. Annie D. Tallent, *The Black Hills, or Last Hunting Grounds of the Dakotahs* (Sioux Falls: Brevet Press, 1974), 16-43.
5. Ibid., 62-69.
6. Martin Luschei, *The Black Hills & the Indians*, 47-49.
7. *New York Times*, "A Topographer and Astronomer to be Added to the Expedition," April 9, 1875.
8. Edward Lazarus, *Black Hills/White Justice*, 77-78.
9. Luschei, *The Black Hills & the Indians*, 70-71.
10. Ibid., 76-77.
11. *New York Times*, June 24, 1875.
12. Tallent, *The Black Hills, or Last Hunting Grounds*, 93.
13. Richard Irving Dodge, *The Black Hills* (Minneapolis: Ross & Haines, Inc.: Minneapolis, 1965) 70-75.
14. Parker, *Gold in the Black Hills*, 27.
15. Luschei, *The Black Hills & the Indians*, 78.
16. Dodge, *The Black Hills*, 104-10.
17. Ibid., 110-12.
18. Ibid., 106-109.
19. Ibid., 137-38.
20. *Deadwood Daily Pioneer-Times*, February 28, 1909.
21. Ibid.
22. Tallent, *The Black Hills, or Last Hunting Grounds of the Dakotahs*, 98-100.
23. *New York Times*, August 11, 1875.
24. Linfred Schuttler, "Bryant Party and Frank S. Bryant Family," *Some History of Lawrence County* (Pierre: The State Publishing Company, 1981), 57-59.
25. Tallent, *The Black Hills, or Last Hunting Grounds*, 124-25.
26. Schuttler, "Bryant Party and Frank S. Bryant Family," 56-57.
27. Ibid.
28. Tallent, *The Black Hills, or Last Hunting Grounds*, 125.

29. Schuttler, "Bryant Party and Frank S. Bryant Family," 57.
30. John S. McClintock, *Pioneer Days in the Black Hills* (Norman: University of Oklahoma Press, 2000), 32-33.
31. Ibid.
32. Ibid.
33. Tallent, *The Black Hills, or Last Hunting Grounds*, 128-29.
34. Cleophas C. O'Harra, "Early Placer Gold Mining in the Black Hills," 356-57.
35. T. A. Rickard, *The Stamp Milling of Gold Ores* (New York: The Scientific Publishing Company, 1897), 75.
36. O'Harra, "Early Placer Gold Mining in the Black Hills," 356-57.
37. Bob Lee, ed., *Gold, Gals, Guns, Guts* (Deadwood: Deadwood-Lead '76 Centennial, Inc., 1976) 16.
38. Mildred Fielder, *The Treasure of Homestake Gold* (Aberdeen: North Plains Press, 1970), 18.
39. Homestake Mining Company, "Pre-Patent Title Summary for the Old Abe, M. S. No. 117."
40. Fielder, *The Treasure of Homestake Gold*, 18.
41. Don C. Clowser, *Deadwood . . . The Historic City*, (Deadwood: Fenwyn Press Books, 1969), 7-8.
42. Tallent, *The Black Hills, or Last Hunting Grounds*, 134-36.
43. Christopher Hills, *Gold Pans and Broken Picks*, 19-20.
44. Ibid., 43-44.
45. P. H. Kellar, *Seth Bullock's The Founding of a County* (Deadwood: Dakota Graphics, 1986), 21.
46. Ibid., 21-22.
47. Parker, *Gold in the Black Hills*, 71-72.
48. Hills, *Gold Pans and Broken Picks*, 23-24.

Chapter Five

1. The Dakota Experience, "Memorial of the Legislature of Dakota Praying that the Black Hills of Dakota be opened for settlement, and the Indian title to the same be extinguished," House of Representatives, Miscellaneous Document No. 33, 43rd Cong., 2nd sess., 1875, South Dakota State Historical Society and Educational Web Adventures, http://dakotaexperience.org/cvfrontier/coll_whitesettlement_text.html (accessed March 23, 2009).
2. *New York Times*, "The Black Hills—Gold Hunting Excursions to be Prevented—The Sioux Treaty to be Strictly Observed," March 13, 1875, http://query.nytimes.com/mem/archive-free/pdf?_r=2&res=9A0CEED9153BEF34BC4B52DFB566838E669FDE&oref=slogin&oref=slogin (accessed March 23, 2009).

3. *New York Times*, "The Black Hills Country, Communications from the President to the Senate," March 18, 1875, http://query.nytimes.com/gst/abstract.html?res=9B04EED8153BEF34BC4052DFB566838E669FDE (accessed March 23, 2009).
4. Ibid.
5. *New York Times*, "The Black Hills, Letter from Lieut. Gen. Sheridan to Gen. Sherman," March 27, 1875, http://query.nytimes.com/mem/archive-free/pdf?res=9F01E1DC153CE63ABC4F51DFB566838E669FDE (accessed March 23, 2009).
6. Ibid.
7. Annie D. Tallent, *The Black Hills, or Last Hunting Grounds*, 95.
8. Martin Luschei, *The Black Hills & the Indians*, 80-83.
9. Edward Lazarus, *Black Hills/White Justice*, 82.
10. Luschei, *The Black Hills & the Indians*, 82-83.
11. Native American Documents Project, *Annual Report of the Commissioner of Indian Affairs to the Secretary for the Year 1875* (Washington, DC: Government Printing Office, 1875), 3-21, NADP Document R875001A, California State University, San Marcos, http://www.csusm.edu/nadp/r875001a.htm (accessed March 23, 2009).
12. Ibid.
13. Lazarus, *Black Hills/White Justice*, 83.
14. Ibid., 84.
15. Mario Gonzalez and Elizabeth Cook Lynn, "The Politics of Hallowed Ground: Wounded Knee and the Struggle for Indian Sovereignty, Appendix H," (Chicago: Univ. of Illinois Press, 1999). 330-357, http://www.geocities.com/lakotastudentalliance/hesapachronologygonzalez.html (accessed March 23, 2009).
16. Lazarus, *Black Hills/White Justice*, 84-85.
17. John G. Neihardt, *Black Elk Speaks*, 90.
18. John T. Woolley and Gerhard Peters, "Ulysses S. Grant: Seventh Annual Message to the Senate and House of Representatives," December 7, 1875, The American Presidency Project, http://www.presidency.ucsb.edu/ws/print.php?pid=29516 (accessed March 23, 2009).
19. John D. McDermott, *Gold Rush: The Black Hills Story*, 12-13.
20. Mary Trotter Kion, "Battle of Little Big Horn: Subdue the Sioux," Suite101.com, http://www.suite101.com/lesson.cfm/17638/1151/6 (accessed March 23, 2009).
21. Ibid.
22. Stanley Vestal, *Sitting Bull: Champion of the Sioux: A Biography* (Norman: University of Oklahoma Press, 1932), 140-41.
23. Ibid.
24. Luschei, *The Black Hills & the Indians*, 88.

25. Ibid.
26. Vestal, *Sitting Bull: Champion of the Sioux: A Biography*, 143-46.
27. Ibid., 148-51.
28. William Glenn Robertson et al., "Atlas of the Sioux Wars," United States Army Combined Arms Center, Command and General Staff College, Fort Leavenworth, Kansas, http://cgsc.leavenworth.army.mil/carl/resources/csi/sioux/sioux.asp#Sheridan's%20Campaign%20Plan (accessed March 23, 2009).
29. Ibid.
30. Vestal, *Sitting Bull: Champion of the Sioux: A Biography*, 154-55.
31. Kion, "Battle of Little Big Horn."
32. Lazarus, *Black Hills/White Justice*, 88-89.
33. Vestal, *Sitting Bull: Champion of the Sioux: A Biography*, 163-64.
34. Ibid., 167-76.
35. Lazarus, *Black Hills/White Justice*, 89.
36. Vestal, *Sitting Bull: Champion of the Sioux: A Biography*, 183-84.
37. Fred H. Werner, *The Slim Buttes Battle*, September 9-10, 1876 (Self-Printed, 1981), 11-18.
38. Ibid., 19-24.
39. Peter Rosen, *Pa-Ha-Sa-Pah: Or the Black Hills*, 409-410.
40. *Black Hills Pioneer*, September 23, 1876.
41. Watson Parker, *Gold in the Black Hills*, 136-37.
42. Watson Parker, *Deadwood: The Golden Years* (Lincoln: University of Nebraska Press, 1981), 52-53; Captain Charles King with an introduction by Don Russell, *Campaigning with Crook* (Norman: University of Oklahoma Press, 1983), 157-58.
43. Ibid., 55.
44. Lazarus, *Black Hills/White Justice*, 89-91.
45. Luschei, *The Black Hills & the Indians*, 89-90.
46. Lazarus, *Black Hills/White Justice*, 91-93.
47. Jerome A. Greene, *Lakota and Cheyenne* (Norman: University of Oklahoma Press, 1994), 97.
48. Lazarus, *Black Hills/White Justice*, 93.
49. Robertson et al., "Atlas of the Sioux Wars."
50. Luschei, *The Black Hills & the Indians*, 92.
51. Ibid., 93-94.
52. Lazarus, *Black Hills/White Justice*, 93-94.
53. Luschei, *The Black Hills & the Indians*, 94-97; Thomas R. Buecker, "Can You Send Us Immediate Relief?: Army Expeditions to the Northern Black Hills, 1876-1878," South Dakota History 25, no. 2 (Summer 1995): 101-113.
54. U.S. Department of the Interior, Bureau of Land Management, "Fort Meade Recreation Area," Bureau of Land Management, http://www.blm.gov/mt/st/en/fo/south_dakota_field/ft_meade.html (accessed March 23, 2009).

55. Lazarus, *Black Hills/White Justice*, 96-97.
56. Luschei, *The Black Hills & the Indians*, 103-104.
57. Weider History Network, "Brulé Sioux Chief Spotted Tail," Weider History Group, http://www.historynet.com/culture/native_american_history/3037121.html?page=2&c=y (accessed March 23, 2009).

Chapter Six

1. Moses Manuel, "Forty-Eight Years in the West," dictated to Mary Sheriff, Typescript, Helena, Montana, 1903, Black Hills Mining Museum, Lead, SD.
2. Homestake Mining Company, "Pre-Patent Title Summary for the So. Segregated Golden Terry, M. S. No. 130."
3. H. C. Harney Affidavit before Moses Liverman, Notary Public, in *Old Abe Mining Company vs. James M. Young et al.*, County of Lawrence, Territory of Dakota, August 29, 1879, Black Hills Mining Museum, Lead, SD.
4. Manuel, "Forty-Eight Years in the West;" Bruce C. Yates, "The Homestake Mine, A Historical Sketch," The Black Hills Engineer XIV, no. 3, (May 1926), 131-133.
5. Fielder, *The Treasure of Homestake Gold*, 26-27.
6. Don C. Clowser, *Deadwood . . . The Historic City*, 7-8; Homestake, "Pre-Patent Title Summary for the Old Abe, M. S. No. 117."
7. Homestake Mining Company, "Pre-Patent Title Summary for the Homestake claim, M. S. No. 121."
8. Samuel McMaster Deposition in *Old Abe Mining Company vs. William H. Bull, James Carney, Hugh McKenna, and George Adams*, County of Lawrence, Territory of Dakota, April 25, 1879, Black Hills Mining Museum, Lead, SD; Homestake, "Pre-Patent Title Summary for the Old Abe, M. S. No. 117."
9. John Schofield Affidavit before Henry Hill, Notary Public, in *Old Abe Mining Company vs. James M. Young et al.*, County of Lawrence, Territory of Dakota, July 3, 1878, Black Hills Mining Museum, Lead, SD; A. G. McShane Affidavit before Moses Liverman, Notary Public, in *Old Abe Mining Company vs. James M. Young et al.*, County of Lawrence, Territory of Dakota, June 26, 1878, Black Hills Mining Museum, Lead, SD.
10. H. C. Harney Affidavit.
11. Manuel, "Forty-Eight Years in the West."
12. Harney Affidavit.
13. *Pioneer News*, June 21, 1876.
14. Manuel, "Forty-Eight Years in the West."
15. Parker, *Gold in the Black Hills*, 104; McClintock, *Pioneer Days in the Black Hills*, 149-57.
16. *Pioneer News*, June 21, 1876.

17. Fred Manuel Affidavit before S. P. Ramicus, Notary Public, in *Old Abe Mining Company vs. James M. Young et al.*, County of Lawrence, Territory of Dakota, July 23, 1879, Black Hills Mining Museum, Lead, SD.
18. John Schofield Affidavit.
19. H. C. Harney and Alex Engh to H. B. Young (June 16, 1876), Elizabeth City, Dakota Territory, deed filed for record with the Lawrence County Register of Deeds, Territory of Dakota, August 13, 1877 before J. A. Hairel, Black Hills Mining Museum, Lead, SD.
20. *Black Hills Pioneer*, August 12, 1876.
21. McShane Affidavit.
22. Fred Manuel Affidavit.
23. *Black Hills Pioneer*, September 30, 1876.
24. Ibid., October 21, 1876.
25. Tallent, *The Black Hills, or Last Hunting Grounds*, 208.
26. *Black Hills Daily Pioneer*, December 2, 1876.
27. Tallent, *The Black Hills, or Last Hunting Grounds*, 208.
28. *Black Hills Daily Times*, May 18, 1878.
29. Willard Larson, *Early Mills in the Black Hills* (Isanti, Minnesota: privately printed, 2001), 1.
30. Fred Manuel Affidavit.
31. Thomas Jones Affidavit before Moses Liverman, Notary Public, *Old Abe Mining Company vs. James M. Young et al.*, County of Lawrence, Territory of Dakota, August 2, 1879, Black Hills Mining Museum, Lead, SD.
32. Fred Manuel Affidavit.
33. *Black Hills Pioneer*, March 10, 1877.
34. Edward Lazarus, *Black Hills/White Justice*, 91-93.
35. Homestake Mining Company, "Pre-patent Title Summary for the Homestake claim, M. S. 121;" Homestake Mining Company, "Pre-Patent Title Summary for the Old Abe claim, M. S. 117."
36. *Black Hills Pioneer*, April 28, 1877.
37. Fred Manuel Affidavit.
38. Thomas Jones Affidavit.
39. Homestake, "Pre-patent Title Summary for the Old Abe claim, M. S. 117."
40. Joel Waterland, *The Spawn & the Mother Lode* (Rapid City: Grelind Photographics & Typesetters, 1987), 125-26.
41. U.S. Department of the Interior, Bureau of Land Management, Survey No. 230, Plat of the claim of J. B. Haggin upon the Segregated Northerly 150 feet of Golden Star Lode, May 21, 1880.
42. Fielder, *The Treasure of Homestake Gold*, 31.
43. U.S. Department of the Interior, Bureau of Land Management, Amended Survey No. 132, Plat of the claim of Samuel McMaster upon the Lincoln Lode, November 28, 1882.

44. L. H. Mallory and Dan Latham, Location Notice for the Homestake No. 2 Claim, October 25, 1876, Black Hills Mining Museum, Lead, SD.
45. Waterland, *The Spawn & the Mother Lode*, 126-27.
46. Fred Manuel Affidavit; Alexander Engh Affidavit before S. P. Ramicus, Notary Public, in *Old Abe Mining Company vs. James M. Young et al.*, County of Lawrence, Dakota Territory, July 23, 1879, Black Hills Mining Museum, Lead, SD.
47. William Filan Affidavit before James F. Nahai, Notary Public, in *Old Abe Mining Company vs. James M. Young et al.*, County of Lawrence, Dakota Territory, June 29, 1878, Black Hills Mining Museum, Lead, SD; Homestake, "Pre-patent Title Summary for the Old Abe claim, M. S. 117."
48. *Black Hills Daily Times*, May 18, 1878.
49. Ibid., July 7, 1877.
50. Ibid., August 11, 1877.
51. Ibid., July 21 and September 27, 1877.
52. Ibid., November 16, 1877.
53. Ibid., May 18, 1878.
54. *Black Hills Pioneer*, July 29, 1876.
55. *Black Hills Daily Times*, August 11, 1877.
56. *Black Hills Pioneer*, August 11, 1877.
57. Yates, "The Homestake Mine, A Historical Sketch," 133.

Chapter Seven

1. George Hearst, memoir (1890), Black Hills Mining Museum, Lead, SD.
2. Ibid.
3. Joseph H. Cash, *Working the Homestake* (Ames: The Iowa State University Press, 1973), 17.
4. Homestake, "Pre-Patent Title Summary for the Homestake claim, M. S. No. 121."
5. Waterland, *The Spawn & the Mother Lode*, 125-26.
6. *Black Hills Daily Times*, August 11, 1877.
7. Ibid., October 2, 1877.
8. Homestake, "Pre-Patent Title Summary for the Homestake claim, M. S. No. 121."
9. Yates, "The Homestake Mine, A Historical Sketch," 133-34.
10. Homestake, "Pre-Patent Title Summary for the Homestake claim, M. S. No. 121."
11. Hearst memoir.
12. Cash, *Working the Homestake*, 17.
13. Hearst memoir.

14. William Bronson and T. H. Watkins, *Homestake: The Centennial History of America's Greatest Gold Mine* (San Francisco: Homestake Mining Company, 1977), 30.
15. Homestake, "Pre-Patent Title Summary for the So. Segregated Golden Terry, M. S. No. 130."
16. Rickard, *The Stamp Milling of Gold Ores*, 75.
17. Homestake, "Pre-Patent Title Summary for the Homestake claim, M. S. No. 121."
18. Hearst memoir.
19. Homestake, "Pre-Patent Title Summary for the Homestake claim, M. S. No. 121."
20. William A. Farish to W. W. Farlow, as communicated by C. S. T. Farish to Donald H. McLaughlin, September 12, 1949, Black Hills Mining Museum, Lead, SD.
21. George Hearst to J. B. Haggin and Lloyd Tevis, November 1, 1877, Homestake Mining Company.
22. Homestake Mining Company, "Articles of Incorporation," Thomas H. Reynolds, Clerk, Office of the County Clerk of the City and County of San Francisco, California, November 5, 1877.
23. Fielder, *The Treasure of Homestake Gold*, 70.
24. Ibid., 48-50.
25. Homestake, "Pre-Patent Title Summary for the Homestake claim, M. S. No. 121."
26. *Black Hills Daily Times*, December 17, 18, and 20, 1877.
27. Fielder, *The Treasure of Homestake Gold*, 43.
28. Homestake, "Pre-Patent Title Summary for the Homestake claim, M. S. No. 121."
29. Agnes Wright Spring, *The Cheyenne and Black Hills Stage and Express Routes* (Lincoln: The University of Nebraska Press, 1948), 83.
30. Fielder, *The Treasure of Homestake Gold*, 50-51.
31. T. J. Grier, personal notes, Black Hills Mining Museum, Lead, SD.
32. Fielder, *The Treasure of Homestake Gold*, 50-51.
33. Homestake Mining Company, Minutes of Board of Directors Meeting, March 5, 1878, D. F. Verdenal, Secretary.
34. George Hearst to J. B. Haggin, May 22, 1878, Homestake Mining Company.
35. Ibid.
36. Ibid., May 23, 1878.
37. Ibid.
38. Fielder, *The Treasure of Homestake Gold*, 58-60.
39. Homestake, "Pre-Patent Title Summary for the Homestake claim, M. S. No. 121."

40. Grier, personal notes.
41. Lounsbery and Haggin, "Report of the Homestake Mining Company from January 1, 1878 to September 1, 1880," Homestake Mining Company.
42. Ibid.
43. Homestake Veterans Association, "Recollections of the Early Days of the Homestake Mining Company," Black Hills Mining Museum, Lead, SD.
44. Lounsbery and Haggin, "Report of the Homestake Mining Company from January 1, 1878 to September 1, 1880."
45. Ibid.
46. Fielder, *The Treasure of Homestake Gold*, 114-37.
47. Grier, personal notes.
48. T. J. Grier, "Superintendent's Report," Homestake Mining Company, June 1, 1891.
49. Application for Patent for the Old Abe Lode Mining Claim, Lot No. 117, United States Land Office, Deadwood, Dakota Territory, July 27, 1878; Homestake, "Pre-Patent Title Summary for the Old Abe claim, M. S. No. 117."
50. Old Abe Mining Company, "Articles of Association," Thomas D. Meade, Clerk, Circuit Court of Keoughton County, Michigan, February 2, 1878.
51. Dick Ross and June Ross, *Portage Lake Mining Gazette*, Houghton, Michigan, January 17, 1878, Dick and June Ross, http://www.mfhn.com/houghton/rosscoll/news/mg1878_1.txt (accessed March 23, 2009).
52. John H. Kehoe Affidavit before Edward Allsmon, Notary Public, in *Old Abe Mining Company vs. James M. Young et al.*, County of Lawrence, Territory of Dakota, July 8, 1878, Black Hills Mining Museum, Lead, SD.
53. Michael Lynch Affidavit before James F. Nahai, Notary Public, in *Old Abe Mining Company vs. James M. Young et al.*, County of Lawrence, Territory of Dakota, June 29, 1878, Black Hills Mining Museum, Lead, SD.
54. James M. Young Affidavit before James F. Nahai, Notary Public, in *Old Abe Mining Company vs. James M. Young et al.*, County of Lawrence, Territory of Dakota, June 21, 1878, Black Hills Mining Museum, Lead, SD.
55. Application for Patent for the Old Abe Lode Mining Claim, Lot No. 117.
56. Homestake, "Pre-Patent Title Summary for the Old Abe claim, M. S. No. 117."
57. Hearst to Haggin, March 6, 1879.
58. Homestake, "Pre-Patent Title Summary for the Old Abe claim, M. S. No. 117."
59. Giant and Old Abe Mining Company, "Articles of Incorporation," Thomas H. Reynolds, Clerk, Office of the County Clerk of the City and County of San Francisco, California, September 23, 1878.
60. Fraser and Chalmers, "Price Quotation to the Giant & Old Abe Mining Company for Hoisting and Pumping Machinery," December 11, 1879, Black Hills Mining Museum, Lead, SD.

61. Giant & Old Abe Mining Company, "Shaft Sinking Bids," January 8, 1879, Black Hills Mining Museum, Lead, SD.
62. Waterland, *The Spawn & the Mother Lode*, 150.
63. Ibid., 158.
64. Homestake, "Pre-Patent Title Summary for the Old Abe claim, M. S. No. 117."
65. Waterland, *The Spawn & the Mother Lode*, 126-27.
66. Grier, personal notes.
67. Waterland, *The Spawn & the Mother Lode*, 126-27.
68. Hearst to Haggin, March 4, 1878.
69. Ibid., May 19, 1878.
70. Highland Mining Company, "Articles of Incorporation," Thomas H. Reynolds, Clerk, Office of the County Clerk of the City and County of San Francisco, California, September 30, 1878.
71. Waterland, *The Spawn & the Mother Lode*, 126, 146.
72. Highland Mining Company, "Bids for the Highland Shaft," November 19, 1878, Black Hills Mining Museum, Lead, SD.
73. Hearst to Haggin, March 9, 1879.
74. Fielder, *The Treasure of Homestake Gold*, 79.
75. Hearst to Haggin, March 23, 1879.
76. Grier, personal notes.
77. Lawrence County Register of Deeds, Deadwood, South Dakota, 279:508-12.
78. Grier, personal notes.
79. Waterland, *The Spawn & the Mother Lode*, 146-48.
80. Homestake Mining Company, "Pre-Patent Title Summary for the No. Segregated Golden Terry, M. S. No. 149."
81. Ibid.
82. U.S. Department of the Interior, Bureau of Land Management, Plat of the Claim of Deadwood Mining Company, Survey No. 149, J. D. McIntyre, U.S. Deputy Surveyor, February 19-26, 1879.
83. Homestake, "Pre-Patent Title Summary for the No. Segregated Golden Terry, M. S. No. 149."
84. Hearst to Haggin, May 19, 1878.
85. Homestake, "Pre-Patent Title Summary for the No. Segregated Golden Terry, M. S. No. 149."
86. Thomas G. Ingham, *Digging Gold Among the Rockies* (Philadelphia: Hubbard Brothers Publishers, 1888), 192.
87. Hearst to Haggin, March 9, 1879.
88. Waterland, *The Spawn & the Mother Lode*, 133.
89. Ibid., 128.
90. Willard Larson, *Early Mills in the Black Hills*, 3.

91. George Hearst to Haggin and Tevis, November 1, 1877.
92. U.S. Department of the Interior, Bureau of Land Management, Plat of the Claim of The Golden Terra Mining Company, Golden Terry Lode, Survey No. 130, J. D. McIntyre, U.S. deputy surveyor, September 2, 1878; Homestake, "Pre-Patent Title Summary for the So. Segregated Golden Terry, M. S. No. 130."
93. Yates, "The Homestake Mine, A Historical Sketch," 139.
94. Hearst to Haggin, May 19, 1878.
95. Ibid., March 6, 1879.
96. Ibid., March 9, 1879.
97. Waterland, *The Spawn & the Mother Lode*, 128.
98. Hearst to Haggin, August 28, 1879.
99. Grier, personal notes.
100. Waterland, *The Spawn & the Mother Lode*, 145-46.
101. *Black Hills Daily-Times*, December 7, 1879.
102. Deadwood-Terra Mining Company, "Articles of Incorporation," Secretary of State, New York, August 23, 1880.
103. Yates, "The Homestake Mine, A Historical Sketch," 139.
104. Waterland, *The Spawn & the Mother Lode*, 139.
105. Donald D. Toms, *In the Midst of Life: Mining Company Fatalities, Black Hills, South Dakota, 1876-1995*, I (1996), 76-77.
106. Grier, personal notes.
107. Mildred Fielder, *A Guide to Black Hills Ghost Mines* (Aberdeen: North Plains Press, January 1974), 47.
108. Ingham, *Digging Gold Among the Rockies*, 194.
109. Waterland, *The Spawn & the Mother Lode*, 128.
110. *Black Hills Pioneer*, October 21, 1876.
111. Ibid., January 27, 1877.
112. *Black Hills Daily Times*, May 18, 1878.
113. Waterland, *The Spawn & the Mother Lode*, 128.
114. Lawrence County Register of Deeds, Deadwood, South Dakota, 139:163.
115. Ibid., 166.
116. Watson Parker, *Gold in the Black Hills*, 195.
117. Lawrence County Register of Deeds, Deadwood, South Dakota, 139:169.
118. Ibid., 171.
119. Waterland, *The Spawn & the Mother Lode*, 129-30.
120. Richmond L. Clow, *Chasing the Glitter* (Pierre: South Dakota State Historical Society Press, 2002), 36-37.
121. Waterland, *The Spawn & the Mother Lode*, 129-30.
122. Wyoming and Dakota Water Company, "Articles of Incorporation," City and County of San Francisco, California, January 25, 1878;

Wyoming and Dakota Ditch, Flume, and Mining Company, "Articles of Incorporation," County of Lawrence, Territory of Dakota, June 23, 1877.
123. DeSmet, Bowie, and Palmer Ditch Water Right, Homestake Mining Company Water Rights files.
124. Waterland, *The Spawn & the Mother Lode*, 130.
125. *Samuel McMaster vs. Wyoming and Dakota Water Company*, Judgment in District Court, Lawrence County, Dakota Territory, April 23, 1881, Homestake Mining Company Water Rights files.
126. Yates, "The Homestake Mine, A Historical Sketch," 139.
127. Lawrence County Register of Deeds, Deadwood, South Dakota, 1:46-47.
128. *Borland vs. Haven et al.*, F. 37 (Circuit Court, N. D., California, December 17, 1888): 398.
129. Waterland, *The Spawn & the Mother Lode*, 130.
130. Fielder, *The Treasure of Homestake Gold*, 110.
131. Waterland, *The Spawn & the Mother Lode*, 130.
132. *Black Hills Daily Times*, January 21, 1880.
133. Fielder, *The Treasure of Homestake Gold*, 86-87.
134. Waterland, *The Spawn & the Mother Lode*, 137-38.
135. Fielder, *The Treasure of Homestake Gold*, 18, 73.
136. Caledonia Gold Mining Company, "Articles of Incorporation," City and County of San Francisco, California (September 5, 1878), 29:288.
137. Louis Janin to G. W. Gashwiler, April 18, 1879, The Newberry Library, Chicago, Illinois, Call Number: Graff 4936.
138. Fielder, *The Treasure of Homestake Gold*, 120.
139. *New York Times*, July 9, 1881.
140. Toms, *In the Midst of Life*, 89-91.
141. Ibid., 172-176.
142. Waterland, *The Spawn & the Mother Lode*, 138, 141-45.
143. Ibid., 137-39.
144. Ibid., 141-45.
145. Lawrence County Register of Deeds, 146:135.
146. Ibid., 154:22.
147. Ibid., 154:509.
148. Ibid., 141:163
149. Ibid., 166:378-84.
150. Deadwood-Terra Mining Company, "Certificate as to Dissolution," county of New York, State of New York, April 25, 1902.
151. Lawrence County Register of Deeds, 166:384.

Chapter Eight

1. *Joseph. E. Cook et al. vs. Robert H. Evans et al.*, SD 8d (January 30, 1918): 12:617.

2. Joel Waterland, *Gold, Silver, Sweat & Tears* (Rapid City: Grelind Photographics & Typesetters, 1988), 211.
3. Black Hills Canal and Water Company, "Renewal Certificate of Location of Water Right," Lawrence County Register of Deeds, Deadwood, South Dakota (November 13, 1899): 146:177.
4. Wyoming and Dakota, Ditch, Flume, and Mining Company, "Articles of Incorporation."
5. Coll Deane and the Wyoming and Dakota, Ditch, Flume, and Mining Company, Agreement dated November 15, 1877, Homestake Mining Company Water Rights files.
6. *Black Hills Daily Pioneer*, December 30, 1877.
7. Wyoming and Dakota Water Company, "Articles of Incorporation."
8. Watson Parker, *Deadwood: The Golden Years*, 196-97.
9. Annie D. Tallent, *The Black Hills, or Last Hunting Grounds*, 356.
10. Wyoming and Dakota Water Company, "Articles of Incorporation."
11. Wyoming and Dakota Water Company, Minutes of the Board of Directors, December 6, 1878, Homestake Mining Company Water Rights files.
12. Homestake Mining Company, abstract notes for DeSmet and Bowie Water Rights.
13. Homestake Mining Company, abstract notes for Foster Ditch Water Right.
14. John H. Davey et al., Quit Claim Deed to the Wyoming and Dakota Water Company, February 14, 1880, Homestake Mining Company Water Rights files.
15. Black Hills Canal and Water Company, "Articles of Incorporation," City and County of San Francisco, California, September 30, 1878.
16. Ibid.
17. Black Hills Canal and Water Company, Minutes of the Board of Directors, October 14, 1878, Homestake Mining Company Water Rights files.
18. Hearst to Haggin, March 6, 1879.
19. *Black Hills Daily Times*, April 10, 1879.
20. *George Atchison et al, vs. George Hearst et al.*, "Findings of Fact," Judge G. C. Moody, District Court of the First Judicial District, Lawrence County, Dakota Territory (July 23, 1880), Homestake Mining Company Water Rights files.
21. Homestake Mining Company, "Ditch Histories and Proof of use of Water," Homestake Mining Company Water Rights files.
22. Homestake Mining Company, "Notes regarding a conversation with Sam Blackstone," August 8, 1936, Homestake Mining Company Water Rights files.
23. Homestake, "Ditch Histories."

24. Homestake Mining Company, abstract notes for Pioneer Ditch and Water Right.
25. Moody, "Findings of Fact," *George Atchison et al. vs. George Hearst et al.*
26. Homestake Mining Company, abstract notes for French Boys Ditch and Water Right.
27. Moody, "Findings of Fact," *George Atchison et al. vs. George Hearst et al.*
28. Homestake, "Notes regarding a conversation with Sam Blackstone."
29. Homestake Mining Company, abstract notes for the Montana Ditch and Water Right.
30. Homestake, abstract notes for the Foster Ditch and Water Right.
31. McClintock, *Pioneer Days in the Black Hills*, 231-32.
32. Moody, "Findings of Fact," *George Atchison et al. vs. George Hearst et al.*
33. Homestake, "Notes regarding a conversation with Sam Blackstone."
34. Moody, "Findings of Fact," *George Atchison et al. vs. George Hearst et al.*
35. Hearst to Haggin, May 19, 1878.
36. *Black Hills Daily Times*, April 10, 1879.
37. Homestake Mining Company, abstract notes on Boulder Ditch and Water Right.
38. Hearst to Haggin, March 19, 1879.
39. Hearst to Haggin, March 29, 1879.
40. McClintock, *Pioneer Days in the Black Hills*, 231-32.
41. Stokes and Driggs, *Deadwood Gold*, 129-32.
42. *Black Hills Daily Pioneer*, May 31, 1879.
43. Ibid., June 4, 1879.
44. Ibid., June 6, 1879.
45. Ibid.
46. *Black Hills Pioneer*, June 15, 1879.
47. Ibid.
48. Stokes and Driggs, *Deadwood Gold*, 132-33.
49. *Black Hills Daily Pioneer*, June 19, 1879.
50. R. M., "Letter to the Editor," *Black Hills Daily Pioneer*, June 13, 1879.
51. *Black Hills Daily Pioneer*, June 19, 1879.
52. The Water Controversy, Homestake vs. Father DeSmet, Who Will Supply Water to Deadwood 1800s, http://deadwoodwatercontroversy.blogspot.com/ (accessed March 21, 2009).
53. *Black Hills Daily Pioneer*, June 21, 1879.
54. Ibid., June 22, 1879.
55. Ibid., June 24, 1879.
56. Tallent, *The Black Hills, or Last Hunting Grounds*, 362.
57. *Black Hills Daily Pioneer*, October 2, 1879.

58. Citizens of Deadwood, South Dakota, Petition to A. J. Bowie, agent for the Wyoming and Dakota Water Company, September 29, 1879, Homestake Mining Company Water Rights files.
59. *Black Hills Daily Times*, October 10, 1879.
60. Wyoming and Dakota Water Company, Agreement with Town of Central City, October 1, 1879, Homestake Mining Company Water Rights files.
61. Lounsbery and Haggin, "Report of the Homestake Mining Company from January 1, 1878, to September 1, 1880."
62. *Black Hills Daily Times*, August 24, 1881.
63. *George Atchison et al. vs. George Hearst et al.*, "Judgment and Decree" of Judge G. C. Moody, District Court, First Judicial District, Lawrence County, Dakota Territory (January 7, 1882), 1:482.
64. Borland, F. 37 at 399.
65. Ibid., 399-402.
66. Wyoming and Dakota Water Company, Minutes of the Board of Directors, March 2, 1881, Homestake Mining Company Water Rights files.
67. *Samuel McMaster vs. Wyoming and Dakota Water Company*, "Sheriff's Certificate of Sale No. 10579," District Court, Lawrence County, Dakota Territory, August 18, 1881, Homestake Mining Company Water Rights files.
68. Homestake, abstract notes for the Foster Ditch and Water Right.
69. Moody, "Findings of Fact," *George Atchison et al. vs. George Hearst et al.*
70. G. C. Moody to T. J. Grier, December 7, 1896, Homestake Mining Company Water Rights files.
71. Homestake Mining Company, *Sharp Bits* 5, no. 7 (August 1954).
72. Black Hills Canal and Water Company, Agreement with the city of Deadwood, August 1893, Homestake Mining Company Water Rights files.
73. Homestake Mining Company, abstract notes for the Columbia Ditch and Water Right.
74. Homestake Mining Company, abstract notes for Martin Sands Water Rights on Whitewood Creek.
75. Homestake Mining Company, abstract notes for Peake No. 1 and No. 2 Water Rights.
76. Homestake Mining Company, abstract notes for Little Rapid Ditch and Water Right.
77. Black Hills Canal and Water Company, Agreement with Homestake Mining Company, January 9, 1901, Homestake Mining Company Water Rights files.
78. Black Hills Canal and Water Company, Agreement with Homestake Mining Company, February 17, 1905, Homestake Mining Company Water Rights files.

79. Richard Blackstone to T. J. Grier, January 16, 1907, Homestake Mining Company Water Rights files.
80. *Deadwood Daily Pioneer-Times*, February 24, 1909.
81. Blackstone to Grier, January 16, 1907.
82. Ibid.
83. Homestake Mining Company, "Peake Ditch Pipeline Notes," Homestake Mining Company Water Rights files.
84. Ibid.
85. Blackstone to Grier, January 16, 1907.
86. South Dakota Water Management Board, "Meeting Minutes," (July 10, 2003), 12-16.
87. Margie Winsel Boorda, director, Corporate Land Management, Barrick Gold of North America, e-mail message to author, April 25, 2008.

Chapter Nine

1. Bruce C. Yates, "Some Features of Mining Operations in the Homestake Mine, Lead, South Dakota" (paper presented at the Black Hills Mining Men's Association, Deadwood, South Dakota, January 19, 1904), 24-26.
2. Bernard McDonald, "Mine Timbering by the Square Set System at Rossland, B. C.," *Mining and Scientific Press* LXXXV, no. 12 (September 20, 1902): 158-61.
3. Yates, "Some Features of Mining Operations," 26-27.
4. A. J. M. Ross, "Ore Extraction and Transportation," *Engineering and Mining Journal* 132, (1931): 324-29.
5. A. J. M. Ross and R. G. Wayland, "A Brief Outline of Homestake Mining Methods," *The Black Hills Engineer* XIV, no. 3 (May 1926): 145-46.
6. Clow, *Chasing the Glitter*, 43-45.
7. Ross and Wayland, "A Brief Outline of Homestake Mining Methods," 147-51.
8. Ibid., 152.
9. Grier, personal notes.
10. Lounsbery and Haggin, Financial Agents, "Report of the Homestake Mining Company from January 1, 1878, to September 1, 1880."
11. Grier, personal notes.
12. Ibid.
13. U.S. Department of the Interior, *Eleventh Census of the United States, Statistics of Mines and Mining*, 1889, Special Schedule 3A, Homestake Mining Company files.
14. *Deadwood Daily Pioneer-Times*, November 26, 1908.
15. Yates, "Some Features of Mining Operations," 15.

16. Donald D. Toms, Kristie L. Schillinger, William J. Stone, eds., *The Gold Belt Cities: The City of Mills* (Lead: Black Hills Mining, 1993), 146.
17. Donald Toms, William J. Stone, and Gretchen Motchenbacher, eds., *The Gold Belt Cities: Lead & Homestake* (Lead: G.O.L.D. Unlimited, 1988), 142.
18. Toms et al., *The Gold Belt Cities: The City of Mills*, 146.
19. *Lead Daily Call*, December 14, 1900.
20. Homestake Mining Company, "Air Locomotives of Homestake Mining Company," Sheet No. 1762-6-59, Homestake Mining Company.
21. Toms et al., *The Gold Belt Cities: The City of Mills*, 184.
22. Homestake, "Air Locomotives of Homestake Mining Company."
23. Ibid.
24. *Deadwood Daily Pioneer-Times*, November 26, 1908.
25. Homestake, "Air Locomotives of Homestake Mining Company."
26. Homestake Mining Company, locomotive repair office records.
27. Fielder, *The Treasure of Homestake Gold*, 378.
28. Homestake Mining Company, "Report on Operations for 1951."

Chapter Ten

1. Will Meyerriecks, *Drills and Mills: Precious Metal Mining and Milling Methods of the Frontier West* (privately printed, 2003), 192-93.
2. Moody, "Findings of Fact," *George Atchison et al. vs. George Hearst et al.*
3. Fielder, *The Treasure of Homestake Gold*, 26-27.
4. Yates, "The Homestake Mine, A Historical Sketch," 133.
5. Waterland, *The Spawn & the Mother Lode*, 49-51.
6. Clow, *Chasing the Glitter*, 69-71.
7. Deadwood-Terra Mining Company, "Articles of Incorporation."
8. Homestake Veterans Association, "Recollections of the Early Days."
9. Ibid.
10. Hearst to Haggin, May 19, 1878.
11. Lounsbery and Haggin, "Report of the Homestake Mining Company from January 1, 1878, to September 1, 1880."
12. H. O. Hofman, "Gold-Mining in the Black Hills," *American Institute of Mining and Metallurgical Engineers Transactions* 17 (February 1889): 502-504.
13. Fielder, *The Treasure of Homestake Gold*, 61-63.
14. *Black Hills Daily Pioneer*, January 19, 1879.
15. Ibid., January 18, 1879.
16. *Black Hills Daily Times*, January 20, 1879.
17. Brown and Willard, *The Black Hills Trails*, 366-69.
18. *Black Hills Daily Pioneer*, January 26, 1879.
19. Ibid., January 28, 1879.

20. Ibid., January 23 and 25, 1879.
21. Ibid., January 29, 1879.
22. Hearst to Haggin, March 6, 1879.
23. Ibid., March 19, 1879.
24. *Black Hills Daily Pioneer*, March 20, 1879.
25. Brown and Willard, *The Black Hills Trails*, 366-69.
26. *Black Hills Daily Times*, February 28, 1879.
27. Ibid., March 7, 1879.
28. Lounsbery and Haggin, "Report of the Homestake Mining Company from January 1, 1878, to September 1, 1880."
29. Fraser and Chalmers, "Contract Proposal for Constructing a 120-Stamp Mill for the Highland Mining Company," July 21, 1879, Black Hills Mining Museum, Lead, SD.
30. Waterland, *The Spawn & the Mother Lode*, 146-48.
31. Fraser and Chalmers, "Contract Proposal for Constructing a 120-Stamp Mill for the Highland Mining Company."
32. Lounsbery and Haggin, "Report of the Homestake Mining Company from January 1, 1878, to September 1, 1880."
33. Hofman, "Gold-Mining in the Black Hills," 505.
34. Hofman, "Gold-Mining in the Black Hills," 507-508.
35. Lounsbery and Haggin, "Report of the Homestake Mining Company from January 1, 1878, to September 1, 1880."
36. Hofman, "Gold-Mining in the Black Hills," 509.
37. Clow, *Chasing the Glitter*, 57.
38. National Inventors Hall of Fame, "Hall of Fame Inventor Profile for Eli Whitney Blake," National Inventors Hall of Fame Foundation, Inc., http://www.invent.org/hall_of_fame/318.html (accessed March 25, 2009).
39. Rickard, *The Stamp Milling of Gold Ores*, 78.
40. Ibid., 78.
41. For Mining Basics, "Gyratory Crusher Design," http://www.miningbasics.com/html/gyratory_crusher_design.php (accessed March 25, 2009).
42. Rickard, *The Stamp Milling of Gold Ores*, 78.
43. Allan J. Clark and W. J. Sharwood, "The Metallurgy of Homestake Ore," *American Institute of Mining and Metallurgical Engineers Transactions* 22 (November 21, 1912): 72.
44. Hofman, "Gold-Mining in the Black Hills," 509-17.
45. T. A. Rickard, "Gold-Milling in the Black Hills, South Dakota, and at Grass Valley, California," *American Institute of Mining and Metallurgical Engineers Transactions* 25 (October 1895): 911-12.
46. Hofman, "Gold-Mining in the Black Hills," 500-504.
47. C. W. Merrill, "The Metallurgy of the Homestake Ore," *American Institute of Mining and Metallurgical Engineers Transactions* 34 (1904): 586.

48. Clark and Sharwood, "The Metallurgy of Homestake Ore," 73.
49. Hofman, "Gold-Mining in the Black Hills," 523-24.
50. Ibid., 526-27.
51. Rickard, "Gold-Milling in the Black Hills, South Dakota, and at Grass Valley, California," 915-16.
52. Hofman, "Gold-Mining in the Black Hills," 527-28.
53. Ibid., 529-38.
54. Clark and Sharwood, "The Metallurgy of Homestake Ore," 76-77.
55. J. F. Wiggert, "Development of the Homestake Electrical System," *The Black Hills Engineer: Homestake Semi-Centennial* 14, no. 3 (May 1926): 183.
56. Clark and Sharwood, "The Metallurgy of Homestake Ore," 76-77.
57. Merrill, "The Metallurgy of the Homestake Ore," 586.
58. Clark and Sharwood, "The Metallurgy of Homestake Ore," 80.
59. Hofman, "Gold-Mining in the Black Hills," 499.
60. Waterland, *The Spawn & the Mother Lode*, 141-45.
61. Ibid., 137.
62. Homestake Mining Company, "Homestake Bullion Product from June 1, 1897, to June 1, 1898."
63. Lawrence County Register of Deeds, Deadwood, South Dakota, 154:22.
64. Fielder, *The Treasure of Homestake Gold*, 180.
65. Richard Blackstone, "Superintendent's Report," Homestake Mining Company, December 31, 1915, and December 31, 1916.
66. B. C. Yates, "Superintendent's Report," Homestake Mining Company, December 31, 1924.
67. Homestake Mining Company, "Metallurgical Summary for 1917."
68. Homestake Mining Company, *Sharp Bits* 19, no. 2 (Summer 1968).
69. Clark and Sharwood, "The Metallurgy of Homestake Ore," 84-87.

Chapter Eleven

1. *Black Hills Daily Pioneer*, March 7, 1879.
2. Lounsbery and Haggin, "Report of the Homestake Mining Company from January 1, 1878, to September 1, 1880;" Lounsbery and Haggin, "Fiscal Report from September 1, 1880, to June 1, 1881, Homestake Mining Company."
3. Ibid.
4. Hearst to Haggin, March 6, 1879.
5. Ibid., March 9, 1879.
6. Ibid., March 19, 1879.
7. Ibid., March 23, 1879.
8. *Black Hills Daily Times*, September 30, 1880.
9. Ibid., April 24, 1881.

10. Mildred Fielder, *Railroads of the Black Hills* (Seattle: Superior Publishing Company, 1964), 38.
11. Rick W. Mills, *Black Hills Railroading* (Hermosa: Battle Creek Publishing Company and Black Hills Central Railroad, 2004), 30.
12. Grier, personal notes.
13. Fielder, *Railroads of the Black Hills*, 38.
14. *Black Hills Daily Times*, December 3, 1881.
15. Martha Linde, *Sawmills of the Black Hills* (Rapid City: Fenske Printing, Inc., 1984), 76.
16. Mills, *Black Hills Railroading*, 30.
17. Irma H. Klock, *Whistle Stops on the Black Hills and Fort Pierre* (Deadwood: Dakota Graphics, 1982), 19-20.
18. B. C. Yates, "Superintendent's Report," Homestake Mining Company, December 31, 1919.
19. Fielder, *Railroads of the Black Hills*, 49-50.
20. Ibid., 41-42.
21. Brown and Willard, 276-77.
22. Ibid., 277-78.
23. Ibid., 277-84.
24. Ibid., 284-86.
25. Klock, *Whistle Stops*, 19-27.
26. Fielder, *Railroads of the Black Hills*, 46-47.
27. Klock, *Whistle Stops*, 22-26.
28. Virginia Driving Hawk Sneve, *South Dakota Geographic Names*, Brevet Press: Sioux Falls, South Dakota, 1973, p. 127.
29. Lawrence County, Deadwood, South Dakota, "Record of Appointment of Postmasters 1885-1900," Vol. 59, line 12, 282-83.
30. Klock, *Whistle Stops*, 27.
31. Irma H. Klock, *Yesterday's Gold Camps and Mines in the Northern Black Hills* (Lead: Seaton Publishing Company, 1975) 75-76.
32. Ibid., 75-76.
33. Klock, *Whistle Stops*, 31-31.
34. Black Hills & Fort Pierre Railroad Company, "Financial Report for the Black Hills and Fort Pierre Railroad Company," June 30, 1889, Black Hills Mining Museum.
35. Klock, *Whistle Stops*, 31-31.
36. Grier, personal notes.
37. Fielder, *Railroads of the Black Hills*, 41-42.
38. Klock, *Whistle Stops*, 33-36.
39. Klock, *Yesterday's Gold Camps*, 77-78.
40. Klock, *Whistle Stops*, 33-36.
41. Klock, *Yesterday's Gold Camps*, 77-78.

42. Fielder, *Railroads of the Black Hills*, 39-40.
43. Elna Artus, "Coffee Pot Link between Old World and Black Hills," *Deadwood Daily Pioneer-Times*, date unknown.
44. Klock, *Whistle Stops*, 39-40.
45. John R. Honerkamp, *At the Foot of the Mountain* (Stickney: Argus Printers, 1978), 148-49.
46. Klock, *Whistle Stops*, 41-43.
47. John R. Honerkamp, *At the Foot of the Mountain*, 156-57.
48. Klock, *Whistle Stops*, 43.
49. Ibid., 45-47.
50. *Rapid City Black Hills Weekly*, May 23, 1890.
51. Ibid.
52. Honerkamp, *At the Foot of the Mountain*, 118.
53. Ibid., 121-22.
54. *Rapid City Black Hills Weekly*, June 6, 1890.
55. Fielder, *Railroads of the Black Hills*, 55.
56. Honerkamp, *At the Foot of the Mountain*, 43-44.
57. Klock, *Whistle Stops*, 45-47.
58. *Rapid City Daily Journal*, September 11, 1890.
59. Klock, *Whistle Stops*, 46.
60. Honerkamp, *At the Foot of the Mountain*, 150.
61. Fielder, *Railroads of the Black Hills*, 13.
62. Joel Waterland, *The Mines Around & Beyond* (Rapid City: Grelind Printing Center, 1991), 198.
63. Charles and Fredricka K. Flormann, "Right of Way Deed" to the Black Hills & Fort Pierre Railroad Company, April 5, 1898.
64. "Record of Appointment of Postmasters 1885-1900," 282-83.
65. Klock, *Yesterday's Gold Camps*, 85-86.
66. Clarence Miller, papers, Colette (Flormann) Bonstead collection.
67. Patricia Lavier Mechling, "Greenwood—Vanished Village," *Black Hills Nuggets* (Rapid City: Rapid City Society for Genealogical Research), 110-115.
68. Lawrence County, Deadwood, South Dakota, "Record of Appointment of Postmasters, 1894-1930" Vol. 89, line 17, 635-636.
69. Miller, papers.
70. Richmond L. Clow, "Homestake Mining Company and the First Regulated Timber Harvest," U.S. Forest Service, Black Hills National Forest—Projects and Plans, http://www.fs.fed.us/r2/blackhills/projects/timber/case1/case1.shtml (accessed March 27, 2009).
71. Ibid.
72. Ibid.

73. Elton and Norma Adams, editors, *Nemo, South Dakota: One Hundred Years, 1889-1989* (Nemo Centennial Association, 1989), 5.
74. Clow, "Homestake Mining Company and the First Regulated Timber Harvest."
75. Tallent, *The Black Hills, or Last Hunting Grounds*, 372.
76. *Black Hills Weekly Times*, January 25, 1891.
77. Joseph H. Cash, *Working the Homestake*, 25.
78. Fielder, *Railroads of the Black Hills*, 44-48.
79. Honerkamp, 49-50.
80. Fielder, *Railroads of the Black Hills*, 49-54.
81. Honerkamp, *At the Foot of the Mountain*, 89.
82. Fielder, *Railroads of the Black Hills*, 13.
83. Fielder, *Railroads of the Black Hills*, 48.
84. Ibid., 53.
85. Ibid.
86. Marilyn Anderson, "Hearst Mercantile Store," *Nemo, South Dakota: One Hundred Years, 1889-1989* (Nemo Centennial Association, 1989), 22-24.
87. Homestake Mining Company, *Sharp Bits* 5, no. 6 (July 1954).
88. Adams, *Nemo, South Dakota: One Hundred Years*, 5-8.
89. Ibid., 10.
90. Homestake Mining Company, "Report on Operations for 1935."
91. Fielder, *The Treasure of Homestake Gold*, 211-71.
92. Homestake Mining Company, "Report on Operations for 1934."
93. Homestake Mining Company, "Report on Operations for 1937."
94. W. D. Beardshear, "Lumbering in the Black Hills," *The Black Hills Engineer* XIV, no. 3 (May 1926): 175-181.
95. Homestake Mining Company, *Sharp Bits* 5, no. 6 (July 1954).
96. Homestake Mining Company, *Sharp Bits* 7, no. 9 (October 1956).
97. Homestake Mining Company, *Sharp Bits* 12, no. 6 (July 1961).
98. Homestake Mining Company, "Annual Report to Shareholders," 1972.
99. Homestake Mining Company, "Annual Report to Shareholders," 1980.

Chapter Twelve

1. Clowser, *Deadwood . . . The Historic City*, 7-8; Homestake Mining Company, "Pre-Patent Title Summary for the Old Abe, M. S. No. 117."
2. *Lead Daily Call*, August 5, 1926.
3. Ibid.
4. *Black Hills Pioneer*, July 27, 1876.
5. *Lead Daily Call*, August 5, 1926.
6. Ibid.
7. *Black Hills Pioneer*, July 27, 1876.

8. *Lead Daily Call*, Homestake edition, May 1953.
9. *Black Hills Pioneer*, July 27, 1876.
10. *Pioneer News*, July 29, 1876.
11. Irma H. Klock, *All Roads Lead to Deadwood* (Lead: privately printed, 1979), 235.
12. *Black Hills Pioneer*, August 12, 1876.
13. *Lead Daily Call*, August 5, 1926.
14. Fielder, *The Treasure of Homestake Gold*, 149.
15. Waterland, *The Spawn & the Mother Lode*, 223-24.
16. Fielder, *The Treasure of Homestake Gold*, 149-51.
17. Chambers Kellar to Donald McLaughlin and Guy Bjorge, June 3, 1947, Homestake Mining Company.
18. Mineral Claimants J. B. Haggin and Homestake Mining Company, Agreement with Lead City Townsite Trustees, Lawrence County Register of Deeds, Deadwood, South Dakota (March 18, 1892), 82:289.
19. Homestake Mining Company et al., Quit Claim Deed to Cyrus H. Enos, Trustee for Lead Townsite Claimants, Lawrence County Register of Deeds, Deadwood, South Dakota (January 5, 1893), 91:160.
20. Ibid.
21. U.S. Land Office, Townsite Patent to Leonard Gordon, Probate Judge in trust for occupants of the Lead City Townsite, Lawrence County Register of Deeds, Deadwood, South Dakota (filed February 1, 1899), 139:374.
22. Albert Gushurst Jr, "Manuel, Fred and Moses and Their Niece Josie," *Some History of Lawrence County* (Pierre: The State Publishing Company, 1981), 7-10.
23. Fred Manuel Affidavit.
24. Gushurst, "Manuel, Fred and Moses and Their Niece Josie," 7-10.
25. Toms et al., *The Gold Belt Cities: The City of Mills*, 68-69.
26. Tallent, *The Black Hills, or Last Hunting Grounds*, 388.
27. Gushurst, "Manuel, Fred and Moses and Their Niece Josie," 7-10.
28. *Lead Daily Call*, "Homestake Edition," May 1953; *Black Hills Daily Times*, February 26, 1878.
29. Ibid.
30. Ibid.
31. Tallent, *The Black Hills, or Last Hunting Grounds*, 390.
32. Toms et al., *The Gold Belt Cities: Lead & Homestake*, 112.
33. Ibid., 112-115.
34. Lead High School, *Goldenlode* 46, 1972.
35. Fielder, *The Treasure of Homestake Gold*, 311.
36. *Deadwood Daily Pioneer-Times*, June 2, 1908.
37. Lee, ed., *Gold, Gals, Guns, Guts*, 238.
38. Toms et al., *The Gold Belt Cities: Lead & Homestake*, 112.

39. Toms et al., *The Gold Belt Cities: The City of Mills*, 18.
40. *Lead Daily Call*, August 5, 1926.
41. Toms et al., *The Gold Belt Cities: Lead & Homestake*, 30-31, 69.
42. Lee, *Gold, Gals, Guns, Guts*, 238.
43. Tallent, *The Black Hills, or Last Hunting Grounds*, 388.
44. *Lead Daily Call*, August 5, 1926.
45. Fielder, *The Treasure of Homestake Gold*, 37, 54.
46. *Lead Daily Call*, August 5, 1926.
47. Homestake Edition, May 1953.
48. Dwayne F. Knight, *Steeples above Stopes: The Churches in the Gold Camps, 1876-1976* (Deadwood: Deadwood-Lead '76 Centennial, 1976), 19-20.
49. Knight, *Steeples above Stopes*, 19-39.
50. Centennial Committee, *First Lutheran Church, 100 Years, 1889-1999* (Deadwood: Dakota Graphics, June 1989).
51. LaVaughn Storsve, *The History of Bethel Lutheran Church, 75th Anniversary Edition* (Lead: privately printed), 1970.
52. Knight, *Steeples above Stopes*, 57-63, 69-70.
53. Fielder, *The Treasure of Homestake Gold*, 234-237.
54. Ibid., 282.
55. Jeanette Larson, "City of Lead, 1876 to 1981," *Some History of Lawrence County* (Pierre: The State Publishing Company, 1981), 583.

Chapter Thirteen

1. Irma H. Klock, *The Gold Camps in Upper Deadwood Gulch* (Deadwood: Dakota Graphics, 1984), 20-21.
2. Lee, *Gold, Gals, Guns, Guts*, 22-23.
3. Ibid., 63.
4. Watson Parker and Hugh K. Lambert, *Black Hills Ghost Towns* (Athens: Swallow Press/Ohio University Press, 1974), 100.
5. Klock, *The Gold Camps in Upper Deadwood Gulch*, 7.
6. Tallent, *The Black Hills, or Last Hunting Grounds*, 267-69.
7. Ibid., 269-70.
8. Fielder, *Railroads of the Black Hills*, 126.
9. Klock, *The Gold Camps in Upper Deadwood Gulch*, 6-8.
10. Waterland, *The Spawn & the Mother Lode*, 62-68, 105-110.
11. Klock, *All Roads Lead to Deadwood*, 229-38.
12. Brown and Willard, *The Black Hills Trails*, 497.
13. Klock, *The Gold Camps in Upper Deadwood Gulch*, 11-14.
14. *South Dakota History* 2, no. 3 (Pierre: South Dakota State Historical Society, 1972): 231-232.
15. Ibid.

16. Parker, *Gold in the Black Hills*, 96-97.
17. *South Dakota History* 2, no. 3, 231-32.
18. Tallent, *The Black Hills, or Last Hunting Grounds*, 396-97.
19. Knight, *Steeples above Stopes*, 27-28.
20. Doane Robinson, "History of South Dakota II," MyFamily.com, Inc., http://ftp.rootsweb.com/pub/usgenweb/sd/biography/doane2/rosenkranz.txt (accessed January 17, 2009).
21. BreweryGems.com, "The Black Hills Brewery of Central City, SD (1878-1927): Illustrated History of the Black Hills Brewing Co.," Gary@BreweryGems.com, http://www.brewerygems.com/black-hills.htm (accessed April 9, 2009).
22. *South Dakota History* 2, no. 3, 244-45.
23. Ibid., 245-48.
24. Lee, *Gold, Gals, Guns, Guts*, 128.
25. Klock, *The Gold Camps in Upper Deadwood Gulch*, 27.
26. Ibid., 30.
27. Knight, *Steeples above Stopes*, 47-49.
28. Waterland, *The Spawn & the Mother Lode*, 128.
29. Ibid., 162.
30. Klock, *The Gold Camps in Upper Deadwood Gulch*, 35-40.

Chapter Fourteen

1. Hofman, "Gold-Mining in the Black Hills," 538-40.
2. Rickard, *The Stamp Milling of Gold Ores*, 98.
3. Clow, *Chasing the Glitter*, 43-46.
4. A. J. Clark, "A Review of Homestake Metallurgy," *The Black Hills Engineer* XIV, no. 3 (May 1926): 160.
5. Waterland, *The Spawn & the Mother Lode*, 164.
6. Clow, *Chasing the Glitter*, 117.
7. *Black Hills Daily-Pioneer*, June 30, 1894.
8. Clow, *Chasing the Glitter*, 82-83.
9. Clark, "A Review of Homestake Metallurgy," 159-60.
10. Rickard, "Gold-Milling in the Black Hills, South Dakota, and at Grass Valley, California," 917.
11. David W. Ryder, *The Merrill Story* (San Francisco: The Merrill Company, 1958), 27-29, 41-42.
12. McClintock, *Pioneer Days in the Black Hills*, 335-36.
13. William A. Remer, personal diary, Adams Museum, Deadwood, South Dakota.
14. Ryder, *The Merrill Story*, 41-42.
15. Fielder, *The Treasure of Homestake Gold*, 168-70.

16. Ryder, *The Merrill Story*, 42.
17. Ibid., 46-47.
18. Merrill, "The Metallurgy of the Homestake Ore," 588.
19. Ibid., 588-90.
20. Ibid., 592-94.
21. Clark and Sharwood, "The Metallurgy of Homestake Ore," 104-108.
22. Fielder, *The Treasure of Homestake Gold*, 171-72.
23. Clark, "A Review of Homestake Metallurgy," 162.
24. Yates, "Some Features of Mining Operations," 45.
25. Waterland, *The Spawn & the Mother Lode*, 164, 176.
26. Clark and Sharwood, "The Metallurgy of Homestake Ore," 104-108.
27. Ryder, *The Merrill Story*, 40.
28. Ibid., 36-37.
29. Ibid., 48.
30. Ibid., 49.
31. Homestake Mining Company, *Sharp Bits* 16, no. 10 (November 1965).
32. Clark and Sharwood, "The Metallurgy of Homestake Ore," 104-106.
33. Waterland, *The Spawn & the Mother Lode*, 164, 176.
34. Homestake Mining Company, "Bullion Production Data, January 1, 1878 through December 31, 2000."
35. C. W. Merrill, "Treatment of Homestake Ores," *Mining and Scientific Press* LXXXVI, no. 10 (March 7, 1903): 150.
36. Homestake Mining Company, "Water Used in Batteries," November 22, 1907.
37. Homestake Mining Company, "Bullion Production Data, January 1, 1878 through December 31, 2000."
38. Merrill, "The Metallurgy of the Homestake Ore," 585-91.
39. Mark Ehle Jr., "The Homestake Slime Plant," *Mines and Minerals* (March 1907): 1.
40. Yates, "Some Features of Mining Operations," 36.
41. Ryder, *The Merrill Story*, 43-44.
42. Allan J. Clark, "Notes on Homestake Metallurgy," *American Institute of Mining and Metallurgical Engineers Transactions* 52 (1915): 14.
43. Clark and Sharwood, "The Metallurgy of Homestake Ore," 111.
44. Allan J. Clark, "Milling Methods and Costs at the Homestake Mine, Lead, South Dakota," U.S. Department of the Interior, Bureau of Mines, Information Circular No. 6408 (February 1931): 16.
45. Ehle, "The Homestake Slime Plant," 1-6.
46. A. W. Allen, "The Homestake Enterprise," *Engineering and Mining Journal* 132, no. 7 (October 12, 1931): 304.
47. Ehle, "The Homestake Slime Plant," 5.
48. Merrill, "Treatment of Homestake Ores," 151-52.

49. Remer diary.
50. Ryder, *The Merrill Story*, 51-52.
51. Clark, "Notes on Homestake Metallurgy," 10.
52. Yates, "Superintendent's Report," December 31, 1924.
53. Merrill, "The Metallurgy of the Homestake Ore," 587.
54. Clark and Sharwood, "The Metallurgy of Homestake Ore," 71.
55. Homestake Mining Company, "Metallurgical Summary by Years (1909-1922)."
56. K. D. Pyle, "The Homestake Tungsten Mill," *The Black Hills Engineer* (Rapid City: The South Dakota State School of Mines, May 1926), 171-174.
57. B. C. Yates, "Superintendent's Report," Homestake Mining Company, December 31, 1919, December 31, 1920, and December 31, 1921.
58. Clark, "Milling Methods and Costs," 3-7.
59. Edward Hodges Robie, "Milling Practice at the Homestake Gold Mine," *Engineering and Mining Journal*, 122, no. 15 (October 9, 1926): 566-67.
60. Ibid.
61. Ibid.
62. Fielder, *The Treasure of Homestake Gold*, 265-75.
63. Clark, "Milling Methods and Costs," 19-21.
64. Ibid., 18.
65. Clark and Sharwood, "The Metallurgy of Homestake Ore," 134-36.

Chapter Fifteen

1. Black Hills Mining Museum, "Edison Dynamo Exhibit," Black Hills Mining Museum, Lead, South Dakota.
2. Wiggert, "Development of the Homestake Electrical System," 182-83.
3. Black Hills Corporation, "A Century of Light: The Early Years," Black Hills Corporation, http://www.blackhillspower.com/chapter2b.htm (accessed April 11, 2009).
4. Wiggert, "Development of the Homestake Electrical System," 184.
5. Homestake Mining Company, "Hydroelectric Power Generation (1906-2000)."
6. Black Hills Corporation, "The Early Years," Black Hills Corporation, http://www.blackhillspower.com/chapter2a.htm#1 (accessed April 11, 2009).
7. Wiggert, "Development of the Homestake Electrical System," 184.
8. Black Hills Corporation, "The Early Years."
9. Richard Blackstone, "The Hydro-Electric Power Plant of the Homestake Mining Co.," *Mining and Engineering World* XLI, no. 1 (July 4, 1914): 3.
10. *Deadwood Daily Pioneer-Times*, December 17, 1908.
11. Blackstone, "The Hydro-Electric Power Plant of the Homestake Mining Co.," 3-10.

12. *Deadwood Daily Pioneer-Times*, March 7, 1909.
13. Ibid., April 13, 1909.
14. Blackstone, "The Hydro-Electric Power Plant of the Homestake Mining Co.," 5.
15. Ibid., 5-7.
16. Ibid.
17. Homestake Mining Company, "Hydroelectric Power Generation (1906-2000)."
18. Wiggert, "Development of the Homestake Electrical System," 186-89.
19. Ibid.
20. Richard Blackstone, "Superintendent's Report," Homestake Mining Company, December 31, 1917.
21. Wiggert, "Development of the Homestake Electrical System," 189.
22. Richard Blackstone, "Superintendent's Report," Homestake Mining Company, December 31, 1914; Richard Blackstone, "Superintendent's Report," Homestake Mining Company, December 31, 1915.
23. S. J. Staple, "The Mechanical Department of the Homestake Mining Company," *The Black Hills Engineer* XIV, no. 2 (May 1926): 202-204.
24. Ibid., 210.
25. Fielder, *The Treasure of Homestake Gold*, 175-76, 258-59.
26. Logan Actuator Company, "A Brief History of Logan Actuator Company and Logan Engineering Company," Logan Actuator Company, http://www.loganact.com/history.htm (accessed April 11, 2009).
27. Fielder, *The Treasure of Homestake Gold*, 258-59.
28. Logan Actuator Company, "A Brief History of Logan Actuator Company."
29. Wiggert, "Development of the Homestake Electrical System," 189.
30. B. C. Yates, "Superintendent's Report," Homestake Mining Company, December 31, 1931.
31. Homestake Mining Company, "Hydroelectric Power Generation (1906-2000)."
32. Staple, "The Mechanical Department of the Homestake Mining Company."
33. Homestake Mining Company, "Hydroelectric Power Generation (1906-2000)."
34. B. C. Yates, "Superintendent's Report," Homestake Mining Company, December 31, 1918; B. C. Yates, "Superintendent's Report," Homestake Mining Company, December 31, 1919.
35. Staple, "The Mechanical Department of the Homestake Mining Company," 207-209.
36. Ibid., 202-203.
37. Staple, "The Mechanical Department of the Homestake Mining Company," 193-95.

38. Fielder, *The Treasure of Homestake Gold*, 257-58.
39. Wiggert, "Development of the Homestake Electrical System," 192-98.
40. B. C. Yates, "Superintendent's Report," Homestake Mining Company, December 31, 1921.
41. Ibid., December 31, 1923.
42. Homestake Mining Company, "Report on Operations," December 31, 1933.
43. B. C. Yates, "Superintendent's Report," Homestake Mining Company, December 31, 1928; B. C. Yates, "Superintendent's Report," Homestake Mining Company, December 31, 1929.
44. Wiggert, "Development of the Homestake Electrical System," 201.
45. John F. Wiggert, "Homestake Mining Company's Carrier-current Shaft Signaling System," Technical Publication No. 1286-A, *Mining Technology* (February 1941).
46. Fielder, *The Treasure of Homestake Gold*, 10, 381.
47. Homestake Mining Company, "Report on Operations," December 31, 1935.
48. Ibid., December 31, 1955.
49. Ibid., December 31, 1956.
50. Homestake Mining Company, "Homestake Mine Operations Annual Report," December 31, 1992.
51. Homestake Mining Company, "Hydroelectric Power Generation (1906-2000)."

Chapter Sixteen

1. Joseph H. Cash, "The Homestake Lockout—1909-1910," paper presented at The American Middle Period and the West seminar, 16:235, Hearst Library, Lead, South Dakota, 4.
2. *Lead Daily Call*, "Homestake . . . 75th Year of Production," May 1953.
3. Ibid.
4. United States Commission on Industrial Relations, *Proceedings of Public Hearing Held at Lead, South Dakota, August 3-4, 1914* (Kansas City, Missouri: Shorthand Reporting Company, 1914), 330-31.
5. Cash, "The Homestake Lockout—1909-1910," 5.
6. Tallent, *The Black Hills, or Last Hunting Grounds*, 436.
7. Cash, "The Homestake Lockout—1909-1910," 6-8.
8. *Deadwood Daily Pioneer-Times*, June 7, 1908.
9. Cash, "The Homestake Lockout—1909-1910," 6-8.
10. U.S. Commission on Industrial Relations, *Proceedings of Public Hearing Held at Lead, South Dakota*, 471-72.
11. Cash, "The Homestake Lockout—1909-1910," 6-8.

12. U.S. Commission on Industrial Relations, *Proceedings of Public Hearing Held at Lead, South Dakota*, 472-73.
13. Ibid., 471-74.
14. Cash, "The Homestake Lockout—1909-1910," 11.
15. *Lead Daily Call*, November 17, 1909.
16. Cash, "The Homestake Lockout—1909-1910," 12-15.
17. Ibid., 15-17.
18. U.S. Commission on Industrial Relations, *Proceedings of Public Hearing Held at Lead, South Dakota*, 357.
19. *Deadwood Daily Pioneer-Times*, July 23, 1910.
20. Cash, "The Homestake Lockout—1909-1910," 19-24.
21. Cash, *Working the Homestake*, 92-93.
22. Cash, "The Homestake Lockout—1909-1910," 26-29.
23. T. J. Grier, "Superintendent's Report," Homestake Mining Company, June 1, 1910.
24. Cash, "The Homestake Lockout—1909-1910," 29-32.
25. Ibid., 32-34.
26. Fielder, *The Treasure of Homestake Gold*, 347-48, 437-38.

Chapter Seventeen

1. The Weekly South Dakotan, "Homesteading and Town Building," MyReportLinks.com, http://www.sd4history.com/Unit6/hsandtblesson3.htm (accessed May 16, 2009).
2. Cash, *Working the Homestake*, 30-31.
3. Donald D. Toms, *Ethnic Heritage, Lead South Dakota I* (Lead, South Dakota: South Dakota Historic Preservation and City of Lead Grant Project): 1.
4. Ibid.
5. *Lead Evening Call*, March 17, 1896.
6. Cash, *Working the Homestake*, 30-33.
7. U.S. Commission on Industrial Relations, *Proceedings of Public Hearing Held at Lead, South Dakota*, 65-75, 270-72.
8. Ibid., 34-61.
9. Ibid.
10. Waterland, *The Spawn & the Mother Lode*, 193-94.
11. Homestake Mining Company, "Metallurgical Summary for 1921."
12. Waterland, *The Spawn & the Mother Lode*, 196.
13. Homestake Mining Company, "Report on Operations for 1938."
14. Homestake Mining Company, "Report on Operations for 1939."
15. Homestake Mining Company, "Report on Operations for 1952."
16. Homestake Mining Company, "Manager's Report on the Homestake Mine and Bulldog Mine for 1973."

17. *Lead Daily Call*, August 5, 1926.
18. U.S. Commission on Industrial Relations, *Proceedings of Public Hearing Held at Lead, South Dakota*, 65-75, 294.
19. Toms et al., *The Gold Belt Cities: Lead & Homestake*, 270, 289.
20. Homestake Mining Company, "Manager's Report on the Homestake Mine and Bulldog Mine for 1974."
21. Toms et al., *The Gold Belt Cities: Lead & Homestake*, 42.
22. Fielder, *The Treasure of Homestake Gold*, 152-53.
23. U.S. Commission on Industrial Relations, *Proceedings of Public Hearing Held at Lead, South Dakota*, 399-406.
24. Ibid., 156-58.
25. Toms et al., *The Gold Belt Cities: Lead & Homestake*, 42.
26. Fielder, *The Treasure of Homestake Gold*, 97, 269.
27. Homestake Veterans Association Committee, *The Homestake Veterans Association* (Lead, South Dakota: Homestake Mining Company, March 1951).
28. Jeri Mykleby, "New Member Induction Count," Homestake Veterans Association, November 2008.
29. *Deadwood Daily Pioneer Times*, June 9, 1908; *Deadwood Daily Pioneer-Times*, June 9, 1908.
30. Toms et al., *The Gold Belt Cities: Lead & Homestake*, 78.
31. *Deadwood Daily Pioneer-Times*, August 21, 1908.
32. *Lead Daily Call*, June 7, 1946.
33. Ibid., September 3, 1946.
34. Ibid., August 31, 1947.
35. Ibid.
36. *Deadwood Daily Pioneer-Times*, April 25, 1909.
37. U.S. Commission on Industrial Relations, *Proceedings of Public Hearing Held at Lead, South Dakota*, 27-29.
38. Homestake Mining Company, "Agreement between Homestake Mining Company and United Steelworkers of America, AFL-CIO-CLC," June 1, 1989.
39. Allen, "The Homestake Enterprise," 296-97.
40. Homestake Mining Company, *Sharp Bits* 4, no. 11 (December 1953).
41. Homestake Mining Company, *Sharp Bits* 6, no. 4 (May 1955).
42. Toms, *In the Midst of Life: Mining Company Fatalities*, 1-20.
43. Toms et al., *The Gold Belt Cities: Lead & Homestake*, 226.
44. The Historic Homestake Opera House, "Homestake Opera House: The Jewel of the Black Hills," The Historic Homestake Opera House http://www.leadoperahouse.org/index.html (accessed May 16, 2009).
45. Senate, *Final Report of the Commission on Industrial Relations*, 64th Cong., 1st sess., S. Doc. 415 (Washington, DC: GPO, 1916): 8-22.

46. U.S. Commission on Industrial Relations, *Proceedings of Public Hearing Held at Lead, South Dakota*, 215-30.
47. Ibid., 430-31.
48. Ibid., 435-37.
49. Senate, *Final Report and Testimony Submitted to Congress by the Commission on Industrial Relations Created by the Act of August 23, 1912* I, 64th Cong., 1st sess., S. Doc. 415 (Washington, DC: GPO, 1916): 77-79.
50. *Lead Daily Call*, August 5, 1926.
51. Homestake Mining Company, *Sharp Bits* 7, no. 8 (September 1956); Toms et al., *The Gold Belt Cities: Lead & Homestake*, 232.
52. Homestake Mining Company, "Report on Operations for 1936."
53. Homestake Mining Company, *Sharp Bits* 14, no. 6 (July 1963).
54. Homestake Mining Company, *Sharp Bits* 14, no. 11 (December 1963).
55. Homestake Mining Company, "Report on Operations for 1973;" Homestake Mining Company, "Report on Operations for 1974;" Homestake Mining Company, "Report on Operations for 1975;" Homestake Mining Company, "Report on Operations for 1976."
56. Homestake Mining Company, *Sharp Bits* 1, no. 1 (February 1950).
57. Homestake Mining Company, *Sharp Bits* 1, no. 2 (March 1950).

Chapter Eighteen

1. Waterland, *The Spawn & the Mother Lode*, 168.
2. Toms et al., *The Gold Belt Cities: The City of Mills*, 178.
3. Fielder, *The Treasure of Homestake Gold*, 162.
4. Toms et al., *The Gold Belt Cities: The City of Mills*, 178.
5. Waterland, *The Spawn & the Mother Lode*, 168.
6. T. J. Grier, "Superintendent's Report," Homestake Mining Company, June 1, 1907.
7. Fielder, *The Treasure of Homestake Gold*, 191-95.
8. Grier, "Superintendent's Report," June 1, 1907.
9. Fielder, *The Treasure of Homestake Gold*, 196-97.
10. *Deadwood Daily Pioneer-Times*, January 16, 1908.
11. Ibid., December 17, 1908.
12. Ibid., February 20, 1909.
13. *Lead Daily Call*, "Homestake Edition," May 1953.
14. Waterland, *The Spawn & the Mother Lode*, 194-96.
15. Fielder, *The Treasure of Homestake Gold*, 252.
16. Homestake Mining Company, "Bullion Production Data, January 1, 1878, through December 31, 2000."
17. Felix Edgar Wormser, "Gold-Mining Developments in the Black Hills," *Engineering and Mining Journal*, 114, no. 18 (October 28, 1922):759-60.

18. Waterland, *The Spawn & the Mother Lode*, 202.
19. Homestake Mining Company, "Report on Operations for 1934."
20. Homestake Mining Company, "Metallurgical Summary for 1926;" Homestake Mining Company, "Metallurgical Summary for 1932."
21. Waterland, *The Spawn & the Mother Lode*, 205-207.
22. Toms et al., *The Gold Belt Cities: The City of Mills*, 58.
23. Homestake Mining Company, "Report on Operations for 1932."
24. Homestake Mining Company, "Report on Operations for 1933."
25. Homestake Mining Company, "Report on Operations for 1934."
26. Waterland, *The Spawn & the Mother Lode*, 216.
27. Homestake Mining Company, "Report on Operations for 1936;" Homestake Mining Company, "Report on Operations for 1937;" Homestake Mining Company, "Report on Operations for 1938."
28. Homestake Mining Company, "Report on Operations for 1955."
29. Homestake Mining Company, "Report on Operations for 1935."
30. Homestake Mining Company, "Report on Operations for 1969."
31. Homestake Mining Company, *Sharp Bits* 17, no. 3 (April 1966).

Chapter Nineteen

1. Klock, *Yesterday's Gold Camps*, 17.
2. Columbus Mining Company, Stock certificate No. 68 issued to R. W. Hamilton for two hundred shares of capital stock, August 3, 1880.
3. Waterland, *The Spawn & the Mother Lode*, 105-107.
4. Ibid., 105-108.
5. George P. Baldwin, ed., *The Black Hills Illustrated* (1904; repr., Deadwood: Dakota Graphics), 139.
6. Waterland, *The Spawn & the Mother Lode*, 107-109.
7. U.S. Department of the Interior, Bureau of Mines, *Black Hills Mineral Atlas, South Dakota: Part 1*, Information Circular 7688 (July 1954): 80.
8. Waterland, *The Spawn & the Mother Lode*, 233-35.
9. Ibid., 100-101.
10. Hidden Fortune Gold Mining Company, "Annual Report Covering the Fiscal Year Ending February 25, 1903," Black Hills Mining Museum.
11. Waterland, *The Spawn & the Mother Lode*, 103.
12. Hidden Fortune Gold Mining Company, "Third Annual Report of the Hidden Fortune Gold Mining Company Covering the Fiscal Year Ending February 20, 1904," Black Hills Mining Museum.
13. Ibid.
14. Hidden Fortune Gold Mining Company, "Important Notices and other General Information to the Stockholders of the Hidden Fortune Mining Company," Black Hills Mining Museum.

15. U.S. Department of the Interior, Bureau of Mines, *Black Hills Mineral Atlas*, 81.
16. Waterland, *The Spawn & the Mother Lode*, 105.
17. K. D. Pyle, "The Homestake Tungsten Mill," *The Black Hills Engineer* XIV, no. 3 (May 1926): 171-174.
18. Waterland, *The Spawn & the Mother Lode*, 228.
19. *Deadwood Daily Pioneer-Times*, "Homestake South Extension Company," July 8, 1908.
20. Waterland, *The Spawn & the Mother Lode*, 225, 228.
21. Ibid., 224.
22. Baldwin, ed., *The Black Hills Illustrated*, 163.
23. Waterland, *The Spawn & the Mother Lode*, 224.
24. Ibid., 225.
25. Ibid., 225-27.

Chapter Twenty

1. Fielder, *The Treasure of Homestake Gold*, 314.
2. Waterland, *The Spawn & the Mother Lode*, 208-13.
3. Homestake Mining Company, "Report on Operations for 1935."
4. Waterland, *The Spawn & the Mother Lode*, 209-12; Homestake Mining Company, "Report on Operations for 1936;" Homestake Mining Company, "Report on Operations for 1937;" Homestake Mining Company, "Report on Operations for 1938."
5. Ibid.
6. Ibid.
7. Homestake Mining Company, "Report on Operations for 1955."
8. Homestake Mining Company, "Report on Operations for 1933;" Homestake Mining Company, "Report on Operations for 1934."
9. Homestake Mining Company, "Report on Operations for 1935."
10. Homestake Mining Company, "Report on Operations for 1936;" Homestake Mining Company, "Report on Operations for 1937."
11. John T. Woolley and Gerhard Peters, "*The American Presidency Project,*" Gerhard Peters—The American Presidency Project http://www.presidency.ucsb.edu/ws/?pid=14611 (accessed May 18, 2009).
12. "Gold Reserve Act of 1934," *The Statutes at Large of the United States of America from March 1933 to June 1934* 48, pt. 1 (Washington, DC: GPO, 1934), 337-44.
13. World Gold Council, "After the Gold Standard, 1931-1999," World Gold Council http://www.gold.org/assets/file/value/reserve_asset/history/monetary_history/vol3pdf/1934jan31.pdf (accessed May 18, 2009).
14. Fielder, *The Treasure of Homestake Gold*, 312-14.

15. Homestake Mining Company, "Report on Operations for 1936."
16. Homestake Mining Company, "Report on Operations for 1937."
17. J. M. Berry, "The Budget of the State of South Dakota for the Biennium 1939 to 1941," prepared for the legislature by the secretary of Finance (November 1938).
18. Homestake Mining Company, "The Truth About Homestake," public information circular, 1938, Black Hills Mining Museum.
19. Homestake Mining Company, "The Ore Tax Should be Reduced," public information circular, 1944, Black Hills Mining Museum.
20. Fielder, *The Treasure of Homestake Gold*, 392, 419, 443.
21. Justia: U.S. Laws, "South Dakota 10-39-43: Severance tax on gold," Justia, Inc. http://law.justia.com/southdakota/codes/10/10-39-43.html (accessed June 21, 2009).
22. Homestake Mining Company, "Report on Operations for 1934."
23. Homestake Mining Company, "Report on Operations for 1936."
24. Waterland, *The Spawn & the Mother Lode*, 220.
25. Homestake Mining Company, "Report on Operations for 1937."
26. Homestake Mining Company, "Report on Operations for 1936."
27. Homestake Mining Company, "Report on Operations for 1935."
28. Homestake Mining Company, "Report on Operations for 1937."
29. Fielder, *The Treasure of Homestake Gold*, 329-30.
30. G. N. Bjorge, "Comparison of Estimate to Cost to Dec. 1, 1942—Yates Shaft Construction," Homestake Mining Company, December 28, 1942.
31. Homestake Mining Company, "Report on Operations for 1938;" Homestake Mining Company, "Report on Operations for 1939."
32. Bjorge, "Comparison of Estimate to Cost to Dec. 1, 1942—Yates Shaft Construction."
33. Waterland, *The Spawn & the Mother Lode*, 218-20.
34. Homestake Mining Company, *Sharp Bits* 4, no. 9 (October 1953).
35. Homestake Mining Company, "Report on Operations for 1954."
36. Homestake Mining Company, "Report on Operations for 1955."
37. Homestake Mining Company, "Report on Operations for 1956."
38. Homestake Mining Company, "Report on Operations for 1952."
39. Homestake Mining Company, "Report on Operations for 1953."
40. Homestake Mining Company, "Report on Operations for 1955."
41. Homestake Mining Company, "Report on Operations for 1956."
42. A. H. Shoemaker, "Report of the General Manager," Homestake Mining Company, February 12, 1957.

Chapter Twenty-one

1. Homestake Mining Company, "Report on Operations for 1953."

2. Homestake Mining Company, "Report on Operations for 1955."
3. Homestake Mining Company, "Report on Operations for 1957."
4. Homestake Mining Company, "Report on Operations for 1959."
5. Homestake Mining Company, *Sharp Bits* 13, no. 11 (December 1962).
6. Homestake Mining Company, "Annual Report of Homestake Mine Operations," 1986; Annual Report of Homestake Mine Operations," 1987; "Annual Report of Homestake Mine Operations," 1988; "Annual Report of Homestake Mine Operations," 1989.
7. Homestake Mining Company, *Sharp Bits* 9, no. 2 (April 1958).
8. Homestake Mining Company, "Report on Operations for 1951."
9. Homestake Mining Company, *Sharp Bits* 11, no. 11 (December 1960).
10. Homestake Mining Company, *Sharp Bits* 15, no. 9 (October 1964).
11. Albert P. Gilles, "Raise Boring at Homestake," technical paper presented at the Seventh Annual Intermountain Minerals Conference in Vail, Colorado, July 29-31, 1971.
12. Ibid.
13. Homestake Mining Company, "Annual Report of Homestake Mine Operations," 1983.
14. Homestake Mining Company, *Sharp Bits* 19, no. 2 (Summer 1968).
15. Homestake Mining Company, "Manager's Report on the Homestake Mine and Bulldog Mine for 1973."
16. Homestake Mining Company, "Bullion Production Data, January 1, 1878, through December 31, 2001."
17. Homestake Mining Company, "Deep Level Project: Engineering and Cost Study, Homestake Mine," July 1969.
18. Homestake Mining Company, "Report on Black Hills Operations for 1971."
19. Homestake Mining Company, "Manager's Report on the Homestake Mine and Bulldog Mine for 1972."
20. Ibid.
21. Homestake Mining Company, "Report of Mine Superintendent," 1975.
22. Homestake Mining Company, "Manager's Report on the Homestake Mine and Bulldog Mine," 1977.
23. Homestake Mining Company, "Manager's Report on the Homestake Mine and Bulldog Mine for 1974."
24. Homestake Mining Company, "Report of Mine Superintendent," 1975.
25. Homestake Mining Company, "Annual Report Summary for Homestake Mine Operations," 1976.
26. Homestake Mining Company, "Annual Report, Homestake Mine Operations," 1978.
27. LeEtta Shaffner to Steven Mitchell, February 25, 2009.

28. Homestake Mining Company, "Deep Level Project: Engineering and Cost Study, Homestake Mine," July 1969.
29. Homestake Mining Company, "Manager's Report on the Homestake Mine and Bulldog Mine for 1972."
30. Homestake Mining Company, "Deep Level Ventilation Data, End Position When Complete," April 7, 1971; Homestake Mining Company, "Manager's Report on the Homestake Mine and Bulldog Mountain Operation for 1973."
31. Ibid.
32. Homestake Mining Company, "Manager's Report on the Homestake Mine and Bulldog Mine for 1972."
33. Homestake Mining Company, "Manager's Report on the Homestake Mine and Bulldog Mine for 1973."
34. Homestake Mining Company, "Report of Mine Superintendent," 1975.
35. Ibid.
36. Homestake Mining Company, "Manager's Report on the Homestake Mine and Bulldog Mine," 1977.
37. Homestake Mining Company, "Annual Report, Homestake Mine Operations," 1979.
38. Homestake Mining Company, "Annual Report, Homestake Mine Operations," 1986.
39. John Marks to LeEtta Shaffner, February 25, 2009.
40. Homestake Mining Company, "Annual Report, Homestake Mine Operations," 1988.
41. John Marks, "Basic Statistics for 6950 Vent. Project," Homestake Mining Company, December 19, 1989.
42. Homestake Mining Company, "Annual Report to Shareholders," 1990.
43. Homestake Mining Company, "Annual Report to Shareholders," 1994.
44. LeEtta Shaffner to Steven Mitchell, February 25, 2009.

Chapter Twenty-two

1. Homestake Mining Company, "Manager's Report on the Homestake Mine and Bulldog Mine for 1974."
2. Ibid.
3. Ibid.
4. Ibid.
5. Homestake Mining Company, "Report on Black Hills Operations for 1971."
6. Homestake Mining Company, "Report on Black Hills Operations for 1972."

7. Homestake Mining Company, "Report on Black Hills Operations for 1973."
8. Homestake Mining Company, "Report on Black Hills Operations for 1974."
9. Harold L. Hinds and Larry L. Trautman, "Metallurgical Practice at Homestake, Lead Operation," *Gold, Silver, Uranium, and Coal: Geology, Mining, Extraction, and the Environment* (New York: The American Institute of Mining, Metallurgical, and Petroleum Engineers, Inc., 1983): 88-98.
10. Ross Ommen and Larry L. Trautman, "New Innovations in Gold Processing using Gravity, Oxygen-Vacuum and Carbon at the Homestake Mine," *Fifth Western Regional Conference on Precious Metals and the Environment* (Rapid City: Black Hills Section, American Institute of Mining, Metallurgical, and Petroleum Engineers, 1993): 97-113.
11. Hinds and Trautman, "Metallurgical Practice at Homestake, Lead Operation," 95-98.
12. Lin Carr, *History of Homestake Gold Mine: 1876 to Present* (Lead: Homestake Mining Company, August 1, 1994).
13. U.S. Environmental Protection Agency, Office of Solid Waste, *Treatment of Cyanide Heap Leaches and Tailings*, EPA530-R-94-037 (September 1994): 15, http://www.epa.gov/waste/nonhaz/industrial/special/mining/techdocs/cyanide.pdf (accessed July 3, 2009).
14. Homestake Mining Company, "Manager's Report on the Homestake Mine and Bulldog Mine," 1981.
15. Homestake Mining Company, "Manager's Report on the Homestake Mine and Bulldog Mine," 1982.
16. URS Corporation, "Operations Manual for Grizzly Gulch Tailing Dam," prepared for Homestake Mining Company, August 2005.
17. H. L. Hinds and R. D. Ommen, "Homestake's Gravity Recovery System," *Proceedings of the Third Western Regional Conference on Precious Metals, Coal, and the Environment*, K. N. Han and C. A. Kliche, eds., (Rapid City: Black Hills Section, Black Hills Section, American Institute of Mining, Metallurgical, and Petroleum Engineers, 1987): 143-49.
18. Homestake Mining Company, "Annual Report, Homestake Mine Operations," 1987.
19. Ommen and Trautman, "New Innovations in Gold Processing using Gravity, Oxygen-Vacuum and Carbon at the Homestake Mine," 97-113.
20. Homestake Mining Company, "Annual Report, Homestake Mine Operations," 1989.
21. Homestake Mining Company, "Annual Report, Homestake Mine Operations," 1992.
22. Ommen and Trautman, "New Innovations in Gold Processing using Gravity, Oxygen-Vacuum and Carbon at the Homestake Mine," 97-113.

23. Ibid.
24. Homestake Mining Company, "Annual Report, Homestake Mine Operations," 1992.

Chapter Twenty-three

1. Homestake Mining Company, "Report on Operations for 1951."
2. Carl Schmuck, "Cablebolting at Homestake," (paper presented at Rocky Mountain Minerals Conference, Vail Colorado, July 24-27, 1979).
3. Homestake Mining Company, "Annual Report of Homestake Mine Operations," 1983.
4. Homestake Mining Company, "Report on Operations for 1951."
5. Homestake Mining Company, "Report on Operations for 1952."
6. Homestake Mining Company, "Report on Operations for 1955."
7. Homestake Mining Company, "Annual Report of Homestake Mine Operations," 1984.
8. Homestake Mining Company, "Report on Operations for 1953."
9. Homestake Mining Company, *Sharp Bits* 9, no. 3 (April 1958).
10. Homestake Mining Company, "Manager's Report on the Homestake Mine and Bulldog Mine for 1972;" Homestake Mining Company, "Manager's Report on the Homestake Mine and Bulldog Mine for 1973;" Homestake Mining Company, "Manager's Report on the Homestake Mine and Bulldog Mine for 1974;" Homestake Mining Company, "Manager's Report on the Homestake Mine and Bulldog Mine for 1975;" Homestake Mining Company, "Manager's Report on the Homestake Mine and Bulldog Mine for 1976;" Homestake Mining Company, "Manager's Report on the Homestake Mine and Bulldog Mine for 1977."
11. Steven T. Mitchell, "VCR Stoping at Homestake," *Mining Engineering*, 32, no. 11 (November 1980): 1581-1586.
12. Homestake Mining Company, "Manager's Report on the Homestake Mine and Bulldog Mine for 1982."
13. Homestake Mining Company, "Annual Report to Shareholders," 1995.

Chapter Twenty-four

1. Homestake Mining Company, "Annual Report, Homestake Mine Operations," 1981.
2. Homestake Mining Company, "Annual Report, Homestake Mine Operations," 1982.
3. Homestake Mining Company, "Annual Report, Homestake Mine Operations," 1983.

4. Jerry Pfarr, "Terraville Test Pit Executive Summary," Homestake Mining Company, November 22, 1985.
5. Ibid.
6. Homestake Mining Company, "Annual Report, Homestake Mine Operations," 1985.
7. Homestake Mining Company, "Annual Report, Homestake Mine Operations," 1987.
8. Homestake Mining Company, "Annual Report, Homestake Mine Operations," 1985.
9. Homestake Mining Company, "Annual Report, Homestake Mine Operations," 1992.
10. Ibid.
11. Homestake Mining Company, "Annual Report, Homestake Mine Operations," 1998.
12. The Minerals and Mining Program, "Summary of the Mining Industry in South Dakota, April 2000," South Dakota Department of Environment and Natural Resources http://denr.sd.gov/des/mm/Goldrpta3.pdf (accessed May 20, 2009).
13. The Minerals and Mining Program, "Summary of the Mining Industry in South Dakota for 2007," South Dakota Department of Environment and Natural Resources, May 2008, http://denr.sd.gov/des/mm/documents/Goldrpt07.pdf (accessed May 20, 2009).

Chapter Twenty-five

1. A. B. Mitchell, "Mine Surveying," *Sharp Bits* 10, no. 11 (December, 1959): 1-6.
2. W. W. Staley, *Introduction to Mine Surveying* (Stanford: Stanford University Press, 1964), 148-153.
3. Homestake Mining Company, "Manager's Report on the Homestake Mine and Bulldog Mine for 1973."
4. Charles Herman Fulton, *A Manual of Fire Assaying* (New York: Hill Publishing Company, 1907), 41.
5. Nathaniel Herz, "Fire and Chemical Analysis, The Homestake Enterprise," *Engineering and Mining Journal* 132, No. 7 (October 12, 1931).
6. Ibid.
7. Hinds and Trautman, "Metallurgical Practice at Homestake, Lead Operation," 88-98.
8. Ibid., 99-107.
9. *Deadwood Daily Pioneer-Times*, December 24, 1908.
10. Charles L. Tesch, "Wood Pattern," Homestake Mining Company, undated report.
11. Homestake Mining Company, "Report on Operations for 1934."
12. Homestake Mining Company, "Report on Operations for 1937."

13. Fielder, *The Treasure of Homestake Gold*, 331-34.
14. Alvin Dyer, pers. comm. to author, April 10, 2008.
15. Homestake Mining Company, *Sharp Bits* 16, no. 8 (September 1965).
16. Homestake Mining Company, *Sharp Bits* 17, no. 5 (June 1966).
17. Homestake Mining Company, *Sharp Bits* 16, no. 8 (September 1965).
18. Homestake Mining Company, *Sharp Bits* 17, no. 5 (June 1966).
19. Homestake Mining Company, *Sharp Bits* 20, no. 1 (Spring 1969).
20. Nobel Prize.org, "The Nobel Prize in Physics, 2002," Nobel Web AB, http://nobelprize.org/nobel_prizes/physics/laureates/2002/ (accessed May 20, 2009).

Chapter Twenty-six

1. Yates, "The Homestake Mine, A Historical Sketch," 143.
2. O'Harra, "Early Placer Gold Mining in the Black Hills," 361.
3. Homestake Mining Company, "Bullion Production Data, January 1, 1878 through December 31, 2002."
4. Yates, "The Homestake Mine, A Historical Sketch," 143.
5. Coinlink Coin Resources, "Coinage Act of 1834," Coinlink, http://www.coinlink.com/Resources/coinage-acts-by-congress/coinage-act-of-1834/ (accessed May 20, 2009).
6. The Privateer Gold Pages, "The Early Gold Wars," *The Privateer Market Letter* (2001), http://www.the-privateer.com/gold2.html (accessed May 20, 2009).
7. Austin Gold Information Network, "History of Gold Prices Since 1793," Austin Gold Information Network, http://www.goldinfo.net/yearly.html (accessed May 20, 2009).
8. Homestake Mining Company, "Bullion Production Data, January 1, 1878, through December 31, 2002."
9. Ibid.
10. *The New York Times*, September 12, 2000.
11. Homestake Mining Company, "Homestake Lead Operations, 2000 Ore Reserve and Resource," December 31, 2000.
12. *The New York Times*, September 12, 2000.

Chapter Twenty-seven

1. Homestake Adams Research and Cultural Center, "Homestake Adams Research and Cultural Center: Overview," Adams Museum and House, Inc., http://theadamsdeadwood.org/HomestakeAdamsOverview.aspx (accessed July 30, 2009).
2. Homestake Mining Company, "Strategy Plan for the Homestake Closure Project" 2001.

3. The Minerals and Mining Program, "Summary of the Mining Industry in South Dakota for 2006," South Dakota Department of Environment and Natural Resources, May 2007, http://denr.sd.gov/des/mm/documents/2006Goldrpt.pdf (accessed July 30, 2009).
4. The Minerals and Mining Program, "Summary of the Mining Industry in South Dakota for 2006," South Dakota Department of Environment and Natural Resources, May 2007, http://denr.sd.gov/des/mm/documents/2006Goldrpt.pdf (accessed July 30, 2009).
5. Homestake Mining Company and Barrick Gold Corporation, "Barrick and Homestake Announce Merger Plan: US$2.3 Billion Share Exchange to Create Industry Leader," joint press release, June 25, 2001.
6. Sanford Underground Laboratory at Homestake, "Timeline," South Dakota Science and Technology Authority, http://www.sanfordundergroundlaboratoryathomestake.org/index.php?option=com_content&view=article&id=11&Itemid=16 (accessed July 30, 2009).
7. Mathematical and Physical Sciences, "Announcement of Awards for Developing Conceptual Designs for a Deep Underground Science and Engineering Laboratory (DUSEL)," The National Science Foundation, http://www.nsf.gov/news/news_summ.jsp?cntn_id=104313&org=PHY&from=news (accessed July 30, 2009).
8. Charlie Raines, "DUSEL Redux," Cascade Chapter of the Sierra Club, June 26, 2006, http://cascade.sierraclub.org/node/559 (accessed July 30, 2009).
9. Sanford Underground Laboratory at Homestake, "Timeline," South Dakota Science and Technology Authority, http://www.sanfordundergroundlaboratoryathomestake.org/index.php?option=com_content&view=article&id=11&Itemid=16 (accessed July 30, 2009).
10. *Rapid City Journal,* "Lab water down to 4,600 feet level," (October 31, 2008): A5.
11. Wendy Pitlick, "4850 is dry," *Black Hills Pioneer,* May 14, 2009.
12. Sanford Underground Laboratory at Homestake, "Hydrostatic Sensors Detect Earth Tides," South Dakota Science and Technology Authority, http://www.sanfordundergroundlaboratoryathomestake.org/index.php?option=com_content&view=article&id=171:hydrostatic-system-detects-qearth-tidesq&catid=22:hydrology-studies&Itemid=56 (accessed July 30, 2009).
13. Wendy Pitlick, "Early science measures ground movement in Homestake," *Black Hills Pioneer,* March 10, 2009.
14. Max Mankin, "Professor shedding light on dark matter," *The Brown Daily Herald,* November 4, 2008, article reprinted by the South Dakota Science and Technology, http://www.sanfordundergroundlaboratoryathomestake.org/index.php?option=com_content&view=article&id=51:prof-shedding-light-on-dark-matter&catid=13:science-catigory-1&Itemid=53 (accessed July 30, 2009).

15. Neutrino Scientific Assessment Group, "Recommendations to the Department of Energy and the National Science Foundation on a United States Program in Neutrino-less Double Beta Decay, Report to the Nuclear Science Advisory Committee and the High Energy Physics Advisory Panel," Pacific Northwest National Laboratory (September 1, 2005): 3, http://majorana.pnl.gov/documents/NuSAG_report_final_version.pdf (accessed July 30, 2009).
16. DUSEL Experiment and Development Coordination (DEDC), "Deep Underground Science and Engineering Laboratory," Lawrence Berkeley National Laboratory, http://www.lbl.gov/nsd/homestake/dedc.html (accessed July 30, 2009).
17. Division of Physics, "Deep Underground Science and Engineering Laboratory (SUSEL S4)," The National Science Foundation, http://www.nsf.gov/funding/pgm_summ.jsp?pims_id=503136 (accessed July 30, 2009).
18. Homestake Deep Underground Science and Engineering Laboratory (DUSEL), "DUSEL Experiment and Development Coordination (DEDC)," Lawrence Berkeley National Laboratory, http://www.lbl.gov/nsd/homestake/dedc.html (accessed July 30, 2009).
19. Kayla Gahagan, "Researchers continuing review in touring mine lab," *Rapid City Journal*, March 26, 2009; Wendy Pitlick, "DUSEL efforts intensify," *Black Hills Pioneer*, October 2, 2009.
20. Division of Physics, "Deep Underground Science and Engineering Laboratory (SUSEL S4)," The National Science Foundation, http://www.nsf.gov/funding/pgm_summ.jsp?pims_id=503136 (accessed July 30, 2009).
21. Kayla Gahagan, "Researchers continuing review in touring mine lab," *Rapid City Journal*, March 26, 2009; Jaci Conrad Pearson, "National, local lab efforts move forward, *Black Hills Pioneer*, August 9, 2009.
22. Homestake Deep Underground Science and Engineering Laboratory (DUSEL), "Principle Aspects of the Homestake Conceptual Design Report," Lawrence Berkeley National Laboratory, http://www.lbl.gov/nsd/homestake/PDFs/Homestake_DUSEL_Concepts-1.pdf (accessed July 30, 2009); Wendy Pitlick, "DUSEL team announces preliminary construction timeline," *Black Hills Pioneer*, October 3, 2009.
23. Ibid.; Syd DeVries, pers. comm. to author, February 11, 2009.
24. Wendy Pitlick, "DUSEL meeting successful," *Black Hills Pioneer*, October 6, 2009.
25. Syd DeVries, pers. comm. to author, February 11, 2009.

Chapter Twenty-eight

1. Ronald J. Baril Sr., "The House of Francois Baril," Ancestry.com, http://www.geocities.com/barilgenealogy/ (accessed August 1, 2009).
2. Manuel, "Forty-Eight Years in the West."

3. Ibid.
4. Baril, "The House of Francois Baril."
5. Manuel, "Forty-Eight Years in the West."
6. Edna Adamson Manuel, "Appendage to 'Forty-Eight Years in the West,'" October 3, 1951, Black Hills Mining Museum.
7. Baril, "The House of Francois Baril."
8. James William Harney Sr., e-mail message to author, May 20, 2008.
9. Ibid., May 21, 2008.
10. Baril, "The House of Francois Baril."
11. Shala Huff, "John Hearst of the Long Canes, Hearst Genealogy," Ancestry.com, http://homepages.rootsweb.ancestry.com/~amcolan/Hearst/hearst.html (accessed August 1, 2009).
12. George Hearst, memoir (1890).
13. Huff, "John Hearst of the Long Canes, Hearst Genealogy."
14. George Hearst, memoir (1890).
15. Ibid.
16. *Deadwood Magazine*, "George Hearst: The Pauperly Prince," (Deadwood: TDG Communications, Inc., July 2008).
17. George Hearst, memoir (1890).
18. Ibid.
19. Steven Schoenherr, "William Randolph Hearst," Steven Schoenherr, http://history.sandiego.edu/gen/media/hearst.html (accessed August 1, 2009).
20. George Hearst, memoir (1890).
21. *Black Hills Visitor Magazine*, "The Hearst Legacy in the Black Hills, Part 1," http://www.blackhillsvisitor.com/main.asp?id=14&cat_id=30276 (accessed May 20, 2009).
22. Huff, "John Hearst of the Long Canes, Hearst Genealogy."
23. Mort Young, "Phoebe Hearst: A Pioneer Woman," *Seattle Post-Intelligencer*, September 12, 1976.
24. Huff, "John Hearst of the Long Canes, Hearst Genealogy."
25. Young, "Phoebe Hearst: A Pioneer Woman."
26. The California Museum, "Phoebe Hearst, California Hall of Fame 2007 Inductee," The California Museum, https://www.californiamuseum.org/Exhibits/Hall-of-Fame/2006/hearst.html (accessed August 1, 2009).
27. Castlewood Country Club, "The Hearsts and Hacienda del Pozo de Verona," Castlewood Country Club, http://www.castlewoodcc.org/Default.aspx?p=DynamicModule&pageid=248496&ssid=105323&vnf=1 (accessed August 1, 2009).
28. Young, "Phoebe Hearst: A Pioneer Woman."
29. Toms et al., *The Gold Belt Cities: Lead & Homestake*, 45.
30. Fielder, *The Treasure of Homestake Gold*, 153-54, 276.
31. Young, "Phoebe Hearst: A Pioneer Woman."

32. Susan Cerny, "Phoebe Hearst was a Major Benefactress to the University," *The Berkeley Daily Planet*, February 2, 2002.
33. Young, "Phoebe Hearst: A Pioneer Woman."
34. Jen Stevenson, "Phoebe Apperson Hearst," Embarcadero Publishing Company, http://www.pleasantonweekly.com/morgue/2001/2001_04_13.phoebe13.html (accessed August 1, 2009).
35. Linda H. Peck, "Biography of James Ben Ali Haggin," Linda H. Peck—Haggin Histories, http://haggin.org/JBAH_Biography.html (accessed August 1, 2009).
36. Ibid.
37. Del Paso Country Club, "Del Paso Country Club History," Del Paso Country Club, http://www.delpasocountryclub.com/delpasohistory/index.cfm?fuseaction=cor_av&artID=2208 (accessed August 1, 2009).
38. Peck, "Biography of James Ben Ali Haggin."
39. Eva Dano, "Purebred Arabian Haggin Cup Winners, Percentages," ME Ranch Arabians, http://www.meranch.com/books/hagin-pc.htm (accessed August 1, 2009).
40. Michael Svanevik and Shirley Burgett, "A 19-Century Power Duo: Special to the *Examiner*," November 18, 2002, Linda H. Peck—Haggin Histories, http://haggin.org/PowerDuo.html (accessed August 1, 2009).
41. Wikipedia, Encyclopedia, "Lloyd Tevis," Wikimedia Foundation, Inc., http://en.wikipedia.org/wiki/Lloyd_Tevis (accessed August 1, 2009).
42. Svanevik and Burgett, "A 19-Century Power Duo: Special to the *Examiner*."
43. U.S. Department of the Interior, Bureau of Land Management, "Desert Land Entries," Bureau of Land Management, http://www.blm.gov/ut/st/en/res/utah_public_room/desert_land_entries.html (accessed August 1, 2009).
44. Peter Barnes, "The Great American Land Grab," *The Progress Report*, Benjamin Banneker Center for Economic Justice and Progress, http://www.progress.org/archive/barnes4.htm (accessed August 1, 2009).
45. E. A. Schwartz, "History 347: California History," The California State University: San Marcos, http://courses.csusm.edu/hist347as/br/h4713t02.htm (accessed August 1, 2009).
46. Barnes, "The Great American Land Grab."
47. Philip L. Fradkin with a foreword by J. S. Holliday, *Stagecoach: Wells Fargo and the American West* (New York: Simon and Schuster, 2002), 81.
48. Ibid., 85-86.
49. Noel M. Loomis, *Wells Fargo: An Illustrated History*, (New York: Bramhall House, 1968), 210-12; Fradkin, *Stagecoach: Wells Fargo and the American West*, 86.
50. Loomis, *Wells Fargo: An Illustrated History*, 215-19.
51. Ibid., 235.

52. Svanevik and Burgett, "A 19-Century Power Duo: Special to the *Examiner.*"
53. David R. Baker, "On the Trail of Black Gold," *San Francisco Chronicle*, October 20, 2004, http://www.sfgate.com/cgi-bin/article.cgi?f=/c/a/2004/10/20/BUGPQ9CAGD22.DTL (accessed August 1, 2009).
54. *The New York Times*, July 25, 1899.
55. David W. Ryder, *The Merrill Story* (San Francisco: The Merrill Company, 1958), 1-12.
56. Ibid., 17-24, 27-29.
57. Ibid., 24-30.
58. Ibid., 28; Wikipedia Encyclopedia, "Charles Washington Merrill," Wikimedia Foundation, Inc., http://en.wikipedia.org/wiki/Charles_Washington_Merrill (accessed August 1, 2009).
59. Ryder, *The Merrill Story,* 26-36, 50-61, 137-38.
60. *Deadwood Daily Pioneer-Times*, October 21, 1911.
61. Ibid., October 20, 1911.
62. Ibid., October 24, 1911.
63. Toms et al., *The Gold Belt Cities: Lead & Homestake*, 230-32.
64. *Deadwood Daily Pioneer-Times*, October 22, 1911.
65. Fielder, *The Treasure of Homestake Gold*, 303-304.
66. Toms et al., *The Gold Belt Cities: Lead & Homestake*, 230.

Epilogue

1. *Rapid City Journal*, December 21, 2007.

Appendix

1. U.S. Department of the Interior, Bureau of Land Management, "1872 Mining Law Index," Energy and Minerals Technical Assistance Program, http://www.blm.gov/iemtap/mineintro.html (accessed July 31, 2009).
2. Jeanine Feriancek, "Minerals & Mining Law," FindLaw for Legal Professionals, http://library.findlaw.com/1999/Jan/1/241491.html, (accessed July 31, 2009).
3. National Archives and Records Administration, "How do I locate a lode or placer mining claim?" U.S. Government Printing Office, http://ecfr.gpoaccess.gov/cgi/t/text/text-idx?c=ecfr;rgn=div5;view=text;node=43%3A2.1.1.3.78;idno=43;cc=ecfr#43:2.1.1.3.78.2.180.1, (accessed July 31, 2009).

Glossary of Geologic and Mining Terms

Source: Modified from U.S. Geological Survey, U.S. Bureau of Mines *Dictionary of Mining, Mineral, and Related Terms* published on CD-ROM in 1996.

adit. A horizontal or nearly horizontal passage driven from the surface for the working or dewatering of a mine. If driven through the hill or mountain to the surface on the opposite side, it would be a tunnel.

alluvium. A general term for all detrital deposits resulting from the operations of modern rivers, thus including the sediments laid down in river beds, flood plains, lakes, fans at the foot of mountain slopes, and estuaries.

amalgam. A general term for alloys of mercury with one or more of the well-known metals (except iron and platinum); esp. an alloy of mercury with gold, containing 40 to 60 percent gold, and obtained from the plates in a mill treating gold ore.

amalgamate. (a) To unite a metal with mercury. (b) To form an amalgam with; as mercury amalgamates gold.

amalgamation. The process by which mercury is alloyed with some other metal to produce an amalgam. It was used at one time for the extraction of gold and silver from pulverized ores, but was superseded by cyanidation.

amalgam retort. The vessel used to separate and recover mercury from gold or silver amalgam.

amphibolite. A crystalloblastic rock consisting mainly of amphibole and plagioclase with little or no quartz. As the content of quartz increases, the rock grades into hornblende plagioclase gneiss.

anticline. A fold that is convex upward or had such an attitude at some stage of development. In simple beds, the beds are oppositely inclined, whereas in more complex types, the limbs may dip in the same direction.

apron plate. A sheet of copper or special alloy placed on a table in front of a stamp battery and coated with mercury to trap and amalgamate gold discharged from the stamp mortar.

arrastra. A circular rock-lined pit in which broken ore is pulverized by drag stones that are attached to horizontal arms that rotate around a central pillar. Also spelled arrastre.

assay. To analyze the proportions of metals in an ore; to test an ore or mineral for composition, purity, weight, or other properties of commercial interest. The test or analysis itself; its results.

assay ton. A sample weight of 29.166 grams; used in assaying to represent proportionately the assay value of an ore. Because it bears the same ratio to one milligram that a ton of 2,000 pounds bears to one troy ounce, one milligram of precious metal obtained from an assay ton of ore equals the one troy ounce of precious metal per ton of ore. Abbrev. AT.

back. The roof or upper part of any underground mine cavity.

backfill. Mill tailings or waste rock placed in a stope to provide support for the walls and back after the ore has been removed.

barring down. Prying loose rock from the back or ribs using a scaling bar to prevent an unplanned fall of rock.

batholith. A large, generally discordant plutonic mass that has more than forty square miles of surface exposure and no known floor.

battery. (a) A blasting machine. (b) A series of stamps, usually five, operated in one box or mortar for crushing ore; also, the box in which they are operated.

Blake jaw crusher. A small historic rock crusher with one moving jaw plate, hinged at one end that cycled toward and away from a stationary jaw plate in a regular oscillatory cycle. The jaw crusher served as a primary crusher to reduce run-of-mine ore to a size small enough to be fed into a secondary crusher, where the rock was again reduced in size.

blockhole. (a) A large rock. (b) A small hole drilled into a rock or boulder into which a small charge or explosive is placed to break the rock into smaller pieces.

blockholing. The breaking of oversize rock by loading and firing small explosive charges in small-diameter drilled holes.

Butters and Mein distributor. A turbo distributor that spreads sand evenly in a circular leaching tank or vat used for gold cyanidation.

button. (a) A globule of metal remaining in an assaying crucible or cupel after fusion has been completed. (b) Globule of lead formed during fire assay of gold or silver ore.

cablebolt. A length of high-strength cable used to reinforce ground prior to mining. A typical cablebolt consists of a ⅝-inch-diameter cable 60 feet long that is inserted into a nominal 2½-inch-diameter drill hole and grouted with cement. A pattern of cablebolts provides extended ground support in stope backs or other mining areas.

cage. An elevator-like conveyance used to move people or supplies up or down a shaft or winze; a hoisting engine and wire rope is used to raise and lower the cage, which slides along wooden or steel guides affixed to the shaft compartment.

cager. One who communicates with the hoist operator and directs the movement of cages used to raise and lower workers, mine cars, and supplies between the various levels and the surface; works at the top of a shaft or at an intermediate level or rides the cage itself. Also called a cage tender.

Cambrian rocks. A geologic period assigned to the oldest sedimentary rock units that were formed between 500 and 570 million years ago.

cap. (a) A detonator or blasting cap. (b) To seal, plug, or cover a raise or borehole. (c) The top piece placed over the two posts of a three-piece timber set used for tunnel support. (d) One of the structural members of a square set of timber.

carbide. A commercial term for calcium carbide formerly used in a miner's lamps.

carbide lamp. A lamp that is charged with calcium carbide and water and burns the acetylene generated to produce a flame for illumination.

carbon adsorption. Recovery of dissolved soluble constituents onto activated carbon due to some form of chemical sorption at the active sites. Carbon adsorption is particularly useful for removing gold and silver from cyanide-leach solutions or dissolved organics from process solutions.

carbon-in-pulp leaching. A precious metals-leaching technique in which granular activated carbon particles much larger than the ground-ore particles are added to the cyanidation pulp after the precious metals have been solubilized. The activated carbon and pulp are agitated together to enable the solubilized precious metals to become adsorbed onto the activated carbon. The loaded activated carbon is mechanically screened to separate it from the barren-ore pulp and processed to remove the precious metals and prepare it for reuse.

channel sample. Material collected from a channel that is cut across an ore exposure to obtain a true cross section of it.

chute. An inclined, three-sided trough, constructed of timber and lined with steel plate, that controls the flow of broken rock from a raise to railcars on the haulage level. The rock flow is controlled by raising and lowering planks or a steel gate at the lower end of the chute.

classifier. (a) A device for separating the constituents of a material according to relative sizes and densities, thus facilitating concentration and treatment. The term "classifier" is used in particular where an upward current of water is used to remove fine particles from coarser material. (b) In mineral beneficiation, the classifier is a device that takes the ball-mill discharge and separates it into two portions—the classifier overflow, which is ground as fine as desired, and the classifier underflow, which consists of oversize material.

compressed-air locomotive. A mine locomotive powered by compressed air. The air is transferred from a compressor to the underground charging stations via pipelines.

conglomerate. The rounded and water-worn pebbles of river or beach gravel that have been cemented together to form a solid rock.

contact. A plane or irregular surface between two types or ages of rock. The surface of delimitation between a vein and its wall or country rock.

cope. The upper section of a flask or mold.

coplaning. The process of moving the head of a transit or theodolite laterally until its vertical axis lies in the same vertical plane common to two plumb lines.

core drill. A lightweight, usually mobile drilling machine that imparts thrust and rotation to a string of drill rods, core barrel, and bit and causes a continuous core of rock to be cut and extracted from the bedrock.

crosscut. A passageway driven at right angles to the main drift, tunnel, or adit that extends across the top, middle, or bottom of an ore body; provides access to stopes and chutes.

crown pillar. An ore pillar at the top of an open stope that is left for ground support or protection of facilities above; was typically mined using a square-set method.

crucible. A heat-resistant container used to heat or melt samples or bullion.

crusher. A machine for crushing rock or other materials.

cupola. A cylindrical vertical furnace for melting metal, esp. gray iron, by having the charge come in contact with the hot fuel, usually metallurgical coke.

cut-and-fill stoping. A mining method in which the ore is excavated by successive flat or inclined slices, working upward from the level. After each slice is blasted down, the broken ore is removed and the stope is filled with waste or sandfill. The term "cut-and-fill stoping" implies a definite and characteristic sequence of operations: (1) breaking a slice of ore from the back, (2) removing the broken ore, and (3) backfilling.

cutoff grade. The lowest grade of mineralized material that qualifies as ore in a given deposit; rock of the lowest assay included in the outline for an ore estimate.

cyanidation. A process of dissolving gold and silver using a solution of potassium or sodium cyanide. The dissolved metals are recovered using other treatment processes.

cyclone. A classifying (or concentrating) separator into which pulp is fed into a vortex of water. Coarser and heavier fractions of solids represent cyclone underflow. Finer particles, called the cyclone overflow, flow from a central vortex.

dark matter. Invisible matter that has not yet been detected, but whose presence is inferred from its effects on visible matter, including the rotation of galaxies; is believed to account for about 90 percent of the mass of the universe.

desorption. A process where adsorbed material, such as gold, is removed from the adsorbent, activated carbon; the reverse process of adsorption.

dike. A tabular igneous intrusion that fills a fissure and cuts across the bedding or foliation of the country rock.

dike swarm. A group of dikes, which may be in radial, parallel, or en echelon arrangement. Their relationship with the parent plutonic body may not be directly observable.

dip. The angle at which a bed, stratum, or vein is inclined from the horizontal, measured perpendicular to the strike and in a vertical plane.

doghouse. An enclosure or small chamber that is used for an office in an underground mine.

dolomite. A limestone that has been transformed by the substitution of magnesium carbonate for a portion of the original calcium carbonate.

dore'. An impure alloy of gold and silver.

drag. The bottom part of a two-piece mold used in sand casting. The cope and drag comprise the framework that holds the sand mold. Metal parts are created or cast by pouring molten metal into the mold cavity that is created after a wood pattern is removed from tightly packed sand.

drift. An adit or horizontal opening that is parallel to the strike or course of the vein or long dimension of the ore body.

dry. A miner's change house, which is typically equipped with showers, lockers, clothes baskets, and a means of drying wet clothing.

dump. (a) The location at which broken rock is discharged from railcars or a loader into a raise or stope. (b) A location on the surface where waste rock is placed.

face. The end of a drift being advanced or the active mining front in a stope.

fill. Mill tailings or waste rock used to fill underground voids after the ore has been removed. Termed "hydraulic fill" if placed as a slurry.

fineness. The proportion of pure silver or gold in jewelry, bullion, or coins; often expressed in parts per thousand, such as "900 fine."

fire assay. A process of fusing a small sample of gold and silver ore to determine metal content; requires furnace heat and involves scorification and cupellation.

foliation. (a) A general term for a planar arrangement of textural or structural features in a rock such as schist. (b). The laminated structure resulting from segregation of different minerals into layers parallel to the schistosity.

footwall. The underlying side of a fault, ore body, or mine working; the wall rock beneath an inclined ore vein or fault.

free gold. (a) Gold that is not chemically combined with other substances. (b) Placer gold.

free-milling. Ore from which the free gold or silver and can be recovered by simple milling methods.

geothermal gradient. The rate of increase of temperature per unit depth in the Earth. The gradient differs from place to place depending on the heat flow in the region and the thermal conductivity of the rocks.

giant powder. A blasting powder consisting of nitroglycerin, sodium nitrate, sulfur, rosin, and sometimes kieselguhr.

glory-hole mining. A surface mining method, usually located lower than the mill tramway level that consists of a small open cut, the bottom of which is connected to an ore raise driven from an underground haulage level. The excavation of the ore begins at the surface. Broken ore is removed by transferring it into the raises and chutes and loading it into cars on the haulage level. The ore is hoisted to the mill tramway level at or near the surface.

grade. The average metal content (e.g., gold) of an ore reserve, stope, face, or ore in process. Usually expressed in troy ounces per short ton or in dollars per ton, based on either the assay-ton system or the dollar-ton system. Under the latter two systems, an assay containing one milligram of gold was equivalent to either

$20.67 per ton or $20 per ton, respectively, based on the historic gold price of $20.67 per troy ounce.

Granby car. A type of mine car that has a "fifth wheel" attached to the side of the car body. As the fifth wheel travels over a ramp at the dumping point, the car body is automatically raised and lowered, permitting the car to dump its load.

grizzly. A series of iron or steel bars spaced to control the size of rock that is allowed to be transferred to a raise, chute, bin, or crusher.

grizzly man. A person who works underground at a grizzly and breaks oversized rocks with a sledge hammer so that they will pass through the grizzly.

ground. The rock mass that surrounds a mine opening such as a drift, raise, shaft, stope, or open cut.

ground control. The act of controlling the rock mass or ground that surrounds a mine opening; may include engineered designs for slopes, walls, or openings, scaling, smooth-wall blasting, or the use of support systems such as rock bolts, stulls, cribs, square sets, cablebolts, shotcrete, backfill, and other means of controlling or stabilizing the rock mass.

hanging wall. The overlying side of an ore body, fault, or mine working, esp. the wall rock above an inclined ore vein or fault.

hoist. (a) A winch that is used to handle, hoist, and lower drill-string equipment, casing, pipe, etc., while drilling, or to move the drill from place to place. (b) A power-driven windlass for raising ore, rock, or other material from a mine and for lowering or raising people and material. (c) A drum on which hoisting rope is wound in the engine house, as the cage or skip is raised in the hoisting shaft.

hydraulic fill. Mill tailings that are transported underground in the form of a slurry and discharged into the stope or other cavity to be filled.

hydrothermal solution. A hot magmatic emanation rich in water that contains dissolved mineral substances.

igneous rock. Rock formed by the solidification of hot molten magma.

jig. A concentrating table covered with water that separates metals or ore from foreign matter in layers according to their specific gravity.

manway. A compartment, vertical or inclined, for the accommodation of ladders, pipes, and utilities. Its purpose is to provide access into a stope. Also called ladderway.

Merrill-Crowe process. A process of recovering gold using cyanidation to dissolve gold and zinc dust to precipitate the gold. The gold precipitate is then filtered out of the solution, mixed with fluxes and smelted to form crude bullion. The crude bullion is refined to separate the gold and silver and any impurities.

mesh size. A phrase used to designate screen size or particle size. The number of openings per lineal inch of screen. A 200-mesh screen has two hundred openings per lineal inch or forty thousand openings per square inch. Particles of material small enough to pass through a 200-mesh screen are referred to as "minus 200-mesh." Particles retained on the screen are referred to as "plus 200-mesh." Mesh fraction represents that percentage of a material that will pass through a specified screen size or is retained by a certain screen size.

metamorphic rock. Rock that has had its mineral content changed by being subjected to extreme heat and pressure, such as a shale to slate or schist or sandstone to quartzite.

milling. The grinding or crushing of ore. The term may include the operation of removing valueless or harmful constituents and preparing the ore for other treatment processes.

miner's inch. The quantity of water that will flow through a 1-inch-square opening in a 2-inch-thick plank under a head of 6 inches. Under South Dakota statute, 50 miner's inches is equivalent to 1 cubic foot per second of flow.

mining claim. (a) That portion of the public mineral lands that a miner, for mining purposes, locates and possesses in accordance with mining laws. (b) A mining claim is a parcel of land containing valuable minerals in the soil or rock. A location is the act of appropriating such a parcel of land pursuant to provisions of the Mining Law of 1872.

motorman. The person who operates a haulage locomotive.

muck. (a) Blasted rock, particularly from underground. (b) To excavate or remove muck.

mucking. The operation of shoveling broken rock by hand or machine.

neutrino. (a) Any of three stable leptons having a mass approaching zero and no charge. (b) Elementary particles that travel close to the speed of light, lack an electric charge, are able to pass through ordinary matter with almost no interaction and are extremely difficult to detect.

open-stope method. Stoping in which no regular method of artificial support is employed other than rock bolting or occasional props or cribs.

ore. The naturally occurring rock from which a mineral or minerals of economic value can be extracted profitably. The term is generally but not always used to refer to metalliferous material and is often modified by the name of the valuable constituent, e.g., gold ore.

ore deposit. A mineral deposit that has been tested and is known to be of sufficient size, grade, and accessibility to yield a profit when mined and processed.

orepass. A vertical or inclined passage for the downward transfer of ore; equipped with gates or other appliances for controlling the flow of rock. An orepass is driven in ore or country rock and connects a level with the skip loader or a lower haulage level.

ore reserve. The total tonnage and grade of ore that are determined and proven in accordance with federal standards.

outcrop. That part of a geologic formation or structure that is exposed at the surface; may include bedrock that is covered only by surficial deposits such as alluvium.

patent. A document from the federal government that conveys title to a previously unpatented mining claim; no further assessment work need be done; however, taxes must be paid.

patented mining claim. A mining claim to which title to ownership of the surface and mineral rights has been transferred to the applicant by the U.S. government in compliance with the laws relating to such claims.

pattern maker. A person who makes wooden patterns for sand castings.

Pelton wheel. An impulse water turbine with buckets bolted to its periphery, which are struck by a high velocity jet of water.

penstock. A closed conduit for supplying water under pressure to a waterwheel or turbine.

phonolite. The extrusive equivalent of nepheline syenite. The principal mineral is soda orthoclase or sanidine. Other major minerals are nepheline and aegirine—diopside, usually with other feldspathoidal minerals such as sodalite or hauyne. Phonolite is an important ore progenitor.

phyllite. A metamorphic rock, intermediate in grade between slate and mica schist. Minute crystals of sericite and chlorite impart a silky sheen to the schistosity.

pillar. (a) A column of ore that has been left to support the back. (b) An unmined area between two stopes.

pilot raise. A small raise driven in the center and in advance of the main shaft, raise, or winze to create a free face for subsequent enlargement by stripping.

placer. A deposit of sand or gravel that contains particles of gold, ilmenite, gemstones, or other heavy minerals of value. The common types are stream gravels and beach sands.

placer claim. (a) A mining claim located upon gravel or ground whose mineral contents are extracted by the use of water by sluicing, hydraulic mining, dry washing, etc. (b) Ground with defined boundaries that contains mineral in the earth, sand, or gravel; ground that includes valuable deposits not fixed in the rock. (c) The maximum size of a placer claim is 20 acres. Association claims of two or more persons may be located up to an area of 160 acres for eight people.

placer gold. Gold that has been eroded from bedrock and now occurs as grains, flakes, or nuggets in alluvial or soil material.

placer mining. The process of separating a heavy mineral such as gold from alluvium by wet or dry gravity concentration methods that may include the use of pans, sluice boxes, rocker boxes, dredges, trommels, concentrating tables, etc. When water under pressure is employed to break down the gravel, the term hydraulic mining is generally employed.

plat. (a) The map of a survey in horizontal projection, such as of a mine, townsite, etc. (b) A diagram drawn to scale showing land boundaries and subdivisions, together with all data essential to the description of the several units. A plat differs from a map in that it does not show additional cultural, drainage, and relief features.

porphyry. An igneous rock of any composition that contains conspicuous phenocrysts in a fine-grained groundmass.

portal. (a) The surface location at which an adit or tunnel is started. (b) The surface entrance to a drift, tunnel, adit, or entry. (c) The log, concrete, timber, or masonry arch or retaining wall erected at the opening of a drift, tunnel, or adit.

powder. A general term for explosives including dynamite, but excluding caps.

precipitate. (a) The operation, act, or process of adding a chemical or chemicals to an aqueous solution to react with a dissolved material in the solution and remove the resulting new solid matter by settling. (b) The solids resulting from the precipitation process.

primary crusher. (a) The first crusher in a series for breaking rocks. (b) In comminution of ore, a heavy-duty dry crushing machine capable of accepting run-of-mine coarse ore and reducing it in size to somewhere between 4 inches and 6 inches.

pulp. A mixture of ground ore and water capable of flowing through suitably graded channels as a fluid. Its dilution or consistency is specified as a percentage of solids by weight.

quicksilver. A common name for mercury.

raising. The process of mining a shaft, winze, or steep tunnel in an upward manner.

reclamation. Restoration of mined land to near-original conditions or some other beneficial use.

recovery. (a) The percentage of valuable constituent derived from ore being processed; a measure of extraction efficiency. (b) The ratio of the footage of core acquired from core drilling to a specific length of borehole, expressed in percent.

refining. To purify metals by removing impurities and separating the desirable metals such as gold and silver.

refractory. (a) An ore from which it is difficult or expensive to recover its valuable constituents. (b) Difficult to treat. (c) Resistant to heat.

retorting. The process of removing mercury from an amalgam by volatizing the mercury in an iron retort, conducting the gas away, and condensing it for reuse.

rhyolite. A fine-grained igneous rock that contains small phenocrysts of quartz and alkali feldspar; the aphanitic equivalent of granite.

rib. The side of a pillar or wall of a mine opening.

rock bolt. A rod, usually constructed of steel, that is inserted into a predrilled hole in rock and secured for the purpose of ground control. Rock bolts are classified according to the means by which they are secured or anchored in rock. In current usage, there are mainly four types: friction, point-anchor, grouted, and resin.

roof bolting. A system of establishing ground support in the back of a stope or drift. Holes up to 16 feet long are drilled upward in the back. Rock bolts of $5/8$-1 inch or more in diameter are inserted into the drill holes and are secured by friction, point anchors, grout, or resin.

run-of-mine. The broken ore in the state that it has been removed from the mining areas and delivered to the crushers.

sample. A volume or mass of material that has been collected by approved methods for subsequent analysis, such as fire assaying, to determine metal content. Individual samples are pulverized, blended, and split into smaller portions in the laboratory. In new mines, bulk samples may be collected during the exploration phase to ascertain a suitable treatment method for the ore body. Channel samples, picks, grabs, and wet-sludge test-hole samples are small samples that are taken on a day-to-day basis in drifts and stopes to guide mining operations in accordance with established grade-control policies.

sandfill. Mill tailings that are slurried underground with water to fill and support the cavities created by the extraction of ore.

scaling. (a) The prying, barring, and removing of loose rocks that are marginally affixed to the back or ribs after a shot has been fired. (b) Any removal of loose rocks from the back and ribs using a scaling bar. Syn.: barring.

schist. A strongly foliated crystalline rock, formed by dynamic metamorphism, that can be readily split into thin flakes or slabs due to the well developed parallelism of more than 50 percent of the minerals present, particularly those of lamellar or elongate prismatic habit, e.g., mica and hornblende.

scorification. Part of the assaying process whereby gold or silver is separated from ore by fusion with lead. The gold and silver are removed from the lead by cupellation.

secondary crusher. Crushing and pulverizing machines next in line after primary crushing to further reduce the particle size of rock.

sedimentary rocks. Rocks formed by the accumulation of sediment in water or from air. The sediment may consist of rock fragments or particles of various sizes (conglomerate sandstone, shale); of the remains or products of animals or plants (certain limestones and coal); of the product of chemical action or of evaporation (salt, gypsum, etc.); or of combinations of these materials. A characteristic feature of sedimentary deposits is a layered structure known as bedding or stratification. Sedimentary beds as deposited lie flat or nearly flat.

shaft. An excavation of limited area compared with its depth that is used to access underground ore, hoist rock, transport workers and material, or ventilate underground workings. A shaft is equipped with a hoisting engine at the surface for handling workers, rock, and supplies, or it may be used only for pumping or ventilating operations.

shrinkage stoping. A vertical, overhand mining method whereby most of the broken ore remains in the stope to form a working floor for the miners. Another benefit of leaving the broken ore in the stope is to provide additional wall support until the stope is completed and ready for drawdown. Stopes are mined upward in horizontal slices. Normally, about 35 percent of the ore derived from the stope cuts (the swell) can be drawn off (shrunk) as mining progresses.

skip. A bucket, usually square or rectangular, that rides on guides in the shaft or winze and is used to transfer ore or waste rock up a shaft or winze. The rock is dumped into a transfer bin at the top of the shaft or winze.

skip loader. The chute used to measure and transfer a load of rock from the raise to the skip.

slime. Extremely fine sediment (minus-200 mesh) produced during crushing and grinding operations.

slurry. A mixture of finely ground rock and water that is fluid enough to flow through by gravity through a pipe or launder.

slushing. A means of transferring broken rock from a stope to a raise by dragging the rock over the floor using a scraper bucket that is attached to a multi-drum hoist using cables. The winch is called a slusher.

smelting. The chemical reduction of a metal from its ore by fusion; the impurities readily separate and form a slag that can readily be removed from the reduced metal.

spad. A flat nail that contains a hole in one end and is used to suspend a plumb bob. The spad is inserted into a wooden plug that has been tightly wedged into a hole drilled in the back of the drift. The installed plug and spad constitute an underground survey-control point.

square-set stoping. A method of stoping in which the walls and back of the excavation are supported by a system of interlocking framed timbers. A full square set of timber consists of four posts, two caps, and two ties. A square set is 6 to 7 feet (1.83 to 2.13 m) high, and 4 to 6 feet (1.22 to 1.83 m) square. The square-set mining method is slow, since only enough ore is normally mined to facilitate installation of one set of timber. Ore in the stope is mined out in floors or horizontal panels. The square sets of each successive floor are framed into the top of the preceding floor sets.

stamp mill. An apparatus in which rock is crushed by the weight of falling pestles (stamps) that are operated by steam power. Amalgamation is usually combined with crushing when gold or silver is the metal to be recovered.

stope. Any excavation in a mine, other than development workings, created for the purpose of extracting ore. The outline of the ore body generally determines the outline of the stope. The term is also applied to the process of breaking ore by drilling and blasting.

stripping a shaft. (a) Taking out the timber and utilities from a shaft. (b) Trimming, enlarging, or squaring the sides of a pilot raise, winze, or shaft.

sublevel stoping. A mining method whereby the ore is excavated as an open stope, by retreating the vertical mining face(s) from one end of the stope to the other. The ore body is developed by driving a series of sublevel drifts above the main haulage level. The sublevels are connected by a break raise at one end of the stope and an access raise at the other end. Starting at the break raise, the ore is drilled and blasted using longholes drilled from each of the sublevels; the broken ore falls into the drawholes or an undercut sill, where it is drawn off through chutes or mucked from drawpoint drifts.

subsidence. The sudden or gradual settling of the ground surface above near-surface mining areas that have not been sufficiently supported or backfilled.

sulfide ore. Ore in which sulfide minerals predominate.

syncline. A major fold in rocks in which the strata dip inward from both sides toward the fold axis.

tailings. The gangue or reject material remaining after the washing, concentration, or treatment of ground ore has been completed.

test hole. A small-diameter hole that is drilled with a percussion drill to a depth of up to 15 feet; wet or dry drill cuttings are collected at intervals and assayed to determine gold content. Longhole test holes are infrequently drilled to a depth of 40 feet to test the ground ahead of the active mining face.

troy ounce. One-twelfth of a pound of 5,760 grains (troy pound) or 480 grains. A troy ounce equals 20 pennyweights, 1.09714 avoirdupois ounces, or 31.1035 grams. It is used in all assay returns for gold, silver, and platinum-group metals.

tube mill. A revolving cylinder, usually lined with silex, that is nearly half-filled with pebbles; used for fine grinding of ores to aid in the liberation of gold and enhance recovery. The material to be ground is mixed with water, fed through one end of the mill, and discharged through the opposite end as a slime.

tunnel. A horizontal or nearly horizontal underground passage that is open at both ends. The term is loosely applied in many cases to an adit. An adit, if continued through a hill, would be a tunnel.

unconformity. A substantial break or gap in the geologic record characterized by a lack of continuity in deposition and corresponding to a period of non-deposition, weathering and erosion prior to the deposition of the younger beds, and often marked by absence of parallelism between the strata; the relationship where the younger overlying stratum does not conform to the dip and strike of the older underlying rocks, as shown specifically by an angular unconformity.

vein. (a) A mineral body, narrow in relation to its other dimensions, which cuts across the bedding and in which the minerals were formed after the country rock. (b) A zone or belt of mineralized rock lying within boundaries, clearly separating it from neighboring rock. (c) The filling of a fissure or fault in a rock, particularly if deposited by aqueous solutions. When metalliferous, a vein is called a lode; when filled with igneous intrusive material, a dike.

vertical crater retreat mining. A variation of the sublevel stoping method that is based on crater blasting theory. Blasting is carried out toward the bottom of near-vertical boreholes, creating horizontal cuts that are advanced upwards. A spherical charge is placed at an optimal distance from the stope back so that a maximum volume of rock is broken in the shape of an inverted cone. Borehole spacing is determined so that overlapping fragmentation cones do not disturb adjacent explosive charges. Abbrev. VCR.

waste rock. Barren development rock or submarginal rock that has been mined, but is not of sufficient value to warrant treatment and is therefore not subjected to milling and treatment processes.

winze. (a) A subsidiary shaft that starts underground. It is usually a connection between two levels. (b) Interior mine shaft.

Bibliography

Adams, Elton and Norma Adams. *Nemo, South Dakota: One Hundred Years, 1889-1989.* Nemo Centennial Association, 1989.

Allen, A. W. "The Homestake Enterprise." *Engineering and Mining Journal* 132, no. 7 (October 12, 1931): 304.

Anderson, Marilyn. "Hearst Mercantile Store." *Nemo, South Dakota: One Hundred Years, 1889-1989.* Nemo Centennial Association, 1989.

Artus, Elna. "Coffee Pot Link between Old World and Black Hills." *Deadwood Daily Pioneer-Times*, date unknown.

Austin Gold Information Network. "History of Gold Prices since 1793." Austin Gold Information Network. http://www.goldinfo.net/yearly.html.

Avalon Project at Yale Law School. "Fort Laramie Treaty, 1868." Lillian Goldman Law Library. http://avalon.law.yale.edu/19th_century/nt001.asp.

Badlands Natural History Association. "Early Indians and Explorers." Badlands, History of Badlands National Monument. http://www.nps.gov/history/history/online_books/badl/sec1.htm.

Baker, David R. "On the Trail of Black Gold." *San Francisco Chronicle*, October 20, 2004. http://www.sfgate.com/cgi-bin/article.cgi?f=/c/a/2004/10/20/BUGPQ9CAGD22.DTL.

Baldwin, George P., ed. *The Black Hills Illustrated.* Deadwood: Dakota Graphics. First published 1904 by Black Hills Mining Men's Association.

Baril, Ronald J. Sr. "The House of Francois Baril." Ancestry.com. http://www.geocities.com/barilgenealogy/.

Barnes, Peter. "The Great American Land Grab." *The Progress Report.* Benjamin Banneker Center for Economic Justice and Progress. http://www.progress.org/archive/barnes4.htm.

Beardshear, W. D. "Lumbering in the Black Hills." *The Black Hills Engineer* XIV, no. 3 (May 1926): 175-181.

Berry, J. M. "The Budget of the State of South Dakota for the Biennium 1939 to 1941." Prepared for the legislature by the secretary of Finance (November 1938).

Bjorge, G. N. "Comparison of Estimate to Cost to Dec. 1, 1942—Yates Shaft Construction." Homestake Mining Company, December 28, 1942.

Black Hills & Fort Pierre Railroad Company. "Financial Report for the Black Hills and Fort Pierre Railroad Company." June 30, 1889. Black Hills Mining Museum.

Black Hills Canal and Water Company. "Articles of Incorporation." City and County of San Francisco, California, September 30, 1878.

———. Agreement with the city of Deadwood, August 1893. Homestake Mining Company Water Rights files.

———. "Renewal Certificate of Location of Water Right." Lawrence County, Deadwood, South Dakota, November 13, 1899, 146:177.

———. Agreement with Homestake Mining Company, January 9, 1901. Homestake Mining Company Water Rights files.

———. Agreement with Homestake Mining Company, February 17, 1905. Homestake Mining Company Water Rights files.

———. Minutes of the Board of Directors, October 14, 1878.

Black Hills Corporation. "A Century of Light: The Early Years." Black Hills Corporation. http://www.blackhillspower.com/chapter2b.htm.

———. "The Early Years." Black Hills Corporation. http://www.blackhillspower.com/chapter2a.htm#1.

Black Hills Mining Museum. "Edison Dynamo Exhibit." Black Hills Mining Museum, Lead, South Dakota.

Black Hills Visitor Magazine. "The Hearst Legacy in the Black Hills, Part 1." http://www.blackhillsvisitor.com/main.asp?id=14&cat_id=30276.

Blackstone, Richard. Papers. Homestake Mining Company Water Rights files.

———. "Superintendent's Report." Homestake Mining Company, December 31, 1914.

———. "Superintendent's Report." Homestake Mining Company, December 31, 1915.

———. "Superintendent's Report." Homestake Mining Company, December 31, 1916.

———. "Superintendent's Report." Homestake Mining Company, December 31, 1917.

———. "The Hydro-Electric Power Plant of the Homestake Mining Co." *Mining and Engineering World* XLI, no. 1 (July 4, 1914): 3.

Borland vs. Haven et al. F. 37, Circuit Court, N. D., California, December 17, 1888.

BreweryGems.com. "The Black Hills Brewery of Central City, SD (1878-1927): Illustrated History of the Black Hills Brewing Co. Gary@BreweryGems.com. http://www.brewerygems.com/black-hills.htm.

Bronson, William and T. H. Watkins. *Homestake: The Centennial History of America's Greatest Gold Mine.* San Francisco: Homestake Mining Company, 1977.

Brown, Bruce. "Winter Count of Crazy Horse's Life." B.F. Communications, Inc. http://www.astonisher.com/archives/museum/crazy_horse_wintercount.html.

Brown, Jesse and A. M. Willard. *The Black Hills Trails.* Rapid City: Rapid City Journal Company, 1924.

Buecker, Thomas R. "Can You Send Us Immediate Relief?: Army Expeditions to the Northern Black Hills, 1876-1878." *South Dakota History* 25, no. 2 (Summer 1995): 101-113.

Caledonia Gold Mining Company. "Articles of Incorporation." City and County of San Francisco, California, September 5, 1878.

Campbell, Thomas J. "Synopsis of Homestake Mine Geology." *Homestake—background material for DUSEL, Deep Underground Science and Engineering Laboratory.* http://homestake.sdsmt.edu/Geology/geology.htm.

Canku Luta. "Treaty of Fort Laramie, September 17, 1851." Red Road, Inc. http://www.canku-luta.org/PineRidge/laramie_treaty.html.

Carr, Lin. *History of Homestake Gold Mine: 1876 to Present.* Lead: Homestake Mining Company, August 1, 1994.

Cash, Joseph H. "The Homestake Lockout—1909-1910." Paper presented at The American Middle Period and the West seminar, 16:235, Hearst Library, Lead, South Dakota.

———. *Working the Homestake.* Ames: The Iowa State University Press, 1973.

Castlewood Country Club. "The Hearsts and Hacienda del Pozo de Verona." Castlewood Country Club. http://www.castlewoodcc.org/Default.aspx?p=DynamicModule&pageid=248496&ssid=105323&vnf=1.

Cerny, Susan. "Phoebe Hearst was a Major Benefactress to the University." *The Berkeley Daily Planet,* February 2, 2002.

Citizens of Deadwood. South Dakota. Petition to A. J. Bowie, agent for the Wyoming and Dakota Water Company, September 29, 1879. Homestake Mining Company.

Clark, Allan J. "Notes on Homestake Metallurgy." *American Institute of Mining and Metallurgical Engineers Transactions* 52 (1915): 14.

———. "A Review of Homestake Metallurgy." *The Black Hills Engineer* XIV, no. 3 (May 1926): 160.

———. "Milling Methods and Costs at the Homestake Mine, Lead, South Dakota." U.S. Department of the Interior, Bureau of Mines, Information Circular No. 6408 (February 1931): 16.

Clark, Allan J. and W. J. Sharwood. "The Metallurgy of Homestake Ore." *American Institute of Mining and Metallurgical Engineers Transactions* 22 (November 21, 1912): 72.

Clark, M. J. "Neiman Enterprises Buys Spearfish Sawmill, Doubles Size." TimberBuySell.com. http://www.timberbuysell.com/Community/DisplayNews.asp?id=2553.

Clow, Richmond L. *Chasing the Glitter.* Pierre: South Dakota State Historical Society Press, 2002.

———. "Homestake Mining Company and the First Regulated Timber Harvest." U.S. Forest Service, Black Hills National Forest—Projects and Plans. http://www.fs.fed.us/r2/blackhills/projects/timber/case1/case1.shtml.

Clowser, Don C. *Deadwood . . . The Historic City.* Deadwood: Fenwyn Press Books, 1969.

Coinlink Coin Resources. "Coinage Act of 1834." Coinlink. http://www.coinlink.com/Resources/coinage-acts-by-congress/coinage-act-of-1834.

Columbus Gold Mining Company, Stock certificate No. 68 issued to R. W. Hamilton for 200 shares of capital stock, August 3, 1880.

Connolly, Joseph P. *The Tertiary Mineralization of the Black Hills*. Rapid City: South Dakota School of Mines, September 1927.

Dano, Eva. "Purebred Arabian Haggin Cup Winners, Percentages." ME Ranch Arabians. http://www.meranch.com/books/hagin-pc.htm.

Davey, John H. et al. Quit Claim Deed to the Wyoming and Dakota Water Company dated February 14, 1880. Homestake Mining Company Water Rights files.

Deadwood Magazine. "George Hearst: The Pauperly Prince." Deadwood: TDG Communications, Inc. (July 2008).

Deadwood-Terra Mining Company. "Articles of Incorporation." New York County, State of New York, August 23, 1880.

———. "Certificate as to Dissolution." County of New York, State of New York, April 25, 1902.

Deane, Coll. Agreement with the Wyoming and Dakota, Ditch, Flume, and Mining Company dated November 15, 1877. Homestake Mining Company Water Rights files.

Del Paso Country Club. "Del Paso Country Club History." Del Paso Country Club. http://www.delpasocountryclub.com/delpasohistory/index.cfm?fuseaction=cor_av&artID=2208.

Dippie, Brian D. "It's Equal I have Never Seen, Custer Explores the Black Hills in 1874." *Columbia Magazine* 19, no. 2 (Summer 2005): 3.

Division of Physics. "Deep Underground Science and Engineering Laboratory (SUSEL S4)." The National Science Foundation. http://www.nsf.gov/funding/pgm_summ.jsp?pims_id=503136.

Dodge, Richard Irving. *The Black Hills*. Minneapolis: Ross & Haines, Inc.: Minneapolis, 1965.

Driscoll, D. G., J. M. Carter, J. E. Williamson, and L. D. Putnam. *Hydrology of the Black Hills Area, South Dakota*. U.S. Geological Survey Water-Resources Investigations Report 02-4094, 2002.

Dyer, Alvin. pers. comm. to author, April 10, 2008.

Ehle, Mark Jr. "The Homestake Slime Plant." *Mines and Minerals* (March 1907): 1.

Engh, Alexander. Affidavit before S. P. Ramicus, Notary Public, in *Old Abe Mining Company vs. James M. Young et al.*, County of Lawrence, Dakota Territory, July 23, 1879. Black Hills Mining Museum, Lead, SD.

Evans-Hatch, Gail and Michael. *Place of Passages: Jewel Cave National Monument Historic Resource Study*. Omaha: Midwestern Region National Park Service, 2006.

Farish, William A. Papers. Black Hills Mining Museum, Lead, SD.

Feriancek, Jeanine. "Minerals & Mining Law." FindLaw for Legal Professionals. http://library.findlaw.com/1999/Jan/1/241491.html.

Field Division of Education. "The History of Scottsbluff, Nebraska: Last Stand of the Indians." National Park Service. http://www.nps.gov/history/history/online_books/berkeley/brand1/brand1n.htm.

Fielder, Mildred. *Railroads of the Black Hills*. Seattle: Superior Publishing Company, 1964.

———. *The Treasure of Homestake Gold*. Aberdeen: North Plains Press, 1970.

———. *A Guide to Black Hills Ghost Mines*. Aberdeen: North Plains Press, January 1974.

———. *Lost Gold*. Aberdeen: North Plains Press, 1978.

Filan, William. Affidavit before James F. Nahai, Notary Public, in *Old Abe Mining Company vs. James M. Young et al.*, County of Lawrence, Dakota Territory, June 29, 1878. Black Hills Mining Museum, Lead, SD.

Flormann, Charles and Fredricka. "Right of Way Deed" to the Black Hills & Fort Pierre Railroad Company, April 5, 1898.

For Mining Basics. "Gyratory Crusher Design," Mining Basics. http://www.miningbasics.com/html/gyratory_crusher_design.php.

Fort Berthold Library. "Treaty with the Arikara Tribe, 1825. The Library Corporation. http://lib.fbcc.bia.edu/FortBerthold/TA25.asp.

Fradkin, Philip L. with a foreword by J. S. Holliday. *Stagecoach: Wells Fargo and the American West*. New York: Simon and Schuster, 2002.

Fraser and Chalmers. "Contract Proposal for Constructing a 120-Stamp Mill for the Highland Mining Company." July 21, 1879. Black Hills Mining Museum.

———. "Price Quotation to the Giant & Old Abe Mining Company for Hoisting and Pumping Machinery," December 11, 1879. Black Hills Mining Museum, Lead, SD.

Froiland, Sven G. *Natural History of the Black Hills*. Iowa: Graphic Publishing Company, 1978.

Fulton, Charles Herman. *A Manual of Fire Assaying*. New York: Hill Publishing Company, 1907.

Gahagan, Kayla. "Researchers continuing review in touring mine lab." *Rapid City Journal* (March 26, 2009).

George Atchison et al. vs. George Hearst et al. "Findings of Fact," Judge G. C. Moody, District Court of the First Judicial District, Lawrence County, Dakota Territory, July 23, 1880.

———. "Judgment and Decree" of Judge G. C. Moody, District Court, First Judicial District, Lawrence County, Dakota Territory, January 7, 1882, 1:482.

Giant & Old Abe Mining Company. "Articles of Incorporation." Thomas H. Reynolds, Clerk, Office of the County Clerk of the City and County of San Francisco, California, September 23, 1878.

———. "Shaft Sinking Bids," January 8, 1879. Black Hills Mining Museum, Lead, SD.

Gilles, Albert P. "Raise Boring at Homestake." Paper presented at the Seventh Annual Intermountain Minerals Conference in Vail, Colorado, July 29-31, 1971.

"Gold Reserve Act of 1934." *The Statutes at Large of the United States of America from March 1933 to June 1934* 48, pt. 1 (Washington, DC: GPO, 1934), 337-44.

Gonzalez, Mario and Elizabeth Cook Lynn. "The Politics of Hallowed Ground: Wounded Knee and the Struggle for Indian Sovereignty, Appendix H." Chicago: Univ. of Illinois Press, 1999. http://www.geocities.com/lakotastudentalliance/hesapachronologygonzalez.html.

Goodspeed, Weston Arthur. "The Province and the States." USGenWeb. http://files.usgwarchives.net/sd/history/province/b-hills.txt.

Grafe, Ernest and Paul Horsted. *Exploring with Custer: The 1874 Black Hills Expedition.* Custer: Golden Valley Press, 2005.

Greene, Jerome A. *Lakota and Cheyenne.* Norman: University of Oklahoma Press, 1994.

Grier, Thomas J. Papers. Black Hills Mining Museum, Lead, SD.

———. "Superintendent's Report." Homestake Mining Company, June 1, 1891.

———. "Superintendent's Report." Homestake Mining Company, June 1, 1907.

———. "Superintendent's Report." Homestake Mining Company, June 1, 1910.

Gries, John Paul. *Roadside Geology of South Dakota.* Missoula: Mountain Press Publishing Company, 1996.

Gushurst, Albert Jr., "Manuel, Fred and Moses and Their Niece Josie." *Some History of Lawrence County.* Pierre: The State Publishing Company, 1981.

Harney, H. C. Affidavit before Moses Liverman, Notary Public, in *Old Abe Mining Company vs. James M. Young et al.*, County of Lawrence, Territory of Dakota, August 29, 1879. Black Hills Mining Museum, Lead, SD.

Harney, H. C. and Alex Engh. Deed to H. B. Young dated June 16, 1876, Elizabeth City, Dakota Territory, deed filed for record with the Lawrence County Register of Deeds, Territory of Dakota, August 13, 1877, before J. A. Hairel. Black Hills Mining Museum, Lead, SD.

Hart, Kathy. "Homestake Mine Area Geology." Homestake Mining Company.

Hasselstrom, Linda. *Roadside History of South Dakota.* Missoula: Mountain Press Publishing Company, 1994.

Hearst, George. Memoir (1890). Black Hills Mining Museum, Lead, SD.

———. Papers. Homestake Mining Company, Lead, SD.

Herz, Nathaniel. "Fire and Chemical Analysis, The Homestake Enterprise." *Engineering and Mining Journal* 132, No. 7 (October 12, 1931).

Hidden Fortune Gold Mining Company. "Annual Report Covering the Fiscal Year Ending February 25, 1903." Black Hills Mining Museum.

———. "Important Notices and other General Information to the Stockholders of the Hidden Fortune Mining Company." Black Hills Mining Museum.

———. "Third Annual Report of the Hidden Fortune Gold Mining Company Covering the Fiscal Year Ending February 20, 1904." Black Hills Mining Museum.

Highland Mining Company. "Articles of Incorporation." Thomas H. Reynolds, Clerk, Office of the County Clerk of the City and County of San Francisco, California, September 30, 1878.

———. "Bids for the Highland Shaft," November 19, 1878. Black Hills Mining Museum, Lead, SD.

Hills, Christopher. *Gold Pans and Broken Picks*. Spearfish: Clark Printing, 1998.

Hinds, Harold L. and Larry L. Trautman. "Metallurgical Practice at Homestake, Lead Operation." *Gold, Silver, Uranium, and Coal: Geology, Mining, Extraction, and the Environment*. New York: The American Institute of Mining, Metallurgical, and Petroleum Engineers, Inc., 1983.

———. and R. D. Ommen. "Homestake's Gravity Recovery System." *Proceedings of the Third Western Regional Conference on Precious Metals, Coal, and the Environment*. K. N. Han and C. A. Kliche, eds. Rapid City: Black Hills Section, Black Hills Section, American Institute of Mining, Metallurgical, and Petroleum Engineers, 1987.

Historic St. Mary's Mission. "Fr. De Smet & Historic St. Mary's Mission." Historic St. Mary's Mission, Inc. http://www.saintmarysmission.org/FatherDeSmet.html.

Hofman, H. O. "Gold-Mining in the Black Hills." *American Institute of Mining and Metallurgical Engineers Transactions* 17 (February 1889): 502-504.

Homestake Adams Research and Cultural Center. "Homestake Adams Research and Cultural Center: Overview." Adams Museum and House, Inc. http://theadamsdeadwood.org/HomestakeAdamsOverview.aspx.

Homestake Deep Underground Science and Engineering Laboratory (DUSEL). "DUSEL Experiment and Development Coordination (DEDC)." Lawrence Berkeley National Laboratory. http://www.lbl.gov/nsd/homestake/dedc.html.

———. "Principle Aspects of the Homestake Conceptual Design Report." Lawrence Berkeley National Laboratory. http://www.lbl.gov/nsd/homestake/PDFs/Homestake_DUSEL_Concepts-1.pdf.

Homestake Mining Company et al. Quit Claim Deed to Cyrus H. Enos, Trustee for Lead Townsite Claimants. Lawrence County Register of Deeds, Deadwood, South Dakota (January 5, 1893), 91:160.

Homestake Mining Company and Barrick Gold Corporation. "Barrick and Homestake Announce Merger Plan: US$2.3 Billion Share Exchange to Create Industry Leader." Joint press release, June 25, 2001.

Homestake Mining Company. Abstract notes for DeSmet and Bowie Water Rights.

———. Abstract notes for Foster Ditch Water Right.

———. Abstract notes for French Boys Ditch and Water Right

———. Abstract notes for Little Rapid Ditch and Water Right.

``````. Abstract notes for Martin Sands Water Rights on Whitewood Creek.
``````. Abstract notes for Peake No. 1 and No. 2 Water Rights.
``````. Abstract notes for Pioneer Ditch and Water Right.
``````. Abstract notes for the Columbia Ditch and Water Right.
``````. Abstract notes for the Montana Ditch and Water Right.
``````. Abstract notes on Boulder Ditch and Water Right.
``````. "Air Locomotives of Homestake Mining Company." Sheet No. 1762-6-59.
``````. "Annual Report of Homestake Mine Operations," 1978.
``````. "Annual Report of Homestake Mine Operations," 1979.
``````. "Annual Report of Homestake Mine Operations," 1981.
``````. "Annual Report of Homestake Mine Operations," 1982.
``````. "Annual Report of Homestake Mine Operations," 1983.
``````. "Annual Report of Homestake Mine Operations," 1985.
``````. "Annual Report of Homestake Mine Operations," 1986.
``````. "Annual Report of Homestake Mine Operations," 1987.
``````. "Annual Report of Homestake Mine Operations," 1988.
``````. "Annual Report of Homestake Mine Operations," 1989.
``````. "Annual Report of Homestake Mine Operations," 1992.
``````. "Annual Report, Homestake Mine Operations," 1998.
``````. "Annual Report Summary for Homestake Mine Operations," 1976.
``````. "Annual Report to Shareholders," 1972.
``````. "Annual Report to Shareholders," 1980.
``````. "Annual Report to Shareholders," 1990.
``````. "Annual Report to Shareholders," 1994.
``````. "Annual Report to Shareholders," 1995.
``````. "Articles of Incorporation." Thomas H. Reynolds, Clerk, Office of the County Clerk of the City and County of San Francisco, California, November 5, 1877.
``````. "Bullion Production Data, January 1, 1878, through December 31, 2000."
``````. "Bullion Production Data, January 1, 1878, through December 31, 2001."
``````. "Bullion Production Data, January 1, 1878, through December 31, 2002."
``````. "Deep Level Project: Engineering and Cost Study, Homestake Mine," July 1969.
``````. "Deep Level Ventilation Data, End Position When Complete," April 7, 1971.
``````. "Ditch Histories and Proof of use of Water."
``````. "Homestake Bullion Product from June 1, 1897, to June 1, 1898."
``````

———. "Homestake Lead Operations, 2000 Ore Reserve and Resource." December 31, 2000.
———. "Homestake Mine Operations Annual Report," December 31, 1992.
———. "Hydroelectric Power Generation (1906-2000)."
———. "Manager's Report on the Homestake Mine and Bulldog Mine for 1972."
———. "Manager's Report on the Homestake Mine and Bulldog Mine for 1973."
———. "Manager's Report on the Homestake Mine and Bulldog Mine for 1974."
———. "Manager's Report on the Homestake Mine and Bulldog Mine for 1975."
———. "Manager's Report on the Homestake Mine and Bulldog Mine for 1976."
———. "Manager's Report on the Homestake Mine and Bulldog Mine," 1977.
———. "Manager's Report on the Homestake Mine and Bulldog Mine," 1981.
———. "Manager's Report on the Homestake Mine and Bulldog Mine," 1982.
———. "Metallurgical Summary by Years (1909-1922)."
———. "Metallurgical Summary for 1917."
———. "Metallurgical Summary for 1926."
———. "Metallurgical Summary for 1932."
———. "Minutes of Board of Directors Meeting held March 5, 1878," D. F. Verdenal, Secretary, Homestake Mining Company.
———. "Notes regarding a conversation with Sam Blackstone," August 8, 1936. Homestake Mining Company Water Rights files.
———. "Peake Ditch Pipeline Notes."
———. "Pre-Patent Title Summary for the Homestake claim, M. S. No. 121."
———. "Pre-Patent Title Summary for the No. Segregated Golden Terry, M. S. No. 149."
———. "Pre-Patent Title Summary for the Old Abe, M. S. No. 117."
———. "Pre-Patent Title Summary for the So. Segregated Golden Terry, M. S. No. 130."
———. "Report of Mine Superintendent," 1975.
———. "Report on Black Hills Operations for 1971."
———. "Report on Black Hills Operations for 1972."
———. "Report on Black Hills Operations for 1973."
———. "Report on Black Hills Operations for 1974."
———. "Report on Operations for 1932."
———. "Report on Operations for 1933."
———. "Report on Operations for 1934."
———. "Report on Operations for 1935."
———. "Report on Operations for 1936."
———. "Report on Operations for 1937."
———. "Report on Operations for 1938."
———. "Report on Operations for 1939."
———. "Report on Operations for 1951."
———. "Report on Operations for 1952."

———. "Report on Operations for 1953."
———. "Report on Operations for 1954."
———. "Report on Operations for 1955."
———. "Report on Operations for 1956."
———. "Report on Operations for 1957."
———. "Report on Operations for 1959."
———. "Report on Operations for 1969."
———. "Water Used in Batteries," November 22, 1907.
———. *Sharp Bits* 4, no. 9 (October 1953).
———. *Sharp Bits* 5, no. 6 (July 1954).
———. *Sharp Bits* 5, no. 7 (August 1954).
———. *Sharp Bits* 7, no. 9 (October 1956).
———. *Sharp Bits* 9, no. 2 (April 1958).
———. *Sharp Bits* 11, no. 11 (December 1960).
———. *Sharp Bits* 13, no. 11 (December 1962).
———. *Sharp Bits* 15, no. 9 (October 1964).
———. *Sharp Bits* 16, no. 8 (September 1965).
———. *Sharp Bits* 16, no. 10 (November 1965).
———. *Sharp Bits* 17, no. 3 (April 1966).
———. *Sharp Bits* 17, no. 5 (June 1966).
———. *Sharp Bits* 19, no. 2 (Summer 1968).
———. *Sharp Bits* 20, no. 1 (Spring 1969).
———. "Strategy Plan for the Homestake Closure Project," 2001.
———. "The Ore Tax Should be Reduced." Public information circular, 1944. Black Hills Mining Museum.
———. "The Truth About Homestake." Public information circular, 1938. Black Hills Mining Museum.
Homestake Veterans Association. "Recollections of the Early Days of the Homestake Mining Company." Black Hills Mining Museum, Lead, SD.
Honerkamp, John R. *At the Foot of the Mountain*. Stickney: Argus Printers, 1978.
Houck, Louis. *The Spanish Regime in Missouri*. Chicago: The Lakeside Press, 1909.
Huff, Shala. "John Hearst of the Long Canes, Hearst Genealogy." Ancestry.com. http://homepages.rootsweb.ancestry.com/~amcolan/Hearst/hearst.html.
Humphries. A. A. *Preliminary Report of Explorations in Nebraska and Dakota in the Years 1855-'56-'57*. Engineer Department, United States Army, Washington, DC: GPO, August 20, 1875.
Ingham, Thomas G. *Digging Gold Among the Rockies*. Philadelphia: Hubbard Brothers Publishers, 1888.
Janin, Louis. Papers. The Newberry Library, Chicago, Illinois, Call Number: Graff 4936.

Jones, Thomas. Affidavit before Moses Liverman, Notary Public, in *Old Abe Mining Company vs. James M. Young et al.*, County of Lawrence, Territory of Dakota, August 2, 1879. Black Hills Mining Museum, Lead, SD.

Joseph E. Cook et al. vs. Robert H. Evans et al. SD, 8d, January 30, 1918, at 12:617.

Justia: U.S. Laws. "South Dakota 10-39-43: Severance tax on gold." Justia, Inc. http://law.justia.com/southdakota/codes/10/10-39-43.html.

Kehoe, John H. Affidavit of John H. Kehoe before Edward Allsmon, Notary Public, in *Old Abe Mining Company vs. James M. Young et al.*, County of Lawrence, Territory of Dakota, July 8, 1878. Black Hills Mining Museum, Lead, SD.

Kellar, Chambers. Papers. Homestake Mining Company.

Kellar, P. H. *Seth Bullock's The Founding of a County*. Deadwood: Dakota Graphics, 1986.

Kingsbury, George W. *History of Dakota Territory I*. Chicago: S. J. Clarke Publishing Company, 1915.

Kion, Mary Trotter. "Battle of Little Big Horn: Subdue the Sioux." Suite101.com. http://www.suite101.com/lesson.cfm/17638/1151/6.

Kion, Mary Trotter. "Forts of the West, An Introduction." Suite101.com. http://www.suite101.com/article.cfm/great_american_plains/112946.

Klock, Irma H. *Yesterday's Gold Camps and Mines in the Northern Black Hills*. Lead: Seaton Publishing Company, 1975.

———. *All Roads Lead to Deadwood*. Lead: privately printed, 1979.

———. *Whistle Stops on the Black Hills and Fort Pierre*. Deadwood: Dakota Graphics, 1982.

———. *The Gold Camps in Upper Deadwood Gulch*. Deadwood: Dakota Graphics, 1984.

Knight, Dwayne F. *Steeples above Stopes: The Churches in the Gold Camps, 1876-1976*. Deadwood: Deadwood-Lead '76 Centennial, 1976.

Larson, Jeanette. "City of Lead, 1876 to 1981." *Some History of Lawrence County*, Pierre: The State Publishing Company, 1981.

Larson, Willard. *Early Mills in the Black Hills*. Isanti, Minnesota: privately printed, 2001.

Laveille, E. *The Life of Father De Smet*. New York: P.J. Kenedy & Sons, 1915.

Lawrence County, Deadwood, South Dakota. "Record of Appointment of Postmasters 1885-1900." Vol. 59.

———. "Record of Appointment of Postmasters, 1894-1930." Vol. 89.

Lazarus, Edward. *Black Hills/White Justice: The Sioux Nation Versus the United States, 1775 to the Present*. New York: HarperCollins Publishers, 1991.

Lead High School. *Goldenlode* 46. 1972.

Lee, Bob. *Gold, Gals, Guns, Guts*. Deadwood: Deadwood-Lead '76 Centennial, Inc., 1976).

Linde, Martha. *Sawmills of the Black Hills*. Rapid City: Fenske Printing, Inc., 1984, 76.

Logan Actuator Company. "A Brief History of Logan Actuator Company and Logan Engineering Company." Logan Actuator Company. http://www.loganact.com/history.htm.

Longwell, Chester R., Richard Foster Flint, and John E. Sanders. *Physical Geology.* New York: John Wiley and Sons, Inc., 1969.

Loomis, Noel M. *Wells Fargo: An Illustrated History.* New York: Bramhall House, 1968.

Lounsbery and Haggin. "Fiscal Report from September 1, 1880, to June 1, 1881," Homestake Mining Company.

Lubragge, Michael T. "Manifest Destiny." From Revolution to Reconstruction. http://www.let.rug.nl/usa/E/manifest/manif1.htm#int.

Luschei, Martin. *The Black Hills & the Indians.* San Luis Obispo: Niobrara Press, 2007.

Lynch, Michael. Affidavit before James F. Nahai, Notary Public, in *Old Abe Mining Company vs. James M. Young et al.*, County of Lawrence, Territory of Dakota, June 29, 1878. Black Hills Mining Museum, Lead, SD.

Making of America Books. "Report of a Reconnaissance of the Black Hills of Dakota, made in the Summer of 1874" by William Ludlow. U.S. Government Printing Office, 1875. Andrew W. Mellon Foundation. http://quod.lib.umich.edu/cgi/t/text/text-idx?c=moa;cc=moa;rgn=main;view=text;idno=AFK4434.0001.001.

Mallory, L. H. and Dan Latham. Location Notice for the Homestake No. 2 Claim, October 25, 1876. Black Hills Mining Museum, Lead, SD.

Mankin, Max. "Professor shedding light on dark matter." *The Brown Daily Herald* (November 4, 2008). Article reprinted by the South Dakota Science and Technology. http://www.sanfordundergroundlaboratoryathomestake.org/index.php?option=com_content&view=article&id=51:prof-shedding-light-on-dark-matter&catid=13:science-catigory-1&Itemid=53.

Manuel, Edna Adamson. "Appendage to 'Forty-Eight Years in the West.'" October 3, 1951, Black Hills Mining Museum.

Manuel, Fred. Affidavit before S. P. Ramicus, Notary Public, in *Old Abe Mining Company vs. James M. Young et al.*, County of Lawrence, Territory of Dakota, July 23, 1879. Black Hills Mining Museum, Lead, SD.

Manuel, Moses. "Forty-Eight Years in the West." Dictated to Mary Sheriff. Typescript. Helena, Montana, 1903. Black Hills Mining Museum, Lead, SD.

Marks, John. "Basic Statistics for 6950 Vent. Project." Homestake Mining Company, December 19, 1989.

Mathematical and Physical Sciences. "Announcement of Awards for Developing Conceptual Designs for a Deep Underground Science and Engineering Laboratory (DUSEL)." The National Science Foundation. http://www.nsf.gov/news/news_summ.jsp?cntn_id=104313&org=PHY&from=news.

McCann, Lloyd E. "The Grattan Massacre." Reprinted from *Nebraska History* XXXVII, no. 1, March 1956.

McClintock, John S. *Pioneer Days in the Black Hills*. Norman: University of Oklahoma Press, 2000.
McDermott, John D. *Gold Rush: The Black Hills Story*. Pierre: South Dakota State Historical Society Press, 2001.
McDonald, Bernard. "Mine Timbering by the Square Set System at Rossland, B. C." *Mining and Scientific Press* LXXXV, no. 12 (September 20, 1902): 158-61.
McLain, Gary. *Indian America*. Santa Fe: John Muir Publications, 1994.
McMaster, Samuel. Deposition in *Old Abe Mining Company vs. William H. Bull, James Carney, Hugh McKenna, and George Adams*, County of Lawrence, Territory of Dakota, April 25, 1879.
McShane, A. G. Affidavit before Moses Liverman, Notary Public, in *Old Abe Mining Company vs. James M. Young et al.*, County of Lawrence, Territory of Dakota, June 26, 1878. Black Hills Mining Museum, Lead, SD.
Mechling, Patricia Lavier. "Greenwood—Vanished Village." *Black Hills Nuggets*, Rapid City: Rapid City Society for Genealogical Research.
Merrill, C. W. "Treatment of Homestake Ores." *Mining and Scientific Press* LXXXVI, no. 10 (March 7, 1903): 150.
———. "The Metallurgy of the Homestake Ore." *American Institute of Mining and Metallurgical Engineers Transactions* 34 (1904): 586.
Meyerriecks, Will. *Drills and Mills: Precious Metal Mining and Milling Methods of the Frontier West*. (privately printed, 2003).
Miller, Clarence. Papers. Colette (Flormann) Bonstead collection.
Mills, Rick W. *Black Hills Railroading*. Hermosa: Battle Creek Publishing Company and Black Hills Central Railroad, 2004.
Mineral Claimants J. B. Haggin and Homestake Mining Company. Agreement with Lead City Townsite Trustees. Lawrence County Register of Deeds, Deadwood, South Dakota (March 18, 1892), 82:289.
Minerals and Mining Program. "Summary of the Mining Industry in South Dakota, April 2000." South Dakota Department of Environment and Natural Resources. http://denr.sd.gov/des/mm/Goldrpta3.pdf.
———. "Summary of the Mining Industry in South Dakota for 2007." South Dakota Department of Environment and Natural Resources (May 2008). http://denr.sd.gov/des/mm/documents/Goldrpt07.pdf.
Mitchell, A. B. "Mine Surveying." *Sharp Bits* 10, no. 11 (December, 1959): 1-6.
Mitchell, Steven T., "VCR Stoping at Homestake." *Mining Engineering* 32, no. 11 (November 1980): 1581-1586.
Monroe County Library System. "Custer in the News, Gold in the Black Hills." Letter from General Custer, Fort Abraham Lincoln, Dakota, December 13, 1874. Monroe County Library System. http://monroe.lib.mi.us/books_movies_music/special_collections/custer_news_black_hills.htm.
Moody, G. C. Papers. Homestake Mining Company.

Munkres, Robert L. "Tales of Old Fort Laramie," Muskingum College. http://www.muskingum.edu/~rmunkres/military/Laramie/Tales.html.

National Archives and Records Administration. "How do I locate a lode or placer mining claim?" U.S. Government Printing Office. http://ecfr.gpoaccess.gov/cgi/t/text/text-idx?c=ecfr;rgn=div5;view=text;node=43%3A2.1.1.3.78;idno=43;cc=ecfr#43:2.1.1.3.78.2.180.1.

National Inventors Hall of Fame. "Hall of Fame Inventor Profile for Eli Whitney Blake." National Inventors Hall of Fame Foundation, Inc. http://www.invent.org/hall_of_fame/318.html.

Native Americans Documents Project. *Annual Report of the Commissioner of Indian Affairs to the Secretary for the Year 1875*. Washington, DC: Government Printing Office, 1875. NADP Document R875001A. California State University, San Marcos. http://www.csusm.edu/nadp/r875001a.htm.

Neihardt, John G. *Black Elk Speaks*. 7th ed. Lincoln: University of Nebraska Press, 1993.

Neutrino Scientific Assessment Group. "Recommendations to the Department of Energy and the National Science Foundation on a United States Program in Neutrino-less Double Beta Decay, Report to the Nuclear Science Advisory Committee and the High Energy Physics Advisory Panel." Pacific Northwest National Laboratory (September 1, 2005): 3. http://majorana.pnl.gov/documents/NuSAG_report_final_version.pdf.

Nobel Prize.org. "The Nobel Prize in Physics, 2002." Nobel Web AB. http://nobelprize.org/nobel_prizes/physics/laureates/2002/.

O'Harra, Cleophas C. "Early Placer Gold Mining in the Black Hills." *The Black Hills Engineer* XIX, no. 4 (November 1931): 343-344.

Old Abe Mining Company. "Articles of Association." Thomas D. Meade, Clerk, Circuit Court of Keoughton County, Michigan, February 2, 1878.

Ommen, Ross and Larry L. Trautman. "New Innovations in Gold Processing using Gravity, Oxygen-Vacuum and Carbon at the Homestake Mine." *Fifth Western Regional Conference on Precious Metals and the Environment*. Rapid City: Black Hills Section, American Institute of Mining, Metallurgical, and Petroleum Engineers, 1993.

Parker, Watson and Hugh K. Lambert. *Black Hills Ghost Towns*. Athens: Swallow Press/Ohio University Press, 1974.

Parker, Watson. *Deadwood: The Golden Years*. Lincoln: University of Nebraska Press, 1981.

———. *Gold in the Black Hills*. 1st ed. Lincoln: University of Nebraska Press, 1982.

Peck, Linda H. "Biography of James Ben Ali Haggin." Linda H. Peck—Haggin Histories. http://haggin.org/JBAH_Biography.html.

Pfarr, Jerry. "Terraville Test Pit Executive Summary." Homestake Mining Company, November 22, 1985.

Progulske, Donald R. and Frank J. Shideler. *Following Custer*. Brookings: South Dakota State University, Agricultural Experiment Station, Bulletin 674.

Pyle, K. D. "The Homestake Tungsten Mill." *The Black Hills Engineer* (Rapid City: The South Dakota State School of Mines, May 1926), 171-174.

Raines, Charlie. "DUSEL Redux." Cascade Chapter of the Sierra Club (June 26, 2006). http://cascade.sierraclub.org/node/559.

Remer, William A. Personal diary. Adams Museum, Deadwood, South Dakota.

Rickard, T. A. "Gold-Milling in the Black Hills, South Dakota, and at Grass Valley, California." *American Institute of Mining and Metallurgical Engineers Transactions* 25 (October 1895): 911-12.

———. *The Stamp Milling of Gold Ores*. New York: The Scientific Publishing Company, 1897.

Rickards, Colin. *Bowler Hats and Stetsons: Stories of Englishmen in the Wild West*. New York: Bonanza Books, 1965.

Robertson, William Glenn, Jerold E. Brown, William M. Campsey, and Scott R. McMeen. "Atlas of the Sioux Wars." U.S. Army Combined Arms Center, Command and General Staff College, Fort Leavenworth, Kansas. http://cgsc.leavenworth.army.mil/carl/resources/csi/sioux/sioux.asp#Sheridan's%20Campaign%20Plan.

Robie, Edward Hodges. "Milling Practice at the Homestake Gold Mine." *Engineering and Mining Journal*, 122, no. 15 (October 9, 1926): 566.

Robinson, Doane. "History of South Dakota II." MyFamily.com, Inc. http://ftp.rootsweb.com/pub/usgenweb/sd/biography/doane2/rosenkranz.txt.

———. *A Brief History of South Dakota*. New York: American Book Company, 1905.

Rootsweb. "Report of Captain W. F. Raynolds' Expedition to Explore the Headwaters of the Missouri & Yellowstone Rivers." American Serial Set, 40th Congress, 2d Session, Senate Executive Document 77, Ancestry.com. http://freepages.history.rootsweb.ancestry.com/~familyinformation/fpk/raynolds_rpt.html.

Rosen, Peter. *Pa-Ha-Sa-Pah or the Black Hills of South Dakota*. St. Louis: Nixon-Jones Printing Company, 1895.

Ross, A. J. M. "Ore Extraction and Transportation." *Engineering and Mining Journal* 132, (1931): 324-29.

Ross, A. J. M. and R. G. Wayland. "A Brief Outline of Homestake Mining Methods." *The Black Hills Engineer* XIV, no. 3 (May 1926): 145-46.

Ross, Dick and June. *Portage Lake Mining Gazette*. Houghton, Michigan, January 17, 1878. Dick and June Ross. http://www.mfhn.com/houghton/rosscoll/news/mg1878_1.txt.

Ryder, David W. *The Merrill Story*. San Francisco: The Merrill Company, 1958.

Samuel McMaster vs. Wyoming and Dakota Water Company. Judgment in District Court, Lawrence County, Dakota Territory, April 23, 1881.

"Sheriff's Certificate of Sale No. 10579," District Court, Lawrence County, Dakota Territory, August 18, 1881.

Sanford Underground Laboratory at Homestake. "Hydrostatic Sensors Detect Earth Tides." South Dakota Science and Technology Authority. http://www.sanfordundergroundlaboratoryathomestake.org/index.php?option=com_content&view=article&id=171:hydrostatic-system-detects-qearth-tidesq&catid=22:hydrology-studies&Itemid=56.

———. "Timeline." South Dakota Science and Technology Authority. http://www.sanfordundergroundlaboratoryathomestake.org/index.php?option=com_content&view=article&id=11&Itemid=16.

Schmuck, Carl. "Cablebolting at Homestake." Paper presented at Rocky Mountain Minerals Conference, Vail Colorado, July 24-27, 1979.

Schoenherr, Steven. "William Randolph Hearst," Steven Schoenherr. http://history.sandiego.edu/gen/media/hearst.html.

Schofield, John. Affidavit before Henry Hill, Notary Public, in *Old Abe Mining Company vs. James M. Young et al.*, County of Lawrence, Territory of Dakota, July 3, 1878. Black Hills Mining Museum, Lead, SD.

Schuttler, Linfred. "Bryant Party and Frank S. Bryant Family." In *Some History of Lawrence County*, 57-59. Pierre: The State Publishing Company, 1981.

Schwartz, E. A. "History 347: California History." The California State University: San Marcos. http://courses.csusm.edu/hist347as/br/h4713t02.htm.

Shoemaker, A. H. "Report of the General Manager." Homestake Mining Company, February 12, 1957.

Sneve, Virginia Driving Hawk. *South Dakota Geographic Names*. Brevet Press: Sioux Falls, South Dakota, 1973.

South Dakota History 2, no. 3. Pierre: South Dakota State Historical Society, 1972.

South Dakota Water Management Board. *Meeting Minutes*, July 10, 2003.

Spring, Agnes Wright. *The Cheyenne and Black Hills Stage and Express Routes*. Lincoln: The University of Nebraska Press, 1948.

Staley, W. W., *Introduction to Mine Surveying*. Stanford: Stanford University Press, 1964.

Staple, S. J. "The Mechanical Department of the Homestake Mining Company." *The Black Hills Engineer* XIV, no. 2 (May 1926): 202-204.

Stevenson, Jen. "Phoebe Apperson Hearst." Embarcadero Publishing Company. http://www.pleasantonweekly.com/morgue/2001/2001_04_13.phoebe13.html.

Stokes, George W. and Howard R. Driggs. *Deadwood Gold: A Story of the Black Hills*. New York: World Book Company, 1927.

Storsve, LaVaughn. *The History of Bethel Lutheran Church, 75th Anniversary Edition*. Lead: privately printed, 1970.

Svanevik, Michael and Shirley Burgett. "A 19-Century Power Duo: Special to the *Examiner*." November 18, 2002. Linda H. Peck—Haggin Histories. http://haggin.org/PowerDuo.html.

Tallent, Annie D. *The Black Hills, or Last Hunting Grounds of the Dakotahs*. Sioux Falls: Brevet Press, 1974.

Tesch, Charles L. "Wood Pattern." Homestake Mining Company, undated report.

The California Museum. "Phoebe Hearst, California Hall of Fame 2007 Inductee." The California Museum. https://www.californiamuseum.org/Exhibits/Hall-of-Fame/2006/hearst.html.

The Dakota Experience. "Memorial of the Legislature of Dakota Praying that the Black Hills of Dakota be opened for settlement, and the Indian title to the same be extinguished." House of Representatives, House Misc. Doc. No. 33, 43rd Cong., 2nd sess., 1875. South Dakota State Historical Society and Educational Web Adventures. http://dakotaexperience.org/cvfrontier/coll_whitesettlement_text.html.

The Minerals and Mining Program. "Summary of the Mining Industry in South Dakota for 2006." South Dakota Department of Environment and Natural Resources (May 2007). http://denr.sd.gov/des/mm/documents/2006Goldrpt.pdf.

The Privateer Gold Pages. "The Early Gold Wars." *The Privateer Market Letter* (2001). http://www.the-privateer.com/gold2.html.

Thomson, Frank. *The Thoen Stone: A Saga of the Black Hills*. Detroit: The Harlo Press, 1966.

Toms, Donald D. *In the Midst of Life: Mining Company Fatalities, Black Hills, South Dakota, 1876-1995* I, 1996.

Toms, Donald D., Kristie L. Schillinger, William J. Stone, eds. *The Gold Belt Cities: The City of Mills*. Lead: Black Hills Mining, 1993.

Toms, Donald D., William J. Stone, and Gretchen Motchenbacher, eds. *The Gold Belt Cities: Lead & Homestake*. Lead: G.O.L.D. Unlimited, 1988.

URS Corporation. "Operations Manual for Grizzly Gulch Tailing Dam." Prepared for Homestake Mining Company, August 2005.

U.S. Congress. *Homestead Act of 1862. Public Acts of the Thirty-Seventh Congress*. 37th Cong., 2nd sess., ch. 75, May 20, 1862. The Library of Congress, American Memory. http://memory.loc.gov/cgi-bin/ampage?collId=llsl&fileName=012/llsl012.db&recNum=423.

———. *Pacific Railway Act of 1862. Public Acts of the Thirty-Seventh Congress*. 37th Cong., 2nd sess., ch. 75, July 1, 1862. The Library of Congress, American Memory. http://memory.loc.gov/cgi-bin/ampage?collId=llsl&fileName=012/llsl012.db&recNum=530.

U.S. Department of the Interior. Bureau of Land Management. "Desert Land Entries." Bureau of Land Management. http://www.blm.gov/ut/st/en/res/utah_public_room/desert_land_entries.html.

———. Bureau of Land Management. "Fort Meade Recreation Area." Bureau of Land Management. http://www.blm.gov/mt/st/en/fo/south_dakota_field/ft_meade.html.

———. Bureau of Land Management. "Introduction: 1872 Mining Law." Energy and Minerals Technical Assistance Program. http://www.blm.gov/icmtap/mineintro.html.

———. Bureau of Land Management. *Plat of the Claim of The Golden Terra Mining Company, Golden Terry Lode, Survey No. 130*, J. D. McIntyre, U.S. Deputy Surveyor, September 2, 1878.

———. Bureau of Land Management. *Plat of the Claim of Deadwood Mining Company, Survey No. 149*, J. D. McIntyre, U.S. Deputy Surveyor, February 19-26, 1879.

———. Bureau of Land Management. *Plat of the claim of J. B. Haggin upon the Segregated Northerly 150 feet of Golden Star Lode, Survey No. 230*, May 21, 1880.

———. Bureau of Land Management. *Plat of the claim of Samuel McMaster upon the Lincoln Lode, Amended Survey No. 132*. November 28, 1882.

———. Bureau of Mines. *Black Hills Mineral Atlas, South Dakota: Part 1*. Information Circular 7688, July 1954.

———. *Eleventh Census of the United States. Statistics of Mines and Mining*, 1889, Special Schedule 3A.

———. National Park Service. "The Great Reconnaissance." National Park Service. http://www.nps.gov/archive/jeff/LewisClark2/Circa1804/WestwardExpansion/EarlyExplorers/OtherExplorers.htm.

———. U.S. Geological Survey. *Geology of the Region around Lead, South Dakota*. Bulletin 765, 1924.

U.S. Commission on Industrial Relations. *Proceedings of Public Hearing Held at Lead, South Dakota, August 3-4, 1914*, 1914.

U.S. Environmental Protection Agency. Office of Solid Waste. *Treatment of Cyanide Heap Leaches and Tailings*. EPA530-R-94-037 (September 1994). http://www.epa.gov/waste/nonhaz/industrial/special/mining/techdocs/cyanide.pdf.

U.S. Land Office. Townsite Patent to Leonard Gordon, Probate Judge in trust for occupants of the Lead City Townsite. Lawrence County Register of Deeds, Deadwood, South Dakota (filed February 1, 1899), 139:374.

Vestal, Stanley. *Sitting Bull: Champion of the Sioux: A Biography*. Norman: University of Oklahoma Press, 1932.

Warren, Lieutenant Gouverneur Kemble. *Preliminary Report of Explorations in Nebraska and Dakota in the Years 1855-'56-'57*. Engineer Department, U.S. Army, Washington, DC: GPO, 1875.

Waterland, Joel. *The Spawn & the Mother Lode*. Rapid City: Grelind PhotoGraphics & Typesetters, 1987.

———. *Gold, Silver, Sweat & Tears*. Rapid City: Grelind Photographics & Typesetters, 1988.

———. *The Mines Around & Beyond*. Rapid City: Grelind Printing Center, 1991.

Weider History Network. "Brulé Sioux Chief Spotted Tail." Weider History Group. http://www.historynet.com/culture/native_american_history/3037121.html?page=2&c=y.

Werner, Fred H. *The Slim Buttes Battle, September 9-10, 1876*. Self-Printed, 1981.

Wiggert, J. F. "Development of the Homestake Electrical System." *The Black Hills Engineer: Homestake Semi-Centennial* 14, no. 3 (May 1926): 183.

———. "Homestake Mining Company's Carrier-current Shaft Signaling System." Technical Publication No. 1286-A, *Mining Technology* (February 1941).

Woolley, John T. and Gerhard Peters. *"The American Presidency Project."* Gerhard Peters—The American Presidency Project. http://www.presidency.ucsb.edu/ws/?pid=14611.

———. "Ulysses S. Grant: Seventh Annual Message to the Senate and House of Representatives." December 7, 1875. The American Presidency Project. http://www.presidency.ucsb.edu/ws/print.php?pid=29516.

World Gold Council. "After the Gold Standard, 1931-1999." World Gold Council. http://www.gold.org/assets/file/value/reserve_asset/history/monetary_history/vol3pdf/1934jan31.pdf.

Wormser, Felix Edgar. "Gold-Mining Developments in the Black Hills." *Engineering and Mining Journal*, 114, no. 18 (October 28, 1922):759-60.

Wyoming and Dakota Ditch, Flume, and Mining Company. "Articles of Incorporation." County of Lawrence, Territory of Dakota, June 23, 1877.

Wyoming and Dakota Water Company. "Articles of Incorporation." City and County of San Francisco, California, January 25, 1878.

———. Minutes of the Board of Directors, December 6, 1878. Homestake Mining Company Water Rights files.

———. Agreement with Town of Central City, October 1, 1879. Homestake Mining Company Water Rights files.

———. Minutes of the Board of Directors, March 2, 1881. Homestake Mining Company Water Rights files.

Wyoming Commentary. "Wyoming Atlas of Historical County Boundaries." The Newberry Library. http://historical-county.newberry.org/website/Wyoming/documents/WY_Commentary.htm.

Yates, A. B. "Structure of the Homestake Ore Body." PhD diss., Harvard Engineering School, March 1931.

Yates, B. C. "Some Features of Mining Operations in the Homestake Mine, Lead, South Dakota." Paper presented at the Black Hills Mining Men's Association, Deadwood, South Dakota, January 19, 1904.

———. "Superintendent's Report." Homestake Mining Company, December 31, 1918.

———. "Superintendent's Report." Homestake Mining Company, December 31, 1919.

———. "Superintendent's Report." Homestake Mining Company, December 31, 1920.

———. "Superintendent's Report." Homestake Mining Company, December 31, 1921.

———. "Superintendent's Report." Homestake Mining Company, December 31, 1924.
———. "Superintendent's Report." Homestake Mining Company, December 31, 1928.
———. "Superintendent's Report." Homestake Mining Company, December 31, 1929.
———. "Superintendent's Report." Homestake Mining Company, December 31, 1931.
Young, James M. Affidavit before James F. Nahai, Notary Public, in *Old Abe Mining Company vs. James M. Young et al.*, County of Lawrence, Territory of Dakota, June 21, 1878. Black Hills Mining Museum, Lead, SD.
Young, Mort. "Phoebe Hearst: A Pioneer Woman." *Seattle Post-Intelligencer*, September 12, 1976.

Index

A

Abe Street, 307
Abt, Frank, 192, 307, 312, 314, 319
Abt Hotel, 319
Adams, Dick, 198
Adams Museum & House, 564
Addie Street, 308, 310, 315–16, 321–22, 410
AFL (American Federation of Labor), 408, 422, 427, 649
AFL-CIO (United Steelworkers of America), 408
Aid Fund, 421–22, 425
Aikey, Josephine A. "Josie," 313
Ainley, John, 320
air locomotives, 14, 24, 233–34, 407, 437, 635, 684
Air Raise, 394
  No. 1, 394
  No. 2, 394
Alameda High School, 597
Alaniva, Charles, 322
Albien, H. A., 106
Alexander, L. L., 178, 204, 606
Ali, Adeline Ben, 591
Ali, Ibrahim Ben, 591
Ali, James Ben Jr., 591
Alimak Raise Climber, 485
Alisal Rancheria, 589
Allen, John W., 104
Allen, Nettie, 333
Allis Chalmers Ball Mill, 363, 366–68, 370, 466, 477, 507, 512. *See also under* mills
Allis Chalmers gyratory crusher, 466, 507
Allison, John P., 459
Allison, William B., 115
Alloys Company, 599–600
Alpha mine, 50, 140
amalgam, 14–15, 238–39, 241–42, 257–62, 264–65, 333, 345, 368, 544, 665, 673
amalgamation, 7, 10, 17, 237, 239, 242, 254, 256–57, 261–62, 264, 345, 349, 355, 364, 368, 370, 503, 506–7, 514, 665, 675
American Davidson centrifugal fan, 21, 499
American Davidson Company, 498
American Dream, 29, 63, 70, 97, 409
American Federation of Labor (AFL), 427
American Flag claim, 165
American Fur Company, 56, 65, 105
American Horse, 62, 83, 85, 123–25
American Indians, 29, 33, 63, 118
American Institute of Mining, 253, 261, 350, 364, 635–36, 644, 656, 679, 683, 689–91
American Mine Services (AMS), 495
American Progress, 29, 32
Amicus Ledge, 46–47
Amicus Mill (*see also under* mills), 15, 259, 263, 348, 363, 367, 370, 372, 407, 441, 444, 467, 549
Amicus Plate House, 444

697

ammonium nitrate with fuel oil (ANFO), 524
amphibolite, 42, 49, 665
Anaconda Mining Company, 389, 586
Anchor City, 8, 328, 336, 344
Ancient Order of United Workmen's (AOUW), 401
Anderson, H. P., 274
Andrews, Ed, 429
Annie's Place, 319
Annual Report of the Commissioner of Indian Affairs, 103, 116, 621, 703
Anson Mill, 124. *See also under* mills
anticlinal ledges, 43
Anticline No. 4, 47
AOUW (Ancient Order of United Workmen's), 401
Apache people, 51
Apperson, Phoebe Elizabeth, 585, 588
apron plates, 242, 257–61, 670
Arapaho tribe, 51, 53–54, 66, 89, 115, 120
Archean period, 35
Argue, John, 111
Argue Party, 111. *See also under* prospectors
Arikara tribe, 51, 54, 66
Armes, Hannah, 170
Armstrong, Moses K., 77, 88
Army Appropriation Act of 1853, 70
Army Corps of Topographical Engineers, 70
arrastra, 7, 138, 140–42, 145, 148, 177, 192, 237–39, 344, 580, 670
arsenopyrite, 42–43, 345, 347, 542, 670
artesian, 40
Arthur, Harold, 429
Ashton, George, 110
assaying, 10, 264, 347, 490, 541, 543, 661, 670–71, 683–84, 693
assay office, 15, 163, 261, 264–65, 319, 329, 347, 360, 394, 436, 468, 490, 519, 542, 600

Assembly Hall, 316
Assembly of God Church, 322
Atchison, George, 193–95, 202, 206, 330, 608, 633–34, 636, 694
Atkinson, Henry, 54
Atkinson and O'Fallon Treaty, 54
Atlas Portland cement, 383
Auburn Boulevard, 592
Aunt Sally, 91, 93, 280. *See also* Campbell, Sarah
Auxiliary Steam-Electric Plant, 8, 387, 390–91, 394, 396, 444
Avalon, 278, 297, 618, 689

B

Babcock, W. H., 108
backfill, 219, 221, 437, 440, 448, 450, 486, 505, 508, 670, 677, 687
Bacon (doctor), 248
Bacon, H. D., 157, 595
Bacon, Henry D., 180
Bad and Cheyenne rivers, 54
Bad Road, 131
Bahcall, John, 556
Bailey, John W., 170
Bailey, J. W., 151
Bald Mountain, 107, 401
Bald Mountain Mining districts, 401
Baldwin, Frank D., 130
Baldwin, Ivy, 323–24
Baldwin Locomotive Works, 227–28
Balf, J. H., 330
Ballantyne, Alex, 167
ball mills, 139, 477–78, 551, 670, 677. *See also under* mills
ballpark, 16, 19, 324–25, 419–20
Baltic Tunnel, 456
Baltimore and Deadwood Mill, 16, 332. *See also under* mills
Baltimore and Deadwood Mining Company, 453

Baltimore Street, 322
Band Shell, 428
Bank of Lead, 320
Banner Associates, 511
Barber, H. W., 191
Barclay, Charles, 307
Barney, Ashbel H., 595
Barrick Gold Corporation, 31, 569–71, 662, 696
Barron, E. H., 198–99
Bartholomew, James S., 336
basalt, 42
batteries, 239, 254, 258, 262, 646, 670, 698
battery, 239
Battle of Cedar Creek, 130
Battle of Rosebud Creek, 122
Battle of Slim Buttes, 125, 127, 130, 622
Battle of the Little Big Horn, 123, 125, 128, 136
Battle of Wolf Mountain, 130
Bauer, Sheaffer, and Lear, 504
Baulis, A. B., 157, 174
Bayliss, A. T., 598
Beakley, John, 176
Beaman, H. B., 200
Bear's Rib tribe, 73
Bear Butte, 37–38, 53, 71, 73–74, 76, 91, 95, 106, 132, 138, 140, 274, 328, 605
Bear Butte Creek, 37, 71, 76, 138, 140, 274, 605
Bear Gulch, 111–12
Bear Lodge Mountains, 60, 132, 301
Bear Mountain, 35
"beat," 420, 523
Beaver Creek, 73, 111–12, 120, 123
Beaver Creek District, 112
Beck, Samuel J., 185
Beemer, George, 146, 312
Begole, George D., 461

Belding, John, 191, 206
Beliveau, Josephine, 581
Belknap, William W., 75, 116
Bell, Aron M., 192
Bell, Ben, 274–75
Bell, Thomas, 166–67
Belle Fourche River, 61, 71, 74, 114, 126
Belle Fourche Roundup, 428
Benham, Alexander, 187
Bennett, Harold "Bud," 481
Benteen (tribe), 123
Berean Baptist Church, 322
Bernal Rancho, 589
Besant Park, 207
Bethel Lutheran Church, 322, 534, 644, 705
Bettleheim, Florence Keats, 57
Bevan, A. D., 145
BHCWC (Black Hills Canal and Water Company), 178, 190–92, 197–203, 205–9, 230, 254, 632
Bidwell, Oliver, 193
Big Bend, 51, 54, 120, 189
Big Four, 344, 594–95, 599
Big Horn and Wind Rivers, 114
Big Horn Mountains, 76, 115, 123, 136–37, 311
Big Horn River, 114, 122
Big House, 430
Big Missouri mine, 50
Big Partisan, 68
Big Piney Creek, 80
Bill Bunney, 437
Bingham claim, 456
Bingham Shafthouse, 456–57
binline, 515, 518–19, 521
biotite, 42–43
Bishop, E. J., 315
Bishop Busch, 426
Bismarck, 78, 91–94, 118, 227, 252, 280
Bismarck Tribune, 91, 93
Bisslinghoff, H., 111

Black Elk, 59, 90, 117, 615, 618, 622, 703
Blackfeet tribe, 115
Black Hills, 29–35, 37–41, 48–49, 51–54, 58–63, 73–77, 79–80, 86–107, 109–18, 124–29, 131–34, 138–40, 142–43, 148–52, 184–85, 190–92, 197–202, 249–54, 268–73, 277–89, 291–301, 327–39, 458–62, 580–83, 613–30, 632–45, 652–55, 662–65, 689–95, 698–709
  expedition of 1874, 29
  the expropriation of, 30, 113, 133
  George Hearst, J. B. Haggin, and Lloyd Tevis acquisition of, 34
  miners ordered out from, 103–5
  the Sioux splitting up, but refusing to give up, 130–33
  stratigraphy, 35
Black Hills & Fort Pierre Railroad Company, 269–70, 280, 289, 295, 297, 586, 640–41, 690, 693
Black Hills and Fort Pierre Railroad, 32, 205, 270–71
Black Hills Brewing Company, 16, 337–39, 644, 690
Black Hills Brigade, 127
Black Hills Canal and Water Company (BHCWC), 178, 190–92, 197–203, 205–9, 230, 254, 632
Black Hills Consolidated Mines, 461–62
Black Hills Daily Pioneer, 187, 199–200, 249, 625–26, 632–34, 637, 639
Black Hills Daily Register, 405
Black Hills Daily Times, 146, 191, 195, 201, 252, 269, 271, 329, 625–27, 630–34, 637, 639, 643
Black Hills Electric Company of Deadwood, 374
Black Hills Exploring and Mining Association, 77, 79
Black Hills Forest Reserve, 292
Black Hills Illustrated, 596, 653–54, 689
Black Hills Medical Center, 416
Black Hills Milling and Mining Company, 142–43, 150, 152, 156
Black Hills Mining and Exploring Association, 88, 97
Black Hills Mining Museum, 31, 109, 138, 143–44, 214–16, 227–30, 232–33, 294–95, 297–98, 300–301, 315–20, 334–35, 338–41, 366–68, 378–79, 382–83, 388–89, 393–95, 414–18, 429–30, 434–38, 445–47, 474–76, 623–29, 647, 654–55, 690, 693–95, 699, 701–2
Black Hills Pioneer, 138–39, 142, 148, 198, 250, 279, 572, 623–26, 630, 633, 642, 663–64
Black Hills Power and Light Company, 397
Black Hills Products, 339
Black Hills Register, 336
Black Hills Reserve, 222
Black Hills Telegraphic Herald, 60
Black Hills Treaty, 5, 128–29
Black Hills Weekly Pioneer, 330
Black Moon, 121
Blackstone, Alex J., 376
Blackstone, Richard, 205, 269, 272, 274–75, 343, 376, 381–82, 404, 406, 419, 430, 537, 609, 635, 639, 647–48
Blackstone, Samuel, 205, 276
Blacktail claim, 109
Blacktail Gulch, 62, 106–7, 109, 278, 327, 329, 331–33, 454
Blake, Eli Whitney, 255, 638, 702
Blake Rock Crusher Company, 255
Blanchard, A. S., 106
Blanchard Party, 105–7. *See also under* prospectors
Bland, John, 428
Blanket House, 345–46

Bleeker Street, 303, 314–15, 317–19, 448
Blind Lode, 180
Bliss, Cornelius N., 292
Blodgett, Samuel, 105
Bloody Knife, 91, 94–95, 122–23
Blue Earth River, 52
Blue Mountains, 120
B&M shafts, 18, 167, 385–89, 396, 433, 447, 464, 549
 No. 1, 388, 443, 464
 No. 2, 19, 443–44, 450, 464
Boat, William J., 336
Bobtail Creek, 137
Bobtail Gulch, 106, 134, 136, 171–72, 174, 180, 192, 246, 306, 327–28, 341, 455, 581
Bogle Mill, 335. *See also under* mills
Boley, D. C., 357
Bolthoff, Henry, 148
Bolthoff pulverizer, 139
Boorda, Margie Winsel, 31, 635
booster pump, 505
Booth, J. C., 314
Bored Raise Drilling, 10, 487
Borglum, Gutzon, 601–2
Borland, Archibald, 177, 188
Borland, Archie, 178, 189–90, 196, 203–4
Boughton, M. V., 161
Boulder Canyon, 194, 276
Boulder Ditch, 6, 192–98, 200–203, 205, 632, 684
Bouyer, Mitch, 123
Bowen spring, 210
Bowie, Augustus J. Jr., 177–78, 189, 193, 196, 198, 201, 203, 255, 633, 679
Bowie Ditch, 178, 189, 193, 196, 205, 630
Bowlan, Archie, 196
Bowman and Flannery Saloon, 333
Boxelder Creek, 95, 134, 289, 581–82

Boxelder Creek, 95, 105, 582
Boylan, F. G., 343
Bozeman Trail, 63, 80, 82, 85
Brablec Construction, 505
Bradley, James H., 122
Brady Consultants, 503
Breen, Thomas H., 307
Brennan, John R., 110–11
Brennan, T., 329
Brick Store, 18, 278, 285, 309, 362, 372, 416–18
Bridger, Jim, 61, 65, 67, 77
Briggs, Charles, 165
Broadwater, Charles C., 349, 599
Brookhaven-Homestake Solar Neutrino Observatory, 553–54
Brookhaven National Laboratory, 31, 553–55, 557
Brookhaven solar neutrino experiment, 22, 31, 553–55
Brookings, Wilmont W., 77
Brown, Andrew Jackson, 58
Brown, David, 278
Brown, Jesse, 249, 615
Brown, Robert, 329
Brown, Tom, 58
Brown Pharmacy, 318
Brownsville, 269, 274, 276–78, 321, 418
Brulé and Oglala Tetons, 53
Brulé Tetons, 54
Brulé tribe, 12, 71, 115
Brush Arc Light Company, 372
Bryant, Frank S., 59, 105–7, 180, 333, 619–20, 692
Bryant, Jerry, 31–32, 275, 301
Bryant Party, 105–6, 619–20, 692. *See also under* prospectors
Bryon and Mike Rossiter, 346
Buchanan (president), 78
Buchanan, J. J., 189

Buck, Charles, 279
Buck's Landing, 272, 279, 289, 291
Buck's Rock, 279
Buckeye District, 112
Buffalo Bill's Wild West Show, 133
bulk-mining methods, 524
Bull, W. H., 191
Bullard, Jim, 277
bullion bars, 544
Bullock, Seth, 127, 200, 314, 620, 687
Bunny, Richard, 401
Burk, William, 480
Burke, Karl, 565
Burke Lovejoy, 390
Burlington and Missouri River, 270, 293–95
Burlington and Missouri River railroad, 270, 293–95
Burn, S. G., 269
Burnett, Ernie, 319
  "My Melancholy Baby," 319
Burns, John H., 200
Busch, Joseph F., 426, 601
Butler, Jack, 406
Butte Miners' Union, 401, 405
Butte Miners' Union No. 1, 405
Butters-Mein sand distributors, 508
Buxton Mill, 347. *See also under* mills
Buxton Mining Company, 347
Byro (beer), 339
Byron Party, 111. *See also under* prospectors

C

cablebolts, 302, 517, 527, 666, 670
Caddey, S. W., 41
Cain, Michael, 307
Calamity Gulch, 392
Calcite, 15, 42, 281–83, 295–97
Caledonia Gold Mining Company, 6, 157, 180, 182–83, 206, 263, 630, 679
Caledonia House, 342

Caledonia ledge, 46, 182
Caledonia Mill, 180–82, 206, 245, 255–57, 263. *See also under* mills
Caledonia Mining Company, 46, 180–83, 424, 559
Caledonia Syncline, 46
Calhoun, James, 91
California State Telegraph Company, 593
California Steam Navigation Company, 593
California Street Cable Car Railway, 593
Calumet House, 165
Cambrian Deadwood Formation, 41, 48
Cambrian period, 41, 49
Cameron, Daniel, 181
Camp 5, 300
Campbell, Sarah, 91, 111, 280
Campbell House, 320, 440
Camp Brown, 580
Camp Jack Sturgis, 132
Camp Robinson, 118, 126–27
Camp Ruhlen, 132
Camp Sheridan, 118, 132
Camp Sturgis, 132
Canadian Mine Services, Inc., 493
C and G Hose Company No.1, 341
carbon adsorption system, 512–13
Carbon-in-Pulp
  method, 370, 506
  plant, 21, 357, 509, 513–14, 544
carbon monoxide, 182, 422–23, 437–38, 440–41, 500–501
Cardwell, L., 159
Carey, Thomas E., 303
Carle Mill, 145. *See also under* mills
Carr, Thomas, 337
Car Repair Shop, 436
Carrington, Henry B., 80
Carrington, W. P. C., 73
Carter, Frank, 297
Carter, Janet M., 32, 39–40

Carter, Tom, 331
Carter Mine, 278
Carty, John R., 330
Carwye camp, 296
Case, Sidney, 377
Cassel and McLaughlin Amalgam Mill, 333. *See also under* mills
Castle Creek, 61, 94, 101–2, 108
Castlewood Country Club, 589, 662, 679
Castonguay, John, 429
Catholic Church, 309, 321
Catholic Sisters of Charity, 320
Cathrine Kaia Dorthea Engh, 583
Cavanaugh, Mike, 167
Cave Gulch, 107
Cave Hills Lutheran Church, 322
CBI (Chicago Bridge and Iron Company), 553
Cedar Creek, 119, 130
cement cap, 451, 520–22
Cenozoic era, 35, 37
Centennial Valley, 503–4
Central Campus, 16, 315–16
Central City, 8, 16, 19, 41, 106, 127, 139, 177, 181, 187, 195, 202, 205–6, 208, 212, 246, 328–29, 331, 333–37, 339–42, 344, 401–3, 452–53, 503, 633, 643, 678, 695
Central City Board of Trustees, 202
Central City Champion, 336
Central City Enterprise, 336
Central City Miners' Union, 401
Central Elementary School, 316
Central Pacific Railroad, 79, 594, 597
Central Safety Committee, 423
Central Steam Plant, 8, 18, 265, 348, 387, 391, 394, 436, 444
Centre College, 591
Cepak, Mike, 32, 565
Cetto, Frank, 429
CH2M Hill, 504–5
Chairman Commons, 426–27

Chandler, Charles F., 543
Chandler, Zacharia, 116
Chaplin, Abram B., 165
Chapline, Curley B., 330
Charles Flormann and his wife Fredricka, 289
Charles Street, 205
Cherry Blossom, 339
Cheshire, Thomas L., 181
Cheyenne and Black Hills Stage and Express Line, 156
Cheyenne Consolidated Mill and Mining Company, 455–56
Cheyenne River, 51, 56, 62, 73–74, 114
Cheyenne tribe, 12, 51, 53–54, 56, 62–63, 66, 73–74, 89, 98–99, 104, 109–10, 114–15, 120–23, 130, 139, 151, 156, 246, 455–57, 580, 622, 626, 682, 692
Chicago, Burlington, and Quincy's subsidiary, 294
Chicago Bridge and Iron Company (CBI), 553–54
Chicago Inter-Ocean, 93
Chicago Tribune, 96
Chief Ice, 121
Chief Little Thunder, 69
Chisholm brothers, 109
Chouteau, Azby A., 206
Christ's Ascension, 322
Christ Church, 321
Christensen, Kevin, 565
Christ Episcopal Church, 316, 600
Christian Science Society of Lead, 322
CIP Plant, 506, 508
City Creek, 106–7, 198, 201–2, 605
City Dump, 22, 510, 565–66
Civilian Conservation Corps (CCC), 283
Civil War, 578, 581
claims
  Baltic claim, 455

Bingham claim, 455
Discovery claim, 13, 106–10
  Claim No. 1, 106, 108–9, 177
  Claim No. 2, 13, 108–9, 605
  Claim No. 4, 107, 193
  Claim No. 5, 109
  Claim No. 6, 109
  Claim No. 9, 107, 109
  Claim No. 22, 109
Eagle claim, 455
Giant claim, 110, 135, 159–60
Gold Run claim, 110, 135, 159–60
Gopher and Golden Terra Extension claims, 174
Harrison and St. Patrick claims, 457
Hidden Fortune claim, 455, 458–59
Highland Chief claim, 167
Homestake and Segregated Homestake claims, 135, 137–38, 140–42, 145, 148, 150–52, 155–56, 158, 161–62, 168, 180, 190, 192, 213, 239, 245–46, 313, 434, 581, 623–26, 685
Hoodlebug claim, 455
Lincoln claim, 77, 79, 91–92, 95, 118–19, 121, 124, 141, 143, 205, 248, 252, 327–28, 341–42, 344, 601, 615, 622, 624, 626, 689–90, 692, 694
Mammoth claim, 135, 165
Marvine claim, 455
Mineral Point claim, 177, 344
Clara (open cut), 46, 166, 214, 304, 349, 353, 362, 559, 596–99
Clara Hill, 166, 304
Clark, Allen J., 347
Clark, Charlotte C., 353
Clark, Edward H., 348
Clark, Joseph, 166–68, 190
Clark, William, 131
Clark-Bowen diversion, 212
Clark Spring, 210
Cleveland, Grover, 292
Cleveland 5D, 226
Cleveland No. 44SW, 226
Cleveland No. 44TW, 226
Cleveland Toll Road, 306
Cliff House, 593
clinometer, 490
Clough (doctor), 416
Cloverleaf Mining Company, 278
Coffin, James, 306
Coffin, James D., 304–5
Cohen, A., 248
Cohen, P., 248
Coinage Act in 1834, 561
Cole, Nathaniel P., 597
Coleman, H. S., 177
Coleman, Nathan, 200
College of Mines, 597
Collins, Charles, 87, 89, 97–98, 336
Collins-Russell Expedition, 5, 12, 97–99, 111
Colorado River, 579
Colorado School of Mines Research Institute, 504
Colorado Siding, 277
Colovin, P. J., 336
Colpitts, Edwin H., 396
Columbia Ditch, 205–6, 606, 633, 684
Columbia School of Mines, 99, 543
Columbia Water Company, 206
Columbus Consolidated Gold Mining Company, 9, 19, 364, 452–54, 458–59
Columbus Consolidated Mill, 19, 332, 454. *See also under* mills
Columbus Gold Mining Company, 9, 242, 452–53, 680
Columbus Shaft, 453–54
Comanche band, 51
Commiskey, John B., 274–75
Committee on Indian Affairs, 113
communications systems, 395
Comstock Lode, 217–18, 584
Conclusion of Law, 202

Cone House, 349–50
Congregational Society Church, 321–22
Conley House, 344
Conners, Thomas, 307
Connolly, Joseph P., 49, 613
Conquering Bear, 68–69
Consolidated Power and Light Company, 374–75
Contract Advisory Committee, 522
Converse County, 276
Cook, B. C., 357
Cook, Charles H., 178, 204
Coolidge, A. L., 429
Cooper Institute, 87
coplaning process, 538, 667
Corey, George W., 111
Cornish pastie, 409, 602
Cornish Pump, 18, 385–86
Corral Creek, 95, 122
Corum, A. J., 421
Corwin, Al, 246
Costa Negra, 56. *See also* Black Hills
Costello, John, 146, 181, 246
Cotton's Bar, 318
Cottonwood Creek, 76
Courtney, Louise Thoen, 57
Cousin Jacks, 409
Cowan, R. B., 116
Cox (secretary), 86–87
Cravath, William B., 183
Crawford, Emmet, 124
Crawford, Jack, 110, 127
Crazy Horse, 71, 73, 80–81, 85, 89–90, 115, 117–18, 120–23, 128, 130–31, 133, 616, 678
Creighton (private), 94
Cretaceous period, 37
Crist, Charles H., 274
Crist, S. A., 330
Cristoforo Colombo Saloon, 318
Crocker, Charles, 594–95
Crofutt, George A., 29

Crook, George, 13, 103, 118–19, 127, 131
Crook City, 107, 126–27, 280, 328
Cousins, Charles 429
Crow Dog, 133
Crowe, Thomas B., 599
Crow Foot, 133
Crow tribe, 51, 53–54, 57, 66–67, 74, 85, 111, 121, 130, 132–33
Cuddy, John, 69
Cummings, Joseph, 132
Curley, Edwin A., 96
Curnow, John M., 165
Curtis, William, 91
Custer, Elizabeth, 95
Custer, George Armstrong, 91–92, 97, 114
    expedition 1874, 90, 94, 97, 111, 114, 133, 280, 333
Custer City, 104–6, 112, 126–27, 139
Custer Mining District, 98
Custer Park Mining Company, 94
Custer Peak, 38, 280
custom stamp mills, 13, 16, 138, 145, 147–48, 239, 333–34. *See also under* mills
cut-and-fill method, 21, 451, 515–19, 528–29, 540, 542
Cutting Mine, 457
cyanidation, 8, 17, 345–46, 349, 354, 364, 514, 665–68, 670
Cyanide Charlie, 345, 353, 599 *See also* Merrill, Charles Washington
cyanide mill, 406, 453–54, 459. *See also under* mills
Cyanide Plant, 8–9, 16–17, 20, 205–6, 257–58, 273, 331–32, 342, 346–47, 349–51, 353–55, 357–59, 362–63, 365, 368, 372, 448–49, 460, 464, 467–69, 508, 568, 596, 598
    No. 1, 16–17, 257–58, 273, 347, 349–51, 353–55, 357–58, 362–63,

365, 368, 370, 375, 386, 448–49, 460, 508, 596
  No. 2, 16–17, 205–6, 331–32, 342, 353–55, 357–59, 362, 370, 372, 375, 386, 467, 568
  No. 3, 20, 449, 467–69, 508
Cyclone Hill, 207, 209–10

D

Dacey, W. H., 440
Dahl, Eline Jensen, 583
Dakota Central Railway, 272
Dakota Historical Society, 79
Dakota Sioux, 16, 51–52, 91
Dakota Southern Railroad, 96
Dakota Territory, 5, 12, 16, 62, 77–79, 88, 90–91, 94, 99, 113, 124, 129, 132–33, 161, 165–66, 187–88, 227, 272, 311, 400, 409, 616–17, 623–25, 627, 630–31, 633, 680–82, 687–89, 691–92, 695–96
Dalkenberger, Charles, 320
Daly, Jack, 305, 312, 319
Daly, Marcus, 585–86
Daly Mine, 585
Dames and Moore (engineering firm), 504
Danforth, A. H., 459
Daniels, Walt, 429
Darton, N. H., 48
Davenport and Black Hills Milling & Mining Company, 161, 168
Davenport Company, 142, 150, 156, 161, 169
Davenport Mill, 13, 143–44, 148, 150. *See also under* mills
Davis, I. C., 330
Davis, Marshal, 330
Davis, Ray Jr., 22, 553, 555–57, 572
Davison, W. J., 343
Davy, P. B., 80

Dawes, A. C., 88
Dawson, A. R. Z., 126
Dawson, J. W., 159
Dawson Hotel, 344
Deadwood, Madison (Paha Sapa), Minnelusa, Lakota, and Fall River formations, 40
Deadwood and Delaware Smelter, 346–48
Deadwood Central Railroad, 278, 458
Deadwood City, 107, 110, 126–27, 131–33, 135–38, 151–52, 166, 173–74, 179, 188, 196–202, 205, 208, 212, 232, 246–47, 281–83, 291, 293–94, 305–6, 311, 330–31, 336–37, 339–40, 346–47, 356–57, 460–61, 503, 564, 600–601
Deadwood Creek, 13, 16, 105–10, 134, 139–40, 154, 176, 178, 189, 193, 196, 205, 303, 327–30, 332, 334, 339–41, 344, 354–55, 357, 441, 458, 503, 568, 605
Deadwood Daily Pioneer-Times, 27, 176, 233, 338, 343, 375, 405, 419, 439, 619, 634–35, 639, 641, 645–50, 652, 658, 664, 677
Deadwood Expedition, 132
Deadwood Formation, 37–38, 41, 48–50, 332, 454–55, 560
Deadwood Gulch, 13, 16, 19, 33, 49, 62, 106–9, 112, 136, 171, 193, 305–6, 312, 319, 327–28, 331–34, 341, 441, 454, 559–60, 642–43, 687
  Deadwood Mill (*see also under* mills), 16, 171, 174–75, 180, 205, 243, 245, 263, 332, 561
Deadwood Mining Company, 6, 157, 168, 170–71, 173–75, 243, 252, 453, 559, 628, 694
Deadwood Pioneer, 198
Deadwood Slime Plant, 506
Deadwood-Terra Mining Company, 6,

157, 174–75, 180, 183, 243, 263, 424, 559, 629–30, 635, 680
Deadwood Water Question, 6, 193, 198, 200
Deane, Coll, 187–88, 487, 631, 680
Deatherage, Perry "Jack," 477
DEDC (DUSEL Experiment and Development Coordination Committee), 573–75, 661
Deeble, Joel, 477
Deen, Pat, 297
Deep-Level Projects, 10, 479, 488, 492
deep level ventilation project, 498
Deep Underground Science and Engineering Laboratory (DUSEL), 23, 31, 33, 569–71, 574–75, 603, 613, 660–61, 679–80, 688
Deer Creek, 63, 111, 579
Deerfield Lake, 94
De Gray, Charley, 105
Deidesheimer, Philip, 217–18
Delano (secretary), 89, 91, 99
Delano, Columbus, 88, 99–100, 113
Delano, Moses, 165
Delano, M. S., 165
de Lassus, Carlos de Hault, 56
Delauncy, John, 333
Delicate, Don, 430
Dell Fockler, 274–75
Del Paso Country Club, 592, 663, 680
Delzer Construction, 505
DeMille, Cecil B., 602
Denver Engineering Mill, 363. *See also under* mills
de Oliver, Frank George, 111
Department of Dakota, 80, 118
Department of the Platte, 98, 118
Depression. See Great Depression
Derby, Kentucky, 592
Desert Land Act, 594
DeSmet Ditch, 178, 189, 193, 196, 205, 630

DeSmet Ledge, 47
DeSmet Mill (*see also under* mills), 13, 16, 139, 179–80, 187, 193, 196, 202, 205, 255, 263, 273, 335, 338, 344, 354–55, 458
DeSmet Saloon, 344
DeSmet Syncline, 47
dewatering process, 145, 367–68, 385, 398, 407, 466, 500, 572, 665
Diamond drilling, 43, 446, 463, 466, 481, 490–91, 524, 542
Dick, Billy, 317
Dickinson, Daniel K., 279, 317, 413–14
Dickinson, W. R., 317, 342
discovery shaft, 135, 137–38, 169
dit Manuel, Pierre Hebert, 577
diversion tunnel, 17, 24, 373, 376–77, 382
Dodge, George S., 100–102, 106, 154
Dodge, Richard Irving, 100, 115, 619
Doherty, Napoleon "Jack," 274
Donaldson, Aris B., 91
dore, 239, 242, 261, 362, 370, 544–45, 561, 668
Dorr, John V. N., 357
Dorr duplex classifier, 368
Dorr slime thickeners, 508
double jacking method, 214, 226
Dowe, John, 248
Doyle, William D., 281
Driscoll, Daniel G., 32, 39–40
Driscoll, R. H., 461
Drumlummon Mine, 347, 598
Duex, Todd, 31
Duffy, John, 111
Dull Knife, 119
Dump Draw, 94
Duncan, John, 165
Dunham-Bush Company, 501
Dunmire, Joe, 469
Dunstan, Jack, 429
Durbin, Thomas, 151, 170

DUSEL (Deep Underground Science and Engineering Laboratory), 23, 31, 569–70, 573–75, 603, 613, 660–61, 679–80, 683, 688, 691
DUSEL Experiment and Development Coordination Committee (DEDC), 573–75, 661
DuSette, Charles, 305
Dye, William, 85, 87
Dyer, A. B., 79
dynamos, 17, 371–72, 375, 645, 678

E

Eagle claim, 455. *See also under* claims
Eagle Drug Store, 318
Early, Pat, 339
Early Science Implementation Program, 572
East Main Street, 265, 305, 307, 321–22
East Sand Plant, 505, 508, 544
East Siding, 277
East Stope, 47
Eckland, Charles, 419
Edgemont, 294
Edison Dynamos, 8, 17, 371–72
Edison United Manufacturing Company, 371
Edmunds, Newton, 77
Edwards, Dolph, 336
Eighth Judicial Circuit Court of South Dakota, 403
Eisenhower, Dwight D., 602
Elk Creek, 15, 35, 95, 106, 186, 189, 205, 208, 210, 222, 269, 272, 278, 280–86, 288–89, 295–96, 418, 606
Elk Creek Canyon, 7, 15, 35, 106, 272, 280–86, 288, 295–96
Elk Creek Station, 278, 280–81, 295
Elkhorn Railroad, 284

Elliott Lumber Company, 177
Elliott Mill, 177. *See also under* mills
Ellison Boiler Plant, 436
Ellison circuit, 495
Ellison Compressor Room, 436
Ellison exhaust circuit, 495
Ellison Fire of 1930, 446
Ellison Formation, 42–43
Ellison Mine Office, 436
Ellison Reservoir, 437
Ellison Shaft, 19, 231–32, 273, 388, 396, 407, 433, 435–37, 439, 441–43, 446, 448–50, 461, 463–65, 471, 474, 487, 495–97, 549, 600–601
Elmendorf Farm, 592
Elrod, William, 297
Elster, Henry P., 428–29
Emigrant Gap Pass, 592
Emmet House, 342
Empire Bakery, 201
Enderby, William "Blackie," 477
Engel, P. M., 73
Engel, Roland, 429
Engh, Aksel Elini, 583
Engh, Alexander "Alf," 33, 134–38, 141, 145, 150, 159, 165, 170, 192, 238–39, 341, 581–82, 625
Engh, Martin Aleksander, 583
Engle, Robert R., 285
Englewood Hydroelectric Plant, 8, 17, 186, 209–12, 277, 373–75, 386, 437, 566
Englewood Plant, 210–11, 375
English Tudor mansion, 430
Enos, Cyrus H., 13, 142–45, 150, 152, 161, 168–69, 307–8, 641, 683
Environmental Protection Agency (EPA), 503
Eocene epoch, 38
Epp, Jon, 565
Erickson, Albin, 298

Esmeralda mill, 278. *See also under* mills
Este, 7, 289, 291–93, 296–97
Estes Creek, 95
Evans, Evan, 283
Evans, Fred T., 161, 271, 277
Evans, Robert H., 185, 630, 687
Evans-Tonn Ditch, 185, 606
Evening Press, 199
Evening Times, 198
Ewert, Theodore, 91
Executive Order 6102, 469
Exhaust Raise System No. 31, 488
Ezra Kind Party, 58. *See also under* prospectors

F

Fahey, Ted, 429
"Failure of a Large Mining Operator" (New York Times), 181
Fair Labor Standards Act (FLSA), 413
Fairview Mining Company, 174
Fall River, 40
False Bottom Creek, 37, 62, 106, 131–32, 184, 206, 606
Fan House and Assay Office, 436
Fantail Gulch, 347
Fantail Junction, 294
Fardele, George, 161
Fardell, George, 161
Fargo, Charles, 595
Fargo, William G., 595
Farish, C. S. T., 151, 626
Farish, William A., 151, 155, 159, 246, 609, 626
Farley, James, 176
Farlow, W. W., 151, 626
Farnum, E. B., 126, 131–32
Father De Smet, 59–60, 66, 68, 97, 114, 157, 616, 687. *See also* De Smet, Pierre Jean

Father DeSmet Consolidated Gold Mining Company, 6, 157, 177–80, 183, 188–89, 193, 196, 203–5, 255, 263, 344, 424
Father DeSmet Mining Company, 6, 176–77, 205, 559
Father Redmond, 321
Faust, Maude, 312
F. Berry, James, 187
Federal Housing and Urban Development grant, 570
Federal Indian Bureau, 86
Federal Reserve Banks, 469–70
Fermi National Accelerator Laboratory, 572, 575
Ferry, Patterson, 579
Fetterman, William, 80
Fielder, Mildred, 31, 157, 244, 295, 609, 615, 620, 629, 638
Filan, William, 145, 625
Fillebrown, H. C., 76
Finnish Evangelical Lutheran Church, 232, 321–22, 436
Finnish Temperance Society, 405
Fire assaying, 541, 658, 674, 681
First Finnish Evangelical Lutheran Church, 232, 321–22, 436
First Finnish Lutheran Church, 533
First Lutheran Church, 322, 534, 642
First National Bank
 of Deadwood, 284, 320
 of Lead, 320
First United Presbyterian Church, 322
Fisher, Cyndi, 31
Fisk, J. F., 318
Flag Rocks, 44, 474
Flaherty, John, 177, 249
Flat Iron Springs, 207, 608
Fleming, Hugh B., 68
Flick, D. W., 103
Flormann, Charles, 289–91

Flormann, Robert, 289
FLSA (Fair Labor Standards Act), 413
foliation, 38, 43, 668–69
Folsom, Henry, 185
Forbes, Lester, 331
Forbes, Lloyd, 331
forebay, 17, 24, 373, 376–77, 379, 382–83
Forest Hill, 201
Forest Reserve Act, 222
Forsythe, George A., 91, 96
Fort Abercrombie, 80
Fort Abraham Lincoln, 91–92, 95, 118–19, 124, 689
Fort Buford, 133, 578
Fort Ellis, 121
Fort Fetterman, 63, 118–19, 121
Fort George A. Meade, 132
Fort John and the Gratiot Houses, 69
Fort Laramie, 12, 29, 33, 53, 58, 60–61, 63, 65–69, 71, 73, 75, 80, 82–86, 89, 93, 98–100, 103, 106, 113, 116, 118, 126, 130, 311, 330, 580, 584, 611, 615–17, 677
Fort Laramie Treaty
    of 1851, 12, 33, 67, 73, 83
    of 1868, 12, 29, 33, 58, 83–84, 89, 99, 116, 611
Fort Leavenworth, 74, 622, 691
Fort Peck Indian Reservation, 130
Fort Phil Kearny, 80, 85
Fort Pierre, 15, 24, 32, 69, 76, 96, 105, 170, 175, 205, 245, 249, 254, 269–73, 275, 277–89, 291–99, 309, 325, 330, 373, 395, 434, 458, 586, 638–39, 677, 681, 687
Fort Pierre Railroad, 15, 24, 32, 170, 175, 205, 245, 254, 269–73, 275, 277–80, 282–85, 287–89, 291, 294–99, 309, 325, 373, 395, 434, 458, 586, 638–39, 677, 681

Fort Randall, 74, 79–80, 118
Fort Ransom, 578
Fort Reno, 121
Fort Robinson, 89, 131–32
Fort Stevenson, 578
Foss, Matilda Caroline, 582–83
Foss, Peter Andreas, 583
Foster, W. M., 189, 193, 605
Foster Ditch, 178, 189, 191, 193, 196, 205, 361, 605, 631–33, 683
Fourche, Belle, 61, 71, 74, 114, 126, 321, 428
Fourth Cavalry, 132
Fowler, E. P., 202
Fox Street, 307
Frankenberg, Aleck, 248–49
Franklin County Mining School, 584
Franklin Hotel, 600
Franklin Saloon, 333
Frank S. Peck, 19, 158, 458
Fraser & Chalmers Company, 14–15, 167, 240, 252–56, 387, 443, 627, 636, 681
Fraser & Chalmers Mill, 252. *See also under* mills
Fraser, Robert, 419
Frawley, Henry, 459
Fredrickson, Gabe, 297
Freeman, John W., 414
Free Timber Act in 1878, 291
Fremont, Elkhorn, and Missouri River Railroad, 331, 457
French, Charles H., 596
French, Clara 597
French Boys Arrastra, 140–42, 148
French Boys Ditch, 6, 141, 191–92, 194, 203, 632
French creek, 62, 92–94, 98, 100–103, 127, 134
Fritz, Orville, 480
Fry, C. H., 111

Fuller, Bert, 437
Fyler, C. C., 329

G

Galena Junction, 278, 297
Galena Street, 306
Gall, 121, 133
Gant, Frederic Dent, 91
Garden, Mary, 23, 601–2
Gardner, Craven Van Horn, 139
Gardner, C. V. "Cap," 283
Garretson, Austin, 427
Gashwiler, John W., 151, 157, 166, 180–81
Gaston, A., 200
Gates, Philetus W., 256
Gay, Alfred H., 104, 107, 110, 135, 303
Gay, William, 107, 110, 135, 158, 161, 303, 329, 331–32
Gay Party, 105, 107, 135. *See also under* prospectors
Gayville camp, 16–17, 31, 106–7, 138, 206, 232, 327–32, 339–40, 344, 354–55, 357, 359, 361, 372, 467
Geary Street, 597–98
Gem Theatre, 274
General Ellison mining claim, 433
General Land Office, 291–92, 307
Gentle Annie mine, 50
geology, 32, 35, 41–42, 45–46, 58, 102, 445, 526, 531, 613, 656, 679, 682–83, 688, 694
George, Frank "Portigee," 111
Giachetto, Antone, 429
Giant & Old Abe and Homestake Mines, 138, 143
Giant & Old Abe Mining Company, 627–28, 681
Giant and Old Abe Mining Company, 166–68, 424, 627

Giant claim, 110, 135, 159–60. *See also under* claims
Giant Lode No. 115, 110, 160, 303
Gibbs, Cook, and Parker Mill, 16, 332. *See also under* mills
Gilbert, C. H., 278
Gill, John W., 330
Gill, William, 181
Gilmer, John T., 156, 167, 171
Gilmer and Salisbury Stage Company, 156, 167, 246
Gilson (superindendent), 245, 247
Girard, R. H., 159
Girdler, Charles A., 202
Goiens camp, 296–97
gold, 19–24, 48–50, 56–64, 73–77, 93–94, 96–98, 100–102, 104–17, 134–40, 142–45, 157–61, 176–80, 193–97, 260–64, 303–8, 344–53, 355–58, 362–66, 452–60, 466–71, 506–8, 512–14, 541–48, 558–63, 617–24, 626–32, 634–46, 648–53, 664–83, 689–95
Gold Belt Cities: The City of Mills, 303, 635, 641–42, 650–51, 693
Gold Confiscation of 1933, 469
Golden Gate City, 8, 13, 24, 127, 140, 148, 157, 171, 176–77, 179, 256, 300, 328, 333, 336, 341, 344, 374, 392, 404, 439
Golden Gate Mine and Timber Company, 300
Golden Gate Mining Company, 176–77
Golden Gate Saloon, 344
Golden Gate Tunnel, 177, 344
Golden Hills Resort, 421
Golden Prospect Shaft, 13, 19, 170, 273, 372, 433, 443. *See also* Highland Shaft
Golden Reward Consolidated Gold Mining and Milling Company, 600

Golden Reward Shaft, 441
Golden Stairway, 20, 473, 496
Golden Star claim, 138, 141–42, 148, 156, 159, 162, 213
Golden Star Lode (No. 186), 160
Golden Star Mill (*see also under* mills), 13, 18, 161, 163, 232, 247–48, 252, 255, 257–58, 264, 355, 369, 384, 434–36, 441, 561
Golden Star Shaft, 162, 165, 244, 246, 309, 385, 434, 439, 464, 530, 540, 601
Golden Terra, 14, 50, 151, 157, 168, 172–75, 180, 190, 192, 243, 245, 252, 255, 257, 263, 269, 273, 331, 413, 561, 629, 694
Golden Terra Mining Company, 6, 151, 168, 172–75, 243, 252, 629, 694
Golden Terry, 34, 134, 137–38, 151–54, 169–72, 341, 581, 623, 626, 628–29, 685, 694
Golden Terry claim, 138, 151–52, 169–72
Golden Terry mining, 34
Golden Terry No. 2 Mine, 170
gold mine, 29, 41, 45, 366, 453, 571, 598, 626, 645, 656, 678–79, 691
Gold Nugget Beer, 337–38
Gold Reserve Act, 464, 469–70, 561, 652, 682
Gold Run claim, 110, 135, 151, 158–61. *See also under* claims
Gold Run Creek, 135–36, 142–45, 150, 155, 180, 191, 193–94, 197, 213, 258, 303, 328, 341, 349–51, 355, 448, 503–4, 510, 566
Gold Run Gulch, 24, 94, 136, 144–46, 232, 273, 303–6, 312–13, 365, 374, 433, 435, 475, 510, 581
Gold Run Lode No. 116, 110, 160, 303
Gold Run Toll Road, 307
gold rush, 5, 49, 56, 60, 96–97, 112, 134, 311, 327, 337, 584, 591, 619, 621, 689
California, 584, 591
Montana, 134
Gold Street, 19, 308, 310, 321, 414, 440
Goose Creek, 121
Gopher and Golden Terra Extension claims, 174. *See also under* claims
Gopher Mining Company, 174, 559
Gordon, John, 98
Gordon, Leonard, 307, 310, 641, 694
Gordon Party, 12, 60, 98–99. *See also under* prospectors
Gordon Quartz Mineral District, 306
Gordon Stockade, 13, 98, 104, 127
Gorman, Tom, 297
Graham, Anna, 315
Graham, George, 159
Graham, Kate, 319
Grand Avenue, 310
Grand Central Hotel, 126
Grand Island and Wyoming Central branch of the Burlington and Missouri River railroad also reached Deadwood from Hill City via Ten-Mile Station, Pennington, and Pluma, 293
Grand Prize Shaft, 248
Grand Prize Tunnel, 180
Grand River, 54, 123
Grant (general), 81
Grant, Ulysses S., 90–91, 99–100, 103, 113, 115, 128, 133, 611, 621, 695
Grant Hamilton's Hardware Store, 322
Grantz, Otto P., 455, 459, 461
Grantz Gold Mining, 344
Grantz Street, 307
graphite, 42–43, 264
Grattan, John L., 68
Grattan and Ash Hollow massacres, 68
Graves, L. R., 177, 200, 203
Gray, Jack, 342
Great Council, 68, 71

Great Depression, 409, 429, 464, 470
Great Plains, 54, 64, 69, 74, 118
Great Reconnaissance, 5, 70–71, 77, 133, 616, 694
Great Sioux Nation, 116, 128
Great Sioux Reservation, 12–13, 29–30, 33–34, 82–83, 86, 89–90, 97, 117–19, 124, 128–30, 132–33, 142, 159, 272, 305
Great Sioux War, 5, 29–30, 118, 133
Great Slave Lake, 579
Green, John, 159
Green, Thomas, 176
Green Front, 319, 325
Green River, 61, 580
Greenwood Camp, 289
Greenwood Min, 289
Gregg, Harrison M., 204
Gregg, Harry M., 178
Gregory, J. Shaw, 77
Grier, Thomas J., 13, 164–65, 182, 215, 232, 276, 279, 283–84, 292, 295, 309, 348, 354, 362, 375, 400–401, 404, 407, 412, 417, 419, 421, 425–26, 430, 560–61, 598, 600–601, 609, 626–27, 633–34
  lockout that precludes a strike, 403
Grier Addition, 429, 431
Grinnell, George Bird, 91
Grizzly Gulch, 21, 398, 483–84, 503–6, 508, 511–14, 564, 566, 656, 693
Grizzly Gulch Tailing Storage Facility, 10, 21, 398, 484, 505–6, 508, 511–13, 564, 566, 569
Grouard, Frank, 121, 124–25
Grummond, George, 80
Gunsolly, J. O., 342, 413
Gushurst, Al, 313
Gushurst, Peter Albert, 16, 304, 307–8, 311–13
Gustin Mine, 50, 327, 332–33
Gwinn, Samuel R., 144–46, 148, 167

Gwinn's Homestake No. 2 claim, 148
Gwinn's Tunnel, 246
Gwinn Avenue, 148, 410
Gwinn Mills, 166. *See also under* mills

H

Haag, Robert, 377
Hacienda del Pozo de Verona, 589, 662, 679
Haggard, E. D., 107, 110, 135, 303, 330
Haggin, B. A., 157, 174
Haggin, James Ben Ali, 11, 14, 23, 149, 152, 155, 157, 161, 163, 166, 168–69, 178, 189–90, 192, 194–95, 203–7, 227–28, 232, 246, 267–69, 271, 279, 284, 308, 591–92, 597, 607, 626, 663, 689–90
Haggin, John, 591
Haggin, Louis T., 271, 592
Haggin Cup, 592, 663, 680
Hagginsville, 277
Hall, Herbert S., 283
Hamburg Bell, 592
Hamilton, L. H., 181
Hancock, W. S., 88
Hangs-Around-the-Fort, 89
Hanna Creek, 210
Hanna Pump Station, 13, 208–10, 212, 386–87, 395
Harding, J. A., 108, 200
Hardinge, 363, 370
Harney, Hank, 33, 135–38, 142–43, 150, 156, 239, 313, 583
Harney, Henry Clay, 31, 33, 134–35, 141, 165, 170, 238, 341, 581–83
Harney, William, 69
Harney Peak, 36–38, 580
Harney-Warren Expedition of 1855, 70
Harqua Hala Mine, 347on of 1855, 70
Harqua Hala Mine, 347

Harris, C., 134
Harris, James R., 342
Harris-Corliss engine, 252
Harrison and St. Patrick claims, 457. See also under claims
Harrison-Durango Tunnel, 458
Hart, Olin, 487
Hartley, William, 194
Hartz jigs, 347
Harvey Ranch, 207
Hauser, George, 105
Haven, George D., 176–78, 204
Hawkeye mine, 50
Hawkins, Pat, 181
Hawley, N. H., 110
Hawley Party, 111. See also prospectors
Hay Creek, 95, 289
Hayden, Ferdinand V., 73, 76, 79, 88–89
Hayes, Rutherford B., 132, 138
Head, A. E., 157, 166–68, 190
Hearst, George, 11, 15, 23, 149, 151, 154, 156, 166, 168, 179, 189–90, 192, 194–96, 202–3, 205–7, 217–18, 227–28, 239, 245–46, 267, 270–71, 294–95, 416–17, 583, 587–89, 606, 625–26, 631–33, 662, 680–81
Hearst, Haggin, Tevis, and Company, 23, 149, 151, 583–86, 591
Hearst, Phoebe Apperson, 32, 316, 321, 324, 388, 404, 417, 434, 588, 590, 662–63, 679, 693, 696
Hearst, William Randolph, 179, 404, 585, 587, 589–90, 662, 692
Hearst Castle, 585
Hearst Corporation, 587
Hearst Ditch, 6, 186, 191, 205–7, 212, 437, 608
Hearst Free Kindergarten, 316, 321, 588, 600
Hearst Free Library, 417–18, 588, 590
Hearst Greek Theater, 590
Hearst Memorial Gymnasium for Women, 590
Hearst Memorial Mining Building, 590
Hearst Mercantile Company, 9, 15, 17–18, 245, 273, 278, 280, 285, 297, 309, 320–21, 342, 371–72, 416–18, 426, 590, 640, 677
Hebert, Anonyme, 577
Hebert, Marie Emilie Amelia, 577
Hebert, Marie Ida "Adelaide," 577
He Dog, 120, 131
Heffron, Michael, 165
Heinrich, George, 330
Heinrich, Gustav J., 337
Heitler, Frank, 419
Helena, Montana Territory, 105, 134, 337, 578–81, 623, 688
Heltibridle, Juliet, 32
Henderson Mine, 570
Hendy Challenge feeder, 256
Henley, William, 331
Hepburn, James, 111
Hercules (open cut), 46, 50, 214
Herkner, Henry F., 453
Hermes Mining Company, 585
Heydenfeldt, Sol, 180
Hiawatha Park, 410
Hickok, James Butler, 188, 333
Hicks, James, 107
Hidatsa tribe, 53–54, 66
Hidden Fortune, claim, No. 2, 456
Hidden Fortune claim, 455, 458–59. See also under claims
Hidden Fortune Gold Mining Company, 19, 454, 456–58, 651, 682–83
Hidden Fortune Mining Company, 9, 344, 364, 452, 455–56, 651, 682
Hidden Treasure Gulch, 13, 16, 139, 334–35
Hidden Treasure Mill, 13, 139. See also under mills
Hidden Treasure Mine, 50, 139, 334–35

Higby, T. C., 202
Higgins (assists the project of 80-stamp mill), 246
Highball drift crews, 483
High Forehead, 68–69
Highland Chief claim, 167. *See also* claims
Highland Mill (*see also under* mills), 13–15, 144, 170, 228, 243–44, 252–53, 257, 263, 269, 561
Highland Mining Company, 6, 13, 19, 147, 167–70, 175, 183, 228, 243, 252, 255, 263, 269, 372, 424, 433, 559, 628, 636, 681, 683
Highland Mining Company's Golden Prospect Shaft, 372
Highland Shaft, 628, 683. *See also* Golden Prospect Shaft
Highland Tunnel, 176
Hildebrand, John, 332
Hildebrand Mill, 106, 108, 329, 332. *See also under* mills
Hill, Henry, 314, 623, 692
Hill, Jack, 322
Hill City, 29, 49, 59–62, 79, 86, 89, 97, 102, 105–6, 111–13, 134, 137, 171, 176, 180, 184, 190, 196, 200, 293–94, 311, 322, 330, 534, 580
Hinch, Jack, 330
Hines, M. C., 76
Hines, Mike, 248
Hinman (reverend), 128
Hinman, Samuel D., 101
H. K. Porter Company, 14, 233, 270–71, 285
HLCP (Homestake Laboratory Conversion Project), 569
Hodges, Clarence 429
Hofman, H. O., 253, 345, 635
Hohnes, Philip W., 174
Holstein, Ben, 200
Holstein, W., 289

Holvey Building, 304
Holvey Drug Store, 313
Holway, Darwin, 429
homestake, 11–24, 29–34, 41–46, 140–45, 147–69, 227–34, 242–49, 254–59, 261–67, 273–84, 296–302, 307–11, 341–50, 362–67, 369–76, 386–90, 392–401, 403–9, 411–26, 428–34, 460–64, 469–71, 503–7, 509–20, 530–37, 541–55, 557–73, 598–603, 623–61, 677–93
  claim, 135, 137–38, 140–42, 148, 150–52, 155–56, 158, 161–62, 168, 180, 190, 192, 213, 239, 245–46, 313, 434, 581, 623–26, 685
  See also under claims
  discovery, 134, 138
  custom stamp mills in Lead City, 6, 145
  legacy, 564
  the need for water and a mill, 137
Homestake's Hearst Mercantile Store, 321
Homestake's Reclamation Bond, 536
Homestake's Water, Surface Construction, and Transportation groups, 387
Homestake Adams Research and Cultural Center (HARCC), 31, 564, 659, 683
Homestake Aid Association, 421
Homestake and Segregated Homestake claims, 145. *See also under* claims
Homestake Band, 9, 19, 428–29, 460, 602
Homestake Belt, 7–9, 13, 24, 34, 41, 50, 152, 156–58, 174, 178, 183, 190, 196, 213–14, 217, 237, 240, 242–43, 255, 258, 263, 267, 327, 424, 452, 559–60, 586
Homestake board of directors, 159, 167, 401, 474, 506, 531, 533
Homestake Collection, 31, 259, 282,

323, 335, 343, 354, 444, 469, 525, 539, 551–52, 556, 564, 577, 583, 601, 626
Homestake Formation, 41–43, 45, 49, 490, 501, 542–43, 560
Homestake Foundry, 22, 367, 552
Homestake Gold Mine, 41, 45, 366, 571, 645, 656, 679, 691
Homestake Hotel, 16, 309, 319, 372
Homestake Hydrostatic Water Level System, 572
Homestake Laboratory Conversion Project (HLCP), 569
Homestake Lode (No. 121), 160
Homestake Mill (*see also under* mills), 17, 159, 182, 227, 240, 246–47, 252, 254–59, 263–64, 267, 269, 320–21, 347–48, 369, 371, 434, 441, 503, 561
Homestake Mine, 11–12, 14–15, 19–22, 24, 33–34, 37–38, 41–43, 45–46, 48, 134–35, 164–65, 208, 216–18, 228–29, 245–46, 255–56, 263–64, 370–72, 407–9, 411–12, 541–43, 553–54, 557–64, 568–70, 598–603, 613, 629–30, 647–49, 654–57, 684–85
  baseball, 9, 419, 428
  capitalists, 6, 34, 149, 156, 165, 171, 178, 203, 207, 243, 255, 263, 269, 273, 279, 284, 559
  claim, 135, 137–38, 140–42, 148, 150–52, 155–56, 158, 161–62, 168, 180, 190, 192, 213, 239, 245–46, 313, 434, 581, 623–26, 685
   See also under claims
  community support, 409
  decision to close, 561
  early VIPs, 577
  early workforce, 9, 409–11
  employee housing, 9, 429
   employees' wages of, 409, 412–13, 426, 522
  famous VIPs as visitors in, 600
  hospitals in, 9, 18, 325, 413–14, 416
  locators of the, 134
  mergers of, 199
  mine safety, 422
  other belt companies of the capitalists of, 165
  people's nicknames in, 9, 431–32
  production statistics of, 10, 24, 558
  recreation building and opera house in, 9, 425
  return to work with nonunion employees in, 406
   subsequent efforts to organize the workforce in, 407
  system of stoping, 7, 222
  tailing-storage facilities, 504–5, 511, 514
  waste-rock facilities, 453, 535–36, 568–69
  wastewater-treatment plant, 21, 505, 510–11, 514
  water collection system, 13, 208, 212
  water rights in, 31, 156, 178, 184–85, 187, 189–92, 194, 203, 205, 207–9, 212, 372, 585–86, 605, 608, 633, 684
Homestake Mining Company, 44, 154–57, 182–83, 208, 224–26, 234–36, 244, 263–64, 307–8, 343–45, 404–6, 418–19, 479–85, 489–91, 497–501, 509–11, 518–23, 526–29, 534–36, 558–60, 565–69, 607–8, 623–24, 626–28, 630–41, 644–59, 677–80, 682–90, 692–93, 695–96
Homestake Mining Company and Barrick Gold Corporation, 569, 660, 683
Homestake No. 2, 144–45, 148, 153, 167–69, 191, 246, 424, 625, 688
Homestake No. 2 claim, 145, 148, 167, 169, 191, 246, 625, 688
Homestake No. 2 tunnel, 144–45
Homestake Open Cut, 138, 213
Homestake Powder House, 273

Homestake Railroad, 269, 271
Homestake South Extension Mine, 462
Homestake South Extension Mining Company, 9, 452, 460–61
Homestake Veterans Association, 9, 411, 419, 627, 635, 649, 659, 686
Homestake Visitor Center, 533–34
Homestead Act of 1862, 409, 617, 693
Hoodlebug (mining company), 344, 428, 455
Hooper, Thomas, 104
Hopkins, A. L., 174
Hopkins, George S., 158, 307
Hopkins, Mark, 594–95
Horney, Leonidas, 581
Horney, William, 581
Horse Creek, 66, 89
Horseshoe Grove, 277
Hose Parlors No. 2, 419
House Bill No. 4, 470
Howe, John
  recreation building and opera house, 320
H-Raise, 450
Hughes, R. B., 330
Hulin, Hilan "Pat," 107
Hull, H. F., 104
Humphreys, A. A., 75
Hunkpapa tribe, 12, 52, 54, 71–74, 87, 115, 117, 120, 122, 130–31, 133
Hunter, Edith, 591
Huntington, Collis P., 594
Hurricane District, 112
Hutton, J. D., 76
Hydroelectric Plant No. 1, 17–18, 24, 373, 376–80, 382–84, 392
Hydroelectric Plant No. 2, 18, 380, 391–92, 566

I

Idaho Territory, 78
igneous granites, 37

Illingworth, William H., 91
ilmenite, 42, 672
Immanuel Lutheran Church, 322
Incline Anticline, 47
Incline Ledge, 47, 162
Incline Shaft, 138, 162, 246
Independence Anticline, 48
Independence Stope No. 3, 440
Indian Affairs, of the Indian Bureau, 81, 83, 88, 99, 101, 103, 113, 116–17, 119, 128, 621, 690
Indian Appropriations Act, 128
Indian Flats, 205
Indian Land Cessions in the United States (Thomas), 129
Indians, 24, 29, 33, 51–52, 55, 57, 59–69, 73–75, 77, 80–82, 84–87, 89–90, 97–98, 101–5, 113–28, 130–33, 136–37, 149, 246, 251, 330, 337, 579–80, 589, 614, 616–19, 621–23, 677, 681, 688
Ingaldsby, M. J., 107, 110, 135, 303
Ingersoll-Rand CCW11, 226
intake dam, 17, 191, 196, 205–6, 376
Inyan Kara, 73–74, 92
Iowa State College, 349
Iron Creek, 111–12
Iron Shell, 82–84, 123–24
Irwin, E. F., 410
Itazipco Lakota tribe, 51
Iverson, Arley, 477

J

Jack Langrishe's theater, 126
Jackson, A. S., 428
Jacob (George Heart's brother), 583
James River Basin, 53
Jane, Calamity, 111
Janin, Alexis, 597
Janin, Henry, 154, 166–67
Janin, Louis, 13, 138, 143, 151, 180,

630
Janklow, William, 569
Japanese Pipe Conveyor, 21, 532
Jefferson, Thomas, 601
Jeffrey Aerodyne Mine Fan, 20–21, 474, 486, 498–99
Jenkins, L. P., 318
Jenney, Walter P., 13, 62, 99–101, 111, 115, 138, 143
Jentges, Paul, 313
JL's Gift Shop, 322, 534
Joe, California, 111
Johnson, Andy, 248
Johnson, B., 277
Johnson Brothers Company, 531
Johnson Gulch, 392
John Telford Ranch, 274
Jones, Charles, 305
Jones, Robert, 148, 304–5
Jones, Thomas, 141, 320, 624
Jones, William E., 139, 281
Judge Moody, 250–51, 307. *See also* Moody, G. C.
Judge Slaughter, 246
Julius Street, 321
Jumping Bull, 121
Junction, 280
Justice Mining Company, 157, 176–77
J. W. Sparks and Company, 462

K

Kaiser, Sam, 336
Kane, John, 107, 194, 330
Karney, James, 191
Keenihan, Mark, 565
Keets mine, 50
Kehoe, John H., 165, 627, 687
Kellar, Chambers 404, 406, 426, 428, 641
Keller, Gus, 206
Kellogg, Ludwig D., 150–51
Kellogg, Mark, 123
Kelly, P. J., 336
Kenyon, Robert, 104, 109
Keough Draw, 207
Kern River, 594
Keyo, J. H., 159
Keystone, 294
Kimberly, Peter, 170
Kincaid Coach Raid, 71
King, Dick, 111
King, John, 305
Kiowa, 51, 53–54, 66–67
Kiowa tribe, 51
Kipp, Noah, 111
Kirk Canyon, 441
Kirk Power Plant, 8, 18, 207, 391, 396–98
Kirk Road, 41
Kirk Timber Yard, 496–97
Kirwan, James, 401, 404, 426
Kissack, J. W., 145, 155, 419
Knife Blade Rock, 281
Knight (sheriff), 276
Knight, Dwayne, 343
Knowles, Freeman, 406–7
Konsella, P., 159
Koontz, Bob, 428–29
Kulpaca, George, 429
Kuykendall, William Littlebury, 187–89

L

Labarge, Tom, 105
Labor Day celebration, 16, 325–26, 420
laccoliths, 35, 38
Laflin, 289
Lake Traverse, 53, 71
Lake Winnipeg, 52
Lakota, 12, 33, 40, 51–52, 54, 66, 72, 86, 89–90, 101, 120, 130, 614, 622, 682
Lakota Sioux, 12, 72, 90
Lamb, Lyman, 98–99

Lancaster, Nimrod, 333
Lancaster City, 8, 327, 332–33
Lande, Kenneth, 557, 568
Lang, Gertrude, 333
Lang, J. C., 297
Lang, William, 419
Langdon, Emma F., 401
Langhoff, Fred, 377
Langrishe Theater, 202
Lantern, 403, 405–7
Laramide orogeny, 37–38
Laramie River, 114
Lardner, Elizabeth C., 206
Lardner, William Jr., 107, 177, 206, 333, 452
Lardner Ditch, 206
Lardner Party, 105–8, 332. *See also under* prospectors
Large Underground Xenon (LUX), 572
Larson, Ariel, 32, 280
Larson, Charles, 280
Larson, Harold, 280
Larson, John "Happy," 298–99
Larson, Willard, 146, 334, 624, 628
Latham, Dan, 145, 625, 688
Laughing Water Creek, 103
Lavier, Charles, 274–75
Lawie, Alexander, 318
Lawler, John J., 336
Lawrence Berkeley National Laboratory, 32, 570, 575–76, 661, 683
Lawrence County, 32, 41, 45, 127, 165, 178, 188, 197, 204, 329, 348, 405–6, 470, 600, 619, 624, 628–31, 633, 637–39, 641–42, 678, 681–83, 687, 689, 691, 694
Lawrence County Board of Commissioners, 197
Lawrence County Commission, 127
Lawson, Andy, 181
Lazarus, Edward, 29, 616, 619, 621, 624
Lead's Post Office, 418
Lead Anticline, 43

Lead City, 13, 15–16, 24, 31–32, 38, 41, 49, 143–45, 147–48, 164–65, 194–96, 206–9, 211–12, 227–28, 230–32, 252, 256, 267–73, 276, 281–82, 284–88, 294, 296–97, 303, 305–10, 312–13, 319, 321–22, 402–3, 641
  celebrations, 323
  churches, 321
  first murder in, 7, 313
  schools, 314
  townsite trustees, 314
Lead City Miners' Union, 269, 400, 402–3
Lead City Miners' Union Hall, 400
Lead Commercial Club, 428
Lead Congregational Society Church, 322
Lead Country Club, 207, 210, 428
Lead Daily Call, 230–31, 303–4, 403, 405, 635, 640–42, 648–50
Lead Deadwood Sanitary District, 208, 212, 503–4, 568
Lead Evening Call, 410, 648
Lead High School, 316, 641, 687
Lead Kiwanis Club, 428
Lead Labor Day parade and celebrations, 428
Lead Miners' Union, 400–401, 403, 405, 407, 421, 426
Lead Opera House, 400
Lead Public School System, 316
Lead Substation, 384, 386–87, 391–92
Lead Syncline, 43, 48
Lead Temperance Society Hall, 321
Lead Townsite Boundary, 158
Lead Window, 5, 38, 41, 43, 46
Lead Womans' Club, 316
LeBeau, J. B., 331
Lecompton Mine, 584
Ledge
  No. 3, 47
  No. 4, 47

Leeper, Jack, 277
Leggett, Thomas H., 598
Legislative Assembly of Dakota Territory, 90
Lennon, John, 426–27
Leroy, Carrie, 284
Leroy, Elnora, 284
Leroy, Henry, 284
Lessell, E. K., 321
Levy, James, 158, 161
Lewis, J., 248
Lillehaug, Gary, 31, 376–78, 380
Lilly, William J., 18, 389–90
Lilly Hoist Safety Controller, 18, 389–90
limekiln, 15, 282
limekilns, 282
Lincoln, Abraham, 77, 79, 91–92, 95, 118–19, 121, 124, 601, 689
Lincoln Avenue, 205
Lincoln House, 342
   Newton's Boarding House, Miner's Home Hotel, Emmet House, and the Caledonia House, 342
Lincoln Lode, 143, 624, 694
Lincoln Shaft, 248
Linn Tractor and Trailer, 298
Linscott, William D., 289
Little Big Horn River, 119, 122–23
Little Big Man, 131
Little Elk Creek, 35, 95, 186, 189
   canyon, 35
   ditch, 6, 189
Little Hawk, 131
Little Maud, 141
Little Missouri River, 51, 123, 136
Little Rapid, 107–8, 186, 205, 207, 210, 212, 373, 606, 633, 683
Little Rapid Ditch, 7, 205, 207, 210, 373, 633, 683
Little Rapid diversions, 212
Little Rapid Springs, 207, 606

Little Rapid Weir House, 207
Little Spearfish, 37, 107, 391, 607
Little Thunder, 68
Little Wolf, 115
Lloyd, Harry, 593
load-haul-dumps, 526
Lobdell, A. C., 330
Locan, Frank, 180
Lockout of 1909-1910, 400, 404–6, 421, 425–26, 600
locomotive no. 10090, 228
"lode gold," 33. *See also under* placer
Lodge Trail Ridge, 80
Logan, B. B., 98–99
Logan, Henry J., 389–90
Logan Engineering Company, 390, 646, 688
Logeman, Jack, 429
Loisel, Regis, 56
London, A. L., 248
Lonergan, John, 336
Long, A. F., 98
Long, James, 319
Long Baseline Neutrino Experiment, 575
Long Valley, 392
Looking-for-Enemy, 124
Lookout Mountain, 57, 62–63
Lost District, 110
Lost Mining District, 107
Loudon, A. H., 333
Louisa Ann Haggard, 581
Louisiana Purchase in 1803, 55
Lounsbery, Richard P., 174
Lowe, Richard, 105
Loyal Legion, 406–7
Lucas, J. W., 343
Lucky Strike Mining Company, 289
Ludlow, William, 12, 91–95, 114, 618, 688
Lundberg, Dorr, and Wilson Mining Company, 600

Lundberg, Dorr, and Wilson Mill, 357. *See also under* mills
Lundin, A. H., 419
Lunt, D. S., 111
Luschei, Martin, 29, 616–17, 619, 621
Lutheran Church, 232, 321–22, 436, 533–34, 642, 692
LUX (Large Underground Xenon), 572, 594
Lyford, Charles, 400
Lyman, William P., 77, 148
Lynch, James, 248
Lynch, J. J., 336
Lynch, Michael, 165, 627

M

MacArthur-Forrest cyanide process, 346
Mackay, John, 592
Mackin, Bernard, 321, 336
Madison Formation, 40, 76
Maggie (horse), 231
magmas, 38, 43
Mahler, J. C., 174
Main Ledge, 43, 47, 236, 444–45, 450, 461–64, 471–73, 481, 497, 537, 540, 542, 551, 561–63
Main Ledge Anticline, 47
Main Street, 13, 18–19, 147, 202, 265, 305, 307–8, 310–11, 313, 315–18, 321–22, 324, 329, 372, 408, 414, 418, 425, 429, 440, 445, 448, 533
Majorana experiment, 573
Major Ledwich, 314
Major Research Equipment and Facilities Construction (MREFC), 574
Mallory, L. H., 145, 148, 625
Mallory, Thomas H., 111
Mallory Gulch, 111–12
Mallory Open Cut, 138, 213
Mallory Party, 112
Mammoth claim, 135, 165. *See also under* claims
Man-Afraid-of-His-Horses, 68, 83, 85
Mandan tribe, 53–54, 66
Manifest Destiny, 29, 63, 65, 133, 615, 688
Manly, Basil M., 427
Mann, George B., 185
Manning, John, 249
Manuel, Fred, 23, 33, 134–35, 138, 141, 145, 165, 170, 192, 238–39, 305, 313, 341, 577, 581–82, 624–25, 641
Manuel, Marie Phelonise, 577
Manuel, Moses, 11, 23, 33, 134–37, 141, 148, 150, 165, 170, 192, 238–39, 313, 341, 577–78, 582, 623
Manuel Brothers Park, 421, 533
"manway," 220, 465, 479, 493, 515, 518, 521, 670
Manypenny, George, 128
Manypenny Agreement, 5, 33, 128–30, 132, 142
Manypenny Commission, 128
Marconi Avenue, 592
Marcy Rod Mill, 477, 507. *See also under* mills
Marks, John, 32, 499, 655
Marshman, Lewis, and Brown Mill, 148. *See also under* mills
Martha (George Heart's sister), 111, 597, 638, 687
Martin, Charles, 333
Martin, E. W., 600
Martin Hotel, 319
Marvine, G. E., 455
Marvine claim, 455. *See also under* claims
Marysville District, 580
Mason, David, 179
Maurice Intake, 17–18, 376, 383
May, Ernest, 307–8, 312, 319
Mayham, H. J., 453, 459, 461

Mayham, Messers, 459
Maynadier, H. E., 76
Mayo, John, 404
McAleer, Jack, 109
McBatt, Alec, 136
McCaffrey, Hugh, 200
McCall, Jack, 188
McCarty, Jerry, 330
McClellan, George D., 406, 421
McDonald, Angus W., 98
McDonald, J. C., 178
McDonald, M. L., 154, 157, 180
McDonald, Pete, 197
McDougal, Alex, 159
McGillycuddy, Valentine T., 100
McGovern Hill, 205, 306
McGregor (millwright), 246
McIntyre, J. D., 13, 155, 160–61, 172, 306, 628–29, 694
McKay, Edward, 107, 333
McKay, William T., 91, 93
McKenzie, Alex, 273–74, 278
McKinney, A. M., 194
McLaughlin, Donald H., 151, 445, 626
McLemore, J. C., 401
McLennan, R., 191, 607
McMakin, W. J., 406, 419
McMaster, John, 167
McMaster, Samuel, 13, 155–56, 159, 164, 167–68, 174, 176, 178–79, 189–90, 192–95, 199, 204–5, 209, 246, 248–49, 267, 269, 271, 345, 605, 609, 623–24, 630, 633, 691, 694
McMaster, Thomas, 174
McMasters, Angus, 159, 248–49
McMillan, James, 73
McNabb Ranch, 428
McShane, A. G., 138, 623
McTeague, John, 314
Meadow Valley Wash, 579
mercury trap, 258, 260
Merrick, Al, 198

Merrill, Beatrice Maud, 599
Merrill, Bruce Robinson, 599
Merrill, Charles Washington, 11, 23, 257, 347–49, 353, 357, 363, 370, 507, 596–600, 664
Merrill, Gregor Charles, 599
Merrill, John Lisgar, 599
Merrill-Crowe precipitation process, 370, 508, 512, 514, 599, 670
Merrill Story, The (Ricker), 353, 643–45, 664, 691
Merritt, Wesley, 120
Mesozoic era, 35, 37–39
Metals Disintegrating Company of Elizabeth, 600
metamorphic rocks, 37
metamorphism, 42, 674
Methodist church, 284, 341, 343
Methodist society, 321
Mexican Cession, 55, 70
Meyer, C. W., 200
Middle Boxelder Creek, 95
Mile-High Mobile Home Park, 421, 533
Miles (colonel), 130
Milk River, 579
Miller, John F., 588
Miller, Joseph, 126
Miller, Newel, 283, 285, 296
Miller's Platform, 280–81, 283
Millet, R. D., 252
Milliken (No. 1) Winze, 446, 473
Milliken, James, 307–8, 461
Milliken, John T., 462
Milliken Shaft, 495
Milliken Winze, 19, 446, 448, 471, 497
milling circuit no. 4, 512–13
Milling Company, 344, 346, 600
Mill Optimization Projects, 10, 512, 514
Mill Reservoir, 505
mills
    Allis Chalmers Ball, 363, 366–68, 370, 466, 477, 507, 512

Amicus Mill, 15, 24, 259, 261, 263, 348, 354, 363, 367, 370, 372, 407, 441, 444, 467, 549
Anson Mill, 124
ball, 139, 477–78, 551
Baltimore and Deadwood Mill, 16, 332
Bogle Mill, 334–35
Buxton Mill, 347
Caledonia Mill, 180–82, 206, 245, 255–57, 263
Carle, 145–47
Cassel and McLaughlin Amalgam Mill, 333
Columbus Consolidated Mill, 19, 332, 454
custom stamp, 13, 16, 138, 145, 147–48, 239, 333–34
cyanide, 406, 453–54, 459
Davenport Mill, 13, 143–44, 148, 150
Deadwood Mill, 16, 171, 174–75, 180, 205, 243, 245, 263, 332, 561
Denver Engineering Mill, 363
DeSmet Mill, 13, 16, 139, 179–80, 187, 193, 196, 202, 205, 255, 263, 273, 334–35, 338, 344, 354–55, 458
Elliott Mill, 177
Esmeralda, 278
Fraser & Chalmers Mill, 252
Gibbs, Cook, and Parker Mill, 16, 332
Golden Star Mill, 7, 13, 18, 161, 163, 232, 244, 247–48, 252, 255, 257–58, 264, 355, 369, 384, 413, 434–36, 441, 561
Gwinn Mills, 166
Hidden Treasure Mill, 13, 139
Highland Mill, 7, 13–15, 144, 170, 228, 243–44, 252–53, 257, 263, 269, 561
Hildebrand Mill, 106, 108, 329, 332
Homestake Mill, 7, 17, 159, 182, 227, 240, 244–47, 252, 254–59, 263–64, 267, 269, 320–21, 347–48, 369, 371, 434, 441, 503, 561
Lundberg, Dorr, and Wilson Mill, 357
Marcy Rod Mill, 477, 507
Marshman, Lewis, and Brown Mill, 148
M. L. White Mill, 246
Monroe Mill, 243, 262–64, 341, 354, 369, 372, 374, 407, 439, 441, 689
Morris and Costello Mill, 145, 147
Pocahontas Mill, 20, 228, 365, 370, 372, 374, 386, 469
Racine Mill, 145, 147–48
Sheldon Edwards Amalgam Mill, 333
Smith and Pringle Mill, 148
Snell Mill, 145
South Mill, 8, 17, 20, 207, 236, 244, 263–64, 317, 365–70, 387, 396–97, 429, 431, 436, 441, 463–65, 467–68, 470, 474–75, 477–78, 505, 507, 512, 514, 531–32, 544–45, 549, 551
Thompson, 145–48, 159, 187–88, 193, 315, 334
Tual, Casey, and Webber Amalgam Mill, 333
Tungsten Mill, 244, 263–64, 364–65, 444, 460, 645, 652, 691
White Stamp Mill, 147
Whitney Amalgam Mill, 333
Mills, B. F., 336
Mills, Louis D., 349, 599
Mine and Engineering Office, 475
Mine Enforcement and Safety Administration, 497
mine fire of 1907, 437
mine fire of 1919, 440, 442
Mine Monitoring System, 467
Miners' Home Hotel, 342
Miners' Magazine, 401
Miners' Union Hall, 282, 321, 400–401, 421

Mineral Point, 177, 179, 244, 263–64, 344, 354, 374, 404, 407, 413, 439, 468
   claim, 177, 344
   See also under claims
Mineral Springs, 207, 608
Mine Rescue, 19, 422–23, 497
Miners and Merchants Bank, 428
Miners' Union Opera House, 590
Minerva mine, 50, 332
mine surveying, 537, 658, 689, 692
mining
   method, 50, 444, 515, 542, 634, 691
     mechanization of, 10, 515
       blasthole sublevel stoping, 10, 524
         mechanized cut-and-fill stoping, 10, 528
       open cut-and-fill, 20–21, 451, 490, 515, 517–22, 524, 527–28, 537, 540
       square-set cut-and-fill stoping, 21, 515–17, 524, 527
       uphole and bench, 527
       vertical crater retreat mining, 525–26, 676
   placer, 61, 94, 102, 107, 109, 134, 325, 664, 672, 690
   See also under placer
   quartz, 102, 325
Mining Enforcement and Safety Administration of the Department of Interior, 504
Mining Law of 1872, 90, 142, 159, 291, 307, 611, 671
Minneapolis Brewing Company, 337
Minnesota River, 52–53
Minnesota Territory, 78
Minniconjou tribe, 51–52, 68, 71, 115, 117, 124, 130, 133
Minute Men, 127
Miocene epoch, 39
Mississippi River, 51–52, 55, 58, 70
Missouri River, 51, 53–54, 56, 62, 73–76, 82, 86, 88–89, 98, 105, 120, 128, 130, 132, 136, 270, 272, 293–95, 331, 457
Mitchell, Albert "Bud," 420, 469, 487, 689
Mitchell, Edwin W., 310
Mitchell, Steven, 565, 689
Mitchell, W. J., 428–29
Mix, John, 98
M. L. White Mill, 246. *See also under* mills
Mogul Mining Company, 600
Monarch-Rockwell furnace, 544
Monheim, John, 278
Monitor mine, 50
Monroe Mill (*see also under* mills), 262–64, 341, 354, 369, 372, 374, 407, 439, 441, 689
Montana Ditch, 6, 190–92, 194–96, 203, 205, 608, 632, 684
Montana Ditch Company, 192
Montana Mining Company, 347–48, 598
Montana Territory, 84, 107, 130, 134, 578–80, 586
Moody, Gideon C., 133, 174, 202–3, 250, 279, 292, 307, 631, 633, 681
Moon, S. M., 330
Moore, George, 193, 357
Moore, John F., 206
Moore, Joseph B., 307
Moore, Thomas, 105
Morcom, John H., 459
Morcom, William, 429
Morgan, Julia, 589–90
Morgan, Reece, 274–75
Morgando, John Sr., 429
Morris, Frank, 16, 310
Morris and Costello Mill, 145, 147. *See also under* mills
Morrow, Stanley, 31, 147

Morse, Samuel F. B., 596
Moskee sawmill, 15, 300–301
Mountain Top ballpark, 18, 325
Mountain Top Field, 420
Mountain Top Park, 533
Mount Rushmore, 37, 428, 601–2
Mowatt, Jack, 281
Mowatt camp, 280–81
Moyer, John, 277
MREFC (Major Research Equipment and Facilities Construction), 574
Mrs. O'Hara's Capital Boarding House, 333
Muckler, Dan, 107, 110, 135, 303
Mudder, Terry, 510
Mudlock (bacteria), 510
Munday, Henry, 200
Murphy, "Spud," 274
Murphy, B. E., 106, 267, 274–76, 296–97
Murphy Gulch, 267
Murray, Dick, 429
Murray, Ed, 429
Murrin, Thomas D., 285, 426
Myers, Ira, 191, 206
"My Melancholy Baby" (Burnett), 319

N

Narcouter, Lephiere, 105
National Academy of Sciences, 291
National Cathedral School of Girls, 590
National Congress of Mothers, 590
National Environmental Policy Act, 503, 506, 514
National Forestry Commission, 291
National Parent-Teacher Association, 590
National Pollutant Discharge Elimination System (NPDES), 510
National Science Foundation (NSF), 569–75, 603, 660–61, 680, 688, 690

Neary, J. L., 426
Neary, Thomas, 307
Nebraska Territory, 70, 78
Neill, Robert, 109, 111
Neiman, McLaughlin, and Nicholson sawmills, 302
Nelson, Chris, 320
Nelson, Mark, 565
Nelson, Tom, 31
Nemo area, 15, 95, 222, 278, 289, 291–93, 297–301, 418, 640, 677
Neubert, Armin, 337–39
neutrino, 22, 31, 553–56, 573, 575–76, 661, 671, 690
Nevada, 105–6, 191, 194, 199, 201, 217, 246, 273, 504, 579, 584–85, 592, 594, 597, 607
Nevada Gulch, 191, 194, 273, 504, 607
Newell, Milton H., 599
Newell town, 428
New French Boys Ditch, 203
Newton, Henry, 100
Newton's Boarding House, 342
Newton-Jenney expedition, 5, 62, 99–101, 103–4, 106, 111, 116, 133, 333
New York, 87, 91, 93, 99, 113–14, 151, 155, 173–75, 181, 294, 336, 371, 453, 569, 590, 592, 613–17, 619–21, 629–30, 656, 658–59, 663–64, 680–81, 683, 687–88, 691
New York Stock Exchange, 155, 569
New York Times, 99, 104, 113–14, 181, 294, 617, 619–21, 630, 659, 664
"Failure of a Large Mining Operator," 181
New York Tribune, 93
Nichols, David E., 176–77
Nichols, Richard, 159
Niemi, John, 321
9 Ledge, 48, 236, 464, 466, 471, 481, 524, 540, 562–63
nine-post raise, 220

Niobrara River, 56, 73, 129
Nix, George M., 461
Nob Hill, 19, 429–30, 593
Noonan, John, 94–95
Nordberg hoist, 441, 446
Nordberg Manufacturing Company, 388, 466, 475
Norris, Samuel, 592
Norris Peak Road, 95
Northam, G. A., 405, 426
North and South Sandfill, 450
North Battery, 365–66
north Bleeker Street, 314, 319, 448
Northern Pacific Railroad, 82, 90, 94, 252
North Fork of Rapid Creek, 94, 210
North Gold Street, 321
North Homestake Exploration Project, 10, 501
North Mill Street, 15, 18, 147, 304, 321, 371, 416–18, 441, 448, 590
North Platte River, 66, 84, 580
North Sand Line, 497
North Sand Raise, 449, 451
Northwestern Stage and Transportation Company, 252
Northwestern Transportation Company, 88, 252
Norton, George A., 319
Novak, John, 336
NSF (National Science Foundation), 569, 573–74
Nutless Mine, 344

O

O'Brien, W. S., 222
O'Connell, James, 426–27
O'Donnell, D. J., 307
O'Fallon, Benjamin, 54
O'Leary
 Tim, 159
 Timothy, 313
O'Neill Pass, 40
O' Reilly, J. J., 336
Ogden, David B., 336
Oglala Lakota tribe, 54, 130
Oglala tribe, 12, 51–54, 59, 65, 67, 69, 71–72, 80, 85–86, 89, 115, 117, 120, 122, 124, 130, 132–33
Oiyuwega, 52
Ojibwa tribe, 52–53
Old Abe Anticline, 47
Old Abe claim, 135, 145–46, 148, 155, 165–67, 624–25, 627–28
Old Abe Mining Company, 6, 165–68, 248, 424, 623–25, 627–28, 680–82, 687–90, 692, 696
Old Abe Shaft, 13, 144, 165, 182, 309, 344, 385, 549
Old Brig Shaft, 175, 180, 245, 273, 441
Old Woman Creek, 73
Oleson, Andrew H., 460
Oligocene epoch, 39
Olson, Freddie, 481
Omaha, 73, 79, 105, 110, 118, 346–47, 595, 614, 680
Omaha Treaty, 595
O. Mills, Darius, 594
Omisis tribe, 51
Ontario Mine, 585
Open Cut, 10, 12–14, 16, 20–22, 24, 32, 38, 43, 46–47, 110, 135, 137–38, 147, 171–72, 180–81, 213–14, 322–23, 416, 421, 438–39, 450–51, 515, 517–22, 527–28, 530–35, 549, 562–64, 566, 568–69, 669–70
 East Waste Rock Facility, 22, 535–36, 568
 exploration work in, 530
 modern mining in, 530
 pit expansion in, 534
  reclamation of Waste-Rock Facilities, 536

Terraville Test Pit, 531
Open Square-Set Method, 7, 217
open timber stoping, 515
Ophir Mine, 217–18, 585
Ord, Edward O. C., 118
ore cars, 14, 19, 162, 213, 216, 219, 227, 229, 232, 435, 518–19, 551
Oregon Territory, 70
Organic Administration Act, 77, 222, 292
Oriental and Occidental Steamship Company, 594
Orman, James B., 461
Oro City, 189, 328, 344, 605
orogeny, 37–38
Oro Hondo Mining Company, 9, 20, 452, 460–63, 486
Oro Hondo Shaft, 10, 20–21, 398, 461, 463, 472, 474, 486–87, 496, 498–500
Ottinger, Charles C., 167
Ottman, A., 318
Otto, Bob, 31–32, 433
Ourth, Frank, 337
Owl Creek, 126
oxidation phase, 362
oxygen-vacuum system, 512, 514

P

Paananen, LeeAnn, 32
Pacific Coast Oil Company, 596
Pacific Express Company, 594–95
Pacific Railway Act, 79, 617, 693
Paddock, Algernon, 128
Padouca band, 51
Pa-ha-Sa-Pah: or the Black Hills of South Dakota (Rosen), 59
Paige, Sidney, 46, 48, 613
Paleocene epoch, 38
Paleozoic era, 35, 37, 49
Palethorpe, Mary Jane, 590

Palmer, George, 111
Palmer, Virgil, 539
Palmer Ditch, 178, 189, 193, 196, 205, 630
Palmer Gulch Mining District, 111
Palmer Party, 111. *See also under* prospectors
Palmetto and American Flag Consolidated Mines, 166
Palms, Francis, 165
Panic of 1873, 133
Park Avenue, 143–45, 394, 410
Parker, E. B., 330
Parker, Ely, 86
Parker, Gifford, 278
Parnell, W. E., 165
Parsons, H. B., 157, 174
Parsons, William, 482
parting process, 544, 546
Pascoe, John, 181
Patterson, Tom, 106
Peak, Terry, 38, 426, 504
Peake
  No. 1, 207, 633, 684
  No. 2, 207
Peake Ditch, 7, 205, 207–8, 210, 373, 634, 685
Peake diversions, 212
Pearson, John B., 105, 107, 109, 135, 160, 180, 303, 341
Pearson Party, 105, 135
Pendo, Esther, 333
Pepin, Marguerite, 577
perchloroethylene, 22, 554–56, 568
permanent camp, 93–94
Peterson, H. J., 322
Phifer, W. D., 321
Phillips, Henry, 428–29
Phillips, Kirk G., 194, 459
Phillips Academy, 597
Phoebe Hearst Free Kindergarten, 321
Phoenix Iron Works, 337

phonolite dikes, 43
phonolite intrusive rocks, 44
Piedmont, 7, 15, 95, 231, 272, 280, 282–89, 293, 296–97, 418
    town of, 283–84
Piedmont Townsite Company, 284
Piedmont Valley, 95
Pierce, J. A., 189, 605
Pierce, No. 1, 47
Pierce, No. 2, 47
Pierce Anticline, 47
Pierce Ledge, 14, 47, 218–19, 448
Pierman, James, 105
Pietila, Eric, 281
Pillar Peak, 38
Pimlico. See Silver City
Pinchot, Gifford, 292
Pine Ridge Agency, 133
Pine Street, 305, 308, 314–15, 320
Pinney, Milton E., 139
Pioneer Ditch, 6, 142, 145, 190–92, 194, 201, 607, 632, 684
Pioneer News, 137, 142, 306, 623, 641
placer
    gold, 33, 62, 109, 154, 303, 341, 580, 618, 620, 659, 669, 672, 690
    miners, 50, 142, 193–94, 312, 327
    mining, 61, 94, 102, 107, 109, 134, 325, 664, 672, 690
    See also under mining
Plains Apache tribe, 51, 54
Platte Agency, 69
Platte River, 51, 54, 56, 63–66, 68–69, 73, 75, 81–82, 84, 89, 98, 100, 111, 118, 129, 580
Pleistocene epoch, 37
Pliocene epoch, 37
Plowman, Adoniram J., 307
Pluma mine, 50, 232, 274, 293, 347, 375, 503
Pocahontas Mill, 20, 228, 365, 370, 372, 374, 469. *See also under* mills

Pock, Deane, 487
Pomeroy, Brick, 127
Ponca tribe, 51
Poorman Anticline, 43, 48
Poorman Formation, 42–43
Poorman Gulch, 8, 189, 192, 273, 306, 325, 328, 344, 455–57, 606
Poorman Gulch Road, 306
Pope and Talbot, Inc., 301–2
Porphyry Dyke, 581
Porter, C. D., 172
Portuguese Siding, 279
Potato Creek, 111–12
Povandra, Ray, 408
Powder River, 38, 51, 76, 81–82, 85–86, 89–90, 114, 117–20, 122–24, 130, 137
Powder River Basin, 38, 81, 85–86, 89–90, 117–18, 120, 130
Power Company, 373–74, 386
powerhouse, 376, 379–80
Prairie, Gillette, 94
Prairie and Pines Parish, 322
Preble, Edward, 189
Precambrian era, 35, 37, 41, 43
Precambrian Homestake Formation, 41
Precambrian rock formations, 12, 35, 37–41, 44, 49
precipitation, 8, 360, 362, 370, 508, 512, 514, 599, 673
Presidential Proclamation 2072, 470, 561
Prince Oscar mining claim, 168
prospectors
    Blanchard Party, 33
    Brennan Party, 111
    Bryant Party, 33, 105–6, 619–20, 692
    Ezra Kind Party, 58
    Gay Party, 33, 105, 107, 135
    Gordon Party, 12, 60, 98–99
    Lardner Party, 33
    Smith Party, 33

Proterozoic
   period, 37, 41
   rocks, 37–38, 41–42
Proterozoic Harney Peak granites, 38
Pryor, Thomas, 314
Public Utilities Commission, 296
pulp stream, 368
pyrite, 42, 49, 345, 347
pyrrhotite, 42–43, 345, 347, 542

Q

quartz, 33, 42–43, 49, 93, 101–3, 105, 110, 134–35, 137–38, 160, 171, 213, 306, 325, 579–80, 587, 665, 673
Quartz Ledge, 137
quartz mining, 102, 325.
   See also under mining
quicksilver, 238, 260, 368, 673
Quit Claim Deed, 153, 308, 310, 631, 641, 680, 683

R

Rabbit Creek, 124
Racine Mill, 145, 147–48.
   See also under mills
Raddick, W. P., 135, 304–5
Raddick Gulch, 187
Raddick Intake, 209
Railroad Avenue, 308, 311, 322, 410
Rainwater, E. M., 320
Raise-boring technology, 487
Raiser, Robe, 103
Raise System No. 52, 488
Ralph Brothers, 429
Ralston, Robert, 111
Ralston, Wilson, 193
Ramsdell, Joseph, 185, 606
Ramsdell Ditch, 185, 606
Rancho Del Paso, 592
Randolph, William, 179, 228, 404, 585, 587, 590, 662, 692
Rapid City, 49, 157, 244, 276, 280, 284–85, 287, 291, 293, 328, 428, 451, 456, 531, 571, 601, 613, 615, 624, 631, 638–39, 645, 656, 660–61, 664, 678, 680–81, 683, 689–91, 694
Rapid City Range Days parade, 428
Rapid Creek, 61, 94, 101–2, 108, 207–8, 210, 606
Rashleigh Ball, 429
Rathburn, Dan, 155
Rawlings District, 112
Raymond, Robert M., 598
Raynolds, W. F., 75, 79, 616, 691
Raynolds Expedition, 5, 75, 77, 92, 97
Raynolds-Hayden Expedition of 1859, 70
RBC (Rotating Biological Contactor), 510
Reagan, Dan, 248
Reausaw, Fred, 289
Reausaw Lake, 95, 289
Recreation Building, 415, 418, 425, 590
Red Canyon, 111, 580
Red Cloud, 12, 59, 63, 69, 71–72, 80–82, 84–87, 89, 98, 101, 103, 115, 120, 128, 130–33
Red Cloud's War, 80
Red Cloud Agency, 63, 89, 98, 101, 115, 120, 128, 130, 132
Red Dog, 101
Redwater River, 131–32, 136, 606
Redwood Manufacturer's Company, 376
Ree people, 51
refining process, 17, 22, 264, 347, 364, 544–46, 548, 673
Regan, Tom, 31, 565
regrind plant, 8, 17, 263, 362–65, 374, 459
Reid, G. W., 185
Remer, William A., 274, 348, 362, 643
Reno, Marcus, 123

Reno Gulch, 7, 192, 273–74, 277, 608
Reno Powder Magazines, 273
Reno Siding, 273–74
Repass, Robert E., 296
Repass camp, 296–97
Republican Fork, 84, 100, 602
retorting process, 7, 239, 242, 260–61, 264, 673
Reva Gap, 124
Reynolds, Charley, 91, 93, 122
Reynolds, Jack, 579
Reynolds, Joseph J., 119
rhyolite dikes, 38, 47
Richards, Gary, 31, 294
Richards, Henry, 165–66
Rickard, T. A., 345, 620, 636
Ricker, Frank H., 349, 353, 599
Rincon Hill, 593
Roach, James, 181
Roberts, John M., 192
Robins Engineers and Constructors, 532
Robinson, Clara Scott 598
Robinson's Wood Camp, 278
Rockhold, Ray, 429
Roderick Dhu, 452
Roosevelt, Franklin D., 469–70, 561
Roosevelt, Theodore, 230, 601
Rosebud Agency, 133
Rosebud Creek, 119, 122, 132
Rosebud River, 114, 120, 122
Rosen, Peter, 58–59, 614, 622
  Pa-ha-Sa-Pah: or the Black Hills of South Dakota, 58–59, 614, 622
Rosenkranz, Henry, 336, 338–39
Rosenkranz Brewery, 336–37
Rosevear, Henry, 181
Rosowitz, Don, 565
Ross, Alec J. M., 415, 464, 634
Ross, Horatio N., 91, 93
Ross complex, 396
Ross Compressor Plant, 466
Rossio, Charles, 429

Rossiter Cyanide Plant, 16, 346
Rossiter Plant, 346
Ross Shaft, 20, 317, 397, 399, 420, 429, 464–67, 469, 471–76, 484, 488, 492, 494, 498, 540, 572
Rotating Biological Contactor (RBC), 510
Roubaix Lake, 95
Rounds, Mike, 569–71
Rowland, Clyde, 429
Rowland, Herbert, 429
Rowland, John, 429
Royal School of Mines, 349
Royce, W. E., 284
Royce, William, 407
Ruby Basin, 306, 401
Ruby Basin Road, 306
Runkle, George, 281
Runkle camp, 280–81, 286
Running Antelope, 133
run-of-mine ore, 241, 256, 441, 507, 666
Russell, John, 246
Russell, Smith, 245
Russell, Thomas H., 88, 97–98
Ruth, Chris, 452
Ryan, Thomas J., 401, 426

S

Saari, Matt, 321
Sacramento Valley Colonization Company, 592
Safe Investment Mine, 289
Safe Investment Mining Company, 291
safety bonus system, 423
Sage Creek, 74
Salisbury, O. J., 156, 171
Salt Lake City, 276, 554, 585
sampling method, 22, 50, 153, 262, 454, 459, 518, 537, 541–42
sand casting process, 10, 548, 669, 672

Sand Creek, 111–12
sand dam, 19, 449, 451, 505, 508, 521
Sanders, Elizabeth Jane, 591
Sanders, Margaret, 591
Sanders, Susan Gano, 591–92
Sanders, William, 593
sandfill, 21, 448–51, 488, 496, 515, 517–18, 521–22, 527, 668, 674
Sand House, 210
Sanford, T. Denny, 571
Sanford Underground Science and Engineering Laboratory, 568, 571–74, 603
San Francisco Chronicle, 594, 664, 677
San Francisco Examiner, 587
San Francisco Stock and Mining Exchange, 155
San Joaquin Valley, 594
Sans Arc tribe, 52, 71, 115, 117, 130, 133
Santa Fe, 56, 63, 614, 689
Santee Sioux, 51–52
Santee tribe, 115
Saone, 53–54
Sapsley, James, 189
Saskatchewan, Canada, 131, 133
Sauders, J., 111
Savage Tunnel, 19, 205, 307, 438–39, 441, 461
Save Centennial Valley Association, 504
sawmills, 101, 163, 170, 267, 273, 278, 280–81, 289, 292, 297–302, 638, 687
Sawpit Gulch, 19, 41, 242, 334, 452–54, 606
Sawpit Waste Rock Facility, 22, 453, 535–36
Sawyer, Fred, 280
Scandinavian Lutheran Church, 322
Schaffer spring, 210
schists, 37–38, 42–43, 48–49, 455, 669, 671, 674

Schlichting, Hubert B., 337, 339
Schmitz, Henry, 307
Schnitzel, J. "Henry," 459
Schwatka, Frederick, 124
Scott, Dan, 88
Scroggin, W. E., 600
Searle, John K., 307
Second Level, 177
Seely Drug Store, 418
seepage, 200, 354, 484, 564, 566, 568
Seepage Collection Pumphouse, 484
Segregated Golden Star, 142, 144, 168–69
Segregated Homestake claims, 144–45, 156, 161, 168. *See also under* claims
Segregated Homestake interest, 161
Seip, William, 185
Senter, John, 165
Service Raise No. 28, 491, 540
Seven Campfires, 71
Seventh Cavalry, 87, 91, 121–22, 125, 132
7-foot standard cone crusher, 532
Shafthouse and Crusher Building, 474
shaft jumbo drill, 484
Shaft No. 5, 20, 483–85, 487–88, 495, 504, 537–38
shallow open cuts, 213
Shankland, Samuel, 104
Sharp Bits, 244, 431, 633, 637, 640, 644, 649–51, 653–54, 657–59, 686, 689
Sharwood, W. J., 261, 348–49, 636, 679
Shaw, Herbert S., 459
Shears, G. W., 165
Sheldon Edwards Amalgam Mill, 333. *See also under* mills
Shepherd of the Hills Lutheran Church, 322, 534
Sheridan, Philip, 90, 97, 114, 118, 132
Sherman, William, 80–81, 83, 90, 96, 113

Sherman Street, 205
Shoemaker Gulch, 245, 342
Shoshone tribe, 66, 121
"Shotgun" Layther, 314
Shuey, Herbert, 599
Shugardt, G., 330
Shule, C. W., 330
Sicangu tribe, 51–52, 65
Siegrist, Jake, 155
Sierra Nevada Mountains, 579, 592, 597
Siever Street, 308, 315–16, 321
Sihasapa tribe, 52, 74
silicate facies (grunerite), 42
siliceous ore, 457
Silver City, 328, 344
Simonton, O. H., 330
Sinking Gardens, 322, 533–34
Sioux City, 73–75, 87–88, 97–98, 336
Sioux City Journal, 88
Sioux City Times, 97
Sioux City Weekly Times, 87
Sioux nation, 12, 52, 67, 73, 83, 116, 128–30, 616, 687
Sisters of St. Benedict, 321
Sitting Bull, 12, 62, 71–72, 83, 85, 87, 89, 115, 117–18, 120–25, 128, 130–31, 133, 621–22, 694
Skidmore, I. V., 333
Slabtown camp, 291
Slag Pile, 346
slates, 37, 42, 48, 671–72
Slavonians, 401, 405, 410–11
Sligo, Matthew, 176
Slim Buttes, 13, 119, 124–25, 127, 130, 622, 694
Slime Plant, 8, 17, 21, 205, 282–83, 350, 355, 357–64, 368, 370, 374–75, 386, 448, 506–7, 544, 566, 644, 680
Sludge Tank Building, 360
Sluice and hydraulic mining, 102
Slurry Pumphouse, 505

Smead, Walter E., 320
smelting process, 10, 346–48, 512–13, 544–45, 548, 585, 675
Smith, A. G., 246
Smith, Byron M., 77
Smith, Edward, 99, 116
Smith, E. P., 101
Smith, Eugene 108
Smith, Henry Weston, 131, 330
Smith, Hoke, 291
Smith, J. B., 185
Smith, John, 248
Smith, Michael, 91
Smith, Palmer, 189, 197
Smith, S. R., 319
Smith, William L., 107, 307
Smith and Pringle Mill, 148. *See also under* mills
Smith Gulch, 107
Smith Party, 105, 107. *See also under* prospectors
Smithsonian Institution, 79, 129
S. Moffit, 73
Smoky Hill River, 84
Smoots Mine, 289
Snell Mill, 145. *See also under* mills
Snowden, J. H., 73
Snyder, Julia, 314–15
Soldier's Town, 89–90, 117
Solution Storage and Precipitation Press Building, 360
Sonza, Nesto, 429
Soule, Charles D., 88
South Battery, 17, 365–66
South Bend, 8, 127, 327–28, 332–33, 344
South Dakota, 16–17, 22–23, 31–32, 36–37, 39–41, 44–45, 48–49, 53–54, 56–59, 78, 138–39, 270–72, 309–10, 424–26, 453–57, 535, 568–74, 590–92, 612–14, 618–20, 628–31, 633–34, 636–45, 647–51,

658, 660, 677–80, 682–83, 687–89, 691–95
South Dakota Board of Minerals and Environment, 535, 568
South Dakota Department of Environment and Natural Resources (SD-DENR), 564
South Dakota legislature, 337, 339, 570
South Dakota Mined Land Reclamation Act, 535
South Dakota Ore Tax, 9, 17, 469
South Dakota Science and Technology Authority, 23, 570–73, 603, 660, 692
South Dakota State Cement Plant, 451
South Dakota State Department of Health, 503
South Dakota Water Management Board, 212, 634, 692
Southern Pacific Railroad, 592, 595
South Fork of Rapid Creek, 94
South Lead (Ellison) Reservoir, 211
South Lead Cemetery, 207, 317
South Lead School, 16, 316–17
South Mill (*see also under* mills), 17, 20, 207, 236, 263–64, 317, 365–70, 387, 397, 429, 431, 436, 441, 463–65, 467–68, 470, 474–75, 477–78, 505, 507, 512, 514, 531–32, 544–45, 549, 551
South Mill complex, 429, 531
South Mill Street, 207, 317
South Pass, 73
South Sand Line, 449
South Sand Raise, 449–50
Spanish Cession in 1819, 55
Spanish flu epidemic, 413, 445
Spargo, J. A., 419
Spargo, John, 389
Spearfish Canyon, 17, 37, 40, 106, 178, 189, 372–73, 375–77, 380–82, 391

Spearfish Creek, 106, 132, 178, 184–85, 187–89, 208–9, 375–76, 383, 391, 606
Spearfish ditch, 178, 187–91, 205, 209–10, 373
Spearfish diversion, 212
Spearfish Electric Light, 373
Spearfish Fish Hatchery, 380
Spearfish Intake, 209
Spearfish Sawmill, 15, 300–302, 420, 679
Spearfish Water Right, 187
Special Order No. 117, 91
Split Tail Gulch, 197
Spotted Tail, 12, 71, 82–86, 101, 103, 115, 131–33, 623, 694
Spotted Tail Agency, 101, 131–32
Spring Creek, 49, 101–2, 119, 130, 205, 607
Springer, Ed, 319
Springer House, 319
Spring Street, 307, 388, 391
Spring Valley Water Company, 593
Spruce Gulch, 106–7, 134
square-set method, 217–18, 222–23, 302, 449, 517, 541, 668
Stagebarn Canyon, 272, 283, 289, 295–96
stamp mills, 7, 13, 16–17, 24, 138–40, 145, 147–48, 166–67, 184, 197, 231, 237, 239–40, 242–43, 245, 252, 254–58, 262–65, 267, 310, 333–35, 339, 345, 349, 354–55, 357–58, 363, 365–66, 464, 467–68
Standard Consolidated Mine, 347, 598
Standard Oil Company, 596
Standing Rock Agency, 130, 133
standpipes and penstocks, 376
Stanford, Leland, 593–94
Stanley, D. S., 80
Staple, Sydney, 389

Star Shaft pillar, 530
Station No. 5. Stream, 383
Steam Tunnel, 387
Steele, Thomas J., 459
Steir, Herbert, 429
Stetler, Larry, 572
Stevens, Kenneth, 480
Stevenson, A. M., 459
Stewart (prisoner), 276
Stillman, Albert, 317
Stillwell, W., 200
St. Joseph's Hospital, 336, 416
St. Louis, 56, 59, 69, 74, 96, 100, 331, 614, 691
St. Louis Globe-Democrat, 59
Stokes, George W., 110–11, 198, 614
Stokes Party, 111. *See also under* prospectors
Stone, Thoen, 57–59, 614–15, 693
Stoneman, George, 588
Stonewall, 104–5
Stope No. 5, 437
Stope North, 222
Stope South, 222
stoping process, 7, 20, 150, 217–18, 220, 222, 224, 302, 437, 450–51, 488, 515–18, 524, 527–28, 542, 657, 668, 671, 674–76, 689
   open cut-and-fill, 10, 20, 490, 517, 524, 527
   square-set cut-and-fill, 515
St. Patrick Tunnel, 458
St. Paul, 91, 94, 118, 130, 228
Strahorn, Robert, 77
Strawberry Springs, 207, 608
"stumpage corporation," 89
Sturgis, 132, 274, 600–601
Suhati tribe, 51
sulfide regrind system, 512–13
Sullivan, John, 159, 176
Summit, Inc., 505
Summit Siding, 273

Sun Dance, 121
Sunnyside claim, 165
Suomi Synod, 322
surveying method, 10, 268, 380, 537, 658, 689, 692
Sutherland, James J., 189
Sutro Tunnel Company, 594
Svoboda, Bob, 576
Swartout (White Mill foreman), 246
Sweatman Art Memorial, 533
Sweeney, W. W., 275–76
Sweet Water River, 114
Swindler, H. P., 274
Symons 7-foot shorthead cone crusher, 477, 507, 532
synclinal ledges, 43
Syncline, 47–48

T

Taft, Robert A., 602
Taft, William Howard, 23, 600–602
Taft-Hartley Act, 407
Tallent, Annie D., 98, 111, 312, 580, 619, 621, 631
Tallent, David G., 98, 580
Tallent, Robert E., 98
Talley, Paul, 482
Tanner, Henrik, 321
Tanner (reverend), 321
tappet, 239, 241, 256
Tapster, J. R., 318
Taylor Street, 593, 597
Telford, John, 274, 276
Ten-Mile Station, 186, 189–90, 192, 194, 206, 277, 293, 608
Tenth Cavalry, 132
Tenth Missouri Infantry, 581
Terah, Louis, 591
Terah Temple Haggin, 591
Terra Incognita Mining Company, 183
Terraville Test Pit, 10, 531, 533, 535, 658,

690. *See also under* Open Cut
Terraville town, 8, 13–14, 16, 20, 172, 175, 181–82, 192, 194, 206, 228, 245–46, 262, 288, 321, 328, 336, 341–44, 355, 372, 374, 384, 404, 414, 418, 421, 439, 458, 530–31, 535
Terry, Alfred H., 80–81, 83, 91, 97, 117–19
Tertiary rhyolite dikes, 38, 43–44, 46–47
Tesch, Charles L., 317, 658, 693
Teton Sioux, 12, 33, 52–54, 71, 132
Tetrault, P. S., 333
Tevis, Lloyd, 11, 23, 34, 149, 151–52, 154–55, 157, 171, 178–79, 189, 204, 269, 584, 586, 591–93, 595–97, 626, 663
Tevis Cup, 592
Texaco Service Station, 322
Texas Annexation in 1845, 55
Thayer, B. B., 347, 598
Third Cavalry, 124, 132
Thirty-sixth Congress, 77
Thoen, Louis, 57–58
Thomas, Cyrus, 129
    Indian Land Cessions in the United States, 129
Thompson, Alexander, 104
Thompson, C. F., 159
Thompson, Jake, 437
Thompson, Moses, 151, 172, 187–88, 453
Thompson, O. B., 187, 193
Thompson mill, 145, 148, 159, 187–88, 193, 315
Thompson Mill, 145, 148, 159, 187–88, 193, 315
Thompson mill, *See also under* mills
Thompson Mill, *See also under* mills
Thomson, Frank, 57, 59, 62–63, 614
Thorpe, A. G., 139

Thunder Hoop, 123
Tilford, 35, 98, 289
Tilson, Walt, 437
Tilson, William, 207, 606
Timber Case No. 1, 292
Tiyospayes tribe, 52
Todd, J. B. S., 77
Tongue River, 51, 114, 117, 120, 123, 130
Tower, E. M., 306
Townsend, Robert "Bob," 565
Townsite Agreement of 1892, 309
Tracy, William E., 401, 404, 406
Trainor, Bill, 330
Trask, Alvah D., 104
Travis, William, 248–49
The Treasure of Homestake Gold, 157, 244, 609, 620, 623–24, 626–28, 630, 635, 637, 640–46, 648, 650, 652–53, 664, 681
Treaty of Fort Laramie in 1851, 84
Treaty of Fort Laramie in 1868, 12, 29, 58, 83–90, 99–100, 113–14, 116, 118, 130, 611
Tretheway, Sam, 429
Treweek, Will, 407
Trojan Mining Company, 600
Troy. *See* Gayville
Tual, Casey, and Webber Amalgam Mill, 333. *See also under* mills
Tullock Automatic feeder, 256
Tullock Creek, 121–22
Tully, Thomas, 400
Tungsten Mill, 244, 263–64, 364–65, 444, 460, 645, 652, 691. *See also under* mills
Turner & Wilson's Saloon, 330
Twin Buttes, 123
Two Moon, 120–21
Two Strike, 101
two-wire method, 538, 540
Tyler screens, 507

U

Uhlig, Otto, 185
Uncle Sam Mine, 15, 164, 278–79
Union Iron Works, 155, 177, 240, 245–46, 255–56, 435
Union Pacific Railroad, 59, 79, 595
Unionville, 337
United Mine Workers, 407
United Presbyterian Church, 321–22
United Steelworkers of America (AFL-CIO), 407–8
University of Minnesota, 91
Upper Brulé Sioux, 89
Upper Louisiana, 56
Uren, Richard, 157, 165
U.S. Bureau of Mines, 370, 423, 665
U.S. Commission on Industrial Relations, 412, 425, 647–50, 694
U.S. Department of Energy, 574
U.S. Geological Survey, 32, 39–41, 45, 48, 55, 88, 375, 381, 597, 613, 665, 680, 694
U.S. House of Representatives, 113

V

Vandal, 101
Van Dyke, James, 267
Vanocker Canyon, 300
VCR (vertical crater retreat), 21, 451, 525–27, 657, 676, 689
Verpont, James, 106
vertical crater retreat (VCR), 21, 451, 525–27, 676
Vertical Shaft, 19, 138, 162, 165, 433–34, 439, 537
Vested Water Rights, 212
Vinaterri, Felix, 91
Voight, Meno, 245
Volk, James, 572
Von Leuttwitz, A. H., 124

Voorhies, Margaret S., 591
Vroom, Peter, 132

W

Wagner, C. H., 200
Wagner ST-2B LHD, 524
Wagon Box incident, 81
Walker, Francis, 88
Wallace, George, 76
Wall Street, 16, 315, 321
Walsh, Frank P., 427
Walsh Report, 427
Wampler, Linus, 408
Warbonnet Creek, 119
Ward Draw, 207
Warner, Dan, 336
Warner, Porter, 198
War Production Board, 471, 551
Warren, G. K., 73, 75, 79
Warren E. Scoggan, 426
Warren Expedition of 1856, 70
Warren-Hayden Expedition of 1857, 70, 92
Washington, George, 601
Washington School, 316
Washington Street, 307, 593
Washoe Valley, 584
Wasichus, 59, 117
Wasp No. 2 Mining Company, 206, 600
waste rock, 22–23, 218–20, 222–26, 248, 444, 449–50, 453, 460, 476, 480–81, 483, 488, 496, 531, 534–36, 542, 565, 567–68, 666, 669, 674, 676
Waterland, Joel K., 530
Wayland, Russell, 365, 429, 634, 691
WDWC (Wyoming and Dakota Water Company), 178, 188–90, 193, 195–204, 629–31, 633, 679–80, 691
Webb, R. A., 191
Webber, George E., 164, 180

Weber, Carolyn, 31
Weier, Anthony, 280
Weisbach method, 538
Welch, E., 158, 161
Welcome Saloon, 318
Wells Fargo & Company, 594–95
West, Elliott M., 183
Western Bank and Trust Company, 284
Western Federation of Miners, 401–2, 405–7, 410, 426, 600
western frontier, 29, 55
Western Pacific Railroad, 589
Western Specialty Company, 329
Western States Trail Ride, 592
West Lead Cemetery, 273
West Lead Reservoir, 211
West Lead School, 316
West Sand Plant, 467, 508
West Siding, 277
Westward Expansion, 5, 29, 51, 63, 65, 70, 79
W. F. M. No. 1, 401
Wheeler brothers, 108–9, 319
Whipple (bishop), 128
Whiskey Gap, 580
White (doctor), 248
White, A. J., 400
White, Frank G., 165
White, T. G., 165
White and Charles Reed, 400
Whitehead, J. R., 187
White River, 54, 56, 86, 115
White Stamp Mill, 147. *See also under* mills
Whitetail, 106–7, 136, 140–42, 191–92, 194–95, 199–201, 211–12, 239, 273, 406, 504, 608
Whitetail Creek, 136, 140–41, 192, 199, 201, 239
Whitetail Gulches, 106, 194
Whitetail Siding, 273
Whitetail Summit, 273
Whitewood Canyon, 37, 107
Whitewood Creek, 19, 41, 105–7, 109–10, 134, 136, 140–41, 190, 192–96, 201–3, 205–8, 239, 277, 303, 306, 339, 357, 361–62, 373, 437–39, 441, 457, 460–62, 503, 510, 607, 633, 684
Whitewood Springs, 207, 608
Whitfield, J. W., 69
Whitlock, Jim, 510
Whitman County, Washington, D. C., 276
Whitney, R. B., 98
Whitney Amalgam Mill, 333. *See also under* mills
Wibaux, Pierre, 279
Widman, Theodore, 203
Wiedenbeck, Sigmund F., 337
Wilfley No. 6 Concentrating Table, 365
Willard, A. M., 249, 251–52, 615, 678
Willes, William, 190
Williams, Archie, 429
Williams, George, 333
Williams, James, 336
Williams, J. J., 108–9
Williams, John T., 276, 347
Willis, William, 168, 190
Williston basin, 38
Willkie, Wendell L., 602
Wilson, John, 274, 276
Wilson, J. P., 307
Winchell, Newton H., 91
Wind Cave National Park, 51
Wind River, 114
winzes, 9–10, 19–20, 22, 180–81, 236, 446, 448, 456, 462, 466, 471–74, 476, 479–85, 487–88, 491–95, 497, 501–2, 537–41, 576, 666, 672–76
No. 1, 446
No. 2, 462, 466, 471–73
No. 3, 22, 466, 471–72, 476, 479, 484, 539–40

No. 4, 20, 236, 479–85, 487, 494, 540
No. 6, 20, 236, 483, 488, 492–95, 501–2, 540–41, 576
No. 7, 494–95, 541
wireline method, 491
Wisconsin River to central Minnesota, 52
Wolf Mountain, 119, 130, 136–37, 311
  stampede, 6, 136–37, 311
Wolzmuth, Dave, 455
Wood, W. H., 104
Woods, James M., 159
Woodward, Avery, 429
Woodward Clyde, 504, 511
Wooley-Peacho mine, 50, 332, 335, 454
Wooley-Peacho Mine, 335, 454
Workmen's Safety Committee, 423
World War I, 316, 319, 365, 413, 419, 445, 463
World War II, 22, 471, 551–52
Worsham, William "Woody," 481
Wray, Samuel, 245
Wright Irrigation Act, 594
Wright-McLaughlin Engineers, 504
Wyman, Phillip, 181
Wyodak Coal and Manufacturing Company, 394
Wyodak Development Resources Corporation, 397
Wyodak Mine, 18, 394, 397
Wyoming, 12, 18, 53, 60, 62, 66, 73, 78–79, 99, 115, 118, 121, 129, 178, 187–90, 193, 195–99, 246, 271, 276, 293, 295, 300–301, 311, 394, 617, 629–31, 633, 679–80, 695
Wyoming and Dakota Ditch, Flume, and Mining Company, 178, 187–88
Wyoming and Dakota Water Company (WDWC), 178, 188–90, 193, 195–204, 629–31, 633, 679–80, 691
Wyoming Territory, 12, 78–79, 99, 104, 115, 129, 188, 246, 271, 311

Y

Yankton Black Hiller, 96
Yankton Daily Press and Dakotaian, 93, 96
Yankton Sioux, 51, 579
Yankton Treaty, 77
Yankton tribe, 115
Yankton Union and Dakotaian, 87
Yates, A. B., 46, 613
Yates, Bruce C., 217, 222, 365, 390, 406, 428–30, 474, 559, 601–2, 609, 623, 634
Yates Complex, 474
Yates Compressor Plant, 475
Yates Member, 42
Yates Shaft, 9, 20, 420, 448, 474–77, 480–81, 484, 494, 553–54, 653, 677
Yates Waste Rock Facility, 22–23, 476, 480–81, 565, 567
Yates Waste Rock Pile, 460
Yellowstone River, 63, 75–76, 97, 114, 117–18, 122, 124, 130, 133, 616, 691
Yellow Tail, 592
Young, H. B., 137–38, 148, 150, 152, 155–56, 170, 200, 624, 682
Young, James M., 165–66, 623–25, 627, 680–82, 687–89, 692, 696
Young's Hole, 246
Yum, Hi, 333

Z

Zeljadt, E. J., 274
Zwaschka, Mark, 31–32, 532

LaVergne, TN USA
09 December 2009
166433LV00003B/2/P